Computational Transport Phenomena for Engineering Analyses

Computational Transport Phenomena for Engineering Analyses

Richard C. Farmer • Gary C. Cheng
Yen-Sen Chen • Ralph W. Pike

CRC Press
Taylor & Francis Group
Boca Raton London New York

CRC Press is an imprint of the
Taylor & Francis Group, an **informa** business

CRC Press
Taylor & Francis Group
6000 Broken Sound Parkway NW, Suite 300
Boca Raton, FL 33487-2742

First issued in paperback 2017

© 2009 by Taylor and Francis Group, LLC
CRC Press is an imprint of Taylor & Francis Group, an Informa business

No claim to original U.S. Government works

ISBN 13: 978-1-138-11429-6 (pbk)
ISBN 13: 978-1-4200-6756-9 (hbk)

Library of Congress Cataloging-in-Publication Data

Computational transport phenomena for engineering analyses / authors, Richard C.
Farmer ... [et al.].
    p. cm.
  "A CRC title."
  Includes bibliographical references and index.
  ISBN 978-1-4200-6756-9 (alk. paper)
  1. Fluid mechanics--Data processing. 2. Transport theory--Mathematics. I. Farmer,
Richard C. II. Title.

TA357.5.D37C665 2009
620.1'06--dc22   2008054551

Visit the Taylor & Francis Web site at
http://www.taylorandfrancis.com

and the CRC Press Web site at
http://www.crcpress.com

DEDICATION

We lovingly dedicate this book to our wives:
Peggy Farmer, Tina Cheng,
Lanny Chen, and Patricia Pike

Supplementary Resources Disclaimer

Additional resources were previously made available for this title on DVD. However, as DVD has become a less accessible format, all resources have been moved to a more convenient online download option.

You can find these resources available here: www.routledge.com/9781138419445

Please note: Where this title mentions the associated disc, please use the downloadable resources instead.

Contents

PREFACE

Modern computer technology has provided a dramatic improvement in the analysis of complex transport phenomena. However, such methodology has not yet been effectively integrated into engineering curricula. The huge volume of literature on a wide variety of transport processes cannot be appreciated or mastered without using innovative tools that allow one to comprehend and study these processes. Computational methodology is the logical basis for such tool development and is the subject of this text.

Transport phenomena include the disciplines of momentum, heat, and mass transfer. The major improvement in their analyses results from having the ability to predict two- and three-dimensional spatial variations of these processes. Such analyses involve solving conservation laws which are nonlinear partial differential equations. The physical similarity of various transport mechanisms suggests, and practice has proven, that methodology for describing one type of process would also be effective in describing the other transports.

Pertinent conservation equations have been stated by nineteenth century scientists such as Reynolds, Navier, and Stokes. However, it was only in the late twentieth century with the advent of high-speed computers that these equations could be effectively solved by numerical methods. Minimal effort has been devoted to providing instructional material for teaching transport phenomena by using the host of numerical solutions available in the literature. This text is intended as a bridge to connect the basic presentations of such books as Bird, Stewart, and Lightfoot's *Transport Phenomena* and Welty, Wicks, Wilson, and Rorer's *Fundamentals of Momentum, Heat, and Mass Transfer* to the methodology of numerical solutions of the conservation laws.

The objective of this text is to describe and provide a computational fluid dynamics code suitable for illustrating transport processes involving real fluids. This code is termed a computational transport phenomena (CTP) code; it is a mature production code which has been in use for

20 years. It was developed and owned by the authors. The CTP code and example problems are supplied in the companion CD to this text. The CTP code is supplied in Fortran 77 as both object and source code. This code does not require a lease and is designed to run on an individual's personal computer (PC). More importantly, the source code is also supplied. Our contention is that unless the source code is available to the user, the user never really knows the strengths and weaknesses of the analysis. The CTP code is written in Fortran which can be downloaded free, and thus the user can make maximum use of a PC. Currently, an individual's PC is vastly underused as an engineering tool. The PC will do far more than play games and serve as a word processor.

The majority of transport phenomena involve the turbulent flow of various Newtonian fluids. These are precisely the processes the CTP code was designed to simulate. A computational analysis requires two steps: (1) the construction of a grid to represent the geometry of the process and (2) the numerical solution of the pertinent conservation laws. The physics and chemistry appropriate for given problems is theoretically described and several example problems are presented in detail to explain how the CTP code can be used to simulate these phenomena. As these simulations are duplicated on a PC, the knowledge and skills necessary to effectively utilize CTP analyses will be developed. More definitive methodology can be developed by the individual or obtained by intelligent use of leased commercial codes. The CTP code is not a universal tool for solving all transport phenomena problems. It is rather an engineering tool for solving a significant number of transport problems on a routine basis and for learning what the current possibilities are for obtaining useful answers from analyses rather than having to resort to more expensive experimental investigations.

The scope of transport phenomena, including the simulation of turbulence, chemical reactions, and multiphase flow, is immense. Just as immense is the variety of methods which have been and are being applied to their simulation and theoretical understanding. An introductory survey of the most prominent of these methodologies is also included herein. Again the availability of high-speed computing capability has promoted the study and development of many of these methods, sometimes even at the expense of common sense. Turbulence modeling, statistical modeling, and multiphase flow simulation are currently being extensively researched and accompanying theoretical understanding developed. The potential of these studies is reviewed herein. One cannot start from an elementary understanding of these phenomena and appreciate their potential. For the three technologies just mentioned, the content of White's *Viscous Flow Theory*, 3rd ed., Launder and Sandham's *Closure Strategies for Turbulent and Transitional Flow*, Pope's *Turbulent Flows*, and Faghri and Zhang's *Transport Phenomena in Multiphase Systems* form the essential basics for addressing current transport analyses.

The potential of computational methodology for addressing practical engineering problems was brought to the attention of one of these authors (Farmer) early on when the base heating to the S-II stage was to be predicted. This second stage of *Saturn V* utilized a cluster of five hydrogen/oxygen engines. The primary base heating was radiation from the highly three-dimensional, supersonic rocket plumes. Radiation could not be accurately predicted from subscale model experiments. The subsonic base flow was predicted with a Navier–Stokes solver, and the supersonic flow was predicted with a three-dimensional method-of-characteristics analysis. Heating from various lines-of-sight could then be estimated by rigorous radiation analysis for the high temperature and concentration gradients along these rays for their actual physical dimensions. This work was done in about 1965 for NASA's Marshall Space Flight Center (MSFC) by aerospace contractors.

As computers became more efficient, the MSFC sponsored an extensive program to develop tools for launch vehicle analyses, which involved government personnel and many contractor participants. SECA, Inc. was founded by Dr. Farmer to be an active player in this development process. Dr. Yen-Sen Chen, who originated the FDNS computational fluid dynamics code, was a principal engineer in producing practical computational codes. While at SECA, he worked with Dr. Gary Cheng, Dr. Ten-See Wang, and Pete Anderson to produce a notable series of analyses related to rocket engines. These analyses and the codes used to make them were supplied to MSFC engineers for their use. This research was reported in NASA conference proceedings to form an integral part of their computational tool development program. This is the genesis of the CTP code described and provided herein. Dr. Chen moved on to start another small business, Engineering Sciences, Inc., thence to the National Space Organization in Taiwan. Dr. Cheng is now an active researcher and professor at the University of Alabama in Birmingham. We wish to thank Dr. Wang and Anderson for developing the finite-rate chemical reaction portions of the CTP code and the thermodynamics modules of the code, respectively. Dr. Pike and Dr. Farmer have been friends and research and teaching colleagues since graduate school days at Georgia Tech. Their long interest in transport phenomena has convinced them that learning about transport needs to change to master even a small part of what has been discovered and practiced during the past half century.

Of course, this type of activity was not supported only by NASA; for example, the much larger effort by Dr. Joe Thompson and his research team at Mississippi State University developed similar methodology for the Air Force and the National Science Foundation. The totality of these and all other efforts have raised computational analyses of transport problems to a level where practical computational engineering analyses for real problems have been created. Such tools should not remain solely in the hands of graduate students who write and exploit computational codes. The host of previously developed codes must be spread around to constitute modern engineering tools.

We greatly appreciate the host of colleagues who continuously rejuvenate our engineering discipline with research, publications, and technical discussions. We wish to thank Professors Andy Hrymak of McMaster University, Jennifer Curtis of the University of Florida, Rodney Cox of Iowa State University, and Chuck Merkle of Purdue University for discussing these concerns with us and for providing moral support by acknowledging this text as a useful endeavor. Professor John Mathews has been most generous and helpful in providing the notes on Fortran, which he developed while teaching at California State University at Fullerton. We also thank Professors Yasushi Ito and Roy Koomullil of the University of Alabama in Birmingham for allowing us to use the figure illustrating hybrid and overlaid grid generation.

We greatly appreciate the obvious enthusiasm demonstrated by Ms. Barbara Glunn of the Taylor & Francis Group as she piloted our text through the publication process. The entire Taylor & Francis editorial team and their SPi India associates have been most helpful and professional in bringing this work to its finished form.

The enormous size of the current technical literature on transport phenomena cannot be overstated. Effective methods for absorbing and evaluating this literature must be developed and made available to practicing engineers to avoid unnecessary spending on investigations, the outcome of which should already be known.

AUTHORS

Richard C. Farmer recieved his PhD in chemical engineering from Georgia Institute of Technology. Since 1986 until his retirement in 2005, he has held various positions as president, director, and senior staff scientist at SECA, Inc., a company which he founded. SECA performed contract R&D in computational fluid dynamics code development, analyses of rocket engine nozzle and plume flows, internal liquid propellant and two-phase engine related flows, and turbomachinery analysis. Previously he worked with Continuum, Inc. and at the chemical technology division of Science Applications, Inc., where he performed similar research. From 1967 until 1980, Dr. Farmer was a professor of chemical engineering at Louisiana State University where he taught graduate and undergraduate fluid mechanics, heat transfer, and mass transfer; originated graduate courses on combustion, turbulence, and ecological modeling; and conducted vigorous, heavily contract-funded graduate research programs. Research topics included numerical simulation of sediment deposited by the Mississippi River, tidal flushing of Mobile Bay, and thermal discharge of several Tennessee Valley Authority (TVA) power plants. From 1964 until 1967, Dr. Farmer was an aerospace engineer with NASA's Marshall Space Flight Center (MSFC), serving as chief of the Base Heating Section, which was responsible for generating the base thermal design criteria for Saturn launch vehicles. This involved planning and supervising an in-house effort supported by a major contracted effort among many aerospace engineering groups. The program was an interrelated experimental and modeling effort resulting in a validated model for turbulent reacting flow with coupled radiation. The model developed continues to serve as a widely used engineering analysis of rocket plumes.

Gary C. Cheng is an associate professor in the mechanical engineering department at the University of Alabama at Birmingham (UAB). He received a PhD in aerospace engineering from the University of Kansas in 1991. Dr. Cheng joined UAB in 2001 as an assistant professor, and was promoted to associate professor in 2006. From 1991 to 2001, he worked as a senior research engineer at SECA, Inc., conducting various computational fluid dynamics (CFD)-related projects to support NASA, the Air Force, and aerospace companies such as Boeing and Lockheed-Martin. Dr. Cheng has over 17 years experience in numerical simulation using CFD. He has twice received new technology awards from NASA (1995, 2002), and was named a NASA summer fellow in 2002. Dr. Cheng has served as the reviewer for AIAA *Journal of Propulsion and Power, Journal of Applied Numerical Mathematics, Journal of Computational and Applied Mathematics, Chemical Engineering Communication*, and he was on the review committee of NSF MRAC/LRAC (Medium Resource Allocations Committee/Large Resource Allocations Committee) from 2004 to 2007. Dr. Cheng has been the principal investigator (PI), co-PI, or an investigator of more than 25 successful proposals since he joined UAB. The agencies sponsoring his research studies include NASA, DoD, DoE, NIH, and various companies. He has mentored several graduate students for MS degrees and PhDs. Dr. Cheng's research activities include various rocket propulsion systems and their components, multiphase flows, chemical reactions, turbulence modeling and transition, reentry aerothermodynamics (radiation and surface ablation), biofluid flows, computational aeroacoustics, integration of continuum and particulate flow solvers for simulation of flows with disparate length scales, overset chimera grid topology for flow around moving bodies with relative motions, and emerging CFD methodologies. Dr. Cheng has published over 20 peer-reviewed journal publications and more than 40 refereed conference papers. For more information, please visit his Web site at www.eng.uab.edu/me/Faculty/gcheng/.

Yen-Sen Chen received his PhD in aerospace engineering from the University of Kansas in 1984. He is the director of systems engineering and a senior research fellow of the National Space Organization (NSPO), Taiwan, and is also in charge of the development of rocket programs of the NSPO. He also has extensive experience in projects acquisition and program management through his working experience as a visiting scientist at NASA/MSFC from 1984 to 1988, then a senior research scientist at SECA, Inc. from 1988 to 1992, and the president of Engineering Sciences, Inc. from 1992 to 2005. Dr. Chen's technical specialties include theoretical, engineering, and computational skills to obtain solutions of complex problems in launch vehicle propulsion systems, aerothermodynamics design analysis, liquid/solid rocket engine performance, and plume radiation. He developed the FDNS and UNIC codes, multidisciplinary computational tools suitable for all-speed flow regimes, which are used in NASA's launch vehicle, propulsion systems and spacecraft design analyses, and he is recognized as the winner of NASA/MSFC 2007 Software-of-the-Year (SOY) Award and a runner up of the NASA SOY in 2008. He has also received awards from NASA/MSFC for technical innovations in the areas of SSME internal flow study, rocket nozzle flow/plume study, reacting flow modeling, combustion instabilities, 3-D three-phase flow modeling, and pump/turbomachinery flow studies. He is a senior member of the American Institute of Aeronautics and Astronautics (AIAA) and a member of the national honorary Sigma Gamma Tau. He has served on the AIAA Solid Rocket Technical Committee since 2006. He has reviewed technical publications for the AIAA and ASME. Dr. Chen was an adjunct professor and a member of the PhD program advisory committee at the University of Alabama in Huntsville from 1991 to 1995. He is a senior and committee member of the American Institute of Aeronautics and Astronautics and the national honorary Sigma Gamma Tau Society. He has served on the AIAA Solid Rocket Technical Committee since 2006 and as a reviewer for the technical publications of the AIAA Journals, ASME papers, and was appointed by the University of Alabama in Huntsville as an adjunct professor and a member of the PhD program advisory committee from 1991 to 1995. Dr. Chen has initiated a series of lectures on rocket design and fabrication at the National Chiao Tung University, Taiwan, in 2007 and 2008. He has published over 100 journal and conference papers.

Ralph W. Pike is the director of the Minerals Processing Research Institute and is the Paul M. Horton Professor of Chemical Engineering at Louisiana State University. He holds doctorate and bachelor degrees in chemical engineering from the Georgia Institute of Technology. He is the author of a textbook on optimization for engineering systems and coauthor of three other books on design and modeling of chemical processes. He has directed 14 doctoral dissertations and 16 master's theses in chemical engineering. He is a registered professional engineer in Louisiana and Texas. His research has been sponsored by federal and state agencies and private organizations with 106 awards totaling $5.4 million and has resulted in over 200 publications and presentations. His research specialties are fluid dynamics with chemical reactions occurring in the flow which include parts of the areas of transport phenomena, chemical kinetics, chemical and biochemical reactor design, applied mathematics, and biological and ecological systems dynamics. He is a fellow of the American Institute of Chemical Engineers and is first vice-chair of the Environmental Division and second vice-chair of the Fuels and Petrochemicals Division. He is an active member of the Institute for Sustainability and the Safety and Chemical Engineering Education (SACHE) committee of the Center for Chemical Process Safety. He was the meeting program chairman for the 74th annual meeting and has cochaired 61 sessions on transport phenomena, reaction engineering, optimization, and sustainability. He has held all of the positions in the Baton Rouge Section of the AIChE. He has served as coeditor-in-chief of *Waste Management*, an international journal, published by Elsevier Science, Ltd., Oxford, United Kingdom devoted to information on prevention, control, detoxification, and disposition of hazardous, radioactive, and industrial wastes. He is a member of the American Chemical Society and Sigma Xi, a scientific society.

CHAPTER 1

Computational Transport Phenomena

1.1 OVERVIEW

Modern computer technology has provided a dramatic improvement in the analysis of complex transport phenomena processes. This methodology has not yet been effectively integrated into the engineering community. The huge volume of literature associated with the wide variety of transport processes cannot be appreciated or mastered without using innovative tools to allow one to comprehend and study these processes. Computational methodology is the logical basis for such tool development and is the subject of this text.

Transport phenomena include the disciplines of momentum, heat, and mass transfer. The improvement in their analyses results from now having the ability (1) to predict two- and three-dimensional spatial variations of these processes and (2) to realistically represent turbulence, real fluid properties, and chemical reactions. Such analyses involve solving conservation laws which are nonlinear partial differential equations (PDEs). The physical similarity of the various transport mechanisms suggests, and practice has proven, that methodology for describing one type of process would also be useful in describing the other transports.

Pertinent conservation equations have been stated by nineteenth century scientists such as Reynolds, Navier, and Stokes. However, it has only been in the late twentieth century with the advent of high-speed computers that these equations could be effectively solved by numerical methods. Minimal effort has been devoted to providing instructional material for teaching transport phenomena by using the host of numerical solutions available in the literature. This book is intended as a bridge to connect the classical presentations of transport phenomena (Welty et al., 2001; Bird et al., 2002) to the methodology of numerical solution of the conservation laws.

1.2 TRANSPORT PHENOMENA

Transport phenomenon is a term originated by chemical engineers to describe the laminar and turbulent flow of momentum, mass, and heat. The applications which created the need for such knowledge were the unit operations of the chemical process industries. Later the aerospace industry required the same technology to describe hypersonic flow and high-speed combustion. Now, environmental and medical applications are becoming numerous. The conservation laws, or the equations of change, as they are otherwise known, are the fundamental physical laws which describe these transport phenomena. But numerous approximate solutions to these equations used for describing individual unit operations do not come close to realizing the potential of these important physical laws. On the other hand, modern numerical solutions yield powerful computational tools and the synergism for integrating the multitude of diverse technologies.

Three major contributions to the understanding of modern transport phenomena are (1) the basic unification of the transport processes as presented in the classic text of Bird, Stewart, and Lightfoot, (2) the aerospace industries' numerical studies of reacting gases in hypersonic flowfield, termed aerothermochemistry (von Karman, 1954), and (3) the immense literature on turbulence, as typified in classic texts (Hinze, 1975; Monin and Yaglom, 1971, 1975). The many thousands of books, journals, papers, and available reports which have now supplemented these basic works have produced an unmanageable literature.

Flows may be turbulent or laminar, compressible or incompressible, continuum or free molecular, free surface or internal, liquid, gas, or multiphase, and a whole bunch more. The preponderance of flows of engineering interest is turbulent. Although laminar flows are better understood, they are controlled by viscosity, which varies 28 orders of magnitude between silica glasses and hydrogen. Not to mention non-Newtonian flows which cannot be described with a simple viscosity coefficient. A fuller array of these processes is listed in Table 1.1. The classes of flows denoted in this table generally indicate the level of complexity of the flow. The Class I flows include a major fraction of important engineering

TABLE 1.1   Typical Transport Processes

Class I

1. Molecular diffusion in stationary fluids
2. Laminar flow and convective heat and mass transfer in simple gases and liquids
3. High and intermediate Reynolds number turbulent flows including convection of heat and mass
4. Flows with varying physical properties
5. Multicomponent and reacting flows including combustion and explosions
6. Internal flows in pipes, meters, reactors, and turbomachinery
7. Free surface flows in ducts, rivers, and oceans
8. Conjugate and moving-interfacial heat and mass transfer
9. Sparse multiphase flows
10. Flow through porous media
11. Unsteady flows

Class II

1. Steady and time-varying non-Newtonian behavior
2. Solidifying flows
3. Flows with complex dependence on multicomponent physical properties
4. Flows of pure fluids and mixtures near the critical point
5. Flows with extreme geometric complexity, i.e., an entire shell-and-tube heat exchanger

Class III

1. Noncontinuum, free-molecular flow
2. Noncontinuum, unmixedness reaction effects
3. Turbulent flows which require very complex turbulence descriptions
4. Flows on a global scale, i.e., meteorological flow predictions
5. Dense multiphase flows, i.e., fluidized bed reactors
6. Flows with radiative transport

---

applications of transport phenomena and may be analyzed with existing computational codes. Class II flows may be analyzed with minor modifications and generalizations to such codes. Class III flows require codes written specifically to describe such phenomena. But a starting place for developing new educational tools must be chosen, if any approach other than studying a myriad of simple examples is to be employed for understanding transport phenomena. This is not meant to belittle other items in the table, as there are certainly important applications involving all of these transport phenomena.

The term unit operations, the forerunner of transport phenomena, were invented to describe the diverse processes which were important to the chemical manufacturing industry. The chemical and phase changes needed to manufacture various chemicals on a large scale were too complex to describe the individual transport processes upon which they depended. This complexity still exists in large part to the flows within the mixers and reactors being noncontinuum, reacting, and highly three dimensional. The chemical engineer resorted to empirically describing various pieces of separation equipment. This equipment was to accomplish evaporation, drying, distillation, absorption, membrane separation, liquid–liquid extraction, ion exchange, liquid–solid leaching, crystallization, physical phase separations, etc. From the 1920s until the present day, empirical data and experience are still the design tools of the chemical industry. As more analytical and theoretical capability is developed, a shift to more definitive designs using the tools of transport phenomena is being accomplished. Today the unit operations are frequently being termed separation processes, which reflect the trend toward introducing more analysis into the design process. Although computational transport phenomenon (CTP) has made enormous progress, it is not yet possible to replace the unit operation concept. Such replacement is desirable and will eventually be accomplished.

1.3 ANALYZING TRANSPORT PHENOMENA

The conservation laws formulated by the scientific giants of the nineteenth century were stated as nonlinear, PDEs. Such equations could not be solved; therefore, for the next 100 years, transport phenomena were analyzed by simplifying the analyses until solutions could be obtained. The closest anyone came to addressing the analysis directly was the classic work on heat conduction in solids (Carlsaw and Jaeger, 1959). The success of these analyses was due to the heat conduction equation being linear so that series solutions involving superpositioning to satisfy boundary conditions could be employed. Even so, the series solutions required the use of a computer to provide timely computations. Reynolds time averaged the equations to provide the first approach to analyzing turbulence (Monin and Yaglom, 1971). To describe process analyses to the chemical industry and analyses of basic pipe and channel flows, spatial averaging was used. Since the spatial averaging lost many of the important features of the flow, it was combined with empirical test data. A prime example of such technology was the friction factor plot for predicting pressure drop in pipe flow. Even in the twenty-first century this methodology is used in transport phenomena texts. Numerical solution methodology is documented in computational fluid dynamics (CFD) specialists' texts. These practices do not provide suitable tools for studying and understanding the immense field of transport phenomena.

Given the transport equations, one may attempt to find an approximate solution by eliminating unnecessary terms. For example, early on it was realized that omitting the viscous terms allowed accurate pressure distributions to be calculated about airfoils. However, the drag on the airfoil was predicted to be zero. This unsatisfactory condition was eliminated by Prandtl introducing the boundary layer concept (Schlichting, 1979). The boundary layer was postulated as being a thin region near a wall which was dominated by viscous effects. This resulted in two solutions: one to calculate pressure distributions and a second to calculate the boundary layer to evaluate friction. Using CFD methodology, one computation is sufficient to evaluate both pressure fields and friction.

The elements of transport phenomena are summarized in Table 1.2. The solution methods for the various approximations to the fundamental conservation equations are also given in this table. Lacking a complete statistical theory of turbulence, various types of averaging are used to represent turbulent flows. This practice results in the turbulent conservation laws being termed semiempirical, which does not imply that they are not useful. When radiation is not important, the most general form of the conservation laws is a coupled set of nonlinear PDEs. Such equations can be solved effectively only by computational means. In addition to the solver, formulation of the equations, control volumes of the flows, grid systems, and auxiliary conditions must be specified. These are not trivial adjuncts to the conservation laws, if other than trivial analyses are to be made. These adjuncts will be described subsequently.

How has computational methodology given us a new, practical tool to investigate transport phenomena? The short answer is to say that the modern computer power has provided the means for solving complex conservation laws. To implement such methodology, more insight is needed to appreciate how the governing PDEs may be solved. For example, consider the hypothetical, hypersonic vehicle powered by a liquid rocket engine shown in Figure 1.1. This axisymmetric example was chosen because it illustrates most important transport processes and how they were analyzed prior to the development of the computational methods described

TABLE 1.2   Transport Phenomena for Multicomponent, Continuum Flows

---

I. Tools

   A. Physical laws

      1. Conservation of mass: PDE[a]

      2. Conservation of momentum: PDE

      3. Conservation of energy: PDE

      4. Conservation of species: PDE

      5. Thermal equation of state: AE

      6. Caloric equation of state: AE

      7. Reaction kinetics: AE

  Results: Dependent variables:

    Density

    Three velocity components

    Temperature

    Enthalpy or internal energy

    Concentration of species

   B. Transport properties

      1. Laminar flow (all are fluid properties)

         Diffusion coefficients

         Viscosity

         Thermal conductivity

         Specific reaction-rate coefficients

      2. Turbulent flow

         All those of laminar flow

         Eddy diffusivity

         Eddy viscosity

         Eddy conductivity

         (All of the eddy properties are flow properties)

      3. Radiation: Fluid properties

         Absorption coefficients

         Scattering coefficients

         Emission coefficients

  Results: An empirical or theoretical definition of all properties which are needed in a given problem. Usually, empirical descriptions are sufficient.

   C. Geometric relationships

      1. Coordinate systems

      2. Vectors

      3. Tensors

  Results: Definition of the independent variables.

TABLE 1.2 (continued)   Transport Phenomena for Multicomponent, Continuum Flows

II. Formulation of the problem

    A. Define the region of interest, i.e., the control volume

    B. Decide if any dependent variables are constant (to determine the number of equations to be solved)

    C. Decide if gradients in any particular direction or in time are negligible (to determine the number of independent variables)

    D. Choose a coordinate system

    E. Evaluate the transport properties

    F. Determine an initial condition for which all dependent variables are known

    G. Determine what is occurring on all sides of the region to fix boundary conditions

    H. Formulated equations fit one of these categories

        1. Unsteady:

            i.   Varies in all three space directions

            ii. Varies in two space directions

            iii. Varies in one space direction

            iv. No spatial variation: ODE in time

        2. Steady:

            i.   Varies in all three space directions

            ii. Varies in two space directions

            iii. Varies in one space direction: ODE in space

            iv. No spatial variation: AE

  Results: A system of equations to represent transport is obtained.

III. Solution methods

    A. Experimentally determine all variables. Usually possible, but extremely expensive.

    B. Experimentally determine all variables in scaled-down experiment. Scale-up by analysis (usually with algebraic equations) is required. Frequently possible.

    C. Analytically solve the ODE. Frequently possible.

    D. Analytically solve the PDE. All but impossible.

    E. Analytically transform the PDE, then solve analytically. Sometimes possible.

    F. Numerically solve the ODE. Always possible.

    G. Numerically solve the PDE. Usually possible.

Results: Predictions of all dependent variables, as a function of independent variables.

[a] Unless otherwise indicated, the solution is for a set of PDEs. Notations are ODE for ordinary differential equations and AE for algebraic equations. If radiation is significant, the energy equation is an integrodifferential equation.

herein while avoiding geometric complexity. The various regions of the indicated flowfield are all described by the conservation equations of mass, species, momentum, and energy. Since these equations are PDEs, they have different characteristics in the various regions of flowfield. The various flow phenomena indicated by letters in Figure 1.1 are identified in

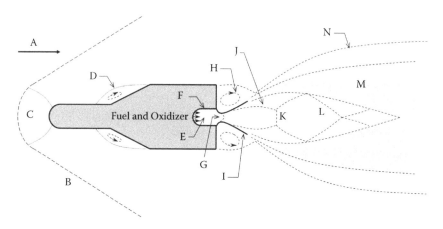

FIGURE 1.1 Hypersonic flow about a rocket propelled vehicle (labels refer to Table 1.3).

TABLE 1.3 Flowfields Denoted by Letters in Figure 1.1

A.	Free-stream hypersonic flow
B.	Bow shock
C.	Subsonic nose cap with ablative surface
D.	Forward facing step with flow separation
E.	Subsonic mixing and combustion
F.	Boundary layer heat transfer to chamber walls
G.	Subsonic/supersonic transition on sonic line
H.	Flow separation on backward facing step
I.	Supersonic nozzle flow with oblique shocks
J.	Oblique shocks from continuation of nozzle shocks
K.	Supersonic/subsonic transition at Mach disc
L.	Oblique shocks within plume
M.	Free shear layer with mixing and combustion
N.	Shock formed by impingement on plume

Table 1.3. Starting at the nose, the approach flow is hypersonic and inviscid. The shock wave is, for all practical purposes, a discontinuity. The subsonic region is viscous, for high altitudes where the atmospheric density is very low, this could be laminar flow. The blunt nose creates a hot region which is depicted as being protected by an ablating surface. This same hot nose cap could produce thermal radiation to the surface. The flow around the body would again be accelerated to become supersonic, except for

the boundary layer along the vehicle surface. As the external flow moves past the vehicle surface to flow around the rocket nozzle it separates from the surface. Eventually it surrounds and contains the rocket plume. The rocket itself mixes fuel and oxidizer in the combustion chamber which burns to form exhaust products. The chamber mixing is not completely efficient; therefore, striations of different mixture ratios will probably still exist. Initial mixing is convective as the propellant jets would be directed toward each other. Subsequent mixing would be accomplished by turbulent transport, which is intermediate between convective and laminar in magnitude. Turbulent boundary layers would be formed along the interior rocket walls. The core flow would be subsonic, choked at the rocket throat, and expanded to supersonic in the diverging nozzle. The core flow and the exiting plume would be supersonic with shock waves to turn the flow as dictated by the pressure field. This flow could even contain imbedded subsonic regions preceded by normal shock waves. The plume would be fuel-rich to maximize performance; therefore, as it contacts air in the free stream it would undergo mixing and further combustion. The hot plume would radiate energy into the environment.

If thermal radiation is neglected, all of these flow processes could be analyzed with the Navier–Stokes form of the conservation laws with suitable averaging and an empirical turbulence model. Why then is obtaining a computational solution so difficult? And it is. The nature of the PDEs changes from hyperbolic, to parabolic, to elliptic depending on the position in the flowfield. With 1950s and 1960s computer methodology, these various flow regimes had to be analyzed separately. The implication is that different numerical algorithms and codes must be used in each type of flow regime. Even with more advanced computer power the conservation laws defied solution. The mixed flow regions were still too complex to resolve, even the simpler nose cap region. This region was successfully analyzed by treating the steady-state problem as time dependent and integrating the equations in time until a steady state was reached (Moretti and Abbett, 1966). Such a treatment made the equations, all of them, behave hyperbolically so that time marching worked! This procedure is still used in many aerodynamics analyses and is recommended for use in this work for general transport phenomena analyses. Furthermore, only one computer code could be written and used to address a wide variety of transport problems. Obviously, all of the code capability would not be required for each analysis attempted. Once the use of the code was mastered, a new learning curve for each application would not be necessary.

1.4 A COMPUTATIONAL TOOL: THE CTP CODE

To appreciate and utilize current transport analyses, one needs a computational tool which is very sophisticated. Ideally, the user would write his own code, but this is not practical or necessary. Many CFD codes are available in the literature and for lease. But the learning curve to effectively use such codes is prohibitive. The purpose of this text is to provide a production quality code and a comprehensive users' manual to serve as an introduction to solving transport problems. The CTP code and example problems are supplied in the companion CD to this text. The CTP code is supplied in Fortran 77 as both object and source code. The example problems are summarized in Appendix A. This code is not touted as the best available or the definitive CTP solver. It is sufficiently general to analyze a wide variety of transport phenomena problems. It was developed and utilized by these authors and their colleagues to address a large number of problems of interest to NASA, the Air Force, and the aerospace industry. Upon mastering the material in this text, the reader will obtain not only an introduction to the computational methodology of transport but also will have a tool which can be used without additional expense or investment in a new learning curve to apply to other industrial problems. It is the further contention of these authors that one must have access to the source code to effectively and wisely use any computational tool.

The development of the personal computer (PC) over the last quarter century has provided the means to solve the partial differential transport equations simply because it calculates incredibly fast. However, the computational methodology has not matured to a single accepted technique. Most practitioners have been trained by accomplishing highly specialized analyses of specific problems. There are commercial CFD codes available, notably the code offered by FLUENT, Inc. of Lebanon, New Hampshire. But such codes are difficult to use and are extremely expensive to lease as source codes. Yet, essentially all undergraduate engineering students have extremely powerful PCs. To provide the bridge between the theoretical statements of the conservation laws and the solving of practical problems, adequate computational methodology designed to run on PCs is desirable. The numerous simplistic codes found in many current textbooks do not serve this purpose. Does an engineer or student really want to learn a different code for each problem he investigates? Fortran code is the workhorse for solving transport problems. Hundreds of millions of dollars have been invested in developing such codes. Such an investment is not likely to be repeated to satisfy software entrepreneurs. Furthermore, Fortran compilers are readily available as shareware which operates on PCs.

There is not a universal CTP solver for all possible transport phenomena analyses, but the CTP code described herein has a wide range of capabilities. The CTP code solves two-dimensional planar and axisymmetric and three-dimensional steady and unsteady, laminar and turbulent conservation equations for mass, momentum, energy, and species. Rectangular Cartesian, orthogonal and nonorthogonal curvilinear coordinates may be employed. Constant density, ideal gas, and real fluid property fluids may be described. Two-equation turbulence models are used for generality and computational efficiency. Multiphase fluids are described as ideal solutions. Reacting fluids are described with detailed reaction mechanisms. Euler–Lagrange tracking models are used to describe solid/fluid flow interactions and multiphase flow phenomena. Property data files are included as part of the code.

However, more importantly, having access to the source code and data files allows capabilities needed but not included in the CTP code to be added by the individual investigator. Other fluids may need to be included in the databases. Other turbulence models may need to be evaluated. It must be realized that without the computational capability inherent in a code like the CTP code, one would not be able to compare predictions made with different turbulence models. Very dense particle/fluid flows might require more elaborated methodology than is currently in the CTP code.

Developing expertise in using the CTP code not only provides education in computational methodology, but gives the user a computational tool to study and apply transport phenomena methodology.

The difficulty in mastering and using the CTP methodology is largely eliminated by an appropriate users' manual. Providing such a manual is a major objective of this text. Since the output data from a CTP analysis are voluminous, another objective is to provide the user with graphics methodology sufficient to visualize and study the output.

1.5 VERIFICATION, VALIDATION, AND GENERALIZATION

How is one assured that the results obtained from a CTP simulation are of sufficient value to be used as design criteria or as the explanation and solution of problems arising in an industrial process? There is not a simple answer to this question as many factors have to be considered. Not the least of which is engineering judgment as to whether or not the simulation is reasonable. As defined by computational analysts (Roach, 1998), "verification" is solving the equations right, and "validation" is solving the right equations. We define "generalization" as determining whether or not the available code contains all of the essential physics and chemistry for the application, or does the code require modification to be appropriate for the analysis. The assurance provided at this point is essential to making CTP analyses a worthwhile endeavor. Roach discusses details of verification and validation for over 400 pages, yet the bulk of the work considered is for constant density and ideal gas fluids. For transport phenomena, real fluid properties and reacting, phase-changing flows have received much less attention and numerical analyses. The studies made to justify the use of the CTP code are discussed in Chapter 10.

1.5.1 VERIFICATION

The selections of the differencing and integration algorithms are the major factors in allowing the CTP code to be verified. Demonstrating the code's grid independence in providing a solution is the major test used to establish that the conservation equations are indeed solved. Details of this verification procedure are presented in Chapter 8.

1.5.2 VALIDATION

Validation of numerical solutions to the conservation equations is practically accomplished by comparing simulated results to experimental data. Historically, constant density and ideal gas flows have been the databases which were used for such comparisons. Transport phenomena include the prediction of mass diffusion, chemical reactions, and phase change for real fluids. The database for such flows is much smaller than for the aerodynamic and water type flows with the associated density restrictions. The most plentiful database for transport type flows is for gaseous combustion. Hence, most of the validation simulations have been made against combusting flows.

A further handicap in providing validation data for transport processes is that many more measurements than those usually reported are required to define the flowfield and to establish the boundary conditions for the analysis. Usually, a validation comparison is required by the user of the simulation. Unfortunately, the experiment is seldom planned in coordination with the computational analyst. This frequently causes the results of the validation exercise to be inconclusive. The proper boundary conditions and the determination of the critical flowfield variables must be determined beforehand. Such prior planning is far less critical in purely aerodynamic and hydrodynamic flowfield simulation validation. The absence of carefully planned experimental data collection is partially mitigated by conducting many simulations for whatever test data are available. For the CTP code supplied herein, the totality of cases analyzed will be referenced and selected cases reviewed. However, since the CTP source code is also supplied, the reader may augment the validation procedure by making additional analyses. Validation is a continuing process, especially for a code which has many options, all of which have seldom been validated.

It cannot be overemphasized that the experiment to be simulated should be as close as possible to process of interest. Likewise, the simulation should be as realistic as possible. For example, if the experiment has pronounced three-dimensional features, the simulation should not be made in two dimensions, so that more numbers could be generated. A further example is the tendency to conduct heat and mass transfer experiments with small heat and mass fluxes so that the fluid properties can be assumed constant and the flowfield can be assumed unaffected by the transport process. After all evaluating such complexities are what make the CTP simulation useful.

1.5.3 GENERALITY

The CTP code was written to solve multidimensional, reacting, real-fluid, multiphase, laminar, or turbulent flows. The speed regime of the flow may be from subsonic to hypersonic. Boundary conditions are such that these speed regimes may be accommodated. Wall functions are used as boundary conditions, to allow larger grid spacing. Most CFD codes are not this general. Some other CFD codes are better suited for analyzing certain of these features, but the code generality is what sets this code apart for simulating transport phenomena.

Turbulence models are limited to two-equation (K–E) models. More elaborate models would reduce the generality of the code. Using wall functions allows for the use of better near-wall simulation without changing the turbulence description far from the wall. But what advantage does using such a turbulence model offer?

1.5.3.1 Example of a Simple Momentum Transport Problem

To determine the essential characteristics which a CTP code must have to be useful for analyzing geometrically and physically complex phenomena, an example problem is given. This example is purposely chosen to be so simple and so basic that even an elementary knowledge of transport phenomena is adequate to understand the conclusions drawn from its consideration.

Consider turbulent pipe flow of a nonreacting, single phase fluid with heat and mass transfer at the pipe wall. Assume that the transfer rate for heat and mass is small, such that the velocity field is uncoupled from the wall boundary conditions and that the fluid properties may remain constant. The eddy transport of momentum, heat, and mass for fully developed flow (FDF) in a smooth pipe from an empirical velocity distribution is estimated as follows. Momentum transport will be considered first.

Spalding's velocity correlation for incompressible, fully developed, turbulent pipe flow from White (2006) is generally considered valid. It is stated as

$$y^+ = u^+ + e^{-\kappa B}\left[e^{\kappa u^+} - 1 - \kappa u^+ - \frac{(\kappa u^+)^2}{2} - \frac{(\kappa u^+)^3}{6} \right] \quad (1.1)$$

$$y^+ = u^+ + 0.1287\left[e^{0.41u^+} - 1 - 0.41u^+ - 0.08405\,(u^+)^2 - 0.01149\,(u^+)^3 \right] \quad (1.2)$$

where (κ, B) are (0.41, 5.0), the values recommended by Coles and Hirst (1968) and used herein. The parameters (κ, B) of (0.40, 5.5) were originally used and are also acceptable. The velocity and distance from the pipe wall are nondimensionalized using the wall shear stress and fluid density and viscosity. These variables are termed the inner law variables. The correlation is very good, except near the pipe centerline. The correlation yields an undesirable slope discontinuity as the predicted profile intersects the pipe centerline. There are another set of variables termed the outer law variables which are useful for analyzing boundary layers, but they are not needed here. The important point is that turbulent, fully developed, pipe flow of a constant density and viscosity fluid is correlated with only the wall shear stress and density and viscosity.

A specific example which has been experimentally studied will be analyzed. Extensive turbulence measurements for fully developed pipe flow have been reported (Laufer, 1954). These measurements were for Reynolds numbers of 5×10^5 (Re_H) and 5×10^4 (Re_L) based on the centerline velocity. The fluid was reported to be incompressible air, but the temperature and pressure were not given. The pipe was 10/in. nominal diameter (9.72/in. or 0.2469/m actual diameter). Unfortunately, insufficient data were presented in the report to fully describe the experiments. An example close to the experimental conditions is used herein to illustrate the desired analysis. Thus, the example cannot be considered as being validated by the experiment. This is frequently the situation when experiments and analyses are not jointly investigated.

To fully specify the example, assume air has a density of $1.18/\text{kg m}^{-3}$ and a kinematic viscosity of $1.505 \times 10^{-5}/\text{m}^2 \text{ s}^{-1}$ giving a viscosity of $1.776 \times 10^{-5}/\text{kg m}^{-1} \text{ s}^{-2}$. These conditions give centerline velocities of

$$Re_H = 5\times10^5 = \frac{DU_{CL}}{\nu} = \frac{0.2469\, U_{CL}}{1.505\times10^{-5}} \quad U_{CL} = 30.48/\text{m s}^{-1} = 100/\text{ft s}^{-1} \quad (1.3)$$

$$Re_L = 5\times10^4 = \frac{DU_{CL}}{\nu} = \frac{0.2469\, U_{CL}}{1.505\times10^{-5}} \quad U_{CL} = 3.048/\text{m s}^{-1} = 10.0/\text{ft s}^{-1} \quad (1.4)$$

The wall shear stress and the mean velocity in the pipe are needed to utilize the well-established velocity profiles from the literature. Laufer reported measured velocity profiles, but did not curve fit the data. Instead of attempting to curve fit data read from small figures, the higher Reynolds number velocity was assumed to be represented by a 1/7.5 power law, and the lower Reynolds number by a 1/7th power law. These profiles were consistent with data from the literature (Schlichting, 1979). Schlichting also offered an equation for relating mean and centerline velocities, namely:

$$U_{av}/U_{CL} = 2n^2/(n+1)(2n+1) \quad (1.5)$$

For $n = 7.5$ and 7, the velocity ratios are 0.827 and 0.817, respectively. These values are also consistent with published graphs (Geankoplis, 2003), although the graphs are too small to be accurately read. The approximate corrections should be adequate, since power law models accurately represent mean turbulent flow in pipe flow. The Reynolds number and mean velocities for the two cases become

$$\begin{aligned} Re_H &= 4.135\times10^5; \quad U_{av} = 25.21/\text{m s}^{-1} \\ Re_L &= 4.085\times10^4; \quad U_{av} = 2.49/\text{m s}^{-1} \end{aligned} \quad (1.6)$$

The relationship between the Reynolds number based on mean velocity and the Darcy friction factor is given by Prandtl's correlation (White, 2006):

$$f_D^{-0.5} = 2.0 \log \left\{ Re\, f_D^{0.5} \right\} - 0.8 \tag{1.7}$$

For the high Reynolds number, $f_D = 0.0136$
For the low Reynolds number, $f_D = 0.0219$
For $Re_H = 413{,}500$, $f_D = 0.0136$:

$$\frac{U_{av}}{u^*} = \left(\frac{8}{f_D} \right)^{1/2} = 24.25 \quad u^* = \frac{25.21}{24.25} = 1.039/\mathrm{m\,s}^{-1}$$

$$y = \frac{y^+ \nu}{u^*} = 1.4527 \times 10^{-5}\, y^+/\mathrm{m} \quad u = u^+ \times u^* = 1.039 u^+/\mathrm{m\,s}^{-1} \tag{1.8}$$

$$\tau_w = \rho u^{*2} = 1.274/\mathrm{kg\,m}^{-1}\mathrm{s}^{-2}$$

For $Re_L = 40{,}850$, $f_D = 0.0219$:

$$\frac{U_{av}}{u^*} = \left(\frac{8}{f_D} \right)^{1/2} = 19.11 \quad u^* = \frac{2.49}{19.11} = 0.130/\mathrm{m\,s}^{-1}$$

$$y = \frac{y^+ \nu}{u^*} = 1.1577 \times 10^{-4}\, y^+/\mathrm{m} \quad u = u^+ \times u^* = 0.130 u^+/\mathrm{m\,s}^{-1} \tag{1.9}$$

$$\tau_w = \rho u^{*2} = 0.01994/\mathrm{kg\,m}^{-1}\mathrm{s}^{-2}$$

At the pipe centerline, $y = 0.1235\,\mathrm{m}$ or $y^+ = 0.1235 \times 1.039/1.505 \times 10^{-5} = 8526$ for the high Reynolds number case, and $y^+ = 0.1235 \times 0.130/1.505 \times 10^{-5} = 1067$ for the low Reynolds number case.

By using the previously defined relationships, the distribution of u, y^+, u^+, and their derivatives can be calculated across the pipe. The distributions of u and u^+ are shown in Figures 1.2 and 1.3 for the high Reynolds number case.

The wall shear stress may also be calculated from the velocity gradient at the wall as calculated from Spalding's correlation equation.

$$\frac{du^+}{dy^+} = \left[1 + 0.41 \left(y^+ - u^+ + 0.001479 u^{+3} \right) \right]^{-1} \tag{1.10}$$

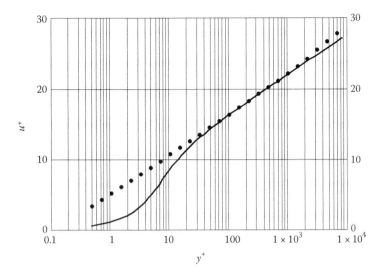

FIGURE 1.2 Dimensionless inner law profile for fully developed pipe flow (Equation 1.2, ——, Equation 1.15, …).

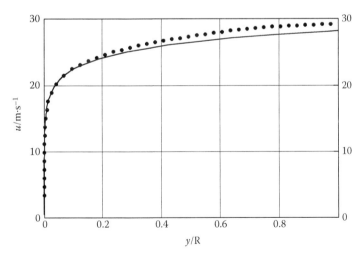

FIGURE 1.3 Dimensional inner law velocity profile for Laufer's high Reynolds number pipe flow experiment (Equation 1.2, ——, Equation 1.15, …).

Converting this velocity gradient to a dimensional form and evaluating it at the wall

$$\tau_w = \mu \frac{du}{dy}\bigg|_w = 1.7759 \times 10^{-5} \times 71722 = 1.274/\text{kg m}^{-1}\text{s}^{-2} \quad (1.11)$$

This confirms that the velocity distribution used gives the correct wall shear stress, for the high Reynolds number case, as the laminar sublayer interacts with the wall.

Also note from Figure 1.3, that the laminar sublayer is very thin. For a numerical solution which integrates the conservation equations all the way to the wall many grid points would be required. For the high Reynolds number example, the layer sublayer which would require resolution is less than 1% of the pipe radius. For a multidimensional analysis, the use of small grid spacing in one direction would also place limitations on the grid sizes that could be used in the other directions. This makes treatment of the wall region critical in any numerical analysis and provides the motivation for developing the more practical wall function solutions.

But how does the shear stress vary across the pipe? For a momentum balance on a cylindrical plug of radius r and a length dx in the FDF section of the pipe

$$P\pi r^2 - \left(P + \frac{dP}{dx}dx\right)\pi r^2 + 2\tau\pi r\,dx = 0$$

$$\tau = \frac{r}{2}\left(\frac{dP}{dx}\right) \quad \text{and} \quad \tau_R = \frac{R}{2}\left(\frac{dP}{dx}\right) = -\tau_w \quad (1.12)$$

Note that the pressure is not a function of r. On eliminating the pressure gradient, $\tau = \tau_R\,(r/R)$, that is the shear stress varies linearly across the pipe.

Using this shear stress variation, an effective viscosity (μ_{eff}) can be defined.

$$\tau = \tau_R(r/R) = \mu_{\text{eff}}\left(\frac{du}{dr}\right)_r = (\mu + \mu_t)\left(\frac{du}{dr}\right)_r \quad (1.13)$$

The effective viscosity and the eddy viscosity (μ_t) can be calculated from these equations. The linear variation of the total shear stress across the pipe is assured by the analysis. By using only the eddy viscosity, the turbulent shear stress distribution across the pipe is predicted. The turbulent and

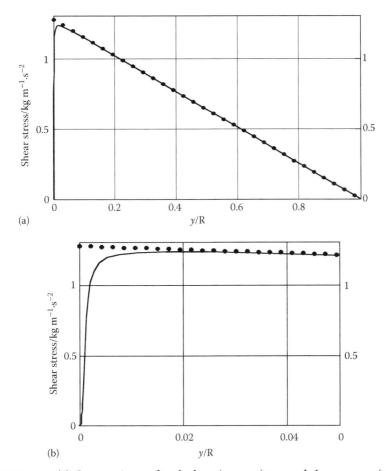

(a)

(b)

FIGURE 1.4 (a) Comparison of turbulent (τ_t, ——) to total shear stress (τ_{total}, ...) for fully developed pipe flow. (b) Comparison of turbulent (τ_t, ——) to total shear stress (τ_{total}, ...) near the wall for fully developed pipe flow.

total shear stress as a function of y is shown in Figure 1.4a. These distributions are shown in an expanded scale near the wall in Figure 1.4b. These curves match the shape of those presented by Laufer. Unfortunately, there are insufficient experimental data to make this comparison quantitative.

The ratio of the turbulent viscosity to the laminar viscosity across the pipe in terms of y^+ is shown in Figure 1.5a. Figure 1.5b shows this ratio as a function of the dimensional position y. Figure 1.5c shows this ratio very near the wall. This method of evaluating eddy viscosity yields a zero value at the pipe centerline. Since the velocity correlation is somewhat erroneous at the centerline, the turbulent viscosity may be subject to this same error. Also, the zero eddy viscosity is aesthetically unappealing. Based on a limited amount of test data, Reichardt (1961) proposed a different correlation:

$$\frac{\mu_t}{\mu} = \frac{\kappa y^+}{6}\left(1+\frac{r}{R}\right)\left[1+2\left(\frac{r}{R}\right)^2\right] \tag{1.14}$$

This correlation has the conceptual advantage that the centerline eddy viscosity is not zero, as shown in Figure 1.5a and b. Wilcox (2006) simulated

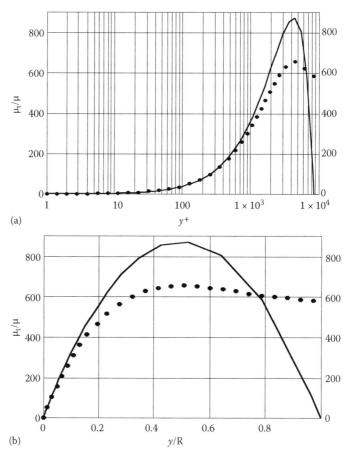

(a)

(b)

FIGURE 1.5 (a) Ratio of turbulent to laminar viscosity (Calculated, ——; Reichardt's correlation, …). (b) Ratio of turbulent to laminar viscosity (Calculated, ——; Reichardt's correlation, …).

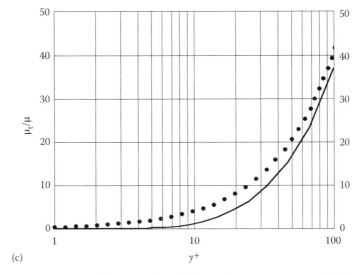

FIGURE 1.5 (continued) (c) Ratio of turbulent to laminar viscosity (Calculated, ——; Reichardt's correlation, …) near the wall.

Laufer's experiment using a two-equation turbulence model and obtained good comparisons to the test data. Wilcox did not calculate and compare the eddy viscosity directly. Hinze (1975) reviewed Laufer's data and discusses the use of Equation 1.13 to represent the eddy viscosity. Hinze showed the Laufer data to behave like Equation 1.14 would predict, but he did not mention this equation or what effect such an equation would imply about the velocity profile. Reichardt's correlation also does not yield good results for predicting wall shear stress, but it is qualitatively correct over most of the pipe cross section. Equation 1.14 can be used to predict the velocity distribution for fully developed pipe flow. By neglecting the molecular viscosity and assuming the linear variation of shear stress across the pipe, the velocity profile which results is (Kays et al., 2005)

$$u^+ = 2.44\ln\left[1.5\,y^+\left(1+r/R\right)\big/\left(1+2\,(r/R)\right)^2\right]+5.0 \qquad (1.15)$$

This profile has zero slope across the centerline. As shown in Figures 1.2 and 1.3, it exhibits a slight wake similar to what is predicted with an outer law profile for boundary layers. In this case, the profile deviates from the inner law above a y^+ value of about 1000. The y^+ values for determining such divergence depends somewhat on Reynolds number, since the friction velocity is used in the definition of y^+. The inner law profile should be used for y^+ values below about 100. Since the velocity profile predicted with Equation 1.15 matches the inner law profile in the overlap region and produces a slight wake, it is recommended for use at y^+ values above several hundreds. This implies that the dotted profiles in Figure 1.5a and b are the better estimate of eddy viscosity near the pipe centerline. Others have proposed similar correlations for the turbulent to laminar viscosity ratio (Longwell, 1966; Wilkes, 2006). In general, these correlations agree qualitatively, but not quantitatively. The velocity profile near a pipe centerline is not well validated with experimental test data. The shear stress at the centerline should be zero, and the profile should have zero slope as it crossed the centerline. Notice that eddy viscosity correlations are frequently stated in terms of kinematic viscosity ratios; however, since these correlations are only for constant density flows, such distinctions are unnecessary.

Regardless of the accuracy questions arising concerning centerline velocity values, the eddy viscosity is seen to increase rapidly from zero in the laminar sublayer to a factor of about 650 and then decrease as the pipe centerline is approached. The low Reynolds number predictions are not presented, because they are qualitatively very similar to Figure 1.5a through c. The maximum eddy viscosity is about an order of magnitude lower than that for the high Reynolds number case. This is roughly the same as the Reynolds number decreases. Extensive investigations of boundary layers and flows in conduits with noncircular cross sections show this same behavior. In fact, the near wall behavior is accurately represented with the same correlation equations.

The conclusions drawn from this example are that if an eddy viscosity model of reasonable accuracy can be devised for boundary layers and internal flows in conduits, the wall friction and velocity field can be predicted fairly well. Such an eddy viscosity model will also predict eddy transport of momentum throughout the flowfield. Of course, its validation further away from the wall must also be established. These issues are discussed in detail in Chapter 4. Furthermore, since the wall effects are controlled by the flow very near the wall, wall functions can be devised such that numerical flowfield predictions can be made using such functions as boundary conditions in lieu of unreasonably dense grids to satisfy zero slip wall conditions. The expectation is that geometrically complex flows can also be so described, as the numerical solutions would provide essentially free-stream conditions with the wall-function boundary conditions accounting for near wall effects. This is the concept used in the CTP code. Its detailed description, verification, and validation are described in the remainder of this text.

1.5.3.2 Simple Heat and Mass Transport Problems

Turbulent exchange processes of heat and mass are so similar that profile data and analyses for fully developed temperature and concentration profiles are interchangeable, provided only that dimensionless parameters are utilized. The following discussion is mainly in terms of temperature, but the results apply equally well to concentrations involving mass transfer.

Fully developed temperature and concentration profile models and experimental data for pipe flow, channel flow, and boundary layer flow have been reported (Kader, 1981). Restrictions placed on fluid properties are extreme. The heat or mass exchange is assumed to be so small that the flowfield and fluid properties are not affected. For boundary layer flow, temperature (or concentration) profiles can be similar from a surface to the edge of the boundary layer. For internal pipe or channel flow, "fully developed" requires more explanation. If the wall temperature is fixed, the entire flow will eventually reach the wall temperature and the profile will be flat. If wall heat flux is specified, the flow will eventually become very hot or very cold, depending on the sign of the specified heat flux. Extensive experiments and analyses have been reported which attempt to define a fully developed temperature (or concentration) profile. Stringent limitations must be imposed on the flowfield analyses for such profiles to be meaningful. For practical problems, such restrictions must be removed. However, to illustrate the importance of turbulent mixing effects, this body of work is informative.

The stringent limitations just mentioned do not apply to velocity profiles. The fully developed velocity profile is obtained when the wall boundary layers grow to reach the pipe or channel centerline. From that point on the profile remains the same as the static pressure drops in response to the wall friction. This situation remains the same until the inlet pressure is insufficient to maintain steady flow in the conduit. The role of pressure does not serve the same purpose in heat or mass transfer; therefore, additional restrictions are needed to realize fully developed temperature and concentration profiles.

Fully developed temperature profiles can be defined as those which exist when the film coefficient is independent of position (Arpaci and Larsen, 1984), thus,

$$h = \frac{q_w}{T_w - T_b} = -\frac{k}{T_w - T_b} \left.\frac{\partial T}{\partial y}\right|_w \neq f\{x\}$$

$$\text{where } T_b = \frac{1}{U_{av} A} \int_A uT \, dA \tag{1.16}$$

or for mass transfer

$$k = \frac{G_{aw}}{(\rho_{aw} - \rho_{ab})} = -\frac{D_{eff}}{(\rho_{aw} - \rho_{ab})} \left.\frac{\partial \rho_a}{\partial y}\right|_w \tag{1.17}$$

T_b and T_w are independent of y, but may depend on x. These conditions imply that a fully developed temperature or concentration profile may be defined as

$$\theta = \frac{T_w - T}{T_w - T_b} \neq f\{x\} \text{ or for mass transfer } \phi = \frac{\rho_{aw} - \rho_a}{\rho_{aw} - \rho_{ab}} \neq g\{x\} \tag{1.18}$$

If q_w = constant,

$$\frac{\partial T}{\partial x} = \frac{\partial T_w}{\partial x} = \frac{\partial T_b}{\partial x} \tag{1.19}$$

If T_w = constant,

$$\frac{\partial T}{\partial x} = \left(\frac{T_w - T}{T_w - T_b}\right)\frac{\partial T}{\partial x} \tag{1.20}$$

Also,

$$Nu = Fn\left\{Re, Pr, \frac{x}{R}\right\} \quad \text{or} \quad Sh = Gn\left\{Re, Sc, \frac{x}{R}\right\} \tag{1.21}$$

for the Nusselt and Sherwood numbers, respectively.

For turbulent flow, the fully developed velocity profile (which we know) is used for the flowfield and the eddy thermal diffusivity (which we do not know) must be determined. Kader provided the following correlation to determine the eddy transport of heat and mass.

Energy balance (von Karman's temperature law of the wall):

$$\frac{d\theta^+}{dy^+} = (Pr^{-1} + \varepsilon/\nu)^{-1} \tag{1.22}$$

$$\text{where} \quad \theta^+\{0\} = 0 \quad \text{and} \quad \theta^+\{\delta\} \quad \text{or} \quad \theta^+\{R\} = \theta_o^+$$

where

δ is a boundary layer thickness

The subscript "o" denotes the pipe centerline value, i.e., the "boundary-layer thickness" in fully developed pipe flow

To avoid specifying an eddy viscosity throughout the entire normal coordinate direction, piecewise portions of the temperature profile were specified. This is the same procedure that was originally used to establish the velocity correlation. The temperature profile was pieced together at different locations than the velocity profile depending on the Pr.

1. At the wall $\varepsilon = 0$, therefore

$$\theta^+ = Pr\, y^+ \tag{1.23}$$

2.1. For the linear sublayer which ends at y_1^+ where $\varepsilon\{y_1^+\} \cong a$ and for the Prandtl number range of $500 \le Pr \le 40 \times 10^3$

$$\varepsilon/\nu = (6E-4)(y^+)^3 \quad \text{and} \quad y_1^+ = 12/Pr^{1/3} \tag{1.24}$$

2.2. An alternative to Equation 1.24 for
$Pr > 1$ is $\theta^+ = Pr\, y^+ - (1.5 \times 10^{-4})Pr^2\, (y^+)^4$

$$\varepsilon/\nu = (6E-4)(y^+)^3 \quad \text{and} \quad y_1^+ = 9/Pr^{1/3} \tag{1.25}$$

2.3. For $Pr \approx 1$, the temperature and velocity profiles are similar

$$y_1^+ \cong 30 \tag{1.26}$$

2.4. For $Pr \ll 1$,

$$\varepsilon/\nu = Pr_t/ky^+ \cong 0.85/ky^+ \quad \text{and} \quad y_1^+ = 2Pr^{-1} \tag{1.27}$$

3. From y_1^+ to δ^+ or R^+ and for $(6 \times 10^{-3}) \le Pr \le (40 \times 10^{-3})$

$$\theta^+ = 2.12\ln y^+ + \beta\{Pr\} \quad \text{and} \quad Pr_t = 0.85$$
$$\text{where} \quad \beta\{Pr\} \equiv (3.85Pr^{1/3} - 1.3)^2 + 2.12\ln Pr \tag{1.28}$$

Putting all the segments of the temperature profile together: boundary layer, pipe flow, and channel flow, temperature profiles for fully developed turbulent flow were obtained. Concentration profiles are described with the same methodology when analogous dimensionless variables are used. The fully developed temperature profile for pipe flow is given as

$$\theta^+ = Pr\, y^+ \exp\{-\Gamma\}$$

$$+ \left\{ 2.12 \ln \left[\left(1 + y^+ \frac{1.5(2 - y/R)}{1 + 2(1 - y/R)^2} \right) \right] + \beta\{Pr\} \right\} \exp\{-1/\Gamma\} \qquad (1.29)$$

where $\quad \Gamma \equiv 0.01(Pr\, y^+)^4 / \left[1 + 5 Pr^3\, y^+ \right]$

Kader showed this equation is valid in the range $0.03 < Pr < 170$. Notice that the friction velocity and kinematic viscosity are needed in this correlation to convert the dimensionless coordinate to a dimensional one.

To consider a specific case, assume water flow in a pipe of 7.62/cm radius (R) at $Re = 2.3 \times 10^4$. Assume the fluid is water with a density of 1000/kg m^{-3} and a viscosity of 1.794×10^{-3}/kg m^{-1} s^{-1}, giving a kinematic viscosity of 1.794×10^{-6}/m^2 s^{-1}. For water at ambient temperature, the Prandtl number is about 7.0. These conditions give a mean velocity of

$$Re = 2.3E4 = \frac{2RU_{av}}{\nu} = \frac{0.1524\,(m)U_{av}\,(m/s)}{1.794 \times 10^{-5}\,(m^2\,s)} \quad U_{av} = 0.2708\,(m/s) \qquad (1.30)$$

From Prandtl's equation for $Re = 23{,}000$, $f_D = 0.02502$

$$\frac{U_{av}}{u^*} = \left(\frac{8}{f_D} \right)^{1/2} = 17.88 \quad u^* = \frac{0.8878}{17.88} = 0.01515\,(m/s) \qquad (1.31)$$

For these flow conditions, the fully developed temperature profile is shown in Figure 1.6. Using the same procedure as used to analyze the Laufer data, the ratio of eddy viscosity to laminar viscosity was calculated and is also shown in Figure 1.6.

A similar analysis using a velocity profile of

$$\frac{u}{u^*} = \frac{1}{0.41} \ln\left\{ \frac{yu^*}{\nu} \right\} + 5.0 \qquad (1.32)$$

a turbulent Prandtl number of 1.0, and a wide range of laminar Prandtl numbers was reported (White, 2006) for boundary layers. Further analyses with similar methodology, but for a more narrow range of laminar Prandtl numbers and a slightly wider range of conduit geometries has also been reported (Kays et al., 2005).

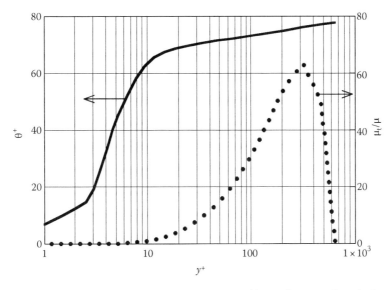

FIGURE 1.6 Fully developed temperature profile and ratio of turbulent to laminar viscosity for pipe flow of water.

- These examples show that the eddy viscosity is much larger than the laminar viscosity, except very near the wall.
- Turbulent Prandtl and Schmidt numbers are essentially constant at a value of 0.9.
- Only trivially small heat and mass transfer rates can be analytically determined.
- The effects of fluid property variations cannot be analytically determined.
- The concept of a fully developed temperature or concentration profile for turbulent pipe flow is valid for a very restricted set of conditions, yet an approximate effect of Prandtl number, Schmidt number, and eddy diffusivity of heat and mass can be established.

Results of this simplistic example are the "theory" upon which many empirical correlations have been developed to produce design information for transport phenomena. Average flows across the wall region are determined to produce *Nu* and *Sh* predictions. Since such severe assumptions have been placed on the analysis, empirical corrections to these "low" transfer rates have been determined to match experimental test data. The goal of CTP is to predict the corrective effects and relate them to the physics of the flow.

1.5.3.3 Potential Prospects of Computational Analyses

All of the severe restrictions employed in the FDF analysis can be removed with computational analysis based on more general solutions of the conservation laws. The conservation equations contain far more information than that derived from subsets of such equations created to expedite solution methodology or simulate particular experiments. Granted, a definitive turbulence model does not exist, reaction kinetics data must mainly be derived from experiments, wall geometries must be faithfully described with computational grids, and practical wall functions must be utilized. A myriad of computational codes have been used to perform transport analyses—yielding various levels of rigor and accuracy. No one CTP code is capable of treating all of the diverse complexities of the chemistry, physics, and geometry involved. But by making a reasonable compromise between what is known, what can be computed, and what accuracy level is required to provide useful results, a fairly general production code has been developed. This is the CTP code described herein.

- Typical restrictions which can be removed from the FDF analysis are general two- and three-dimensional steady and unsteady flows can be analyzed.
- Coupled and conjugate effects of momentum, heat, and mass transfer can be predicted.
- Practical simulations of turbulent flows can be made with two-equation turbulence models and appropriate wall function boundary conditions.
- Real fluid thermodynamic properties for gaseous and liquid fluid mixtures can be described.
- The extent of finite-rate chemical reactions can be determined.
- Several types of multiphase flows can be considered.
- Flow speeds from slow to hypersonic can be predicted.
- Mixed subsonic/supersonic flows can be predicted.
- Flows in porous media can be analyzed.
- Flows in rotating machinery can be predicted.

Application of computed simulations to a wider range of situations requires extensive validation; much of which has been accomplished and much of which remains to be done.

The CTP code is an engineering tool which can be used to provide new technical information. Without such a tool, the multitude of technical literature cannot be effectively utilized. It cannot even be realistically evaluated.

1.6 SUMMARY

A detailed derivation of the conservation equations is to be presented, the necessary mathematics to appreciate these equations will be reviewed, methods of applying these equations to specific problems will be demonstrated, and methods of solving these problems with the CTP code will be explained. Physical property data and elementary problem examples to form an adequate background for using the CTP methodology are also presented. A limited, although extensive, class of CTP analyses will be addressed. Having become familiar with these examples and the codes used to predict such processes, one can use the CTP code to work similar problems, modify the source code to extend its capability, or realize the problem under consideration is beyond the resources available for further code development. In the latter case, one would resort to a commercial code which has the required capability or endeavor to develop such a code.

Upon providing the background for analyzing complex transport phenomena, the remainder of the text is devoted to illustrative examples. These examples were chosen to represent practical flows and to provide an acceptance of the computed results. The experimental data which were used to validate each example are described as appropriate. The CTP code may be used directly to solve the Class I problems in Table 1.2. With minor modifications, once the user is familiar with the code, the Class II processes may be analyzed with the CTP code. Class III processes should be analyzed with special purpose codes designed specifically to address these phenomena. The CTP code is provided as a source code and with a mechanism for producing computational grids. Our experience is that these features are essential to the successful use of the CTP (or any other) solver.

1.7 NOMENCLATURE

Nomenclature is a knotty problem. When complicated analyses are performed, many symbols are used to represent the numerous variables. No satisfactory standard exists, or is likely to be invented. Because of the many systems already in the literature, one must be flexible enough to read and use any of them. This problem is only becoming worse, as the literature is exponentially increasing.

In general, variables that can be typed with the Microsoft Word equation editor will be used. The idea is that another piece of software will not be required to duplicate the equations used in this chapter. Thus the boldness of the characters will not be given a mathematical significance. Single arrows over the variable will indicate a vector quantity. Double arrows will indicate a tensor (usually second order) quantity.

Since the variety of topics covered defeat the purpose of trying to establish one consistent set of nomenclature, each chapter will be supplied with its own nomenclature table. Care must be taken to use the proper table, but this practice allows the maximum use of familiar symbols in any given discussion.

The CTP tool is presented in Fortran. No other language is practical since the millions of dollars of CFD code development, study, and documentation have not been and, in all probability, will never be available for duplicating this technology.

1.7.1 LIST OF SYMBOLS

1.7.1.1 English Symbols

a	thermal diffusivity
A	cross-sectional area of pipe
B	empirical constant in inner law velocity profile (5.5 or 5.0)
C_p	constant pressure heat capacity
D	inside pipe diameter
D_{eff}	effective diffusion coefficient
E	turbulent energy dissipation
f_D	Darcy friction factor
f, Fn	unspecified functions
g, Gn	unspecified functions
G_{aw}	mass flux of species a at the wall
h	heat transfer coefficient
k	mass transfer coefficient
K	turbulent kinetic energy
n	empirical constant in power-law velocity profile
$Nu = hD/\kappa$	Nusselt number
P	pressure

$Pr = \nu/a$	Prandtl number
$Pr_t = \varepsilon/a = C_p\mu_t/k_t$	turbulent Prandtl number
q_w	wall heat flux
r	radial coordinate
R	pipe radius
$Re = D\,U_{av}/\nu$	Reynolds number
$Sh = kD/D_{eff}$	Sherwood number
$Sc = \nu/D_{eff}$	Schmidt number
T	temperature
$T^* = q_w/\rho C_p\, u^*$	characteristic temperature
u	local time-averaged velocity
U_{av}	spatial average of time-averaged velocity
U_{CL}	centerline time-averaged velocity
$u^+ = u/u^*$	inner law velocity
$u^* = \sqrt{\tau_w/\rho_w}$	friction velocity
x	axial coordinate in cylindrical coordinates
$y = R - r$	coordinate normal to the wall
$y^+ = yu^*/\nu$	dimensionless normal from the wall
y_1^+	inner law variable at the edge of laminar sublayer

1.7.1.2 Greek Symbols

β	function in turbulent Prandtl number correlation
Γ	function in turbulent Prandtl number correlation
δ	boundary layer thickness
ε	eddy kinematic viscosity
$\theta = (T_w - T)/(T_w - T_b)$	temperature parameter
$\theta^+ = (T_w - T)/T^*$	dimensionless temperature variable
$\theta_o^+ = (T_w - T_o)/T^*$	dimensionless temperature at pipe centerline
κ	von Kaman constant (0.40 or 0.41) for the inner law velocity profile, otherwise
κ	thermal conductivity
κ_t	eddy thermal conductivity
μ	molecular viscosity
μ_t	eddy viscosity
ν	kinematic viscosity
ρ	density
ρ_i	partial density of species i
τ	shear stress
φ	dimensionless concentration variable

1.7.1.3 Subscripts

a   species a
b   bulk value
av   average value
CL   centerline value
eff   effective value
H   high Reynolds number value
L   low Reynolds number value
o   centerline or edge condition
p   constant pressure quantity
R   fluid value at the wall
t   turbulent quantity
w   wall property

1.7.1.4 Superscripts

*   dimensionless variable
+   inner law variable

Acronyms

AE   algebraic equations
CFD   computational fluid dynamics
CTP   computational transport phenomena
FDF   fully developed flow
ODE   ordinary differential equations
PC   personal computer
PDE   partial differential equations

REFERENCES

Arpaci, V. S. and P. S. Larsen. 1984. *Convective Heat Transfer*. Englewood Cliffs, NJ: Prentice-Hall.

Bird, R. B., W. E. Stewart, and E. N. Lightfoot. 2002. *Transport Phenomena*, 2nd ed. New York: John Wiley & Sons.

Carslaw, H. S. and J. C. Jaeger. 1959. *Conduction of Heat in Solids*, 2nd ed. Oxford: Clarendon Press.

Coles, D. E. and E. A. Hirst. 1968. Computation of turbulent boundary layers—1968 AFOSR—IFP-Stanford Conference. *Proc. 1968 Conf.*, Vol. 2. Stanford University, Stanford, CA.

Geankoplis, C. J. 2003. *Transport Processes and Separation Process Principles*, 4th ed. Upper Saddle River, NJ: Prentice-Hall.

Hinze, J. O. 1975. *Turbulence*, 2nd ed. New York: McGraw-Hill.

Kader, B. A. 1981. Temperature and concentration profiles in fully turbulent boundary layers. *Int. J. Heat Mass Transfer*. 24(9):1541–1544.

Kays, W. M., M. E. Crawford, and B. Weigand. 2005. *Convective Heat and Mass Transfer*, 4th ed. Boston, MA: McGraw-Hill.

Laufer, J. 1954. The structure of turbulence in fully developed pipe flow. NACA Rept. 1174.

Longwell, P. A. 1966. *Mechanics of Fluid Flow*. New York: McGraw-Hill.

Monin, A. S. and A. M. Yaglom. 1971. *Statistical Fluid Mechanics: Mechanics of Turbulence*, Vol. 1. Cambridge, MA: The MIT Press.

Monin, A. S. and A. M. Yaglom. 1975. *Statistical Fluid Mechanics: Mechanics of Turbulence*, Vol. 2. Cambridge, MA: The MIT Press.

Moretti, G. and M. Abbett. 1966. A time-dependent computational method for blunt body flows. *AIAA J.* 4(12):2136–2141.

Reichardt, H. 1961. The principles of turbulent heat transfer. Translated by P. A. Schoeck. In *Recent Advances in Heat and Mass Transfer*, J. P. Hartnett (Ed.), pp. 223–280. New York: McGraw-Hill.

Roach, P. J. 1998. *Verification and Validation in Computational Science and Engineering*. Albuquerque, NM: Hermosa.

Schlichting, H. 1979. *Boundary-Layer Theory*, 7th ed. Translated by J. Kestin. New York: McGraw-Hill.

von Karman, T. 1954. Fundamental equations in aerothermochemistry. *First AGARD Combustion Colloquium*. London: Butterworths.

Welty, J. R., C. E. Wicks, R. E. Wilson, and G. Rorrer. 2001. *Fundamentals of Momentum, Heat, and Mass Transfer*, 4th ed. New York: John Wiley & Sons.

White, F. M. 2006. *Viscous Fluid Flow*, 3rd ed. Boston, MA: McGraw-Hill.

Wilcox, D. C. 2006. *Turbulence Modeling for CFD*, 3rd ed. La Canada: DCW Industries.

Wilkes, J. O. *Fluid Mechanics for Chemical Engineers*, 2nd ed. 2006. Upper Saddle River, NJ: Prentice-Hall.

CHAPTER 2

The Equations of Change

2.1 INTRODUCTION

The equations of change for an open system are a coupled set of partial differential equations that describe the transport of mass, momentum, and energy at a point in the flow field. These equations are integrated to determine the velocity, temperature, and concentration fields in the geometry of the vessel using the appropriate initial and boundary conditions. These equations represent the conservation of mass, momentum, and energy, and they include terms that represent rate equations and equilibrium relations which also require description.

The equations of change are the result of applying three physical laws to an open system. The laws are the law of conversation of mass, Newton's second law of motion, and the first law of thermodynamics. The resulting partial differential equations are referred to as the continuity equation, the equations of motion or the Navier–Stokes equations, and the general energy equation. The conservation laws are applied to an open system, a finite control volume fixed in space with material flowing through it is used, or they are applied to an arbitrary control volume moving with the mass average velocity of the fluid. Unless specifically stated to the contrary, all of the flows described in this chapter are assumed to be laminar. The treatment of turbulent flow is described in Chapters 4 and 5.

A control volume fixed in space can be a cube where the flow of mass, momentum, and energy enter three of the faces and leave through the other three faces as illustrated in Figure 2.1. The dimensions can be finite or differential lengths. The flows shown in Figure 2.1 represent momentum

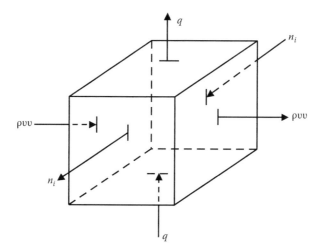

FIGURE 2.1 Control volume as a cube fixed in space.

($\rho \upsilon \upsilon$), mass flux of species i(n_i), and energy (q). The control volume fixed in space can be an arbitrary shape, and a surface element is used to describe the transport in and out of the surface as illustrated in Figure 2.2. In this figure, j_i is the diffusion flux of species i and Ψ is the flux of a property. An arbitrary shaped control volume can be used that is moving with an arbitrary (usually local) fluid velocity.

The advantage of deriving the conservation equations to a cube fixed in space is the visualization of the transport processes through the surfaces.

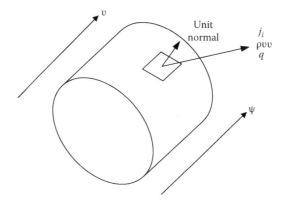

FIGURE 2.2 Control volume as an arbitrary body moving with the local velocity.

This cube fixed in space has the disadvantage in that the resulting equations are in the rectangular coordinate system, and generality to other coordinate systems is not obvious. If the cube fixed in space has finite dimensions, an average velocity, temperature, and concentration are implied at the surfaces of the cube. If differential dimensions are used for the cube, an argument has to be made that the differential dimensions are infinitesimally small but are sufficiently large to not invalidate the treatment of the fluid as a continuum.

The advantage of deriving the conservation equations using an arbitrary control volume is a specific coordinate system is not required. If the control volume is moving with the mass average velocity of the fluid, only diffusive transport of mass, momentum, and energy are included in the balances. If the fluid is not treated as a continuum, the motion of the molecules in the control volume has to be averaged to obtain the local fluid velocity, temperature, and concentration, and this formulation is required when the dimensions of the vessel are of the same order of magnitude as the mean free path of the molecules as might be encountered in the pores of a catalyst. The fluid is treated as a continuum here, and if the equations are needed for the case where the fluid is not a continuum the comparable derivations are provided in Chapman and Cowling (1970).

The equations of change are derived initially using a cube fixed in space, then a general properties balance is used with a control volume moving with the fluid velocity to demonstrate that the equations of change are independent of coordinate systems and of the same form mathematically. This formulation is frequently used in the numerical solution of the equations. The rate equations and equilibrium relations are described for Newtonian and non-Newtonian fluids. Convenient forms of the equations of change are described and tabulated in terms of the rate equations for fluxes of mass momentum and energy and for a Newtonian fluid in several coordinate systems.

2.2 DERIVATION OF THE CONTINUITY EQUATION

The law of conservation of mass is applied to the control volume shown in Figure 2.3. The dimensions of the control volume (system) are Δx, Δy, and Δz, and it is orientated in the flow field such that all of the flow enters through the three faces touching the point (x, y, z) and leaves through the other three faces.

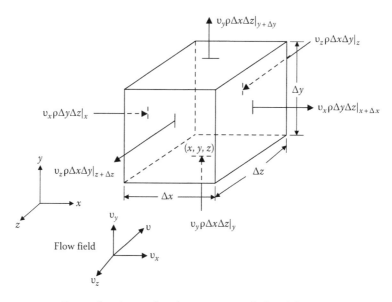

FIGURE 2.3 Control volume fixed in space with fluid flowing through the volume.

The law of conservation of mass can be expressed in the following mnemonic representation:

$$
\begin{vmatrix} \text{Rate of accumulation} \\ \text{of mass in the} \\ \text{control volume} \end{vmatrix} = \begin{vmatrix} \text{Mass flow rate} \\ \text{into the} \\ \text{control volume} \end{vmatrix} - \begin{vmatrix} \text{Mass flow rate} \\ \text{from the} \\ \text{control volume} \end{vmatrix} \quad (2.1)
$$

The flow entering the control volume passes through the faces touching point (x, y, z), and leaving through faces touching point $(x + \Delta x, y + \Delta y, z + \Delta z)$. The mass flow rate into the control volume consists of the sum of the mass flow rate entering the three faces $\Delta x \Delta y$, $\Delta x \Delta z$, and $\Delta y \Delta z$. The mass flow rate entering face $\Delta y \Delta z$ is the product of the area, $\Delta y \Delta z$; the density of the fluid, ρ; and the component of the velocity vector perpendicular to the face, υ_x. This is $\rho \upsilon_x \Delta y \Delta z|_x$ as shown in Figure 2.3. Similar terms can be obtained for the mass flow rates entering and leaving the other faces.

The accumulation of mass in the control volume is the partial derivative with respect to time of the mass in the control volume. The mass in the control volume is the product of the density, ρ, and the volume $\Delta x \Delta y \Delta z$, of the control volume. The law of conservation of mass, Equation 2.1, can be written as

$$\frac{\partial}{\partial t}\left[\Delta x\Delta y\Delta z\rho\right]=\upsilon_x\rho\Delta y\,\Delta z\Big|_x+\upsilon_y\rho\Delta x\Delta z\Big|_y+\upsilon_z\rho\Delta x\Delta\,y\Big|_z-\upsilon_x\rho\Delta y\Delta z\Big|_{x+\Delta x}$$

$$-\upsilon_y\rho\Delta x\Delta z\Big|_{y+\Delta y}-\upsilon_z\rho\Delta x\Delta\,y\Big|_{z+\Delta z} \tag{2.2}$$

Since $\Delta x\Delta y\Delta z$ are fixed in space and not a function of time Equation 2.2 can be written as

$$-\frac{\partial\rho}{\partial t}=\frac{\upsilon_x\rho\Big|_{x+\Delta x}-\upsilon_x\rho\Big|_x}{\Delta x}+\frac{\upsilon_y\rho\Big|_{y+\Delta y}-\upsilon_y\rho\Big|_y}{\Delta y}+\frac{\upsilon_z\rho\Big|_{z+\Delta z}-\upsilon_z\rho\Big|_z}{\Delta z} \tag{2.3}$$

The terms on the right-hand side of the Equation 2.3 are in the form of the definition of partial derivatives. The result of letting Δx, Δy, and Δz approach zero is

$$-\frac{\partial\rho}{\partial t}=\frac{\partial\rho\upsilon_x}{\partial x}+\frac{\partial\rho\upsilon_y}{\partial y}+\frac{\partial\rho\upsilon_z}{\partial z} \tag{2.4}$$

which can be written in vector notation in terms of the vector differential operator ∇ and the velocity vector denoted by an arrow over the symbol as

$$-\frac{\partial\rho}{\partial t}=\nabla\cdot\rho\vec{\upsilon} \tag{2.5}$$

The term $\nabla\cdot\rho\vec{\upsilon}$ is the net mass flux per unit volume through the control volume. The decrease in density with time at a point in the flow field is equal to the net rate of mass efflux per unit volume. This equation is given in Table 2.1 for rectangular and cylindrical coordinates. The equations of

TABLE 2.1 Continuity Equation in Rectangular and Cylindrical Coordinates

Rectangular coordinates (x, y, z)

$$\frac{\partial\rho}{\partial t}+\frac{\partial\rho\upsilon_x}{\partial x}+\frac{\partial\rho\upsilon_y}{\partial y}+\frac{\partial\rho\upsilon_z}{\partial z}=0 \tag{A}$$

Cylindrical coordinates (r, θ, z)

$$\frac{\partial\rho}{\partial t}+\frac{1}{r}\frac{\partial\rho r\upsilon_r}{\partial r}+\frac{1}{r}\frac{\partial\rho\upsilon_\theta}{\partial\theta}+\frac{\partial\rho\upsilon_z}{\partial z}=0 \tag{B}$$

change in cylindrical coordinates can be obtained from the coordinate transformations given in Appendix B or from any of several texts, for example, Bird et al. (1960).

The substantial derivative form of the continuity equation is obtained by expanding Equation 2.5 and rearranging to give the following equation using $\nabla \cdot \rho \vec{\upsilon} = \rho \nabla \cdot \vec{\upsilon} + \vec{\upsilon} \cdot \nabla \rho$.

$$-\frac{\partial \rho}{\partial t} - \vec{\upsilon} \cdot \nabla \rho = \rho \nabla \cdot \vec{\upsilon} \qquad (2.6)$$

The substantial derivative of the density is $D\rho/Dt$, the term on the left-hand side of Equation 2.6.

$$-\frac{D\rho}{Dt} = \rho \nabla \cdot \vec{\upsilon} \qquad (2.7)$$

An important form of Equation 2.7 is for an incompressible fluid (ρ = constant) which is applicable to the flow of most liquids, and Equation 2.7 becomes

$$\nabla \cdot \vec{\upsilon} = 0 \qquad (2.8)$$

2.3 DERIVATION OF THE SPECIES CONTINUITY EQUATION

The law of conservation of mass is applied to species i of a fluid of n species, some or all of which are undergoing chemical reactions, as shown in Figure 2.4. The law of conservation of mass for species i can be expressed in the following mnemonic representation:

$$\begin{bmatrix} \text{Rate of accumulation} \\ \text{of } i \text{ in the control} \\ \text{volume} \end{bmatrix} = \begin{bmatrix} \text{Flow rate of} \\ i \text{ into the} \\ \text{control volume} \end{bmatrix} + \begin{bmatrix} \text{Rate of formation of } i \\ \text{by chemical reaction} \\ \text{in the control volume} \end{bmatrix} - \begin{bmatrix} \text{Flow rate of} \\ i \text{ from the} \\ \text{control volume} \end{bmatrix}$$

$$(2.9)$$

The mass flow of species i entering the control volume is the mass flux of species i, n_i, times the cross section of the face perpendicular to component of the mass flux. The mass flow rate entering the control volume is through the faces touching point (x, y, z), and leaving through faces touching point $(x + \Delta x, y + \Delta y, z + \Delta z)$ as shown in Figure 2.4. The species continuity equation can be written as

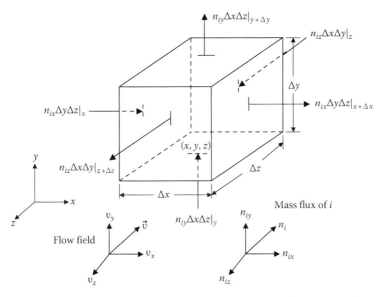

FIGURE 2.4 Control volume fixed in space with mass flow rate of species i flowing through the volume.

$$\frac{\partial}{\partial t}\left[\rho_i \Delta x \Delta y \Delta z\right] = n_{ix}\Delta y \Delta z\big|_x + r_i \Delta x \Delta y \Delta z - n_{ix}\Delta y \Delta z\big|_{x+\Delta x} + n_{iy}\Delta x \Delta z\big|_y$$
$$- n_{iy}\Delta x \Delta z\big|_{y+\Delta y} + n_{iz}\Delta x \Delta y\big|_z - n_{iz}\Delta x \Delta y\big|_{z+\Delta z} \qquad (2.10)$$

where r_i is the rate of formation of species i by chemical reaction.

Simplifying the above and taking the limit as Δx, Δy, Δz go to zero gives the continuity equation for the ith species.

$$\frac{\partial \rho_i}{\partial t} = -\nabla \cdot \vec{n}_i = r_i \quad \text{for} \quad i = 1, 2,\ldots, n \qquad (2.11)$$

Equation 2.11 is given in Table 2.2 A in rectangular and cylindrical coordinates.

The sum of Equation 2.11 over n species gives the continuity equation, Equation 2.5, using

$$\rho = \sum_{i=1}^{n}\rho_i, \quad \rho\vec{v} = \sum_{i=1}^{n}\vec{n}_i, \quad \sum_{i=1}^{n}r_i = 0 \qquad (2.12)$$

TABLE 2.2 Various Forms of the Species Continuity Equation

Continuity equation in terms of the mass flux, n

Rectangular coordinates

$$\frac{\partial \rho_i}{\partial t} + \frac{\partial n_{ix}}{\partial x} + \frac{\partial n_{iy}}{\partial y} + \frac{\partial n_{iz}}{\partial z} = r_i \tag{A}$$

Cylindrical coordinates

$$\frac{\partial \rho_i}{\partial t} + \frac{1}{r}\frac{\partial}{\partial r}(rn_{ir}) + \frac{1}{r}\frac{\partial n_{i\theta}}{\partial \theta} + \frac{\partial n_{iz}}{\partial z} =_i \tag{B}$$

Species continuity equation for constant ρ and D_{AB}

Rectangular coordinates

$$\frac{\partial \rho_A}{\partial t} + \upsilon_x \frac{\partial \rho_A}{\partial x} + \upsilon_y \frac{\partial \rho_A}{\partial y} + \upsilon_z \frac{\partial \rho_A}{\partial z} = D_{AB}\left(\frac{\partial^2 \rho_A}{\partial x^2} + \frac{\partial^2 \rho_A}{\partial y^2} + \frac{\partial^2 \rho_A}{\partial z^2}\right) + r_A \tag{C}$$

Cylindrical coordinates

$$\frac{\partial \rho_A}{\partial t} + \upsilon_r \frac{\partial \rho_A}{\partial r} + \frac{\upsilon_\theta}{r}\frac{\partial \rho_A}{\partial \theta} + \upsilon_z \frac{\partial \rho_A}{\partial z} = D_{AB}\left(\frac{1}{r}\frac{\partial}{\partial r}\left(r\frac{\partial \rho_A}{\partial r}\right) + \frac{1}{r^2}\frac{\partial^2 \rho_A}{\partial \theta^2} + \frac{\partial^2 \rho_A}{\partial z^2}\right) + r_A \tag{D}$$

The mass flux of species i is difficult to evaluate except in a one-dimensional system with no convection. Rather, a diffusion flux (j_i) with respect to a mass (or number) average velocity (υ) is defined and evaluated with a diffusion coefficient and a concentration gradient. Also, a diffusion velocity (υ_{ds}) could be defined by dividing the diffusion flux by the partial mass (or molar) density, The diffusion velocity would therefore be the difference in the velocity of species i and the mass (or molar) average velocity. Using mass-averaged values the following relationships result.

$$\text{Diffusion velocity} = \vec{\upsilon}_{di} = \vec{\upsilon}_i - \vec{\upsilon} \tag{2.13}$$

The diffusion flux of component i relative to the mass average velocity is related to the mass flux of i by

$$\vec{j}_i = \rho_i \vec{\upsilon}_{di} = \rho_i(\vec{\upsilon}_i - \vec{\upsilon}) = \vec{n}_i - \rho_i \vec{\upsilon} \tag{2.14}$$

All of the fluxes and velocities just defined are vectors, which will be denoted by an arrow over the symbol.

Using Equation 2.14 and the species continuity equation, Equation 2.11 can be written in terms of the mass flux relative to the mass average velocity.

$$\frac{\partial \rho_i}{\partial t} + \nabla \cdot (\rho_i \vec{\upsilon}) = -\nabla \cdot \vec{j}_i + r_i \quad \text{for} \quad i = 1, 2, \ldots, n \tag{2.15}$$

The substantial derivative form of Equation 2.15 is obtained by expanding the term $\nabla \cdot (\rho_i \vec{v})$ to obtain

$$\frac{D\rho_i}{Dt} = -\rho_i(\nabla \cdot \vec{v}) - \nabla \cdot \vec{j}_i + r_i \quad \text{for} \quad i = 1, 2, \ldots, n \quad (2.16)$$

2.3.1 BINARY SYSTEMS

Fick's first law of diffusion in a binary mixture is the definition of the binary diffusion coefficient D_{AB}.

$$\vec{j}_A = -\rho D_{AB} \nabla m_A = -\left(\frac{c^2}{\rho}\right) M_A M_B D_{AB} \nabla y_A \quad (2.17)$$

There are equivalent forms of Equation 2.17 in terms of the mass flux $\vec{n}_A$ and molar flux, $\vec{N}_A$, of component A where it must be remembered that the use of a mass (or molar) velocity has been implied in the use of these fluxes. Such a velocity must be evaluated before the species equations can be combined with the other conservation laws. Some applications are so geometrically simple that this step is not necessary. Other forms of Equation 2.17 are:

$$\vec{n}_A = m_A(\vec{n}_A + \vec{n}_B) - \rho D_{AB} \nabla m_A \quad (2.18)$$

$$\vec{N}_A = y_A(\vec{N}_A + \vec{N}_B) - c D_{AB} \nabla y_A \quad (2.19)$$

The species continuity equation for binary diffusion has Equation 2.17 substituted into Equation 2.15 to give an equation in terms of the fluid local mass average velocity and concentration gradients.

$$\frac{\partial \rho_A}{\partial t} + \nabla \cdot (\rho_A \vec{v}) = \nabla \cdot (\rho D_{AB} \nabla m_A) + r_A \quad (2.20)$$

This equation can be used to compute concentration profiles in binary systems, steady or unsteady state, with variable total density and diffusivity. Restrictions are that thermal, pressure, and forced diffusion must be absent. The equations to account for these effects are described in Chapter 3.

For constant density and diffusivity Equation 2.20 can be simplified to the following form using $\nabla \cdot \rho_A \vec{v} = \vec{v} \cdot \rho_A + \rho_A \nabla \cdot \vec{v}$ where $\nabla \cdot \vec{v} = 0$ by continuity.

$$\frac{\partial \rho_A}{\partial t} + \vec{v} \cdot \nabla \rho_A = \frac{D\rho_A}{Dt} = +D_{AB} \nabla^2 \rho_A + r_A \quad (2.21)$$

Equation 2.21 is given in Table 2.2, Equation B, in rectangular and cylindrical coordinates from Bird et al. (1960).

Dividing Equation 2.21 by the molecular weight M_A gives an equation in terms of the molar concentration c_A that is used for dilute liquid solutions.

$$\frac{Dc_A}{Dt} = +D_{AB}\nabla^2 c_A + R_A \tag{2.22}$$

Equation 2.22 can have the same form as the energy equation with constant properties. If the boundary conditions for mass and energy transfer are put in the same form, the solution to the energy equation has the same form as species continuity equation. It is said that there is an analogy between heat and mass transfer. This will be illustrated in the description of flow in the boundary layer involving simultaneous heat and mass transfer.

The equation called Fick's Second Law of Diffusion is obtained from Equation 2.22 for diffusion only ($\upsilon = 0$) and no chemical reaction. This is

$$\frac{\partial c_A}{\partial t} = +D_{AB}\nabla^2 c_A \tag{2.23}$$

Equation 2.23 is used for diffusion in solids and in stationary liquids. It has the same form as the equation for heat conduction. If the boundary and initial conditions are the same, any solution of Equation 2.23 is a solution to the heat conduction equation. The text by Carslaw and Jaeger (1959) contains a number of solutions to this equation for various boundary and initial conditions.

A convenient form of Equation 2.11 in molar units for a binary system is obtained by dividing by the molecular weight M_A:

$$\frac{\partial c_A}{\partial t} + \nabla \cdot \vec{N}_A = R_A \tag{2.24}$$

where
  c_A is the molar concentration
  $\vec{N}_A$ is the molar mass flux
  R_A is the reaction rate

To solve binary diffusion problems using Equation 2.24, information about the diffusion mechanism or concentration is used in analyzing the form of $(N_A + N_B)$ in Equation 2.19. Several cases are encountered in physical systems which are summarized below.

- Species A is slightly soluble in species B, $y_A = 0$, and $y_A(\vec{N}_A + \vec{N}_B) = 0$, and Equation 2.19 becomes

$$\vec{N}_A = -cD_{AB}\nabla y_A \qquad (2.25)$$

and Equation 2.24 has the form

$$\frac{\partial c_A}{\partial t} - \nabla \cdot \left(cD_{AB}\nabla y_A \right) = R_A \qquad (2.26)$$

- Diffusion of species A through a stagnant film of species B, i.e., $N_B = 0$, and Equation 2.19 becomes

$$\vec{N}_A = x_A\vec{N}_A - cD_{AB}\nabla y_A \qquad (2.27)$$

or

$$\vec{N}_A = -\frac{c}{(1-x_A)}D_{AB}\nabla y_A \qquad (2.28)$$

and Equation 2.24 has the form

$$\frac{\partial c_A}{\partial t} - \nabla \cdot \left[\frac{c}{(1-y_A)}D_{AB}\nabla y_A \right] = R_A$$

- Diffusion of species A through species B with rapid chemical reaction on the surface $A \rightarrow \frac{1}{2}B$ to have $N_B = -\frac{1}{2}N_A$ diffusing from the surface. For this case $y_A(N_A + N_B) = y_A(N_A - \frac{1}{2}N_A) = \frac{1}{2}y_A N_A$. Substituting into Equation 2.24 gives

$$\vec{N}_A = \frac{1}{2}y_A\vec{N}_A - cD_{AB}\nabla y_A \qquad (2.29)$$

or

$$\vec{N}_A = -\frac{c}{(1-\frac{1}{2}y_A)}D_{AB}\nabla y_A \qquad (2.30)$$

and Equation 2.24 has the form

$$\frac{\partial c_A}{\partial t} - \nabla \cdot \left[\frac{c}{(1-\frac{1}{2}y_A)}D_{AB}\nabla y_A \right] = R_A \qquad (2.31)$$

- For a homogeneous, first-order, irreversible reaction in the fluid $A \to B$, the reaction rate is $R_A = -kc_A$. With species A being slightly soluble in species B and $c_A = y_A c$, Equation 2.26 has the form:

$$\frac{\partial c_A}{\partial t} = D_{AB} \nabla^2 c_A - kc_A \tag{2.32}$$

The approach described here can be applied to other binary diffusion problems with chemical reactions. For example, a surface reaction could occur at a finite rate rather than instantaneously as shown above. There would be a rate equation for the surface reaction that describes the disappearance of A at the surface by $N_A = kc_{A,\text{surface}}$, where $c_{A,\text{surface}}$ is a surface concentration. The procedure would be the same as used above, and Bird et al. (2002) describes this case and some others that have analytical solutions.

2.3.2 MULTICOMPONENT SYSTEMS

In general, a mass flux is caused by all of the diffusing species. These multi-component diffusion coefficients mean that coupled solutions for the diffusing species are required. The equations governing this behavior are given in Section 3.4.2. Methods of treating such coupling are also discussed in the same chapter.

In addition to mass fluxes of diffusing components resulting from concentration gradients, a mass flux of species with respect to the mean fluid motion can be caused by pressure gradients, temperature gradients, and external force differences. These effects also are discussed in Section 3.4.2.

2.3.3 GENERALIZED CHEMICAL REACTIONS AND SIMULTANEOUS REACTION RATES

Solution of the continuity equation, Equation 2.11, requires an expression for r_i to describe the chemical reaction rate. Chemical reactions involving n chemical species and m chemical reactions are given by

$$\sum_{i=1}^{n} r_{ji} A_i \underset{k_{rj}}{\overset{k_{fj}}{\Longleftrightarrow}} \sum_{i=1}^{n} p_{ji} A_i \quad \text{for} \quad j = 1, 2, \ldots, m \tag{2.33}$$

where

A_i represents the chemical symbol of the ith species

r_{ji} and p_{ji} are the stoichiometric coefficients of the reactants and products, respectively

Based on the nomenclature in Equation 2.33, a general expression can be written for the rate of reaction of the ith species for m simultaneous reactions.

$$r_i = \sum_{j=1}^{m} (p_{ij} - r_{ji}) \left\{ k_{fj} \prod_{i=1}^{n} c_i^{r_{ji}} - k_{rj} \prod_{i=1}^{n} c_i^{p_{ji}} \right\} \tag{2.34}$$

where

k_{fj} and k_{rj} is the forward and reverse rate constants, respectively

c_k is the concentration of component k in appropriate units

The rate constants are a function of temperature given by the Arrhenius equation (Hirschfelder, 1954).

In Chapter 3, a description is given of chemical equilibrium and reaction kinetics for computational transport computations that employs Equation 2.34. Chapter 3 includes methods used to describe equilibrium constants to evaluate the reverse rate constant and free-energy minimization to predict equilibrium concentrations for very fast reactions.

2.4 DERIVATION OF THE EQUATION OF MOTION

The result of applying Newton's second law of motion to an open system—a control volume fixed in space with fluid flowing through—is referred to as the equations of motion or the momentum equation. For a control volume fixed in space, Newton's second law for an open system can be written mnemonically as

Sum of forces acting on the control volume	=	Rate of momentum leaving the control volume	−	Rate of momentum entering the control volume	+	Rate of accumulation of the momentum in the control volume

$$\tag{2.35}$$

The equation of motion is a vector equation with x-, y-, and z-components in rectangular coordinates. The x-component is indicated by

$$\sum F_x = M_{out,x} - M_{in,x} + \frac{\partial M_{cv}}{\partial t} \tag{2.36}$$

2.4.1 FORCES AND STRESSES

Reviewing the forces or stresses (force per unit area) that act on the control volume, they include surface forces and body forces. Body forces are the ones that are proportional to the volume or mass of the body and comprise those forces that involve action at a distance. Examples are gravitational attraction, magnetic forces, and electrodynamic forces.

Surface forces are ones that exert force on the control surface by material outside the control volume. Such forces are exerted by pressure and viscous effects. Normal and shear stresses cause these forces. Pressure is one component of the normal stresses. It is taken as positive in compression. The normal viscous stresses are τ_{xx}, τ_{yy}, τ_{zz}; and they are positive in tension with respect to the positive normals to the surfaces they act upon. The viscous shear stresses act parallel to the surfaces they act upon; they have the same sense as the surface they act upon. The components of the stress tensor comprise a matrix. The first subscript on the matrix element indicates the surface being stressed; the second is the direction that the normal or shear stress acts. Surfaces are vectors; the positive direction of which is the outward normal to the control surface of interest. For the unit cube, three of these normals are in the positive coordinate direction, and three are in the negative coordinate direction. If the positive surface normal points are in a negative coordinate direction, then all three of the viscous stresses on that surface are also positive in all three of the negative coordinate directions. Figure 2.5 indicates the direction of these stress components. For pressure to be compressive it must act in the negative direction of each of the surface normals. This sign convention is customary and it results in the following relationships for the normal stresses.

$$\sigma_{xx} = -P + \tau_{xx} \tag{2.37a}$$

$$\sigma_{yy} = -P + \tau_{yy} \tag{2.37b}$$

$$\sigma_{zz} = -P + \tau_{zz} \tag{2.37c}$$

where
  σ will be used to represent total surface stresses
  τ will be used to represent viscous normal and shear stresses

A different sign convention is to have the normal stresses be positive in compression to be consistent with the pressure and have the shear stresses

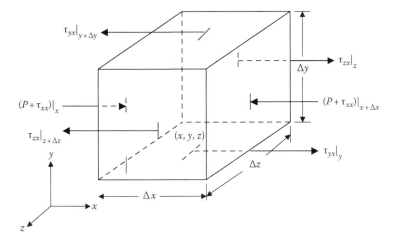

FIGURE 2.5 Shear and normal stresses acting on the control volume in the *x*-direction.

positive on faces in contact with the point (x, y, z) as shown in Figure 2.6. For the fluids at rest or in motion such that the velocity is everywhere the same, each normal stress is equal to the pressure. The resulting stresses are shown in Figure 2.7. The argument is that this sign convention is the same as that for the convective momentum $(\rho \vec{v} \vec{v})$ and that it gives the same form of the equations relating shear stress to shear rate as Fick's diffusion law and Fourier's heat conduction law. These laws for mass and thermal diffusion will be discussed in detail in Chapter 3. The users of this sign convention are very much in the minority; notably Bird et al. (2002) and Brodkey (1967). The remainder of this book will use the customary sign convention exclusively.

An interesting feature, which may be observed by carefully comparing Figures 2.5 and 2.7, is that every component of the stress tensor is exactly opposite in sign when the two conventions are compared. The proponents of the second convention method also carry it a step further by assigning a negative value to the coefficient of viscosity. Viscosity will also be discussed further, but it is simply the scalar coefficient which relates the shear stress to the rate of strain. When this sign change is made and the shear stress is eliminated in favor of the rate of strain (which consists of velocity derivatives), the resulting momentum equations become identical.

Shear stress and the rate of strain are second-order tensors. Tensors will be discussed at length in Appendix B. However, a working definition

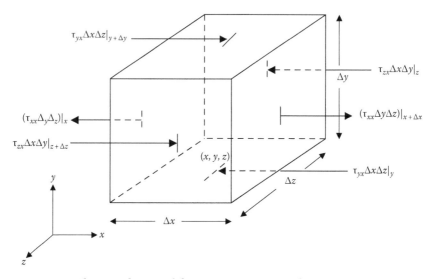

FIGURE 2.6 Shear and normal forces acting on a surface in the x-direction.

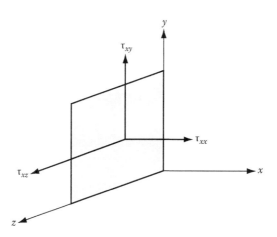

FIGURE 2.7 Shear and normal stresses using a sign convention which is the same as that for the convective momentum flux.

is as follows. Vectors are terms which have three scalar components; one in each of the three coordinate directions which are defined by base vectors. Addition and several types of multiplication of vectors are defined operations. But what is a vector divided by a vector and what is the product of two vectors without a dot or cross multiplication sign. Such operations indicate a simplification of the multiplication process. The division process and a multiplication without the dot or cross simplification produce second-order tensors. These terms have nine scalar coefficients multiplied by pairs of base vectors. For the present these definitions are sufficient. The rectangular (orthogonal) Cartesian coordinate system will be used to discuss the conservation laws and their component parts first because the base vectors are of unit magnitude and always point in the same direction. Other coordinate systems do not have such a simple construction.

In summary, a stress acts on each face of the control volume and has one component normal to the surface and two tangential to the surface, as shown in Figure 2.7, for the face in the y–z plane. The sign conventions are that normal and shear stresses are considered positive when acting on control volume surfaces which are positive and negative when the surfaces are negative. As shown in Figure 2.5, the first subscript on a stress gives the surface on which the stress acts by giving the outward normal to that surface and the second subscript gives the direction that the stress acts. Stresses with repeated subscripts are normal stresses, and those with mixed subscripts are shear stresses.

2.4.2 DERIVATION OF THE X-COMPONENT OF THE EQUATION OF MOTION

For the sum of forces term in Equation 2.36, the x-components of the shear and normal stresses acting on the surfaces of the control volume are shown in Figure 2.5. The x-component of the body force is represented by $\sum \rho_i g_{ix}$, and each chemical species can be acted upon by a different external force per unit mass, g_{xi} such as electric fields (Chapman, 1969). If gravity is the only body force acting on the control volume, then $\sum \rho_i g_{ix} = \rho g_x = \rho g \cos \beta$ where g is the gravitational constant (9.8066 m/s^2) and β is the angle between the x axis and the direction that gravity acts.

The sum of the forces in the x-direction is the sum of the body, shear, and normal forces:

$$\sum_x F_x = \sum_i \rho_i g_{xi} \Delta x \Delta y \Delta z + (+P - \tau_{xx}) \Delta y \Delta z \big|_x - \tau_{yx} \Delta x \Delta y \big|_y - \tau_{zx} \Delta x \Delta y \big|_z$$

$$+ (-P + \tau_{xx}) \Delta y \Delta z \big|_{x+\Delta x} + \tau_{yx} \Delta x \Delta y \big|_{y+\Delta y} + \tau_{zx} \Delta x \Delta z \big|_{z+\Delta z} \qquad (2.38)$$

Rearranging gives

$$\sum F_x = \Delta x \Delta y \Delta z \left\{ \sum \rho_i g_{xi} + \left[\frac{\left(-P + \tau_{xx}\right)\big|_{x+\Delta x} - \left(-P + \tau_{xx}\right)\big|_x}{\Delta x} \right] \right\}$$
$$+ \Delta x \Delta y \Delta z \left[\frac{\tau_{yx}\big|_{y+\Delta y} - \tau_{yx}\big|_y}{\Delta y} + \frac{\tau_{zx}\big|_{z+\Delta z} - \tau_{zx}\big|_z}{\Delta z} \right] \tag{2.39}$$

In Figure 2.8, the control volume is shown with the convective momentum entering and leaving the control volume in the x-direction only. Rate of momentum is defined as the product of the mass flow rate and the velocity in the direction of the mass flow rate.

$$M_{x,\text{in}} = W \upsilon_x = \rho[\upsilon_x \Delta y \Delta z + \upsilon_y \Delta x \Delta z + \upsilon_z \Delta x \Delta y]\upsilon_x \tag{2.40}$$

The net momentum flux through the control volume is

$$M_{x,\text{out}} - M_{x,\text{in}} = \rho \Delta y \Delta z\, \upsilon_x \upsilon_x\big|_{x+\Delta x} - \rho \Delta y \Delta z\, \upsilon_x \upsilon_x\big|_x + \rho \Delta x \Delta z\, \upsilon_y \upsilon_x\big|_{y+\Delta y}$$
$$- \rho \Delta x \Delta z\, \upsilon_y \upsilon_x\big|_y + \rho \Delta x \Delta y\, \upsilon_z \upsilon_x\big|_{z+\Delta z} - \rho \Delta x \Delta y\, \upsilon_z \upsilon_x\big|_z$$

$$\tag{2.41}$$

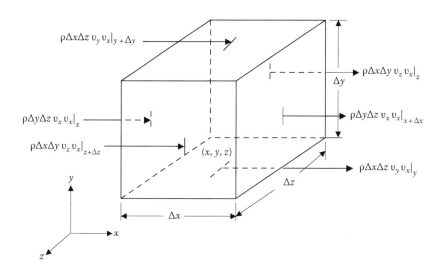

FIGURE 2.8 The rate of convective momentum through control volume in the x-direction only.

Rearranging Equation 2.41 gives

$$\left(M_{outx} - M_{inx}\right) = \Delta x \Delta y \Delta z \left[\frac{\rho \upsilon_x \upsilon_x \big|_{x+\Delta x} - \rho \upsilon_x \upsilon_x \big|_x}{\Delta x} + \frac{\rho \upsilon_y \upsilon_x \big|_{y+\Delta y} - \rho \upsilon_y \upsilon_x \big|_y}{\Delta y} + \frac{\rho \upsilon_z \upsilon_x \big|_{z+\Delta z} - \rho \upsilon_z \upsilon_x \big|_z}{\Delta z} \right]$$

(2.42)

The rate of change of the momentum in the control volume in the x-direction is given by

$$\frac{\partial M_{cv}}{\partial t} = \frac{\partial \left(\Delta x \Delta y \Delta z\right) \rho \upsilon_x}{\partial t} = \Delta x \Delta y \Delta z \frac{\partial \rho \upsilon_x}{\partial t}$$

(2.43)

Substituting Equations 2.39, 2.42, and 2.43 into Equation 2.36 and taking the limit as Δx, Δy, and Δz go to zero, the result is the x-component of the momentum equation.

$$\sum P_i g_{ix} + \left[-\frac{\partial P}{\partial x} + \frac{\partial \tau_{xx}}{\partial x} + \frac{\partial \tau_{yx}}{\partial y} + \frac{\partial \tau_{zx}}{\partial z} \right]$$
$$= \frac{\partial}{\partial x}(\rho \upsilon_x \upsilon_x) + \frac{\partial}{\partial y}(\rho \upsilon_y \upsilon_x) + \frac{\partial}{\partial z}(\rho \upsilon_z \upsilon_x) + \frac{\partial}{\partial t}(\rho \upsilon_x)$$

(2.44)

The same procedure can be used to obtain the y- and z-components of the momentum equation shown below.

y-component

$$\sum P_i g_{iy} + \left[-\frac{\partial P}{\partial y} + \frac{\partial \tau_{xy}}{\partial x} + \frac{\partial \tau_{yy}}{\partial y} + \frac{\partial \tau_{zy}}{\partial z} \right]$$
$$= \frac{\partial}{\partial x}(\rho \upsilon_x \upsilon_y) + \frac{\partial}{\partial y}(\rho \upsilon_y \upsilon_y) + \frac{\partial}{\partial z}(\rho \upsilon_z \upsilon_y) + \frac{\partial}{\partial t}(\rho \upsilon_y)$$

(2.45)

z-component

$$\sum P_i g_{iz} + \left[-\frac{\partial P}{\partial x} + \frac{\partial \tau_{xx}}{\partial x} + \frac{\partial \tau_{yz}}{\partial y} + \frac{\partial \tau_{zz}}{\partial z} \right]$$
$$= \frac{\partial}{\partial x}(\rho \upsilon_x \upsilon_z) + \frac{\partial}{\partial y}(\rho \upsilon_y \upsilon_z) + \frac{\partial}{\partial z}(\rho \upsilon_z \upsilon_z) + \frac{\partial}{\partial t}(\rho \upsilon_z)$$

(2.46)

The above three components of the equation of motion can be written in the concise form of vector notation as

$$\sum \rho_i \vec{g}_i - \nabla P + \nabla \cdot \vec{\vec{\tau}} = \nabla \cdot \rho \vec{v}\vec{v} + \frac{\partial}{\partial t} \rho \vec{v} \qquad (2.47)$$

The first term represents the body force acting per unit volume, the second term represents pressure forces per unit volume, the third represents the viscous shear and normal forces acting per unit volume, the fourth represents the rate of change of convective momentum per unit volume, and the last term represents the rate of change of momentum with time per unit volume. The term $\vec{v}\vec{v}$ is a dyadic product. The terms $\nabla \cdot \vec{\vec{\tau}}$ and $\nabla \cdot \rho \vec{v}\vec{v}$ are single dot products of a vector ∇ and tensors $\vec{\vec{\tau}}$ and $\vec{v}\vec{v}$; they are vectors.

Equation 2.47 can be written in terms of the substantial derivative with the aid of the continuity equation as follows:

$$\rho \frac{D\vec{v}}{Dt} = -\nabla P + \nabla \cdot \vec{\vec{\tau}} + \sum \rho_i \vec{g}_i \qquad (2.48)$$

In the language of computational fluid dynamics (CFD), this form of the momentum equation is said to be the nonconservative form of the equation, because it has already been combined with the continuity equation.

The components of Equation 2.48 are given in Table 2.3 in rectangular and cylindrical coordinates. A description of the stress tensor, $\vec{\vec{\tau}}$, is needed to apply the equation of motion.

The stress has been termed a tensor, but it has not yet been demonstrated that it is a tensor. Its tensor character will be demonstrated subsequently. To this point, it should be considered only a three by three matrix of elements.

2.4.3 RATE OF DEFORMATION

As a fluid element moves it undergoes: (1) translation, (2) rotation, (3) dilatation, and (4) shear strain. Conceptually, a fluid cube could deform with time as shown in Figure 2.9. Translation and rotation moves the fluid as though it were a solid body, so no strain is associated with these motions. Dilatation is associated with linear motions which cause expansive or compressive strains on the element. Shear strain cause angular distortions of the element.

TABLE 2.3 Momentum Equation in Terms of the Shear and Normal Stresses

Rectangular coordinates

x-component

$$\rho\left(\frac{\partial v_x}{\partial t}+v_x\frac{\partial v_x}{\partial x}+v_y\frac{\partial v_x}{\partial y}+v_z\frac{\partial v_x}{\partial z}\right)=-\frac{\partial p}{\partial x}+\left(\frac{\partial \tau_{xx}}{\partial x}+\frac{\partial \tau_{yx}}{\partial y}+\frac{\partial \tau_{zx}}{\partial z}\right)+\sum\rho_i g_{xi} \qquad (A)$$

y-component

$$\rho\left(\frac{\partial v_y}{\partial t}+v_x\frac{\partial v_y}{\partial x}+v_y\frac{\partial v_y}{\partial y}+v_z\frac{\partial v_y}{\partial z}\right)=-\frac{\partial p}{\partial y}+\left(\frac{\partial \tau_{xy}}{\partial x}+\frac{\partial \tau_{yy}}{\partial y}+\frac{\partial \tau_{zy}}{\partial z}\right)+\sum\rho_i g_{yi} \qquad (B)$$

z-Component

$$\rho\left(\frac{\partial v_z}{\partial t}+v_x\frac{\partial v_z}{\partial x}+v_y\frac{\partial v_z}{\partial y}+v_z\frac{\partial v_z}{\partial z}\right)=-\frac{\partial p}{\partial z}+\left(\frac{\partial \tau_{xz}}{\partial x}+\frac{\partial \tau_{yz}}{\partial y}+\frac{\partial \tau_{zz}}{\partial z}\right)+\sum\rho_i g_{zi} \qquad (C)$$

Cylindrical coordinates

r-component (where $\rho v_\theta^2/r$ is the centrifugal "force")

$$\rho\left(\frac{\partial v_r}{\partial t}+v_r\frac{\partial v_r}{\partial r}+\frac{v_\theta}{r}\frac{\partial v_r}{\partial \theta}-\frac{v_\theta^2}{r}+v_z\frac{\partial v_r}{\partial z}\right)$$

$$=-\frac{\partial p}{\partial r}+\left(\frac{1}{r}\frac{\partial}{\partial r}\left(r\tau_{rr}\right)+\frac{1}{r}\frac{\partial \tau_{r\theta}}{\partial \theta}-\frac{\tau_{\theta\theta}}{r}+\frac{\partial \tau_{rz}}{\partial z}\right)+\sum\rho_i g_{ri} \qquad (D)$$

θ-component (where $\rho v_r v_\theta/r$ is the Coriolis "force")

$$\rho\left(\frac{\partial v_\theta}{\partial t}+v_r\frac{\partial v_\theta}{\partial r}+\frac{v_\theta}{r}\frac{\partial v_\theta}{\partial \theta}+\frac{v_r v_\theta}{r}+v_z\frac{\partial v_\theta}{\partial z}\right)$$

$$=-\frac{1}{r}\frac{\partial p}{\partial \theta}+\left(\frac{1}{r^2}\frac{\partial}{\partial r}\left(r^2\tau_{r\theta}\right)+\frac{1}{r}\frac{\partial \tau_{\theta\theta}}{\partial \theta}+\frac{\partial \tau_{\theta z}}{\partial z}\right)+\sum\rho_i g_{\theta i} \qquad (E)$$

z-component

$$\rho\left(\frac{\partial v_z}{\partial t}+v_r\frac{\partial v_z}{\partial r}+\frac{v_\theta}{r}\frac{\partial v_z}{\partial \theta}+v_z\frac{\partial v_z}{\partial z}\right)$$

$$=-\frac{\partial p}{\partial z}+\left(\frac{1}{r}\frac{\partial}{\partial r}\left(r\tau_{rz}\right)+\frac{1}{r}\frac{\partial \tau_{\theta z}}{\partial \theta}+\frac{\partial \tau_{zz}}{\partial z}\right)+\sum\rho_i g_{zi} \qquad (F)$$

Using a cube as the fluid element comparable to the one shown in Figure 2.9, the normal strain (S_{ii}) is the change in length to the original length of a side of the fluid element as shown in Figure 2.10. The strain rate is given by the following equations.

$$\frac{dS_{xx}}{dt}=\frac{d}{dt}\left[\frac{\Delta x_{t+dt}-\Delta x_t}{\Delta x_t}\right]; \quad \frac{dS_{yy}}{dt}=\frac{d}{dt}\left[\frac{\Delta y_{t+dt}-\Delta y_t}{\Delta y_t}\right]; \quad \frac{dS_{zz}}{dt}=\frac{d}{dt}\left[\frac{\Delta z_{t+dt}-\Delta z_t}{\Delta z_t}\right]$$

$$(2.49)$$

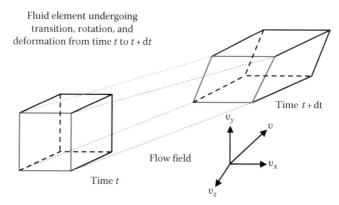

FIGURE 2.9 Diagram of a fluid element undergoing transition, rotation, and deformation from time t to $t + dt$.

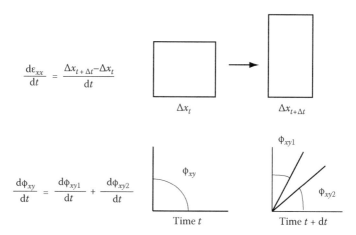

FIGURE 2.10 Diagrams illustrating the rates of normal and shearing strain.

To determine the normal rate of strain (e_{ii}), the limit of Equations 2.49 is taken as the time increment approaches zero.

$$e_{xx} = \lim_{dt \to 0} \frac{d}{dt}\left[\frac{\Delta x_{t+\Delta t} - \Delta x_t}{\Delta x_t}\right] = \frac{\partial v_x}{\partial x} \tag{2.50}$$

$$e_{yy} = \lim_{dt \to 0} \frac{d}{dt}\left[\frac{\Delta y_{t+\Delta t} - \Delta y_t}{\Delta y_t}\right] = \frac{\partial v_y}{\partial y} \tag{2.51}$$

$$e_{zz} = \lim_{dt \to 0} \frac{d}{dt}\left[\frac{\Delta z_{t+\Delta t} - \Delta z_t}{\Delta z_t}\right] = \frac{\partial v_z}{\partial z} \tag{2.52}$$

The velocity gradients shown in Figure 2.11 will cause a distortion of the fluid element to a shape like that shown in Figure 2.12. If $d\Phi_{xy2}$ and $d\Phi_{xy1}$ are in opposite directions, only a rotation of the element would result, thus

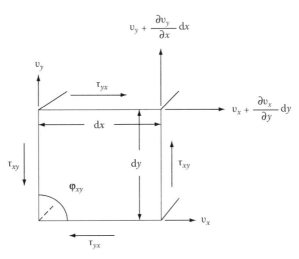

FIGURE 2.11 Shear stresses from a velocity gradient acting on the fluid element causing shearing strain.

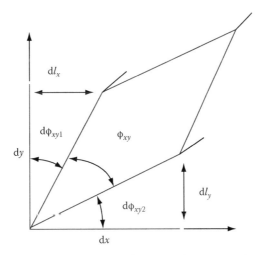

FIGURE 2.12 Shear strain as seen from a observer riding on the fluid element.

$$d\Omega_z = \frac{1}{2}(d\Phi_{xy2} - d\Phi_{xy1}) \tag{2.53}$$

In terms of velocity gradients:

$$d\Phi_{xy2} = \lim_{dt \to 0} \left(\tan^{-1} \left\{ \frac{\dfrac{\partial v_y}{\partial x} dx\, dt}{\left[dx + \dfrac{\partial v_x}{\partial x} dx\, dt \right]} \right\} \right) = \frac{\partial v_y}{\partial x} dt \tag{2.54}$$

$$d\Phi_{xy1} = \lim_{dt \to 0} \left(\tan^{-1} \left\{ \frac{\dfrac{\partial v_x}{\partial y} dy\, dt}{\left[dy + \dfrac{\partial v_y}{\partial y} dy\, dt \right]} \right\} \right) = \frac{\partial v_x}{\partial y} dt \tag{2.55}$$

$$\frac{d\Omega_z}{dt} = \frac{1}{2}\left(\frac{\partial v_y}{\partial x} - \frac{\partial v_x}{\partial y} \right) \tag{2.56}$$

The one-half is used to represent an average value over the incremental time step. This average value represents the angular rotation of the major diagonal of the rhombus shown in Figure 2.12.

By considering a permutation of the coordinates, the other components of the angular velocity vector are determined.

$$\frac{d\Omega_x}{dt} = \frac{1}{2}\left(\frac{\partial v_z}{\partial y} - \frac{\partial v_y}{\partial z} \right) \tag{2.57}$$

$$\frac{d\Omega_y}{dt} = \frac{1}{2}\left(\frac{\partial v_x}{\partial z} - \frac{\partial v_z}{\partial x} \right) \tag{2.58}$$

The vorticity vector is defined by

$$\vec{\omega} = 2\frac{d\vec{\Omega}}{dt} = \nabla \times \vec{v} \tag{2.59}$$

If the vorticity is zero, the flow is said to be irrotational.

The shear strain is determined with the same angular displacement combined in a different manner. From Equation 2.53

$$\frac{d\Phi_{xy}}{dt} = \frac{1}{2}\left(\frac{d\Phi_{xy2}}{dt} - \frac{d\Phi_{xy1}}{dt}\right) \tag{2.60}$$

$$\frac{d\Phi_{xy}}{dt} = \frac{1}{2}\left(\frac{\partial v_y}{\partial x} + \frac{\partial v_x}{\partial y}\right) = e_{xy} = e_{yx} \tag{2.61}$$

For a permutation of the coordinates the other strain rates become

$$\frac{1}{2}\left(\frac{\partial v_z}{\partial y} + \frac{\partial v_y}{\partial z}\right) = e_{yz} = e_{zy} \tag{2.62}$$

$$\frac{1}{2}\left(\frac{\partial v_x}{\partial z} + \frac{\partial v_z}{\partial x}\right) = e_{zx} = e_{xz} \tag{2.63}$$

To this point, two matrices have been defined: the stress matrix $(\hat{\tau})$ and the rate of strain (also called the rate of deformation) $(\hat{e})$. These matrices represent the elements of two second-order tensors, but the proof of this is offered in Section 2.4.4.

$$\hat{\tau} = \begin{bmatrix} \tau_{xx} & \tau_{xy} & \tau_{xz} \\ \tau_{yx} & \tau_{yy} & \tau_{yz} \\ \tau_{zx} & \tau_{zy} & \tau_{zz} \end{bmatrix} \tag{2.64}$$

$$\hat{e} = \begin{bmatrix} e_{xx} & e_{xy} & e_{xz} \\ e_{yx} & e_{yy} & e_{yz} \\ e_{zx} & e_{zy} & e_{zz} \end{bmatrix} \tag{2.65}$$

2.4.4 RELATIONSHIP BETWEEN STRESS AND THE RATE OF STRAIN

To establish that the stress and rate of strain are second-order tensors, two criteria need to be met. First, two directions need to be associated with each of these terms. Since rectangular Cartesian coordinates are only being considered at this time, the base vectors are of unit magnitude and always point in the same direction. This means that such vectors can be added at the conclusion of our analysis without introducing any complexity. The second criterion for the quantity to be a tensor is to demonstrate that it remains invariant under a coordinate transformation. Being invariant does not mean that the elements of the quantity remain constant, only that the elements defined in one coordinate system can be transformed into elements in the other coordinate system likely with different values. For example, if a pitcher throws a ball toward home plate at 45 m/s, a coordinate system originating at the pitcher's mound extending toward home plate would indicate that the ball was traveling at +45 m/s. From the batter's perspective, a coordinate system originating at home plate extending toward the pitcher's mound would indicate that the ball was traveling at −45 m/s. The ball speed would be invariant in either case.

The stress and rate of strain have been defined in an x-, y-, z-coordinate rectangular system originating at an arbitrary point (x, y, z). Yuan (1967) presents this transformation in detail and shows that each element of the stress matrix transforms by the same geometric rules as each element of the rate of strain matrix. This proves that the stress and rate of strain are both second-order tensors.

Before considering transforming the matrix elements further, note that the e_{ij} equals e_{ji}. Assume that $\tau_{ij} = \tau_{ji}$. This assumption will be justified subsequently. This means that both matrices are symmetric. For symmetric tensors, eigenvectors can be defined. To accomplish this construct a tetrahedron $(o–a–b–c)$ located such that its apex is at the origin (o) of the incremental cube for which the stress and rate of strain were defined. Denoting the axes which form the tetrahedron as (a, b, c), the normals to the surfaces $a–b$, $b–c$, $c–a$ point in the directions of the eigenvectors, say 1, 2, and 3. This coordinate system has coordinate lines O–a, O–b, and O–c. These are also called the principal axes. For symmetric tensors, there are only normal stresses and only normal deformations on the surface $a–b–c$. The axes a, b, c constitute a rectangular Cartesian coordinate system. Now consider a transformation from (x, y, z) to (a, b, c). The stresses on the abc surface are called the principal stresses and the strain rates are the principal strain rates. Lamb in 1879 described this analysis and attributed it to Stokes in 1845 (Lamb, 1945). It has also been described by Rouse (1959). It was described in the language of tensors by White (2006) and Sokolnikoff (1964). This provides a background for relating the stress to the rate of strain, but no relationship.

Since $\vec{\vec{\tau}}$ and $\vec{e}$ are now second-order symmetric tensors, scalar invariants can now be determined. A scalar invariant for a vector is its magnitude.

For a second-order tensor, say the rate of strain, three invariants can be defined as

$$I_1 = e_{11} + e_{22} + e_{33} = \nabla \cdot \vec{\upsilon} \tag{2.66}$$

$$I_2 = e_{xx}e_{yy} + e_{yy}e_{zz} + e_{zz}e_{xx} - e_{xy}^2 - e_{yz}^2 - e_{zx}^2 \tag{2.67}$$

$$I_3 = \begin{vmatrix} e_{xx} & e_{xy} & e_{xz} \\ e_{yx} & e_{yy} & e_{yz} \\ e_{zx} & e_{zy} & e_{zz} \end{vmatrix} \tag{2.68}$$

For the principal axes:

$$I_1 = e_1 + e_2 + e_3 = \nabla \cdot \vec{\upsilon} \tag{2.69}$$

$$I_2 = e_1 e_2 + e_2 e_3 + e_3 e_1 \tag{2.70}$$

$$I_3 = \begin{vmatrix} e_1 & 0 & 0 \\ 0 & e_2 & 0 \\ 0 & 0 & e_3 \end{vmatrix} = e_1 e_2 e_3 \tag{2.71}$$

The stress tensor components will be related to the rate of strain components by making the following assumptions which are similar in spirit to the relationships between stress and strain made to describe elastic solids (Daily and Harleman, 1966).

1. Stress components will be expressed as linear functions of the rate of strain components.
2. These relationships will be invariant with respect to rotation of one rectangular Cartesian coordinate system to another. This means that the fluid is isotropic such that its properties are independent of direction. The fluid being isotropic implies that the principal stress axes and the principal rate of strain axes must be identical.
3 The stress components will reduce to hydrostatic pressure when all velocities are zero.

Linear functions of the rate of strain components are made to represent each of the stress components. Negative pressure is added to the normal stress component to account for hydrostatic pressure.

$$\sigma_{11} = -P + \tau_{11} = -P + (A+B)\,e_{11} + Be_{22} + Be_{33} \tag{2.72}$$

$$\sigma_{22} = -P + \tau_{22} = -P + Be_{11} + (A+B)\,e_{22} + Be_{33} \tag{2.73}$$

$$\sigma_{33} = -P + \tau_{33} = -P + Be_{11} + Be_{22} + (A+B)\,e_{33} \tag{2.74}$$

The constants A and B must be supplied to complete this analysis. Only two constants are needed because the fluid is assumed to be isotropic. Based on experimental test data from the middle nineteenth century, $A = 2\mu$. Despite the age of the data, there is no indication that this constant needs to be changed. To evaluate the constant B, sum Equations 2.72 through 2.74 and divide the result by three.

$$\left(\sigma_{11} + \sigma_{22} + \sigma_{33}\right)/3 = -P + (B)(e_{11} + e_{22} + e_{33}) + 2\mu(e_{11} + e_{22} + e_{33})/3$$
$$= -P + (B + 2\mu/3)(\nabla \cdot \vec{\upsilon}) \tag{2.75}$$

If the velocity is zero, the pressure is equal to the average of the sum of the normal stresses. Since pressure is compressive and the normal stress component is in tension, the pressure is correctly represented. The right-hand side of this equation reduces to $-P$ when the velocity is zero. Equation 2.75 is transformed back to the (x, y, z) coordinate system, and B is taken to be $-2\mu/3$ so that the equation would be valid when $\nabla \cdot \vec{\upsilon} \neq 0$. The resulting normal and shear stress equations in terms of the rate of strain components become

$$\sigma_{xx} = -P + 2\mu e_{xx} - \frac{2\mu}{3}(e_{xx} + e_{yy} + e_{zz}) = -P + \tau_{xx} \tag{2.76}$$

$$\sigma_{yy} = -P + 2\mu e_{yy} - \frac{2\mu}{3}(e_{xx} + e_{yy} + e_{zz}) = -P + \tau_{yy} \tag{2.77}$$

$$\sigma_{zz} = -P + 2\mu e_{zz} - \frac{2\mu}{3}(e_{xx} + e_{yy} + e_{zz}) = -P + \tau_{zz} \tag{2.78}$$

$$\tau_{xy} = \mu\left(\frac{\partial \upsilon_x}{\partial y} + \frac{\partial \upsilon_y}{\partial x}\right) = \tau_{yx} \tag{2.79}$$

$$\tau_{yz} = \mu\left(\frac{\partial \upsilon_y}{\partial z} + \frac{\partial \upsilon_z}{\partial y}\right) = \tau_{zy} \tag{2.80}$$

$$\tau_{zx} = \mu\left(\frac{\partial \upsilon_z}{\partial x} + \frac{\partial \upsilon_x}{\partial z}\right) = \tau_{xz} \tag{2.81}$$

Some experiments have suggested that the constant $B = -(2\mu/3)$ should be changed to $B = -[(2\mu/3) - \kappa]$ where κ is the second coefficient of viscosity and B is the bulk viscosity. Be advised, there is no uniformity of the names given to these two terms. For monoatomic gases κ is zero. For thermal nonequilibrium in polyatomic gases and chemically reacting flows with rapid density changes, κ might be significant. For constant density the term it multiplies is zero, so its value is immaterial. For bubbly liquid flows, it might be significant. A single component bubbly mixture possesses a very nonlinear speed of sound which has a strong impact on numerical solutions. This is indirectly related to the second coefficient of viscosity. Further discussion of multicomponent speed of sound will be presented in Chapter 3. The second coefficient of viscosity is believed to be a function of nonequilibrium density changes and not a fluid property at all (see Landau and Lifshitz, 1959; Bird et al., 2002). When it doubt, leave it out.

The viscous stress tensor components are related to the shear rate components in an indicial format by

$$\tau_{ij} = +\left(\kappa - \frac{2}{3}\mu\right)(\nabla \cdot \vec{\upsilon})\,\delta_{ij} + \mu\left(\frac{\partial \upsilon_i}{\partial x_j} + \frac{\partial \upsilon_j}{\partial x_i}\right) \tag{2.82}$$

or in a tensor format by

$$\vec{\vec{\tau}} = +\mu(\nabla\vec{\upsilon} + (\nabla\vec{\upsilon})^{\mathrm{T}}) - \left(\frac{2}{3}\mu - \kappa\right)\nabla \cdot \vec{\upsilon} \cdot \vec{\vec{I}} \tag{2.83}$$

where
  $\nabla\vec{\upsilon}$ is the dyadic product of ∇ and $\vec{\upsilon}$
  $(\nabla\vec{\upsilon})^{\mathrm{T}}$ is the transpose of the dyadic product
  $\vec{\vec{I}}$ is the identity tensor

The Kronecker delta $\delta_{ij} = 1$ if $i = j$ and $= 0$ if $i \neq j$. Beware, some authors redefine the normal viscous stress components to include the pressure term. The components of the stress tensor for a Newtonian fluid are given in Table 2.4 in rectangular and cylindrical coordinates. The transformations given in Appendix B can be used to convert the stress–strain relationship and the momentum equation from rectangular Cartesian coordinates to cylindrical

TABLE 2.4 Components of the Stress Tensor for a Newtonian Fluid

Rectangular Coordinates (x, y, z)		*Cylindrical coordinates (r, θ, z)*	
$\tau_{xx} = \mu\left[2\dfrac{\partial v_x}{\partial x} - \dfrac{2}{3}\nabla\cdot\vec{v}\right]$	(A)	$\tau_{rr} = \mu\left[2\dfrac{\partial v_r}{\partial r} - \dfrac{2}{3}\nabla\cdot\vec{v}\right]$	(H)
$\tau_{yy} = \mu\left[2\dfrac{\partial v_y}{\partial y} - \dfrac{2}{3}\nabla\cdot\vec{v}\right]$	(B)	$\tau_{\theta\theta} = \mu\left[2\left(\dfrac{1}{r}\dfrac{\partial v_\theta}{\partial \theta} + \dfrac{v_r}{r}\right) - \dfrac{2}{3}\nabla\cdot\vec{v}\right]$	(I)
$\tau_{zz} = \mu\left[2\dfrac{\partial v_z}{\partial z} - \dfrac{2}{3}\nabla\cdot\vec{v}\right]$	(C)	$\tau_{zz} = \mu\left[2\dfrac{\partial v_z}{\partial z} - \dfrac{2}{3}\nabla\cdot\vec{v}\right]$	(J)
$\tau_{xy} = \tau_{yx} = \mu\left[\dfrac{\partial v_x}{\partial y} + \dfrac{\partial v_y}{\partial x}\right]$	(D)	$\tau_{r\theta} = \tau_{\theta r} = \mu\left[r\dfrac{\partial}{\partial r}\left(\dfrac{v_\theta}{r}\right) + \dfrac{1}{r}\dfrac{\partial v_r}{\partial \theta}\right]$	(K)
$\tau_{xz} = \tau_{zx} = \mu\left[\dfrac{\partial v_x}{\partial z} + \dfrac{\partial v_z}{\partial x}\right]$	(E)	$\tau_{\theta z} = \tau_{z\theta} = \mu\left[\dfrac{\partial v_\theta}{\partial z} + \dfrac{1}{r}\dfrac{\partial v_z}{\partial \theta}\right]$	(L)
$\tau_{yz} = \tau_{zy} = \mu\left[\dfrac{\partial v_y}{\partial z} + \dfrac{\partial v_z}{\partial y}\right]$	(F)	$\tau_{zr} = \tau_{rz} = \mu\left[\dfrac{\partial v_z}{\partial r} + \dfrac{\partial v_r}{\partial z}\right]$	(M)
$\nabla\cdot\vec{v} = \dfrac{\partial v_x}{\partial x} + \dfrac{\partial v_y}{\partial y} + \dfrac{\partial v_z}{\partial z}$	(G)	$\nabla\cdot\vec{v} = \dfrac{1}{r}\dfrac{\partial r v_r}{\partial r} + \dfrac{1}{r}\dfrac{\partial v_\theta}{\partial \theta} + \dfrac{\partial v_z}{\partial z}$	(N)

coordinates. The transformed equations are also given in many basic texts, for example, White (2006), Bird et al. (2002), Kuo (1986), Wilkes (2006), etc. Care must be taken to determine which sign convention is used for the stress components before using such equations from a given book.

2.4.5 NAVIER–STOKES EQUATIONS

To illustrate the form and complexity of the momentum equation for a Newtonian fluid with varying density and viscosity, the equations for the shear and normal stresses given in Table 2.4 are substituted into the Equations 2.44 through 2.46 in rectangular coordinates. The results are the three components of the general equation of motion given below.

$$\rho\frac{Dv_x}{Dt} = -\frac{\partial P}{\partial x} + \frac{\partial}{\partial x}\left[2\mu\frac{\partial v_x}{\partial x} - \frac{2}{3}\mu\nabla\cdot\vec{v}\right] + \frac{\partial}{\partial y}\left[\mu\left(\frac{\partial v_x}{\partial y} + \frac{\partial v_y}{\partial x}\right)\right]$$

$$+ \frac{\partial}{\partial z}\left[\mu\left(\frac{\partial v_z}{\partial x} + \frac{\partial v_x}{\partial z}\right)\right] + \sum \rho_i g_{ix} \tag{2.84}$$

$$\rho\frac{D\upsilon_y}{Dt} = -\frac{\partial P}{\partial y} + \frac{\partial}{\partial x}\left[\mu\left(\frac{\partial\upsilon_y}{\partial x} + \frac{\partial\upsilon_x}{\partial y}\right)\right] + \frac{\partial}{\partial y}\left[2\mu\frac{\partial\upsilon_y}{\partial y} - \frac{2}{3}\mu\nabla\cdot\vec{\upsilon}\right]$$

$$+ \frac{\partial}{\partial z}\left[\mu\left(\frac{\partial\upsilon_z}{\partial y} + \frac{\partial\upsilon_y}{\partial z}\right)\right] + \sum\rho_i g_{iy} \tag{2.85}$$

$$\rho\frac{D\upsilon_z}{Dt} = -\frac{\partial P}{\partial z} + \frac{\partial}{\partial x}\left[\mu\left(\frac{\partial\upsilon_z}{\partial x} + \frac{\partial\upsilon_x}{\partial z}\right)\right] + \frac{\partial}{\partial y}\left[\mu\left(\frac{\partial\upsilon_z}{\partial y} + \frac{\partial\upsilon_y}{\partial z}\right)\right]$$

$$+ \frac{\partial}{\partial z}\left[2\mu\frac{\partial\upsilon_z}{\partial z} - \frac{2}{3}\mu\nabla\cdot\vec{\upsilon}\right] + \sum\rho_i g_{iz} \tag{2.86}$$

Important simplifications of Equations 2.84 through 2.86 are the case of constant density and viscosity, which yield the Navier–Stokes equations:

$$\rho\frac{D\vec{\upsilon}}{Dt} = -\nabla P + \mu\nabla^2\vec{\upsilon} + \sum\rho_i\vec{g}_i \tag{2.87}$$

These equations are given in rectangular and cylindrical coordinates in Table 2.5.

For the case of $\tau = 0$, an ideal (inviscid) fluid,

$$\rho\frac{D\vec{\upsilon}}{Dt} = -\nabla P + \sum\rho_i\vec{g}_i \tag{2.88}$$

which is the Euler equation or Bernoulli equation if the flow is one dimensional.

TABLE 2.5 Navier–Stokes Equations (Momentum Equation with Constant Properties) in Rectangular and Cylindrical Coordinates

---

Rectangular coordinates

x-component

$$\rho\left(\frac{\partial\upsilon_x}{\partial t} + \upsilon_x\frac{\partial\upsilon_x}{\partial x} + \upsilon_y\frac{\partial\upsilon_x}{\partial y} + \upsilon_z\frac{\partial\upsilon_x}{\partial z}\right) = -\frac{\partial p}{\partial x} + \mu\left(\frac{\partial^2\upsilon_x}{\partial x^2} + \frac{\partial^2\upsilon_x}{\partial y^2} + \frac{\partial^2\upsilon_x}{\partial z^2}\right) + \sum\rho_i g_{ix} \tag{A}$$

y-component

$$\rho\left(\frac{\partial\upsilon_y}{\partial t} + \upsilon_x\frac{\partial\upsilon_y}{\partial x} + \upsilon_y\frac{\partial\upsilon_y}{\partial y} + \upsilon_z\frac{\partial\upsilon_y}{\partial z}\right) = -\frac{\partial p}{\partial y} + \mu\left(\frac{\partial^2\upsilon_y}{\partial x^2} + \frac{\partial^2\upsilon_y}{\partial y^2} + \frac{\partial^2\upsilon_y}{\partial z^2}\right) + \sum\rho_i g_{iy} \tag{B}$$

z-component

$$\rho\left(\frac{\partial\upsilon_z}{\partial t} + \upsilon_x\frac{\partial\upsilon_z}{\partial x} + \upsilon_y\frac{\partial\upsilon_z}{\partial y} + \upsilon_z\frac{\partial\upsilon_z}{\partial z}\right) = -\frac{\partial p}{\partial z} + \mu\left(\frac{\partial^2\upsilon_z}{\partial x^2} + \frac{\partial^2\upsilon_z}{\partial y^2} + \frac{\partial^2\upsilon_z}{\partial z^2}\right) + \sum\rho_i g_{iz} \tag{C}$$

(continued)

TABLE 2.5 (continued)   Navier–Stokes Equations (Momentum Equation with Constant Properties) in Rectangular and Cylindrical Coordinates

Cylindrical coordinates

r-component (where $\rho v_\theta^2/r$ is the centrifugal "force")

$$\rho\left(\frac{\partial v_r}{\partial t}+v_r\frac{\partial v_r}{\partial r}+\frac{v_\theta}{r}\frac{\partial v_r}{\partial \theta}-\frac{v_\theta^2}{r}+v_z\frac{\partial v_r}{\partial z}\right)$$

$$=-\frac{\partial p}{\partial r}+\mu\left[\frac{\partial}{\partial r}\left(\frac{1}{r}\frac{\partial}{\partial r}(rv_r)\right)+\frac{1}{r^2}\frac{\partial^2 v_r}{\partial \theta^2}-\frac{2}{r^2}\frac{\partial v_\theta}{\partial \theta}+\frac{\partial^2 v_r}{\partial z^2}\right]+\Sigma\rho_i g_{ir} \tag{D}$$

θ-component (where $\rho v_r v_\theta/r$ is the Coriolis "force")

$$\rho\left(\frac{\partial v_\theta}{\partial t}+v_r\frac{\partial v_\theta}{\partial r}+\frac{v_\theta}{r}\frac{\partial v_\theta}{\partial \theta}+\frac{v_r v_\theta}{r}+v_z\frac{\partial v_\theta}{\partial z}\right)$$

$$=-\frac{1}{r}\frac{\partial p}{\partial \theta}+\mu\left[\frac{\partial}{\partial r}\left(\frac{1}{r}\frac{\partial}{\partial r}(rv_\theta)\right)+\frac{1}{r^2}\frac{\partial^2 v_\theta}{\partial \theta^2}+\frac{2}{r^2}\frac{\partial v_r}{\partial \theta}+\frac{\partial^2 v_\theta}{\partial z^2}\right]+\Sigma\rho_i g_{i\theta} \tag{E}$$

z-component

$$\rho\left(\frac{\partial v_z}{\partial t}+v_r\frac{\partial v_z}{\partial r}+\frac{v_\theta}{r}\frac{\partial v_z}{\partial \theta}+v_z\frac{\partial v_z}{\partial z}\right)$$

$$=-\frac{\partial p}{\partial z}+\mu\left[\frac{1}{r}\frac{\partial}{\partial r}\left(r\frac{\partial v_z}{\partial r}\right)+\frac{1}{r^2}\frac{\partial^2 v_z}{\partial \theta^2}+\frac{\partial^2 v_z}{\partial z^2}\right]+\Sigma\rho_i g_{iz} \tag{F}$$

2.5 DERIVATION OF THE GENERAL ENERGY EQUATION

The result of applying the first law of thermodynamics to an open system—a control volume fixed in space with fluid flowing through—is referred to as the general energy equation. This law of conservation of energy can be expressed in the following mnemonic representation:

$$\begin{bmatrix}\text{Rate of accumulation}\\ \text{of internal and}\\ \text{kinetic energy in}\\ \text{the control volume}\end{bmatrix}+\begin{bmatrix}\text{Net efflux of internal}\\ \text{and kinetic energy}\\ \text{through the}\\ \text{control volume}\end{bmatrix}$$

$$=\begin{bmatrix}\text{Net rate of heat addition}\\ \text{to the control volume}\\ \text{by conduction and}\\ \text{internal generation}\end{bmatrix}-\begin{bmatrix}\text{Net rate of work}\\ \text{done by the}\\ \text{control volume}\\ \text{on the surroundings}\end{bmatrix} \tag{2.89}$$

By the usual sign convention work done by the control volume (system) on the surroundings is positive. Energy transport by thermal, nuclear, and other electromagnetic radiation is not specifically described, but an energy flux term is included. The internal energy per unit mass, U, is the energy associated with molecular motion and is a function of the material, the temperature, and pressure. The kinetic energy per unit mass, $v^2/2$, is the energy associated with bulk fluid motion $\left(v^2 = \left|\vec{v}\right|^2\right)$.

To evaluate each term in Equation 2.89 consider the fluxes shown on the control volume of Figure 2.13. For the first term, the rate of accumulation of internal and kinetic energy in the control volume is given by the product of the mass of the control volume and the internal and kinetic energy per unit mass in the control volume.

$$\frac{\partial}{\partial t}\left[\left(U + \frac{v^2}{2}\right)\rho\,\Delta x\,\Delta y\,\Delta z\right] = \Delta x\,\Delta y\,\Delta z\,\frac{\partial\rho\left[U + v^2/2\right]}{\partial t} \qquad (2.90)$$

The control volume is orientated in the flow field such that the internal and kinetic energy associated with the flow enters the control volume, passes

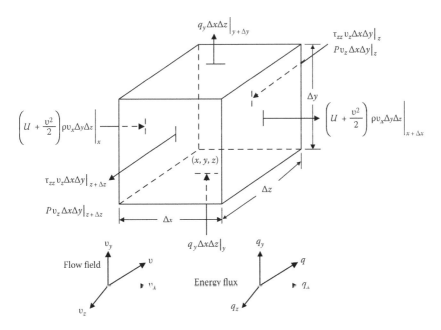

FIGURE 2.13 Typical components of fluxes of internal and kinetic energy, heat transfer and work done by the control volume on the surroundings.

through the faces touching point (x, y, z), and leaves through faces touching point $(x + \Delta x, y + \Delta y, z + \Delta z)$. The internal and kinetic energy entering and leaving through faces $\Delta y \Delta z$ are shown. The combined net flux of internal and kinetic energies (second term of Equation 2.89) can be written as

$$
\begin{aligned}
& \rho \upsilon_x (U + \upsilon^2/2) \Delta y \Delta z \big|_{x+\Delta x} - \rho \upsilon_x (U + \upsilon^2/2) \Delta y \Delta z \big|_x \\
& + \rho \upsilon_y (U + \upsilon^2/2) \Delta x \Delta z \big|_{y+\Delta y} - \rho \upsilon_y (U + \upsilon^2/2) \Delta x \Delta z \big|_y \\
& + \rho \upsilon_z (U + \upsilon^2/2) \Delta x \Delta y \big|_{z+\Delta z} - \rho \upsilon_z (U + \upsilon^2/2) \Delta x \Delta y \big|_x
\end{aligned}
\tag{2.91}
$$

Rearranging the above gives a more convenient form:

$$
\Delta x \Delta y \Delta z \left[+
\begin{array}{c}
\dfrac{\rho \upsilon_x (U + \upsilon^2/2)(U + \upsilon^2/2)\big|_{x+\Delta x} - \rho \upsilon_x (U + \upsilon^2/2)(U + \upsilon^2/2)\big|_x}{\Delta x} \\[2ex]
\dfrac{\rho \upsilon_y (U + \upsilon^2/2)(U + \upsilon^2/2)\big|_{y+\Delta y} - \rho \upsilon_y (U + \upsilon^2/2)(U + \upsilon^2/2)\big|_y}{\Delta y} \\[2ex]
\dfrac{\rho \upsilon_z (U + \upsilon^2/2)(U + \upsilon^2/2)\big|_{z+\Delta z} - \rho \upsilon_z (U + \upsilon^2/2)(U + \upsilon^2/2)\big|_z}{\Delta z}
\end{array}
\right]
\tag{2.92}
$$

The rate of heat addition to the control volume by conduction and diffusion is discussed in Section 3.4.2 and is given by Fourier's law and the heat of mixing according to Equation 2.93.

$$
\vec{q} = -k\nabla T + \sum_i H_i \vec{j}_i / M_i
\tag{2.93}
$$

where
  H_i is the partial molar enthalpy of species i
  k is the thermal conductivity

The net rate of heat addition by conduction and diffusion is given by

$$
q_x \Delta y \Delta z \big|_{x+\Delta x} - q_x \Delta y \Delta z \big|_x + q_y \Delta x \Delta z \big|_{y+\Delta y} - q_y \Delta x \Delta z \big|_y + q_z \Delta x \Delta y \big|_{z+\Delta z} - q_z \Delta x \Delta y \big|_z
\tag{2.94}
$$

Rearranging the above gives a more convenient form

$$
-\Delta x \Delta y \Delta z \left[\frac{q_x\big|_{x+\Delta x} - q_x\big|_x}{\Delta x} + \frac{q_y\big|_{y+\Delta y} - q_y\big|_y}{\Delta y} + \frac{q_z\big|_{z+\Delta z} - q_z\big|_z}{\Delta z} \right]
\tag{2.95}
$$

If q_r is the heat flux by radiation as described in Section 3.5, rate of heat addition to the control volume is given by an equation comparable to Equation 2.95 as follows:

$$-\Delta x \Delta y \Delta z \left[\frac{q_{rx}|_{x+\Delta x} - q_{rx}|_x}{\Delta x} + \frac{q_{ry}|_{yx+\Delta y} - q_{ry}|_y}{\Delta y} + \frac{q_{rz}|_{z+\Delta z} - q_{rz}|_z}{\Delta z} \right] \qquad (2.96)$$

Rate of work is force times the velocity in the direction of the force. The evaluation of the rate of work term in Equation 2.89 consist of two parts—rate of work done by the control volume against volume forces and rate of work done against surface forces. Volume forces are different external forces acting on individual species such as the force on an ion from an electric field. If only gravity is acting, the rate of work done by the control volume per unit volume on the surroundings is given by

$$-\Delta x \Delta y \Delta z \rho (g_x \upsilon_x + g_y \upsilon_y + g_z \upsilon_z) \qquad (2.97)$$

The negative sign arises since work is done against gravity (i.e., on the surrounding) is when υ and g are in the opposite direction. This can also be thought of as the change in potential energy where the potential energy increases when υ and g are opposed.

For different forces acting on individual species, the rate of work is given by

$$\Delta x \Delta y \Delta z \sum \rho_i g_i \cdot \upsilon_i = \Delta x \Delta y \Delta z \sum g_i \cdot n_i \qquad (2.98)$$

In Figure 2.5, the shear and normal forces in the x-direction of the control volume are shown to oppose those forces from the surroundings. The net rate of work done associated with υ_x for the flow field shown is

$$\upsilon_x \tau_{xx} \Delta y \Delta z|_{x+\Delta x} - \upsilon_x \tau_{xx} \Delta y \Delta z|_x + \upsilon_y \tau_{yx} \Delta x \Delta z|_{y+\Delta y} - \upsilon_y \tau_{yx} \Delta x \Delta z|_y$$
$$+ \upsilon_z \tau_{zx} \Delta x \Delta y|_{z+\Delta z} - \upsilon_z \tau_{zx} \Delta x \Delta y|_z \qquad (2.99)$$

which can be rearranged as

$$\Delta x \Delta y \Delta z \left[\frac{\tau_{xx} \upsilon_x|_{x+\Delta x} - \tau_{xx} \upsilon_x|_x}{\Delta x} + \frac{\tau_{yx} \upsilon_x|_{y+\Delta y} - \tau_{yx} \upsilon_x|_y}{\Delta y} + \frac{\upsilon_x \tau_{zx}|_{z+\Delta z} - \upsilon_x \tau_{zx}|_z}{\Delta z} \right]$$
$$(2.100)$$

There are terms associated with υ_y and υ_z which are comparable to Equation 2.100.

For the work done by the control volume against pressure, the following result is obtained since pressure is positive in compression.

$$\Delta x \Delta y \Delta z \left[\frac{P\upsilon_x \big|_{x+\Delta x} - P\upsilon_x \big|_x}{\Delta x} + \frac{P\upsilon_y \big|_{y+\Delta y} - P\upsilon_y \big|_y}{\Delta y} + \frac{P\upsilon_z \big|_{z+\Delta z} - P\upsilon_z \big|_z}{\Delta z} \right]$$

(2.101)

All of the terms for Equation 2.95, the first law of thermodynamics, have been developed. Each is multiplied by $\Delta x \Delta y \Delta z$ and has been put in the form of the definition of a derivative. Canceling the $\Delta x \Delta y \Delta z$ coefficients, taking the limit as Δx, Δy, Δz go to zero, and performing some rearranging, the result is the general energy equation for a multicomponent chemically reacting system.

$$
\frac{\partial \rho (U + \upsilon^2/2)}{\partial t} + \frac{\partial \rho (U + \upsilon^2/2)\upsilon_x}{\partial x} + \frac{\partial \rho (U + \upsilon^2/2)\upsilon_y}{\partial y} + \frac{\partial \rho (U + \upsilon^2/2)\,\upsilon_z}{\partial z}
$$
$$
= -\left(\frac{\partial q_x}{\partial x} + \frac{\partial q_y}{\partial y} + \frac{\partial q_z}{\partial z} \right) - \left(\frac{\partial q_{rx}}{\partial x} + \frac{\partial q_{ry}}{\partial y} + \frac{\partial q_{rz}}{\partial z} \right) + \sum \vec{g}_i \cdot \vec{n}_i
$$
$$
- \left[\frac{\partial(\tau_{xx}\upsilon_x + \tau_{xy}\upsilon_y + \tau_{xz}\upsilon_z)}{\partial x} + \frac{\partial(\tau_{yx}\upsilon_x + \tau_{yy}\upsilon_y + \tau_{yz}\upsilon_z)}{\partial y} + \frac{\partial(\tau_{zx}\upsilon_x + \tau_{zy}\upsilon_y + \tau_{zz}\upsilon_z)}{\partial z} \right]
$$
$$
- \left(\frac{\partial P\upsilon_x}{\partial x} + \frac{\partial P\upsilon_y}{\partial y} + \frac{\partial P\upsilon_z}{\partial z} \right)
$$

(2.102)

Equation 2.102 in vector–tensor notation is

$$
\frac{\partial \rho (U + \upsilon^2/2)}{\partial t} + \nabla \cdot \rho(U + \upsilon^2/2)\,\vec{\upsilon} = -\nabla \cdot \vec{q} - \nabla \cdot \vec{q}_r + \sum \vec{g}_i \cdot \vec{n}_i - \nabla \cdot (\vec{\bar{\tau}} \cdot \vec{\upsilon}) - \nabla \cdot P\vec{\upsilon}
$$

(2.103)

The first term represents the rate of gain in internal and kinetic energy per unit volume. The second term is the net rate of internal and kinetic energy efflux per unit volume. The third is the rate of heat transfer by conduction and diffusion. The fourth term is the rate of heat transfer by radiation. The fifth term is the rate of work done by the fluid on external body forces. The sixth term is the rate of work done by the fluid on viscous forces per unit volume, and the seventh term is the rate of work done by the fluid on pressure forces. Bird et al. (2002) obtains the same equation as Equation 2.103, except that Equation 2.103 includes a radiation flux term.

The rate of work done by the fluid on external body forces can be written in terms of the mass flux of component i relative to the mass average velocity, j_i, using Equation 2.14,

$$\sum \vec{g}_i \cdot \vec{n}_i = \sum \vec{g}_i \cdot \left(\vec{j}_i + \rho_i \vec{v}\right) \tag{2.104}$$

Equation 2.103 can be written in terms of the substantial derivative as given below by using the continuity equation to simplify the term on the left-hand side.

$$\rho \frac{D(U + v^2/2)}{Dt} = -\nabla \cdot \vec{q} - \nabla \cdot \vec{q}_r + \sum \vec{g}_i \cdot \vec{n}_i - \nabla \cdot \left(\vec{\tau} \cdot \vec{v}\right) - \nabla \cdot P\vec{v} \tag{2.105}$$

The thermal energy equation can be obtained by subtracting the dot product of the momentum equation, Equation 2.48, and using the identity $\vec{\tau}:\nabla\vec{v} = \nabla \cdot \left(\vec{\tau}\cdot\vec{v}\right) - \vec{v}\cdot\left(\nabla\cdot\vec{\tau}\right)$ with no loss in generality.

$$\rho \frac{DU}{Dt} = -\nabla \cdot \vec{q} - \nabla \cdot \vec{q}_r + \sum \vec{g}_i \cdot \vec{j}_i - \vec{\tau}:\nabla\vec{v} - P\nabla \cdot \vec{v} \tag{2.106}$$

A sometimes more convenient form of this equation is in terms of the enthalpy, $H = U + P/\rho$. The substantial derivative of the definition of the enthalpy is

$$\frac{DH}{Dt} = \frac{DU}{Dt} + \frac{D\,P/\rho}{Dt} \tag{2.107}$$

Substituting Equation 2.107 into Equation 2.106, expanding, rearranging, and using the continuity equation gives

$$\rho \frac{DH}{Dt} = -\nabla \cdot \vec{q} - \nabla \cdot \vec{q}_r + \sum \vec{g}_i \cdot \vec{j}_i - \vec{\tau}:\nabla\vec{v} + \frac{DP}{Dt} \tag{2.108}$$

A convenient and important starting point for analysis is having the energy equation with temperature as the dependent variable and the reaction rate appearing explicitly in the equation. The following equations are used to obtain this form.

Substantial derivative of enthalpy as a function of temperature, pressure, and concentration

$$\frac{DH}{Dt} = C_p \frac{DT}{Dt} + \frac{1}{\rho}\left[1 - \left(\frac{\partial \ln \hat{V}}{\partial \ln T}\right)_{P,m_i}\right]\frac{DP}{Dt} + \sum_{i=1}^{n} \frac{H_{xi}}{M_i}\frac{Dm_i}{Dt} \tag{2.109}$$

Species continuity equation

$$\frac{\rho}{M_i}\frac{Dm_i}{Dt} = -\nabla \cdot J_i + R_i \qquad (2.110)$$

Continuity equation

$$\left(\nabla \cdot \vec{\upsilon} + \frac{1}{\rho}\frac{D\rho}{Dt}\right) = 0 \qquad (2.111)$$

Substituting Equations 2.109 through 2.111 into Equation 2.108 and simplifying gives

$$\rho C_p \frac{DT}{Dt} = -\nabla \cdot \vec{q} - \nabla \cdot \vec{q}_r + \left(\frac{\partial \ln \hat{V}}{\partial \ln T}\right)_{P,m_i}\frac{DP}{Dt} + \sum_{i=1}^{n}\vec{g}_i \cdot \vec{j}_i - \vec{\tau}:\nabla\vec{\upsilon} + \sum_{i=1}^{n}H_{xi}\left[\nabla \cdot \vec{J}_i - R_i\right]$$

$$(2.112)$$

The term in the energy equation that is yet to be evaluated is $\vec{\tau}:\nabla\vec{\upsilon}$ which represent energy generated by viscous dissipation. Substituting Equation 2.83 for a Newtonian fluid gives the viscous dissipation in terms of the viscosity and velocity gradients as shown below.

$$\left[\mu\left(\nabla\vec{\upsilon} + (\nabla\upsilon)^{\mathrm{T}}\right) - \left(\frac{2}{3}\mu - \kappa\right)\nabla \cdot \vec{\upsilon} \cdot \vec{I}\right] : \nabla\vec{\upsilon} = \mu\Phi_v + \kappa\Psi \qquad (2.113)$$

The term $\left(\vec{\tau}:\nabla\vec{\upsilon}\right)$ is always positive, and Equation 2.113 serves to define the quantities Φ_v and Ψ. The equation for Φ_v is given below in rectangular coordinates and is the sum of velocity gradients squared.

$$\Phi_v = 2\left[\left(\frac{\partial \upsilon_x}{\partial x}\right)^2 + \left(\frac{\partial \upsilon_y}{\partial y}\right)^2 + \left(\frac{\partial \upsilon_z}{\partial z}\right)^2\right] + \left(\frac{\partial \upsilon_x}{\partial y} + \frac{\partial \upsilon_y}{\partial x}\right)^2 + \left(\frac{\partial \upsilon_z}{\partial y} + \frac{\partial \upsilon_y}{\partial z}\right)^2$$

$$+ \left(\frac{\partial \upsilon_x}{\partial z} + \frac{\partial \upsilon_z}{\partial x}\right)^2 - \frac{2}{3}\left(\frac{\partial \upsilon_x}{\partial x} + \frac{\partial \upsilon_y}{\partial y} + \frac{\partial \upsilon_z}{\partial z}\right)^2 \qquad (2.114)$$

The equation for Ψ is

TABLE 2.6   Energy Equation for a Pure Fluid in Terms of the Energy Flux and Stress Tensor

Rectangular coordinates

$$
\rho C_v \left(\frac{\partial T}{\partial t} + \upsilon_x \frac{\partial T}{\partial x} + \upsilon_y \frac{\partial T}{\partial y} + \upsilon_z \frac{\partial T}{\partial z} \right) = -\left[\frac{\partial q_x}{\partial x} + \frac{\partial q_y}{\partial y} + \frac{\partial q_z}{\partial z} \right]
$$

$$
- T \left(\frac{\partial p}{\partial T} \right)_\rho \left(\frac{\partial \upsilon_x}{\partial x} + \frac{\partial \upsilon_y}{\partial y} + \frac{\partial \upsilon_z}{\partial z} \right) + \left\{ \tau_{xx} \frac{\partial \upsilon_x}{\partial x} + \tau_{yy} \frac{\partial \upsilon_y}{\partial y} + \tau_{zz} \frac{\partial \upsilon_z}{\partial z} \right\}
$$

$$
+ \left\{ \tau_{xy} \left(\frac{\partial \upsilon_x}{\partial y} + \frac{\partial \upsilon_y}{\partial x} \right) + \tau_{xz} \left(\frac{\partial \upsilon_x}{\partial z} + \frac{\partial \upsilon_z}{\partial x} \right) + \tau_{yz} \left(\frac{\partial \upsilon_y}{\partial z} + \frac{\partial \upsilon_z}{\partial y} \right) \right\} \tag{A}
$$

Cylindrical coordinates

$$
\rho C_v \left(\frac{\partial T}{\partial t} + \upsilon_r \frac{\partial T}{\partial r} + \frac{\upsilon_\theta}{r} \frac{\partial T}{\partial \theta} + \upsilon_z \frac{\partial T}{\partial z} \right) = -\left[\frac{1}{r} \frac{\partial}{\partial r} (r q_r) + \frac{1}{r} \frac{\partial q_\theta}{\partial \theta} + \frac{\partial q_z}{\partial z} \right]
$$

$$
- T \left(\frac{\partial p}{\partial T} \right)_\rho \left(\frac{1}{r} \frac{\partial}{\partial r} (r \upsilon_r) + \frac{1}{r} \frac{\partial \upsilon_\theta}{\partial \theta} + \frac{\partial \upsilon_z}{\partial z} \right) + \left\{ \tau_{rr} \frac{\partial \upsilon_r}{\partial r} + \tau_{\theta\theta} \frac{1}{r} \left(\frac{\partial \upsilon_\theta}{\partial \theta} + \upsilon_r \right) + \tau_{zz} \frac{\partial \upsilon_z}{\partial z} \right\}
$$

$$
+ \left\{ \tau_{r\theta} \left[r \frac{\partial}{\partial r} \left(\frac{\upsilon_\theta}{r} \right) + \frac{1}{r} \frac{\partial \upsilon_r}{\partial \theta} \right] + \tau_{rz} \left(\frac{\partial \upsilon_z}{\partial r} + \frac{\partial \upsilon_r}{\partial z} \right) + \tau_{\theta z} \left(\frac{1}{r} \frac{\partial \upsilon_z}{\partial \theta} + \frac{\partial \upsilon_\theta}{\partial z} \right) \right\} \tag{B}
$$

$$
\Psi = (\nabla \cdot \vec{\upsilon})^2 \tag{2.115}
$$

There are several comparable forms of the energy equation that can be obtained by various manipulations. Bird et al. (2002) provides several tables of such forms for the energy equation along with the continuity, species continuity, and momentum equations. In Table 2.6 the energy equation is given in rectangular and cylindrical coordinates for a pure fluid in terms of the shear and normal stresses from Bird et al. (1960), as modified to reflect the accepted sign convention on the stress tensor. Assuming that the fluid is of constant density and viscosity and replacing the viscous stress terms with the appropriate rate of strain terms, the energy equations shown in Table 2.6 become those shown in Table 2.7.

It should be appreciated that while many simplified forms of the equations of change may be presented for various assumed properties, not all of them are useful. For example, to assume that the flow reacts but that the density is constant is very likely to be a bad assumption. Remember, do not throw out the baby with the bath water. Assumptions must be justified.

TABLE 2.7   Energy Equation in Terms of the Transport Properties for a Newtonian Fluid with Constant Properties and No Chemical Reactions

Rectangular coordinates

$$\rho C_p \left(\frac{\partial T}{\partial t} + \upsilon_x \frac{\partial T}{\partial x} + \upsilon_y \frac{\partial T}{\partial y} + \upsilon_z \frac{\partial T}{\partial z} \right) = \kappa \left[\frac{\partial^2 T}{\partial x^2} + \frac{\partial^2 T}{\partial y^2} + \frac{\partial^2 T}{\partial z^2} \right]$$

$$+ 2\mu \left\{ \left(\frac{\partial \upsilon_x}{\partial x} \right)^2 + \left(\frac{\partial \upsilon_y}{\partial y} \right)^2 + \left(\frac{\partial \upsilon_z}{\partial z} \right)^2 \right\}$$

$$+ \mu \left\{ \left(\frac{\partial \upsilon_x}{\partial y} + \frac{\partial \upsilon_y}{\partial x} \right)^2 + \left(\frac{\partial \upsilon_x}{\partial z} + \frac{\partial \upsilon_z}{\partial x} \right)^2 + \left(\frac{\partial \upsilon_y}{\partial z} + \frac{\partial \upsilon_z}{\partial y} \right)^2 \right\} \qquad \text{(A)}$$

Cylindrical coordinates

$$\rho C_p \left(\frac{\partial T}{\partial t} + \upsilon_r \frac{\partial T}{\partial r} + \frac{\upsilon_\theta}{r} \frac{\partial T}{\partial \theta} + \upsilon_z \frac{\partial T}{\partial z} \right) = \kappa \left[\frac{1}{r} \frac{\partial}{\partial r} \left(r \frac{\partial T}{\partial r} \right) + \frac{1}{r^2} \frac{\partial^2 T}{\partial \theta^2} + \frac{\partial^2 T}{\partial z^2} \right]$$

$$+ 2\mu \left\{ \left(\frac{\partial \upsilon_r}{\partial r} \right)^2 + \left[\frac{1}{r} \left(\frac{\partial \upsilon_\theta}{\partial \theta} + \upsilon_r \right) \right]^2 + \left(\frac{\partial \upsilon_z}{\partial z} \right)^2 \right\}$$

$$+ \mu \left\{ \left(\frac{\partial \upsilon_\theta}{\partial z} + \frac{1}{r} \frac{\partial \upsilon_z}{\partial \theta} \right)^2 + \left(\frac{\partial \upsilon_z}{\partial r} + \frac{\partial \upsilon_r}{\partial z} \right)^2 + \left[\frac{1}{r} \frac{\partial \upsilon_r}{\partial \theta} + r \frac{\partial}{\partial r} \left(\frac{\upsilon_\theta}{r} \right) \right]^2 \right\} \qquad \text{(B)}$$

2.6 NON-NEWTONIAN FLUIDS

Rheology is the science of deformation and flow, and materials can be classified as shown in Figure 2.14. On one side of the diagram is an inviscid fluid which has zero viscosity and on the other is an inelastic solid that

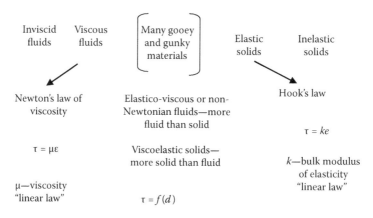

FIGURE 2.14   Classification of materials.

ruptures under an applied stress. In an elastic solid, the applied stress is proportional to the deformation, ε, and the proportionality constant is the bulk modulus of elasticity, k. In a viscous Newtonian fluid the applied stress is proportional to the rate of deformation, e, and the proportionality constant is the viscosity, μ. Between a Newtonian fluid and an elastic solid are "many gunky and gooey materials" according to Professor R. B. Bird that are widely used in industrial applications. These include polymer melts and solutions, solid suspensions, foods, and body fluids, among others. In Figure 2.14 ε is the deformation tensor in Hook's law, and e is the rate of deformation in Newton's law of viscosity.

As indicated in Figure 2.14, there are many fluids where the stress tensor is not proportional to the rate of deformation tensor. Many commercially important materials exhibit this non-Newtonian behavior such as polymer melts and solutions, solid suspensions, foods and body fluids, among others.

Non-Newtonian fluids are classified as fluids with properties that are independent of time or rate of shear (simplest) and fluids with properties that are a function of the length of time that the shear rate is applied. The time-independent fluids are shear-thinning, shear-thickening, and a Bingham plastic which has a finite yield stress and then flows like a Newtonian fluid. Then there are materials that have many characteristics of a solid primarily that of elastic recovery from larger deformations, such as jello, silly putty, and napalm.

For shear-thinning or pseudoplastic fluids, the apparent viscosity, μ_a, decreases with increasing shear rate. Examples include polymer melts and solutions, paint, paper pulp, and mayonnaise. A picture of this behavior is a fluid consisting of long slender molecules that are aligned under shear as the fluid layers slide over each other. The greater the shear is the greater the alignment. These fluids can be described by a power law model with two parameters as discussed below.

For shear-thickening or a dilatant fluid the apparent viscosity, μ_a, increases with increasing shear rate. Examples are starch solutions, potassium silicate solutions, ammonium oleate suspensions, and iron oxide suspensions. A picture of this behavior is a fluid consisting of densely packed particles in which the voids are small and are filled with liquid sufficient to fill the voids. At low shear rates the particles move past on another, and there is sufficient liquid to act as a lubricant. As the shear rate increases there is not sufficient liquid for lubrication and have smooth flow. As a result there is an increase in apparent viscosity. These fluids can be described by a power law model.

A Bingham plastic is a fluid that requires a finite stress to initiate flow but then it flows somewhat like a Newtonian fluid. Examples are drilling mud, suspensions of chalk, sewage sludge, paint, and whole blood. A picture of this behavior is a fluid consisting of a three-dimensional structure which is able to withstand a finite shear stress, τ_0, but crumbles when the stress is exceeded. These fluids can be described by a model with two parameters as discussed below.

A thixotropic or time-thinning fluid has the property that the shear stress decreases with time at a constant shear rate. After standing, the original viscosity is recovered (thixotropic). A picture of this behavior is that a finite time is required for alignment of molecules in a shear-thinning fluid. Examples are paint, ketchup, and some polymer solutions. These fluids can be described by a model with three parameters as discussed below.

A rheopetic or time-thickening fluid has the property that the shear stress increases with time at a constant shear rate. A picture of this behavior is that a build-up of structure with time induced by shear or a finite time is required for the dilatant fluid effect. These fluids can be described by a model with three parameters.

Understanding non-Newtonian flow behavior has taken two directions: development of design procedures and studies of the phenomena of non-Newtonian flow. Design procedures have been semiempirical methods that are based on simple models and are used for scale-up of specific systems. Studies of the phenomena of non-Newtonian flow include rheological equations of state, molecular theory, and the constitutive equations of continuum mechanics that are based on principle of "material indifference."

An evaluation of the simplest form for the stress tensor as a function of the rate of deformation tensor, $\vec{\vec{\tau}} = f(\vec{\vec{d}})$, is obtained from an analysis from the constitutive equations of continuum mechanics (Brodkey, 1967; Bird et al., 1960). The symbol d will be used in the following discussion instead of e to emphasize that the deformation is not like that of a Newtonian fluid. A linear equation relating the stress tensor and the rate of deformation tensor is

$$\vec{\vec{\tau}} = a\vec{\vec{I}} + b\vec{\vec{d}} + c\vec{\vec{d}} \cdot \vec{\vec{d}} \qquad (2.116)$$

where

a, b, and c are scalar functions of $\vec{\vec{d}}$

$\vec{\vec{I}}$ is the identity tensor

Referring to Equation 2.83 for a Newtonian fluid:

$$a = -\left(\frac{2}{3}\mu - \kappa\right)\nabla \cdot \vec{\upsilon} \qquad (2.117)$$

$$b = +\mu \qquad (2.118)$$

$$c = 0 \qquad (2.119)$$

For this analysis the rate of deformation tensor is for a liquid (constant density) and is a simplification of Equation 2.83.

$$\overset{\leftrightarrow}{\tau} = \mu\left(\nabla\vec{\upsilon} + (\nabla\vec{\upsilon})^{\mathrm{T}}\right) \qquad (2.120)$$

For a non-Newtonian fluid, $a(\vec{d}), b(\vec{d}),$ and $c(\vec{d})$ are scalar functions of $\vec{d}$, and this requirement limits the functional form to the following:

$$\overset{\leftrightarrow}{d}:\overset{\leftrightarrow}{I} \quad \text{double dot product of } \overset{\leftrightarrow}{d} \text{ and } \overset{\leftrightarrow}{I} \qquad (2.121)$$

$$\overset{\leftrightarrow}{d}:\overset{\leftrightarrow}{d} \quad \text{double dot product of } d \text{ with itself} \qquad (2.122)$$

$$\left|\overset{\leftrightarrow}{d}\right| \quad \text{determinant of } \overset{\leftrightarrow}{d} \qquad (2.123)$$

Equations 2.121 through 2.123 are called the invariants of $\overset{\leftrightarrow}{d}$, and it can be seen from Equation 2.69 that

$$\overset{\leftrightarrow}{d}:\overset{\leftrightarrow}{I} = 2\nabla \cdot \vec{\upsilon} = 0 \quad \text{for an incompressible fluid} \qquad (2.124)$$

and $\left|\overset{\leftrightarrow}{d}\right| = 0$ for many simple flows. Which leaves $a(\overset{\leftrightarrow}{d}:\overset{\leftrightarrow}{d}), b(\overset{\leftrightarrow}{d}:\overset{\leftrightarrow}{d}),$ and $c(\overset{\leftrightarrow}{d}:\overset{\leftrightarrow}{d})$ or

$$\overset{\leftrightarrow}{\tau} = a(\overset{\leftrightarrow}{d}:\overset{\leftrightarrow}{d})I + b(\overset{\leftrightarrow}{d}:\overset{\leftrightarrow}{d})\overset{\leftrightarrow}{d} + c(\overset{\leftrightarrow}{d}:\overset{\leftrightarrow}{d})\overset{\leftrightarrow}{d} \cdot \overset{\leftrightarrow}{d} \qquad (2.125)$$

A description is provided by Brodkey (1967) that shows $a(\overset{\leftrightarrow}{d}:\overset{\leftrightarrow}{d})I = 0$ and $c(\overset{\leftrightarrow}{d}:\overset{\leftrightarrow}{d})\overset{\leftrightarrow}{d}$ is small compared to $b(\overset{\leftrightarrow}{d}:\overset{\leftrightarrow}{d})\overset{\leftrightarrow}{d}$ and can be neglected. This gives the final form as the Equation 2.126:

$$\overset{\leftrightarrow}{\tau} = f(\overset{\leftrightarrow}{d}:\overset{\leftrightarrow}{d})\overset{\leftrightarrow}{d} \qquad (2.126)$$

The power law model is the two parameter model that is used to describe shear-thinning and shear-thickening non-Newtonian fluids, and it has the following form of Equation 2.126.

$$\vec{\vec{\tau}} = + \left[m \left| \sqrt{1 \big/ 2(\vec{\vec{d}} : \vec{\vec{d}})} \right|^{n-1} \right] \vec{\vec{d}} \tag{2.127}$$

where m and n are parameters that are determined from experimental data (Bird et al., 1960). For a shear-thinning fluid, $n < 1$ and for a shear-thickening fluid, $n > 1$. The equation reduces to Newton's law of viscosity for $m = \mu$ and $n = 1$. The double dot product term can be evaluated using $\frac{1}{2}d{:}d = \Phi_v$, where $\Phi_v = (\nabla \upsilon + (\nabla \upsilon)^{\mathrm{T}})$, Equation 2.83 with the $-(\frac{2}{3}\mu - \kappa)\nabla \cdot \vec{\upsilon}$ term omitted.

There are numerous non-Newtonian flow models that describe a variety of material. The text by Bird et al. (1987) provides a comprehensive discussion of these models, their development and experimental methods to evaluate parameters in the models. One of the more detailed models is the Carreau–Yasuda model shown below.

$$\frac{\mu_a - \mu_\infty}{\mu_0 - \mu_\infty} = \left\{ 1 + \left[\lambda \left(\vec{\vec{d}} : \vec{\vec{d}} \right) \right]^a \right\}^{(n-1)/n} \tag{2.128}$$

This model has five parameters: μ_0 is the zero shear rate apparent viscosity, μ_∞ is the infinite shear rate apparent viscosity, n is a "power-law region" exponent, a is a dimensionless parameter that describes the transition region and the power-law region, and λ is a time constant. The term $\vec{\vec{d}} : \vec{\vec{d}} = \left(\nabla \vec{\upsilon} + (\nabla \vec{\upsilon})^{\mathrm{T}} \right) : \vec{\vec{I}}$ is a scalar derived from the rate of deformation tensor for a non-Newtonian fluid (liquid, $\nabla \cdot \vec{\upsilon} = 0$). Oldrod has a six constant model—there are others.

2.7 GENERAL PROPERTY BALANCE

A general property balance to derive the equations of change serves two purposes. One is to show that the preceding equations of change are applicable to coordinate systems other than rectangular coordinates. The other is to show that the equations of change have the same mathematical form with different values of the dependent variables. In a general property balance, a property is defined along with a flux of this property and a generation of this property. These definitions are used with a conservation law to obtain the continuity equation, species continuity equation, momentum equation, and energy equation. Consider the species continuity equation, Equation 2.15, written in the flowing form:

$$\frac{\partial \rho_i}{\partial t} + \nabla \cdot (\rho_i \vec{v}) + \nabla \cdot \vec{j}_i - r_i = 0 \quad \text{for} \quad i = 1, 2, \ldots, n \qquad (2.129)$$

The first term is the accumulation of the property per unit volume (concentration of species i) in a control volume. The second and third terms are the convective and diffusive flux of the property (species i) through a control volume. The fourth term is the generation of the property per unit volume (chemical reaction rate of species i) in a control volume.

The conservation law used with a control volume moving with the mass average velocity of the fluid, $\vec{v}$, as shown in Figure 2.2, is stated mnemonically and mathematically as

$$\begin{vmatrix} \text{Accumulation of the} \\ \text{property in the} \\ \text{control volume} \end{vmatrix} = \begin{vmatrix} \text{Net flux (out-out)} \\ \text{across the control} \\ \text{surface} \end{vmatrix} + \begin{vmatrix} \text{Generation of the} \\ \text{property in the} \\ \text{control volume} \end{vmatrix}$$

$$(2.130)$$

$$\frac{d}{dt} \int_v \psi \, dV = -\oint_S \vec{\Psi} \cdot d\vec{S} + \int_v \psi_g \, dV \qquad (2.131)$$

where
ψ is the property
$\vec{\Psi}$ is the flux of the property
ψ_g is the generation of the property per unit volume (Brodkey, 1967)

The general property balance, Equation 2.131, is a scalar equation for the continuity, species continuity, and energy equations, and it is a vector equation for the momentum equation. The surface integral has the sign convention of negative for inflow and positive for outflow, and the negative sign accounts for the conservation equation using out–in as positive.

To obtain the differential form of the general property balance, the order of integration and differentiation is interchanged on the term on the left-hand side, and the surface integral is converted to a volume integral. Using

Leibnitz's rule for a multiple integral to change the order of differentiation and integration gives

$$\frac{d}{dt}\int_V \psi \, dV = \int_V \frac{\partial \psi}{\partial t} dV + \oint_S \psi \vec{v} \cdot d\vec{S} \qquad (2.132)$$

Gauss' theorem is used for the surface integrals in Equations 2.131 and 2.132.

$$\oint_S \vec{\Psi} \cdot dS = \int_V \nabla \cdot \vec{\Psi} \, dV \qquad (2.133)$$

$$\oint_S \psi \vec{v} \cdot d\vec{S} = \int_V \nabla \cdot \psi \vec{v} \, dV \qquad (2.134)$$

Substituting Equations 2.133 and 2.134 into Equation 2.132 gives

$$\frac{d}{dt}\int_V \psi \, dV = \int_V \frac{\partial \psi}{\partial t} dV + \int_V \nabla \cdot \psi \vec{v} \, dV \qquad (2.135)$$

Substituting Equations 2.133 and 2.134 into Equation 2.131 gives

$$\int_V \frac{\partial \psi}{\partial t} dV + \int_V \nabla \cdot \psi \vec{v} \, dV = -\int_V \nabla \cdot \vec{\Psi} \, dV + \int_V \psi_g \, dV \qquad (2.136)$$

Equation 2.137 can be written as

$$\int_V \left[\frac{\partial \psi}{\partial t} + \nabla \cdot \psi \vec{v} + \nabla \cdot \vec{\Psi} - \psi_g \right] dV = 0 \qquad (2.137)$$

For the integral to be zero, the term in the brackets must be zero, and the resulting partial differential equation is the differential form of the general property balance.

$$\frac{\partial \psi}{\partial t} + \nabla \cdot \psi \vec{v} + \nabla \cdot \vec{\Psi} - \psi_g = 0 \qquad (2.138)$$

The equations of change are obtained by substituting the property, the flux of the property, and generation of the property for the continuity, species continuity, momentum, and energy equations. For the continuity equation, the property is the density (mass per unit volume). There is no flux of the property since the control volume is moving with the mass average velocity of the fluid, and there is no generation of the property from the conservation of mass. Substituting the density, $\rho = \psi$, into Equation 2.138 gives the same equation as Equation 2.5.

$$\frac{\partial \rho}{\partial t} + \nabla \cdot \rho \vec{v} = 0 \tag{2.139}$$

For the species continuity equation, the property is the concentration (mass of i per unit volume) $\rho_i = \psi$, the flux of the property is the mass flux of species i relative to the mass average velocity $\vec{j}_i = \vec{\Psi}$, and the generation of the property is the chemical reaction rate $r_i = \psi_g$. Substituting into Equation 2.138 gives the same equation as Equations 2.15 and the following equation.

$$\frac{\partial \rho_i}{\partial t} + \nabla \cdot (\rho_i \vec{v}) + \nabla \cdot \vec{j}_i - r_i = 0 \quad \text{for} \quad i = 1, 2, \dots, n \tag{2.140}$$

For the momentum equation, the property is the rate of momentum per unit volume, $\rho \vec{v} = \vec{\psi}$. The flux of momentum is the stress tensor $\vec{\vec{\tau}}$ and pressure written as a tensor $\vec{\vec{PI}}$, i.e., $\vec{\vec{\Psi}} = \vec{\vec{PI}} - \vec{\vec{\tau}}$. The generation of momentum is given by the rate of momentum generated from body forces per unit volume, $\Sigma \rho_i g_{xi} = \psi_g$. Substituting into Equation 2.138 gives the same equation as Equation 2.103.

$$\frac{\partial \rho \vec{v}}{\partial t} + \nabla \cdot \rho \vec{v} \vec{v} + \nabla \cdot \left(\vec{\vec{PI}} - \vec{\vec{\tau}} \right) - \sum \rho_i \vec{g}_i = 0 \tag{2.141}$$

or

$$\frac{\partial \rho \vec{v}}{\partial t} + \nabla \cdot \rho \vec{v} \vec{v} + \nabla P - \nabla \cdot \vec{\vec{\tau}} - \sum \rho_i \vec{g}_i = 0 \tag{2.142}$$

For the energy equation, the property is the internal and kinetic energy per unit volume given by $\rho \left(U + \frac{1}{2} v^2 \right) = \psi$. The energy flux is given by $\vec{q}$ for conduction and diffusion and $\vec{q}_r$ for radiant energy flux. The generation of energy is by work done by the control volume on body and surface (pressure and viscous) forces is $-\Sigma \vec{g}_i \cdot \vec{n}_i + \nabla \cdot \left(\vec{\vec{\tau}} \cdot \vec{v} \right) + \nabla \cdot P \vec{v} = \psi_g$ as described in the derivation of these terms in the energy equation. Substituting into

Equation 2.138 gives the same equation as Equation 2.103 and using $\Sigma \vec{g}_i \cdot \vec{n}_i = \Sigma \vec{g}_i \cdot (\vec{j}_i + \rho_i \vec{v})$.

$$\frac{\partial}{\partial t}\rho\left(U + \frac{v^2}{2}\right) + \nabla \cdot \rho\left(U + \frac{v^2}{2}\right)\vec{v} + \nabla \cdot \vec{q} + \nabla \cdot \vec{q}_r - \sum g_i \cdot (\vec{j}_i + \rho_i \vec{v})$$
$$- \nabla \cdot (\vec{\bar{\tau}} \cdot \vec{v}) + \nabla \cdot P\vec{v} = 0 \tag{2.143}$$

Brodkey (1967) obtained the same equation as Equation 2.143 without the radiation flux term.

In Figure 2.15, the property, flux of the property, and generation of the property are listed for the continuity, species continuity, momentum, and energy equations. The equations of change have the same mathematical form with different values of the dependent variable. This formulation is used for the discretization of the conservation equations to affect their solution. Only the initial and boundary conditions for these equations are different. There are other conservation equations that are used to describe turbulent transport such as the turbulent kinetic energy, dissipation of turbulent kinetic energy, and the Reynolds's stress equations. These equations have the same format as the conservation equations, and there is an increase in the number of dependent variables, but the format used for the solution remains the same. These equations are discussed in detail in Chapter 4.

$$\frac{\partial \psi}{\partial t} + \nabla \cdot \psi \vec{v} + \nabla \cdot \vec{\Psi} - \psi_g = 0$$

	Property ψ	Flux of the property $\vec{\Psi}$	Generation of the property ψ_g
Continuity equation	ρ	0	0
Species continuity equation	ρ_i	$\vec{j}_i$	r_i
Momentum equation	$\rho\vec{v}$	$P\vec{\bar{I}} - \vec{\bar{\tau}}$	$\sum \rho_i \vec{g}_i$
Energy equation	$\rho(U + \frac{1}{2}v^2)$	$\vec{q} = -k\nabla T$ $+\sum H_i \vec{j}_i/M_i$	$-\sum \vec{g}_i \cdot (\vec{j}_i + \rho_i \vec{v})$ $+\nabla \cdot (\vec{\bar{\tau}} \cdot \vec{v}) + \nabla \cdot P\vec{v}$

FIGURE 2.15 The general property balance.

2.8 ANALYTICAL AND APPROXIMATE SOLUTIONS FOR THE EQUATIONS OF CHANGE

2.8.1 INTRODUCTION

There are a wealth of analytical and approximate solutions to the equations of change. Velocity profiles result from the solution of the continuity and momentum equations. For simple geometries the equations of change are usually posed as second-order ordinary differential equation. Most analytical solutions are for incompressible, one-dimensional flows. The concentration and temperature profiles from solutions of the species continuity and energy equations posed as partial differential equation are less abundant since it is necessary to include the convective transport of species and energy in the direction of flow. DeSouza-Santos (2008) presented the analytical and approximate solutions to over 70 problems and classified them according to the number of independent variables, order of the ordinary or partial differential equation, and type of boundary and initial conditions. Extensive collections of analytical and approximate solutions are given by Bird et al. (2002), which are classified by transport mechanism: momentum, energy, mass and some simultaneous momentum, heat, and mass transfer.

Some geometries that have analytical and approximate solutions to the continuity and momentum equations are shown in Figure 2.16 for a fully developed flow with constant properties. For flow between flat plates, Brodkey (1967) describes the solution of the second-order ordinary differential equation from the simplification of the continuity and momentum equations for the cases of the lower plate stationary and moving with a velocity **V** for a negative and positive pressure gradient. The diagram in Figure 2.16 sketches the velocity profile with the lower plate moving and a positive pressure gradient. Other solutions between flat plates include fluid injected and removed at the wall which has a nonzero but constant velocity component in the vertical direction (approximates membrane dialysis) and flow of two immiscible liquids (Bird et al., 1960). For flow in a tube, a parabolic velocity profile is obtained for steady, laminar flow of a Newtonian fluid as discussed below and is given in standard texts. The solution for start-up of flow, stationary liquid in a tube with imposed negative pressure gradient has the liquid accelerating to steady state, and the solution is obtained by the method of separation of variables, as described by Brodkey (1967). Flow in the annulus with a rod moving in the center with a velocity, **V**, approximates coating a wire with a polymer, and

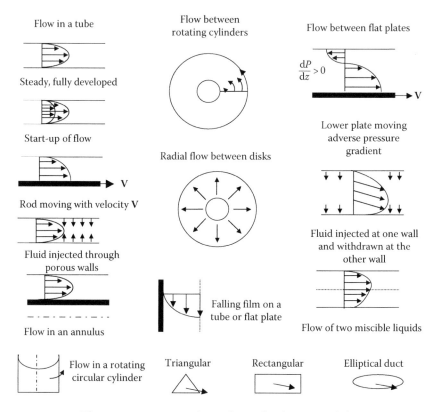

FIGURE 2.16 Flow geometries with analytical solutions of the continuity and momentum equations.

the solution is logarithmic (Bird et al., 1960). For liquid injected through the porous walls of a tube as shown in Figure 2.16 (approximate the flow associated with transpiration cooling), the solution uses the stream function to satisfy the continuity equation, and the momentum equation is solved by separation of variables (Longwell, 1966). For flow in an annulus between two circular tubes (approximates flow in a heat exchanger) the solution for the velocity profile is given by Bird et al. (1960). For a falling film (wetted wall column), the velocity profile is determined using a no-shear at the liquid surface boundary condition, and it is the sum of a quadratic and logarithmic terms (Bird et al., 1960). Flow between rotating cylinders is used in a viscometer, and the solution for the velocity profile is used to evaluate the viscosity as described by Brodkey (1967) and Bird et al. (1960). For radial flow between parallel disks, the velocity varies inversely as the square of the radius, and the radial pressure distribution varies inversely as the fourth power of the radius (Bird et al., 1960). Liquid rotates as a solid body in a rotating cylinder (beaker), and the shape of the liquid surface has the cross section of a parabola (Bird et al., 1960). Flows in ducts with triangular and elliptical cross-sections have closed-form analytical solutions, and flow in a duct with a rectangular cross-section has an infinite series solution (Knudsen and Katz, 1958). Although the diagram of a nozzle is not shown, flow of a compressible gas through a properly designed converging–diverging nozzle is isentropic with subsonic flow in the converging section and either subsonic or supersonic flow in the diverging section. Details of compressible flow of an ideal gas in nozzle ducts and shock waves are given in the text by Shapiro (1953), among others.

The above description is not comprehensive, and a large body of literature is available for solutions of the continuity and momentum equations for velocity profiles, volumetric flow rates, drag, and friction factors. At one point in the history of fluid dynamics, before the advent of numerical solutions, it was said that when a mathematician obtained a new solution to a second-order partial differential equation, a fluid dynamicist found a geometry and flow that matched this new solution.

Two important classes of transport phenomena are not represented by the examples just mentioned. The first is truly one-dimensional flows, like those described in basic thermodynamic texts for single component fluids. These processes consist of piping systems where the flow passes through pumps or turbines, heat exchangers, and chemical reactors and the elevation may change for different parts of the system. Similar analyses may also be made for multicomponent fluids with variable density and physical properties. These analyses are referred to as macroscopic balances and described very adequately by Bird et al. (2002). Flows are taken as average, one-dimensional streams with the processes described generally with empirical test data. Such analyses are particularly useful for process analysis and for constructing lumped parameter models for control systems. This modeling technology constitutes the bulk of basic transport phenomena literature.

The second class of transport phenomena which have been studied and utilized extensively are inviscid flows. All laminar and turbulent viscous terms are omitted from the conservation laws and the remaining terms in the laws are solved to represent the flow. Historically, such methodology was first used to calculate the lift from an airplane wing so that its airfoil shape could be optimized. This practice was very successful; however, it lead to the contradiction that if the pressure field around an airfoil could be accurately calculated, why was the predicted drag zero? Prandtl explained this phenomenon by observing that the effect of friction was limited to a very thin region close to the body surface, i.e., the boundary layer. This treatment of fluid mechanics was indeed useful and eventually extended to the transfer of heat and mass as well. Furthermore, it showed that inviscid flow theory was still a useful technology. In addition to describing pressure forces on bodies moving through fluids, its representation of vortex motions and wave phenomena was worthwhile. Methods of analyzing inviscid flows produced extensive analytical simulations of these flow phenomena as described in Lamb (1945), Kaufmann (1963), and Milne-Thompson (1968). This methodology required extensive mathematical manipulation, but before ready availability to high-speed computers such effort was necessary. This is no longer the case and inviscid flow analysis is seldom used.

One-dimensional flow with heat transfer requires including the convective transport of energy in the direction of bulk flow, and the energy equation is a partial differential equation. As used herein, one-dimensional means that the flow is parallel but that velocity gradient between streamlines may exist. For constant properties, the solution of the continuity and momentum equations are not coupled to the energy equation, and the solution of these equations for the velocity profile can be used in the energy equation. One must be aware of this serious restriction on the solution. This approximation is frequently referred to as being limited to small heat or mass transfer rates. For fully developed flow in a tube and between flat plates, the energy equation can be solved by the method of separation of variables. Initial and boundary conditions are uniform temperature entering the conduit, and either constant wall temperature (condensing steam) or constant heat flux at the wall (resistance heating at a constant rate). Details of these solutions of the classical problems (Graetz–Nusselt problem) are given by Bird et al. (1960).

One-dimensional flow with mass transfer and chemical reaction requires including the convective transport of mass in the direction of bulk flow and the species continuity equations. The species equations are partial differential equations. For chemical reactions, a rate equation is required. For constant properties, the velocity profile can be used from the solution of the momentum and continuity equation. The initial condition is a uniform entering concentration. Boundary conditions can be constant wall concentration, constant mass flux to the wall, and surface reaction at the wall with reactants diffusing to the wall and products diffusing from the wall. A limited number of solutions are available for binary systems given in Bird et al. (2002) and transport in chemical reactors such as DeSouza-Santos (2008).

One must realize that the solutions of the equations of change, before the advent on extensive computational solutions, were very restrictive with respect to the specification of fluid properties and the flow geometry.

2.8.2 ONE-DIMENSIONAL TRANSPORT AND WALL FUNCTIONS

Important insight can be gained by examining analytical solutions to the conservation equations, and these solutions are used in developing the wall functions described in Chapters 1 and 4. These analytical solutions are based on flow in a tube of a binary fluid with constant properties, without viscous dissipation. The continuity equation and momentum equation in terms of the shear stresses can be integrated without having to specify the character of the flow: laminar or turbulent, Newtonian or non-Newtonian. Then having specified the characteristics, the velocity profile, volumetric flow rate, and friction factor are evaluated. The energy and species continuity equation use the velocity profile in the convective term to obtain the temperature and species concentration profiles and corresponding energy and mass transferred. To obtain the wall functions, it is necessary but not realistic to neglect or make a constant of the convective transport in the bulk flow to obtain the wall functions, but these functions have proved useful in CFD codes when the concept of a laminar sublayer, buffer zone, and turbulent core are applied.

The use of computational solutions removes the constraints on the convective solution.

2.8.2.1 Fully Developed Transport in a Tube

In Figure 2.17, a diagram is shown of the fully developed velocity profile in a tube. For this case the velocity component v_z is a function of r only and there is no flow in the θ and r directions, $v_z = v_r = 0$. The continuity equation in cylindrical coordinates (Table 2.1) simplifies to

$$\frac{\partial v_z}{\partial z} = 0 \tag{2.144}$$

Laminar flow, Newtonian fluid parabolic velocity profile

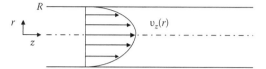

Laminar flow, non-Newtonian fluid limiting cases of $n = 0$ and $n = \infty$

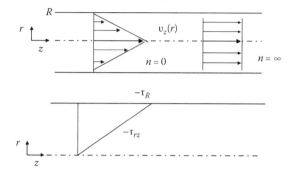

FIGURE 2.17 Tube flow for Newtonian and non-Newtonian fluids.

The velocity component v_z does not change with z, which is described as fully developed flow as compared to changes of v_z with z in the entrance region.

For these conditions the momentum equation in terms of the stresses for cylindrical coordinates (Equation F in Table 2.3) simplifies to

$$0 = -\frac{\partial P}{\partial z} + \frac{1}{r}\frac{\partial}{\partial r}(r\tau_{rz}) \tag{2.145}$$

The pressure gradient term can be transferred to the left-hand side of Equation 2.145 to give an equation that is partially differentiated with respect to z on one side and partially differentiated with respect to r on the other side. For the equation to hold, both sides must be equal to a constant, and the pressure gradient is said to be uniform which is borne out experimentally. Equation 2.145 can be written as

$$\int_0^{r\tau_{rz}} d(r\tau_{rz}) = +\frac{dP}{dz}\int_0^r r\,dr \tag{2.146}$$

Integrating with a lower limit of $r = 0$ where $\tau_{rz} = 0$, no-shear at the centerline, gives an equation for τ_{rz} in which it has not been necessary to specify if the flow is laminar or turbulent, Newtonian or non-Newtonian.

$$\tau_{rz} = +\frac{dP}{dz}\frac{r}{2} \qquad (2.147)$$

Evaluating Equation 2.147 at the wall where $r = R$ and $\tau_{rz} = \tau_R$ gives

$$\tau_R = \frac{dP}{dz}\frac{R}{2} \qquad (2.148)$$

The shear stress τ_R is tending to retard the motion of the fluid, therefore it is negative. This is consistent with the fact that the pressure gradient is negative.

Dividing Equation 2.147 by Equation 2.148 gives an equation that shows the shear stress varies linearly with the radius. This equation is the starting point for determining the wall functions discussed in Chapter 1 and the velocity profiles, volumetric flow rates, and friction factors for Newtonian and non-Newtonian flows.

$$\tau_{rz} = \tau_R \frac{r}{R} \qquad (2.149)$$

For a Newtonian fluid, Equation 2.147 and the definition of the shear stress component in terms of the velocity gradient becomes

$$\mu\left(\frac{d\upsilon_z}{dr}\right) = \left(\frac{dP}{dz}\right)\frac{r}{2} \qquad (2.150)$$

Noting that the pressure gradient is a constant and imposing the boundary condition that $\upsilon_z = 0$ at $r = R$, integration of Equation 2.150 gives the parabolic velocity profile for laminar flow in a tube.

$$\upsilon_z = \frac{R^2}{4\mu}\left(\frac{-dP}{dz}\right)\left[1-\left(\frac{r}{R}\right)^2\right] \qquad (2.151)$$

To obtain the volumetric flow rate Q, the following integral is evaluated which sums all of the differential elements of dimensions $2\pi r\, dr$ over the tube cross-section for r from 0 to R.

$$Q = \int_0^R \upsilon_z\, 2\pi r\, dr = \frac{\pi R^4}{8\mu}\left(\frac{-dP}{dz}\right) \qquad (2.152)$$

This equation is called the Hagen–Poiseuille formula for Newtonian, laminar flow. One important application of this equation is the measurement of viscosity in a Cannon–Fenske viscometer.

Equation 2.152 is used to evaluate the average velocity $\upsilon_{avg} = Q/\pi R^2$, and the Fanning friction factor is defined as $f = \tau_w / \frac{1}{2}\rho\upsilon_{avg}^2$. The wall shear stress, τ_w, is the stress acting on the wall by the fluid, it is equal to $-\tau_R$ the stress on the fluid by the presence of the wall. By combining Equation 2.152 with the definition of the friction factor, Equation 2.153 is obtained for the friction factor in terms of the Reynolds number $N_{Re} = D\upsilon_{avg}\rho/\mu$.

$$f = \frac{16}{N_{Re}} \tag{2.153}$$

The Fanning friction factor is also called the skin friction coefficient. The Darcy or Moody friction factor is four times the Fanning factor. The definition in terms of wall shear stress has to be adjusted accordingly.

For laminar flow of a non-Newtonian fluid, the power law model, Equation 2.127, has the following form for fully developed flow in a tube.

$$\tau_{rz} = m\left(\frac{d\upsilon_z}{dr}\right)^n \tag{2.154}$$

The parameter m is the apparent viscosity. For $n < 1$ the flow behavior is pseudoplastic, and for $n > 1$ dilatant.

Combining Equations 2.149 and 2.154, integrating with the no-slip lower limit that $\upsilon_z = 0$ at $r = R$ gives the velocity profile for flow of a power law fluid in a tube.

$$\upsilon_z = \left[\frac{-1}{2m}\frac{dP}{dz}\right]^{1/n}\frac{R^{1+1/n}}{1+1/n}\left[1-\left(\frac{r}{R}\right)^{1+1/n}\right] \tag{2.155}$$

Using the same procedures as was used to obtain Equation 2.154, the volumetric flow rate for a power law fluid is

$$Q = \frac{\pi R^3}{(3+1/n)}\left[\frac{R}{2m}\frac{dP}{dz}\right]^{1/n} \tag{2.156}$$

Following the same procedure to determine the friction factor, this equation is

$$f = \frac{16}{N_{Re\,n}} \quad \text{where} \quad N_{Re\,n} = \frac{\rho\upsilon_{avg}^{2-n}D^n/m}{(3+\frac{1}{n})2^{n-3}} \tag{2.157}$$

Additional details about the description of power law models and comparable results for Bingham plastics are given by Brodkey (1967).

The conservation equations for energy and mass transport in a fully developed, laminar flow with constant properties in a tube are partial differential equations with a convective transport term in the direction of flow and a conductive transport term in the direction perpendicular to the flow. Initial and boundary conditions are required, and boundary conditions can be constant temperature and concentration at the wall or constant heat flux and mass flux at the wall. These solutions are discussed below in context of obtaining wall functions for use in numerical solutions.

2.8.2.2 Wall-Functions for Momentum, Energy, and Mass Transfer

For fully developed, "turbulent flow of a Newtonian fluid" in a tube, Equation 2.147 applies; and the total shear stress is the sum of the time-averaged turbulent and time-averaged laminar stresses as shown in Figure 2.18 and as discussed in Chapter 1 as given by Equation 2.158.

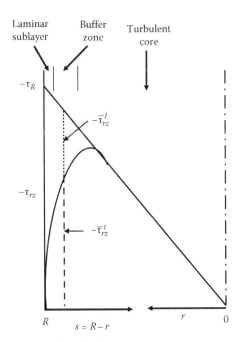

FIGURE 2.18 Laminar and turbulent stresses.

$$\tau_{rz} = \overline{\tau}^t_{rz} + \overline{\tau}^l_{rz} = \tau_R \frac{r}{R} \qquad (2.158)$$

As shown in Figure 2.18, the concept of a laminar sublayer implies a region very near the wall where the flow is laminar and is described by Newton's law of viscosity. There is a transition region from the laminar sublayer to the turbulent core called the buffer zone. In Equation 2.158, the time-averaged laminar shear stress can be approximated by Newton law of viscosity as shown below. A turbulence model is required to describe the time-averaged turbulent stress term, and Prandtl's mixing length shown below is one of many models (see Chapter 4).

$$\overline{\tau}^l_{rz} = \mu \left(\frac{d\overline{\upsilon}_z}{dr} \right) \qquad (2.159)$$

$$\overline{\tau}^t_{rz} = \rho\kappa_v^2 s^2 \left| \frac{d\overline{\upsilon}_z}{dr} \right| \left(\frac{d\overline{\upsilon}_z}{dr} \right) \qquad (2.160)$$

Referring to Figure 2.18 in the laminar sublayer region, the time-averaged, laminar stress dominates, and the time-averaged, turbulent stress term in Equation 2.158 can be omitted. Equations 2.159 and 2.160 are combined to give

$$\mu \left(\frac{d\overline{\upsilon}_z}{dr} \right) = \tau_R \frac{r}{R} \qquad (2.161)$$

Changing the radius r to the distance from the wall $s = R - r$ and using $(1 - s/R) \doteq 1$ near the wall, Equation 2.161 can be written as

$$\int_0^{\overline{\upsilon}_z} d\overline{\upsilon}_z = \frac{\tau_R}{\mu} \int_0^s ds \qquad (2.162)$$

This equation is integrated to give $\overline{\upsilon}_z = (\tau_R/\mu)s$. Defining a friction velocity as $\upsilon^{*2} = \tau_R/\rho$, dimensionless velocity $\upsilon^+ = \overline{\upsilon}_z/\upsilon^*$ and dimensionless distance as $s^+ = \upsilon^* s/(\mu/\rho)$, the resulting equation is called a "wall function." It is used to describe turbulent flow very near the wall as discussed in Chapters 1 and 4.

$$\upsilon^+ = s^+ \qquad (2.163)$$

Referring to Figure 2.18 in the region away from the wall (buffer zone and turbulent core), the time-averaged, turbulent stress dominates, and the time-averaged, laminar stress term in Equation 2.158 can be omitted. Equations 2.158 and 2.160 are combined to give

$$\overline{\tau}_{rz}^t = \rho \kappa_v^2 s^2 \left(\frac{d\overline{\upsilon}_z}{dr}\right)^2 = \tau_R \frac{r}{R} \tag{2.164}$$

Using Equation 2.162 to obtain wall values for a lower limit, Equation 2.164 can be used to obtain

$$\int_{\overline{\upsilon}_\delta}^{\overline{\upsilon}_z} d\overline{\upsilon}_z = \frac{(\tau_R/\rho)^{1/2}}{\kappa} \int_\delta^s \frac{ds}{s} \tag{2.165}$$

A point away from the wall is used as the lower limit where $\upsilon_z = \upsilon_\delta$ at $s = \delta$. Equation 2.165 is integrated and put in the following form:

$$\upsilon^+ = \frac{1}{\kappa_v} \ln s^+ + B \tag{2.166}$$

In this equation $B = -(1/\kappa_v)\ln \delta^+ + \overline{\upsilon}_\delta^+$ and κ_v are determined from turbulent flow measurements in the region away from the wall as discussed in Chapters 1 and 4. Values of κ_v of 0.36 and B of 3.8 have been reported by Bird et al. (1960). Over the past 50 years better experimental data and improved modeling equations have been developed. Equation 1.2 gives currently accepted values of the velocity profile for incompressible, fully developed pipe flow. The summarized analysis illustrates the use of a wall function to develop a velocity profile at a distance removed from the wall.

Equation 2.162 is used to provide a wall function so that Equation 2.164 can be integrated. This concept is used to describe the steep velocity gradient at the wall in turbulent flow. These functions are used in place of a very fine grid that would be required in a numerical solution to have the necessary accuracy to describe the flow. There are similar equations from the solutions of the energy and species continuity equations, and the background to obtain these wall functions is summarized next.

The "energy equation for heating a fluid" that enters a tube section where the velocity profile is given by Equation 2.151, fully developed laminar flow, is obtained by simplifying Equation B in Table 2.7.

$$\rho C_p \upsilon_z \frac{\partial T}{\partial z} = \frac{\kappa}{r}\frac{\partial}{\partial r}\left(r\frac{\partial T}{\partial r}\right) \tag{2.167}$$

The solution of this partial differential equation given the velocity υ_z is called the Graetz–Nusselt problem. Note this analysis assumes that the velocity profile is unaffected by temperature. Also, the viscous dissipation of energy is neglected. Initial conditions of a uniform temperature entering the heated section are used, and boundary conditions are either a constant wall temperature or a constant heat flux at the wall. Jacob (1949) presents a detailed solution for both of theses cases which involve a separation of variables to obtain two ordinary differential equations that require infinite series solutions. Jacob includes the solution for the case of a uniform velocity profile.

For turbulent flow, the energy equation is written in terms of the time-averaged laminar and turbulent heat flux terms as described in Chapter 4. The energy equation has a form similar to Equation 2.167 with the convective transport of energy on the left-hand side, and laminar and turbulent transport of energy on the right-hand side (see Equation B in Table 2.7). Fourier's law describes the laminar transport and a turbulence model is required to describe the turbulent energy transport.

$$\rho C_p \bar{\upsilon}_z \frac{\partial \bar{T}}{\partial z} = -\frac{1}{r}\frac{\partial}{\partial r}\left(r\left(\bar{q}_r^l + \bar{q}_r^t\right)\right) \tag{2.168}$$

To obtain a wall function, it is necessary to consider the convective energy transport term on the left-hand side to be a constant A. Equation 2.168 can be written as

$$\int_{\left(r\left(\bar{q}_r^l+\bar{q}_r^t\right)=0\right)}^{\left(r\left(\bar{q}_r^l+\bar{q}_r^t\right)\right)} d\left(r\left(\bar{q}_r^l+\bar{q}_r^t\right)\right) = -A\int_0^r r\,dr \tag{2.169}$$

Integrating and rearranging gives

$$\bar{q}_r^l + \bar{q}_r^t = -\frac{A}{2}r \tag{2.170}$$

Evaluating Equation 2.168 at the wall ($r = R$) gives $q_R = -(A/2)R$, the heat flux at the wall. Equation 2.170 can be written in a form comparable to Equation 2.158 for the shear stress.

$$\bar{q}_r^l + \bar{q}_r^t = q_R \frac{r}{R} \tag{2.171}$$

Referring to Figure 2.18, Equation 2.171 can be applied to the laminar sublayer using Fourier's law of heat conduction $\bar{q}_r^l = -\kappa d\bar{T}/dr = -\rho C_p \alpha d\bar{T}/dr$ where $\alpha = \kappa/\rho C_p$ and to the buffer zone and turbulent core using Prandtl's equation $\bar{q}_r^t = \rho C_p \kappa_v^2 s^2 (d\bar{v}_z /ds)(d\bar{T}/ds)$. With the definition of $s = R - r$ and $T^+ = \rho C_p v^*(\bar{T} - T_R)/q_R$, the equation for the laminar sublayer is comparable to Equation 2.163.

$$T^+ = s^+ \tag{2.172}$$

The equation using Prandtl's turbulence model is comparable to Equation 2.166.

$$T^+ = \frac{1}{\kappa_v} \ln s^+ + D \tag{2.173}$$

The constant $D = T_\delta^+ - (1/\kappa_v)\ln s_\delta^+$. Details concerning the manipulations and additional wall functions for heat transfer are given in Chapter 1 and Bird et al. (1960). Currently accepted values of turbulent flow temperature profiles which may be used to establish wall functions are discussed in Chapter 1.

The procedure described above for the energy equation is applicable to the species continuity equation for turbulent flow in tube. The species continuity equation is written in terms of the time-averaged laminar and turbulent heat flux terms as described in Chapter 4, and has a form similar to Equation 2.167 with the convective transport of mass on the left-hand side and laminar and turbulent transport of energy on the right-hand side. Fick's law describes the laminar transport, and a turbulence model is required to describe the turbulent energy transport. Defining $\rho_A^+ = v^*$ $(\bar{\rho}_A - \rho_{AR})/n_A$, for the laminar sublayer $\rho_A^+ = s^+$, and Prandtl's equation for the buffer zone and turbulent core is $\rho_A^+ = (1/\kappa_v)\ln s^+ + E$ where $E = \rho_\delta^+ - (1/\kappa_v)\ln s_\delta^+$. Again, recent concentration data are discussed in Chapter 1.

The procedures outlined above that are used to obtain wall functions have required using simplifications and approximations to obtain analytical solutions of the conservation equations. Numerous other methods have been used to obtain analytical solutions that attempt to give better descriptions of experimental measurements as described in Chapter 1. This discussion was provided for the background required to select appropriate wall functions for the particular transport analysis being performed. In general, the better the experimental data for near wall phenomena and the closer to the wall the grid points used in the computational solution, the more accurate the resulting simulation of the entire flow field will become.

2.8.3 REACTING FLOWS IN POROUS MEDIA AND DARCY'S LAW

Transpiration cooling is used to protect surfaces and reduce heat transfer when these surfaces are exposed to extremely high or low temperatures. Very high temperatures are encountered by heat shield of reentering space vehicles, and very low temperatures are encountered in the storage of cryogenic liquids. Transportation cooling has a gas flowing through the walls of the container which is a porous medium, and the gas transports heat by convective flow in the opposite direction of the conductive transport of heat.

Typical results can be obtained describing the flow rate, pressure, and temperature distributions, and energy absorbed by considering the transportation cooling of a porous slab of porosity ε and permeability α by steady flow of an ideal gas with superficial velocity v_0. One side is maintained at T_0 ($z = 0$) and the other at T_L ($z = L$) as is shown in Figure 2.19.

This discussion will illustrate some of the complexities introduced when variable fluid properties are required. Also, the equation of motion for flow in a porous medium, Darcy's law, must be used for transportation cooling in place of the momentum equation. Darcy's law defines the permeability of a porous media, a property of the media.

The continuity equation for flow in a porous medium is given in terms of the porosity of the porous media, ε, and the superficial velocity, v_0, which

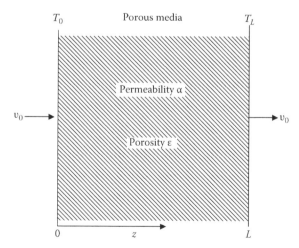

FIGURE 2.19 Diagram of flow through a porous medium subjected to a temperature gradient.

is based on the total cross-sectional area rather than the cross-sectional area of the pores.

$$\varepsilon \frac{\partial \rho}{\partial t} = -\nabla \cdot \rho \upsilon_0 \qquad (2.174)$$

For steady, one-dimensional flow in the z-direction, the above reduces to

$$\frac{d \rho \upsilon_0}{dz} = 0 \qquad (2.175)$$

Integrating gives

$$\rho \upsilon_0 = \text{constant} = \varepsilon W = \varepsilon \rho \upsilon_z \qquad (2.176)$$

where
W is the mass flux of the gas in the pores
υ_z is the mean velocity of the gas in the pores

For flow in a porous medium, Darcy's law is used in place of the equation of motion and is

$$\upsilon_0 = -\frac{\alpha}{\mu}(\nabla P - \rho g) \qquad (2.177)$$

where the superficial velocity $(= \upsilon_z \varepsilon)$ and α in the permeability of the porous media. Equation 2.177 simplifies to the following equation for one-dimensional horizontal flow.

$$\upsilon_0 = -\frac{\alpha}{\mu}\frac{dP}{dz} = \varepsilon \upsilon_z \qquad (2.178)$$

To obtain the pressure distribution for steady flow ($W = $ constant), Equations 2.176 and 2.178 are combined with the ideal gas law $\rho = PM_w/RT$, and rearrangement gives

$$\int_{P_0}^{P} P \, dP = \frac{-WR}{\alpha M_w} \int_{0}^{z} \mu T \, dz \qquad (2.179)$$

Integrating Equation 2.179 gives Equation 2.180 in a form that shows the solution of the energy equation is required.

$$P = \left[P_0^2 - \frac{2WR}{\alpha M_w} \int_{T_0}^{T} \frac{\mu T}{dT/dz} dT \right]^{1/2}$$ (2.180)

To integrate the right-hand side of Equation 2.180, T and dT/dz are needed from the solution of the energy equation, and μ may be a function of T which requires a correlation equation.

The energy equation for an ideal gas is

$$\rho C_p \frac{DT}{Dt} = \nabla \cdot \kappa_g \nabla T + \frac{DP}{Dt} - \vec{\tau} : \nabla \vec{\upsilon}$$ (2.181)

For steady, one-dimensional flow in the pores neglecting viscous dissipation, the previous equation is

$$\rho C_p \upsilon_z \frac{dT}{dz} = \frac{d}{dz} \left(\kappa_g \frac{dT}{dz} \right) + \upsilon_z \frac{dP}{dz}$$ (2.182)

At this point one must resort to a simultaneous numerical solution of Equations 2.180 and 2.182, as they are coupled.

To continue and obtain an approximate but analytical solution, constant thermal conductivity and viscosity are used. The P–V work term is neglected which is small compared to the convective term. Then the energy equation for the gas flowing in the pores, Equation 2.182, combined with the continuity equation, Equation 2.176, gives

$$C_p W \frac{dT}{dz} = \kappa_g \frac{d^2 T}{dz^2}$$ (2.183)

The energy equation for the solid porous medium with constant thermal conductivity simplifies to

$$0 = \kappa_s \frac{d^2 T}{dz^2}$$ (2.184)

The gas and solid are in intimate contact such that at any cross section both are at the same temperature, Equations 2.183 and 2.184 can be combined to give

$$\varepsilon C_p W \frac{dT}{dz} = \left[\varepsilon \kappa_g + (1-\varepsilon)\kappa_s \right] \frac{d^2 T}{dz^2}$$ (2.185)

Letting $\kappa_e = \varepsilon\kappa_g + (1-\varepsilon)\kappa_s$, an effective thermal conductivity for the solid and gas, Equation 2.185 can be put in the form:

$$\frac{d^2 T}{dz^2} = \frac{C_p \varepsilon W}{\kappa_e} \frac{dT}{dz} = 0 \qquad (2.186)$$

For constant properties, the solution of Equation 2.186 is

$$\frac{T - T_0}{T_L - T_0} = \frac{1 - e^{Nz/L}}{1 - e^N} \qquad (2.187)$$

where $N = WC_p\varepsilon L/\kappa_e$. To integrate Equation 2.187 for variable properties, C_p and κ_e, a numerical solution would be required. Using the above with Equation 2.180 the pressure distribution could be obtained.

In the above description of transpiration cooling, the composition of the gas flowing through the porous medium was constant. In cases such as the char zone of a charring ablator where the porous medium is subjected to a steep temperature gradient, chemical reactions will occur in the gas and between the gas and the porous matrix (April 1969). To accurately predict the energy absorbed during reentry of a space vehicle (for proper heat shield design) an accurate description of the reacting flow in a porous medium is needed. Most of the reactions that occur in the char zone during ablation are endothermic, and the heat absorbed by the gases can be bounded by two limiting cases, frozen flow and equilibrium flow.

The results for frozen flow (no chemical reaction) are applicable, except for the constant properties assumption, to give the minimum energy absorbed by the gases flowing through the porous medium. The maximum energy potentially absorbed by the gases flowing through the porous medium is obtained by predicting the changes in composition from the chemical reactions considering the species are in thermodynamic equilibrium (equilibrium flow) at any cross section. Depending on the mass flux of gas, chemical species present, and the temperature gradient, either of these two limiting cases could approximate the energy transferred. To predict more precisely the energy absorbed in the region bounded by these two limiting cases, the composition changes must be evaluated using the kinetics of the reactions (nonequilibrium flow). Differential equations are obtained for the prediction of the energy absorbed in the porous medium for all three cases, considering variable physical and thermodynamic properties. An analytical solution is not feasible, and the results of numerical solutions are presented, that were obtained by simulating the flow in the char zone of a low-density phenolic-nylon ablative composite (April 1969).

The same procedure is followed as described above to obtain equations to describe the energy absorbed at the heated surface of the porous media. The conservation equations are simplified for one-dimensional flow of a reacting gas in the pores of the medium, including the species continuity equation. The form of Darcy's law only requires modification. The energy equation is applied to the porous medium and the gas phase separately, and then these two equations are combined to obtain an equation that predicts the temperature profile in the medium. The material has sufficiently small diameter pores that the gas and solid are at the same temperature at any cross section normal to the flow.

For k gas species, the species continuity equation for the individual species from Table 2.2 reduces to the following form for steady, one-dimensional flow.

$$\frac{dn_i}{dz} = r_i, \quad i = 1, 2, \ldots, k \tag{2.188}$$

This equation shows that the increase in mass flux of component i with distance, z, is from the formation of component i by chemical reaction, r_i.

For steady flow the overall continuity equation (Equation 2.176) is the same as that obtained previously, neglecting changes in gas flow rate due to reaction with the solid. The overall continuity equation is the sum over the species equations. It can vary with position if there are chemical reactions between the gas and the char.

To predict the pressure distribution, the procedure is the same as that discussed above. The mass flux of gas, W, changes due to reaction with the solid phase, and it is included under the integral in Equation 2.180 as shown in

$$P = \left[P_0^2 - \frac{2R}{\alpha M_w} \int_{T_0}^{T} \frac{\mu W T}{dT/dz} dT \right]^{1/2} \tag{2.189}$$

The energy equation for steady, one-dimensional flow of an ideal gas is

$$\rho C_p \upsilon_z \frac{dT}{dz} = \frac{d}{dT}\left(\kappa_g \frac{dT}{dz}\right) - \left(\frac{\partial \ln \rho}{\partial \ln T}\right)_{P,x_i} \frac{dP}{dz} - \sum_{i=1}^{n} H_{xi} R_i \tag{2.190}$$

where $\sum_{i=1}^{k} H_{xi}R_i$ represents the heat absorbed or released by chemical reactions.

The simplifications omitted potential energy changes, viscous dissipation, and internal heat generation. Heat transferred by diffusion in the direction of flow is very small compared to heat transfer by convection and is not included.

For the solid matrix, there is no flow and the energy equation is

$$0 = \frac{d}{dz}\left(\kappa_s \frac{dT}{dz}\right) - \sum_{i=k}^{n} H_{xi}R_i \tag{2.191}$$

where $\sum_{i=1}^{k} H_{xi}R_i$ represents the heat absorbed or released by reactions occurring with the solid.

Multiplying Equation 2.190 by the porosity, ε, Equation 2.191 by $(1 - \varepsilon)$ and adding, the result is an equation that describes the energy transfer at any cross section in the z-direction in the solid and gas

$$W_g \varepsilon C_p \frac{dT}{dz} = \frac{d}{dz}\left(\kappa_e \frac{dT}{dz}\right) - \sum_{i=1}^{n} H_{xi}R_i \tag{2.192}$$

where the overall continuity equation was used along with defining an effective thermal conductivity $\kappa_e = [\varepsilon\kappa_g + (1 - \varepsilon)\kappa_s]$. The term $\sum_{i=1}^{n} H_{xi}R_i$ represents the heat absorbed or released by chemical reactions in the gas phase, between the gas and solid, and in the solid phase.

A numerical solution is required to solve Equations 2.189 and 2.192 with boundary conditions for the temperature, pressure, mass flux, and composition of the species entering the porous media at $z = 0$. The solution is the temperature and mass flux of the species flowing through the length L of the porous media. The pressure distribution is computed from the solution of these equations by Equation 2.189.

The solution to Equations 2.189 and 2.192 was obtained by using a fourth-order Runge–Kutta procedure for the chemical reactions shown in Figure 2.20 (April, 1969). The surfaces of the porous media were maintained at 260°C (500°F) and 1371°C (2500°F). The solid matrix (char) was pure carbon with a porosity of 0.8 and permeability of 1.02×10^{-8} m^2. The gas composition entering the porous media was approximately that which would result from the degradation of a phenolic-nylon heat shield composite. The chemical reactions occurring in the char zone are shown

1. $CH_4 = 1/2C_2H_6 + 1/2H_2$
2. $C_2H_6 = C_2H_4 + H_2$
3. $C_2H_4 = C_2H_2 + H_2$
4. $C_2H_4 = 2C + H_2$
5. $CH_4 = 1/2C_2H_2 + 3/2H_2$
6. $CH_4 = C + 2H_2$
7. $C_6H_6 = 3C_2H_2$
8. $C + CO_2 = 2CO$
9. $NH_3 = 1/2N_2 + 3/2H_2$
10. $CH_4 + 3/2O_2 = CO_2 + H_2O$

FIGURE 2.20 Chemical reactions occurring in the char zone of a phenolic-nylon composite.

in matrix form in Figure 2.21. The inlet composition is given in Figure 2.22. The composition of the degradation products leaving the high-temperature surface for nonequilibrium (finite rate) flow and equilibrium flow is also shown in Figure 2.22. The equilibrium flow case predicted that extensive reactions occurred, and the method of computing the composition of a reacting gas–solid mixture used to predict the equilibrium flow composition used free-energy minimization and was reported by Del Valle (1975). The extent that reactions occur predicted by nonequilibrium

$$
\begin{bmatrix}
0 & 1 & 0 & 0 & 0 & 0 & 0 & 0 & 0 & 0 & 0 & 0 & 0 \\
0 & 0 & 1 & 0 & 0 & 0 & 0 & 0 & 0 & 0 & 0 & 0 & 0 \\
0 & 0 & 0 & 1 & 0 & 0 & 0 & 0 & 0 & 0 & 0 & 0 & 0 \\
0 & 0 & 0 & 0 & 1 & 0 & 0 & 0 & 0 & 0 & 0 & 0 & 0 \\
0 & 1 & 0 & 0 & 0 & 0 & 0 & 0 & 0 & 0 & 0 & 0 & 0 \\
0 & 1 & 0 & 0 & 0 & 0 & 0 & 0 & 0 & 0 & 0 & 0 & 0 \\
0 & 0 & 0 & 0 & 0 & 0 & 0 & 1 & 0 & 0 & 0 & 0 & 0 \\
0 & 0 & 0 & 0 & 0 & 1 & 0 & 1 & 0 & 0 & 0 & 0 & 0 \\
0 & 0 & 0 & 0 & 0 & 0 & 0 & 0 & 0 & 1 & 0 & 0 & 0 \\
0 & 1 & 0 & 0 & 0 & 0 & 0 & 0 & 0 & 0 & 0 & \frac{3}{2} & 0
\end{bmatrix}
\begin{bmatrix}
H_2 \\ CH_4 \\ C_2H_6 \\ C_2H_4 \\ C_2H_2 \\ C \\ C_6H_6 \\ CO_2 \\ CO \\ NH_3 \\ N_2 \\ O_2 \\ H_2O
\end{bmatrix}
=
\begin{bmatrix}
\frac{1}{2} & 0 & \frac{1}{2} & 0 & 0 & 0 & 0 & 0 & 0 & 0 & 0 & 0 & 0 \\
1 & 0 & 0 & 1 & 0 & 0 & 0 & 0 & 0 & 0 & 0 & 0 & 0 \\
1 & 0 & 0 & 0 & 1 & 0 & 0 & 0 & 0 & 0 & 0 & 0 & 0 \\
1 & 0 & 0 & 0 & 0 & 2 & 0 & 0 & 0 & 0 & 0 & 0 & 0 \\
\frac{3}{2} & 0 & 0 & 0 & \frac{1}{2} & 0 & 0 & 0 & 0 & 0 & 0 & 0 & 0 \\
2 & 0 & 0 & 0 & 0 & 1 & 0 & 0 & 0 & 0 & 0 & 0 & 0 \\
0 & 0 & 0 & 0 & 3 & 0 & 0 & 0 & 0 & 0 & 0 & 0 & 0 \\
0 & 0 & 0 & 0 & 0 & 0 & 0 & 0 & 2 & 0 & 0 & 0 & 0 \\
\frac{3}{2} & 0 & 0 & 0 & 0 & 0 & 0 & 0 & 0 & 0 & \frac{1}{2} & 0 & 0 \\
0 & 0 & 0 & 0 & 0 & 0 & 0 & 1 & 0 & 0 & 0 & 0 & 1
\end{bmatrix}
\begin{bmatrix}
H_2 \\ CH_4 \\ C_2H_6 \\ C_2H_4 \\ C_2H_2 \\ C \\ C_6H_6 \\ CO_2 \\ CO \\ NH_3 \\ N_2 \\ O_2 \\ H_2O
\end{bmatrix}
$$

FIGURE 2.21 Matrix form of the reaction set.

Mass flux - 0.05 lb per ft²-s, char characteristics: porosity—0.8, permeability—1×10^{-9} ft²
Pyrolysis gas composition (mole percent)

	Entering Char at 500°F	Leaving Front Surface at 2500°F	
		Nonequilibrium Flow	Equilibrium Flow
CH_4	56.56	3.24	0.01
C_2H_6	0	0.16	0
C_2H_4	0	0.52	0
C_2H_2	0	13.73	0
H_2	1.45	58.27	77.03
H_2O	32.49	18.47	0.01
N_2	8.06	4.61	4.21
CO_2	1.52	0.87	trace
CO	0	0.12	18.65
NH_3	0.02	0.01	trace

	Heat flux at surface (Btu/ft²-s)	Pressure drop (lb_f/ft²)
Frozen flow	63.12	8.9
Nonequilibrium flow	93.79	18.7
Equilibrium flow	271.25	9.8

FIGURE 2.22 Comparison of the high temperature surface heat flux, composition of degradation products and pressure drop for frozen, equilibrium, and nonequilibrium flow.

flow is intermediate between frozen and equilibrium flow as shown in Figures 2.22 and 2.23.

Comparing the composition changes for equilibrium and nonequilibrium flow, it is seen in Figure 2.22 that the flow is between frozen and equilibrium flow. Due to the porous matrix characteristics the pressure drop is very small, and the term neglected in the energy equation could be safely ignored. Chemical reactions do take place as is seen by comparing the heat flux at the surface. This is equivalent to the energy absorbed in the char zone if there is no heat conducted away from the low-temperature surface. In Figure 2.23 the temperature profiles are given for the three cases. The nonequilibrium flow curve is closer to the frozen flow curve because most of the chemical reactions occur in the range from 2000°F to 2500°F.

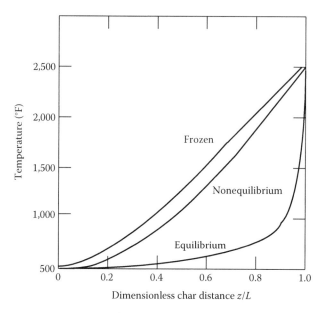

FIGURE 2.23 Comparison of temperature distributions for frozen, equilibrium and nonequilibrium (finite rate) flow in the char zone, 2500°F surface temperature.

The equations for the chemical reactions in the char zone were shown in Figure 2.20. To facilitate the numerical solution, they were put in matrix form using the format of Equation 2.193 as shown in Figure 2.21

$$\sum_{i=1}^{n} r_{ji} A_i \underset{k_{rj}}{\overset{k_{fj}}{\Longleftrightarrow}} \sum_{i=1}^{n} p_{ji} A_i \tag{2.193}$$

where

r_{ji} and p_{ji} represent the stoichiometric coefficients of the reactants and products for species A_i in reaction j

k_{fj} and k_{rj} are forward and reverse reaction rate constants

Using the format of Equation 2.193, the rate of reaction of the ith species, r_i, is given by

$$r_i = \sum_{j=1}^{m} (p_{ji} - r_{ji}) \left\{ k_{fj} \prod_{i=1}^{n} c_i^{r_{ji}} - k_{rj} \prod_{i=1}^{n} c_i^{p_{ji}} \right\} \tag{2.194}$$

There are 10 reactions and 13 chemical species. The matrices are the stoichiometric coefficients r_{ji} and p_{ji}.

To illustrate the use of Equation 2.194, the rate of reaction of methane (component 2) is given by the following:

$$r_2 = \sum_{j-1}^{10} (p_{j2} - r_{j2}) \left[k_{fj} \prod_{i=1}^{13} c_i^{r_{ji}} - k_{rj} \prod_{i=1}^{13} c_i^{p_{ij}} \right]$$

(2.195)

or expanding

$$r_2 = (0-1)\left\{ k_{f1}c_2 - k_{r1}c_1^{1/2}c_3^{1/2} \right\} + (0-1)\left\{ k_{f5}c_2 - k_{r5}c_1^{3/2}c_5^{1/2} \right\}$$
$$+ (0-1)\left\{ k_{f6}c_2 - k_{r6}c_1^2 c_6 \right\}$$

(2.196)

In Equation 2.196, there are five other terms in the expanded form, but these are not included since their coefficients were zero. Modifications of this general form may be required. For example, the powers on the compositions for the actual rate expressions can be different than the stoichiometric coefficients in some cases. In the carbon–water reaction the surface area of carbon is used which is not a "concentration" of carbon. To take the surface area of the char into account the reaction rate constant was modified, and the exponent on the carbon "concentration" was taken equal to zero.

In summary, this analysis demonstrated the development of the appropriate differential equation to describe the one-dimensional reacting flow in a porous medium. The equations were sufficiently complicated to require a numerical solution, and some results were given that applied to the char zone of an ablative heat shield. Three cases were considered: frozen, equilibrium, and nonequilibrium. Computationally, frozen flow is the easiest to solve, followed by equilibrium flow. For equilibrium flow the compositions were a function to temperature only and were represented by algebraic equations. For nonequilibrium flow, 13 species continuity equations were solved simultaneously with the energy equation to predict the mass flux, concentration, and temperature which required an order of magnitude increase in computations. Complete details are given by April (1969) and Del Valle (1975).

2.8.4 SIMULTANEOUS MOMENTUM, HEAT, AND MASS TRANSFER IN THE BOUNDARY LAYER

Flow in the boundary layer refers to flow past a surface where the changes in velocity, temperature, and concentration are confined to a very thin region near the surface. This flow was first identified in studies of flow

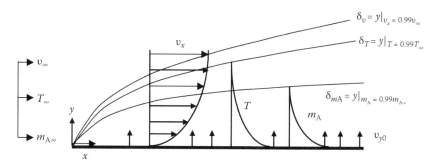

FIGURE 2.24 Boundary layer flow along a flat plate with heat and mass transfer.

around airfoils and other slender bodies. When there is heat and mass transfer, temperature and concentration gradients are very thin and steep from the wall to the free stream. The term thin is relative; the boundary layer at the trailing edge of the Saturn V was about $1\,m^{-1}$. These profiles are shown in an exaggerated form in Figure 2.24 for flow across a flat plate. An order analysis of the terms in the conservation equations was made to obtain the equations that describe this flow (Blasius, 1950; Mickley et al., 1954; Schlichting, 1955).

The boundary layer thicknesses for momentum, δ_v, heat, δ_T, and mass, δ_{Ab}, transfer are shown in Figure 2.24. The boundary layer thicknesses are defined as the distances in the y-direction from the surface where the values of the velocity, temperature, and concentration reach 99% of their free stream values. For example, with air having a free stream velocity of 25.8 ft/s flowing over a flat plate, the momentum boundary layer thickness is 0.026 in. thick at 3.6 in. from the front of the plate and grows to 0.322 in. thick at 96.4 in. from the front of the plate (Mickley et al., 1954).

Knowing that changes in velocity, temperature, and concentration are confined to a very thin region near the surface; an order analysis was performed on the continuity, species continuity, momentum, and energy equations to eliminate terms that are small compared to others. This order analysis is discussed by Blasius (1950), Mickley et al. (1954), and Schlichting (1955); and the result is the following set of partial differential equations for steady, laminar flow of an incompressible binary fluid across a flat plate.

$$\text{Continuity:} \frac{\partial v_x}{\partial x} + \frac{\partial v_y}{\partial y} = 0 \qquad (2.197)$$

$$\text{Momentum: } \upsilon_x \frac{\partial \upsilon_x}{\partial x} + \upsilon_y \frac{\partial \upsilon_y}{\partial y} = \nu \frac{\partial^2 \upsilon_x}{\partial y^2} \qquad (2.198)$$

$$\text{Energy: } \upsilon_x \frac{\partial T}{\partial x} + \upsilon_y \frac{\partial T}{\partial y} = \alpha \frac{\partial^2 T}{\partial y^2} \qquad (2.199)$$

$$\text{Mass: } \upsilon_x \frac{\partial m_A}{\partial x} + \upsilon_y \frac{\partial m_A}{\partial y} = D_{AB} \frac{\partial^2 m_A}{\partial y^2} \qquad (2.200)$$

The velocity components, υ_x and υ_y, temperature, T, and mass fraction $m_A = \rho_A/\rho$ are functions of x and y. The coefficient $\nu(=\mu/\rho)$ is the kinematic viscosity, α is the thermal diffusivity, and D_{AB} is the mass diffusivity. A solution has been obtained to the boundary layer equations using the following initial and boundary conditions.

At $y = 0$ and $x > 0$	$\upsilon_x = 0$	No slip
Surface of plate	$\upsilon_y = \upsilon_{y0}$	Velocity at surface in y-direction
	$T = T_0$	Uniform surface temperature
	$m_A = m_{A0}$	Uniform surface concentration
At $y = \infty$ or $x \le 0$	$\upsilon_x = \upsilon_\infty$	Free stream velocity
Far from plate	$T = T_\infty$	Free stream temperature
	$m_A = m_{A\infty}$	Free stream concentration
	$n_B = 0$	A diffuse through stagnant B

These initial and boundary conditions were specified: free stream velocity, υ_∞; free stream and surface temperatures, T_∞ and T_0; and concentrations, $m_{A\infty}$ and m_{A0}. The surface can be porous, and there is flow with a velocity, υ_{y0}, into or from the boundary layer from this surface. Flow to or from the boundary layer can be interpreted as condensation or evaporation from the surface. There is only mass transfer of component A to or from the surface ($n_B = 0$).

The solution to Equations 2.197 through 2.200 was first obtained by Blasius (1950) for $\upsilon_y|_{y=0} = \upsilon_{y0} = 0$. The solution was obtained by using a dimensionless stream function, $f(\eta)$, to satisfy the continuity equation and reduce the momentum equation from a partial differential equation to an ordinary differential equation. The following equations are the important relations where $\psi(x, y)$ is the stream function.

$$v_x = \frac{\partial \psi(x, y)}{\partial y} \tag{2.201}$$

$$v_y = -\frac{\partial \psi(x, y)}{\partial x} \tag{2.202}$$

$$f(\eta) = \frac{\psi(x, y)}{(vxv_\infty)^{\frac{1}{2}}} \tag{2.203}$$

$$\eta = y \left(\frac{v_\infty}{vx} \right)^{1/2} \tag{2.204}$$

Performing the appropriate partial derivation and substituting into Equation 2.198 the following ordinary differential equation is obtained.

$$ff'' + 2f''' = 0 \tag{2.205}$$

Blasius (1950) solved Equation 2.205 by the Method of Forbenius to obtain the infinite series solution given below.

$$f(\eta) = \sum_{n=0}^{\infty} (-1)^n \frac{c_n \alpha^{n+1}}{(3n+2)!} \eta^{3n+2} \tag{2.206}$$

where

$$\frac{v_x}{v_\infty} = f'(\eta) \tag{2.207}$$

$$\frac{v_y}{v_\infty} = \frac{1}{2} \left(\frac{vv_\infty}{x} \right)^{\frac{1}{2}} \left[\eta f'(\eta) - f(\eta) \right] \tag{2.208}$$

The coefficients in the series solution were determined to have the following recursion formula.

$$c_n = \sum_{l=0}^{n-1} \left(\frac{3n-1}{3l} \right) c_l c_{n-1-l} \tag{2.209}$$

Blasius (1950) reported numerical values for the coefficients, and the first six coefficients have the following values:

$c_0 = 1;\ c_1 = 1;\ c_2 = 11;\ c_3 = 375;\ c_4 = 27{,}897,\ c_5 = 3{,}817{,}137;\ c_6 = 865{,}874{,}115$

The value of α in Equation 2.206 was determined using the boundary condition $v_x = v_\infty$ at $y = \infty$ for $x \le 0$ to be 0.332.

Bennett and Myers (1962) give the first four terms in the infinite series solution as follows:

$$f(\eta) = 0.16603\eta^2 - 4.5943 \times 10^{-4}\eta^5 + 2.4972 \times 10^{-6}\eta^8 - 1.4277 \times 10^{-8}\eta^{11}$$

(2.210)

A graphical representation of the solution for v_x/v_∞ and v_y/v_∞ is shown in Figures 2.25 and 2.26 using the data reported by Brodkey (1967). The velocity v_x reaches 0.99155 of the free stream velocity at a value of $\eta = 5.0$ which is used as the definition of the boundary layer thickness. The limiting value for $v_y(\eta = \infty) = 0.865v_\infty(v/v_\infty x)^{1/2}$ demonstrates that the plate retards the flow, and there is a net flow in the y-direction.

The solution given by Equation 2.210 is applicable to heat and mass transfer for the special case where $v = \alpha = D_{AB}$. The Prandtl number, $Pr = v/\alpha = 1$ and the Schmidt number, $Sc = v/D_{AB} = 1$. The momentum, energy, and species continuity equations have the same form using the following dimensionless temperature and concentration.

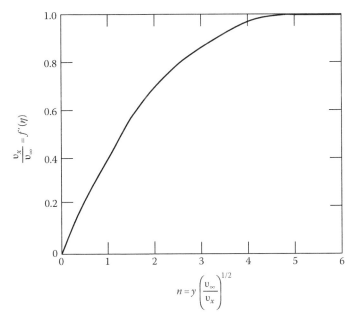

FIGURE 2.25 Blasius solution for flow over a flat plate for velocity v_x.

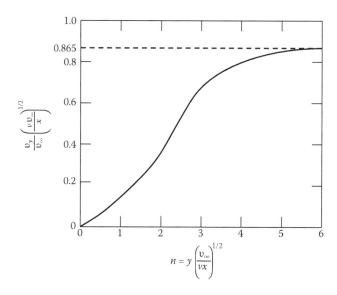

FIGURE 2.26 Blasius solution for flow over a flat plate for velocity v_y.

$$\overline{T} = \frac{T - T_0}{T_\infty - T_0} \tag{2.211}$$

$$\overline{m}_A = \frac{m_A - m_{A0}}{m_{A\infty} - m_{A0}} \tag{2.212}$$

The initial and boundary conditions are the same if the dependent variables are given by the following equation. The velocity, temperature, and concentration profiles are the same.

$$\frac{v_x}{v_\infty} = \frac{T - T_0}{T_\infty - T_0} = \frac{m_A - m_{A0}}{m_{A\infty} - m_{A0}} = f'(\eta) \tag{2.213}$$

The initial and boundary conditions are

At $y = 0$ and $x > 0$	$v_x = 0$; or $v_x/v_\infty = f'(0) = 0$
Surface of plate	$T = T_0; \overline{T} = 0$
$\eta = 0$	$m_A = m_{A0}; \overline{m}_A = 0$
At $y = \infty$ or $x \le 0$	$v_x = v_\infty$ or $v_x/v_\infty = f'(\infty) = 1$
Far from the surface	$T = T_\infty; \overline{T} = 1$
$\eta = \infty$	$m_A = m_{A\infty}; \overline{m}_A = 1$

There is the limitation on the mass transfer solution because $v_y|_{y=0} = v_{y0} = 0$, and there is mass transfer from the surface by diffusion only. This implies that the rate of mass transfer from the surface is small.

The Blasius solution can be extended to allow for convective transport to or from the solid surface, i.e., $v_y = v_{y0}$, a finite velocity at surface in y-direction This transport is referred to as sucking or blowing, sucking has found application in boundary layer control on the wings of high-speed aircraft to permit them to fly at low speeds without stalling. It includes having condensation on a surface or vaporization from a surface.

To extend the solution for a finite value of the velocity, v_{y0}, at the surface, the following boundary condition is used.

$$v_y = v_{y0} \quad \text{at} \quad y = 0 (\eta = 0) \tag{2.214}$$

instead of

$$v_y = 0 \quad \text{at} \quad y = 0 (\eta = 0) \tag{2.215}$$

is used to evaluate the constants in the series solution used to obtain the following equation.

$$f(\eta) = A_0 + A_1\eta + A_2\eta^2 + \cdots \tag{2.216}$$

For $v_x = 0$ and $v_{y0} = 0$, Equations 2.215 and 2.216 were used to determine that $f(0) = A_0 = 0$ and $f'(0) = A_1 = 0$. For $v_x = 0$ and $v_{y0} =$ finite and constant, Equations 2.214 and 2.216 require that $f(0) = A_0 =$ constant and $f'(0) = A_1 = 0$ for the solution given by Equation 2.216 to satisfy the continuity and momentum equations (Equations 2.197 and 2.198). Equation 2.216 must be modified since $A_0 =$ constant and not zero for a finite velocity in the y-direction at the surface.

Solving for $f(0)$ using Equation 2.208 with $f'(0) = 0$ gives

$$f(0) = -2v_{y0} \left(\frac{x}{v v_\infty} \right)^{\frac{1}{2}} = -2 \frac{v_{y0}}{v_\infty} \left(\frac{v_\infty x}{v} \right)^{\frac{1}{2}} = -2 \frac{v_{y0}}{v_\infty} Re_x^{\frac{1}{2}} \tag{2.217}$$

For convective transport from the surface, Equation 2.210 is modified as follows:

$$f(\eta) = -2 \frac{v_{y0}}{v_\infty} \left(\frac{v_\infty x}{v} \right)^{\frac{1}{2}} + \sum_{n=0}^{\infty} (-1)^n \frac{c_n \alpha^{n+1}}{(3n+2)!} \eta^{3n+2} \tag{2.218}$$

The first term in the series above is a constant, and this requirement places a restriction on υ_{y0}. The velocity at the surface, υ_{y0}, must vary to have the product υ_{y0} and $1/x^{1/2}$ be a constant.

It is convenient to let $f(0) = -K$, a constant and solve Equation 2.217 for υ_{y0}.

$$\upsilon_{y0} = \frac{\upsilon_\infty K}{2}\left(\frac{\nu}{\upsilon_\infty x}\right)^{\frac{1}{2}} \tag{2.219}$$

If K is positive, then υ_{y0} is positive, and there is mass transfer into the stream from the surface (blowing or vaporization). If K is negative, then υ_{y0} is negative, and there is mass transfer from the stream to the surface (sucking or condensation).

The solution of the momentum equation, $f(\eta)$ is used to obtain the solution of the energy (Equation 2.199) and species continuity Equation 2.200, for Pr and $Sc \neq 1$. The dimensionless temperature and mass fraction given by Equations 2.211 and 2.212 are substituted into the energy and species continuity Equations 2.199 and 2.200, along with Equation 2.213 for η to obtain Equations 2.220 and 2.221 for energy and species continuity.

$$\frac{d^2\overline{T}}{d\eta^2} + \frac{Pr}{2}f(\eta)\frac{d\overline{T}}{d\eta} = 0 \tag{2.220}$$

$$\frac{d^2\overline{m}_A}{d\eta^2} + \frac{Sc}{2}f(\eta)\frac{d\overline{m}_A}{d\eta} = 0 \tag{2.221}$$

For the case where the Prandtl number and Schmidt number are different than one, Equations 2.220 and 2.221 can be formally integrated where $f(\eta)$ is a known function given by Equation 2.208. Numerical integration is required. Defining $\overline{T}' = d\overline{T}/d\eta$ and integrating Equation 2.220, the result is indicated by

$$\ln\left[\frac{\overline{T}'}{\overline{T}'(\eta=0)}\right] = -\frac{Pr}{2}\int_0^\eta f(\eta)d\eta \tag{2.222}$$

where $\overline{T}'(\eta = 0)$ is evaluated using a boundary condition. Equation 2.222 is written as

$$\overline{T}' = \frac{d\overline{T}}{d\eta} = \overline{T}'(\eta=0)\exp\left[\frac{-Pr}{2}\int_0^\eta f(\eta)d\eta\right] \tag{2.223}$$

Integration is indicated again by integrating again from $\overline{T} = 0$ at $\eta = 0$ to $\overline{T}$ at η to obtain

$$\overline{T}(\eta) = \overline{T}'(\eta = 0) \int_0^\eta \left[\exp \frac{-Pr}{2} \int_0^\eta f(\eta) d\eta \right] d\eta \qquad (2.224)$$

The constant $\overline{T}'(\eta = 0)$ is evaluated using the boundary condition $\overline{T}'(\eta = \infty) = 1$ at $\eta = 0$ substituted into Equation 2.224 which gives

$$\overline{T}(\eta = \infty) = 1 = \overline{T}'(\eta = 0) \int_0^\infty \left[\exp \frac{-Pr}{2} \int_0^\eta f(\eta) d\eta \right] d\eta \qquad (2.225)$$

and the solution to the energy equation is

$$\overline{T}(\eta) = \frac{\int_0^\eta \left[\exp\left(\frac{-Pr}{2} \int_0^\eta f(\eta) d\eta \right) \right] d\eta}{\int_0^\infty \left[\exp\left(\frac{-Pr}{2} \int_0^\eta f(\eta) d\eta \right) \right] d\eta} \qquad (2.226)$$

By the same procedure, the solution to Equation 2.221 is in terms of the known function, $f(\eta)$, and is of the same form.

$$\overline{m}_A(\eta) = \frac{\int_0^\eta \left[\exp\left(-\frac{Sc}{2} \int_0^\eta f(\eta) d\eta \right) \right] d\eta}{\int_0^\infty \left[\exp\left(-\frac{Sc}{2} \int_0^\eta f(\eta) d\eta \right) \right] d\eta} \qquad (2.227)$$

where the only difference is the Schmidt number replaced the Prandtl number.

The solutions to Equations 2.226 and 2.227 are present on Figure 2.27, where the parameters of the solution of the equations are the Prandtl number, Pr, or Schmidt, Sc, and K. From this figure concentration, temperature, and velocity profiles can be obtained for values of the Prandtl and Schmidt numbers of 0.72, 1.0, and 2.0 and for values of the dimensionless mass transfer parameter, K, of 1, 0, and −5. The accuracy of these profiles was confirmed experimentally with good agreement between theory and experiment for air flowing over a porous flat plate with air injected or withdrawn from the boundary layer (Mickley et al., 1954).

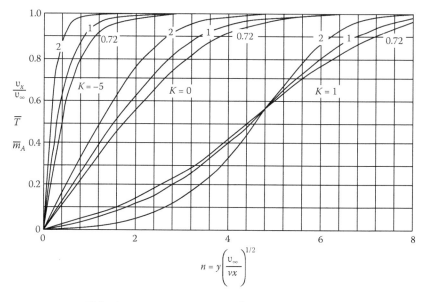

FIGURE 2.27 Velocity, temperature, and concentration profiles in the laminar boundary layer over a flat plate. Mass transfer to the surface, condensation, sucking ($K = -5$). Mass transfer from the surface, evaporation, blowing ($K = 1$). No convective mass transfer from the surface ($K = 0$). Prandlt number is 0.72 and the Schmidt number is 2.0 for air. (Modified from Mickley, H.S., Ross, R.C., Squyers, A.L., and Stewart, W.E., Heat, mass and momentum transfer over a flat plate with blowing or sucking, NASA-TN 3208, National Aeronautics and Space Administration, Washington, D.C., 1954.)

Just as important as the concentration, temperature, and velocity profiles are the fluxes of mass, energy, and momentum at the surface. The expressions for these are given by Bird et al. (1960) along with tabulated values for various values of the dimensionless groups. These results lead to correlations for heat and mass transfer for flow in the boundary layer. The reader is referred to this reference for this information or to the original work, NASA-TN 3208 (Mickley et al., 1954) for the details of both the experimental and theoretical work.

The solution obtained for this example was formidable, yet the restrictions assumed to solve the problem severe. This is precisely the reason that computational methods are needed to solve engineering transport phenomena problems.

2.9 SUMMARY

The equations of change have been described for variable properties, reacting flow that are used in computational transport phenomena. A derivation of these equations was given using a cubic control volume fixed in space to ensure an understanding of the application of the conservation of mass, momentum, and energy applied to an open system. The apparent limitation to a rectangular coordinate system was removed by deriving the conservation equations with a arbitrary control volume moving with the local mass average velocity of the fluid using a general property balance. The general property balance showed the conservation equations were of the same mathematical form with different dependent variables for the property, flux of the property, and generation of the property. The integral formulation can be used in the numerical solution of the equations or the conservation equations in rectangular Cartesian coordinates can be formally transformed into other more appropriate coordinate systems.

An overview of analytical and approximate solutions was described to show the limitations of these solutions and when it would be necessary to move to a numerical solution of the conservation equations. The analytical and approximate solutions served to demonstrate the use of an order analysis to identify important terms and those that could be neglected. They provided the basis for wall functions used in the place of fine grids near the wall in numerical solutions. They provided examples of frozen, equilibrium, and nonequilibrium chemistry used in computational transport phenomena and the concepts associated with simultaneous momentum, heat, and mass transfer in the boundary layer. The primary limitations of the solutions discussed were that they were only appropriate for laminar flow and simple geometries.

2.10 NOMENCLATURE

2.10.1 ENGLISH SYMBOLS

a, b, c, d	parameters in rate of deformation expression for non-Newtonian fluids
A, a	Helmholtz free energy; specific Helmholtz free energy
A_i	symbol for chemical compound in Equation 2.33
A, B, C, D	constants in various property correlations
B	empirical constant in Equation 2.166; bulk viscosity in Equation 2.82 and 2.117

c	total molar concentration
c_A	molar concentration of species A
c_n	coefficients in Blasius boundary layer solution Equation 2.209
C_p	constant pressure heat capacity
C_v	constant volume heat capacity
d_{ij}	rate of strain (deformation) tensor component for non-Newtonian fluids
D	inside pipe diameter
D	empirical constant in wall functions
D_{AB}	binary diffusion coefficient
D_{ij}	diffusion coefficient of binary pair
D_{ij}^*	diffusion coefficient of binary pair in multicomponent mixture
e_{ij}	rate of strain (deformation) tensor component for Newtonian fluids
E	empirical constant in wall functions
f	friction factor
$f(\eta)$	nondimensional stream function given by Equation 2.206
F	force
g_{ix}	external force acting on species i in the x-direction
g	gravitational acceleration constant
G, g	Gibbs free energy; specific Gibbs free energy
H, h	enthalpy; specific enthalpy
H_{xi}	partial molar enthalpy
I, I_i	identity tensor; tensor invariant
j_i	mass diffusion flux of species i
J_i	molar diffusion flux of species i
k	reaction rate in Equation 2.32; bulk modulus for elastic solid
k_{fi}	forward reaction rate constant in Equation 2.34
k_{ri}	reverse reaction rate constant in Equation 2.34
K	equilibrium constant; blowing parameter in Equation 2.219
K_f, K_b	forward and backward specific rate constants
l	arc length
L	width of porous slab
m	parameter in the power law model
$\overline{m}_A$	dimensionless mass fraction given by Equation 2.212

m_i, x_i	mass fraction of species i
M_w	molecular weight
M	momentum
n	number of species
n	parameter in the power law model
n_i	mass flux of species i
N_i	molar flux of species i
$N_{Re} = Dv_{avg}\rho/\mu,$	Reynolds number for tube flow
$N = WC_p \varepsilon\, L/\kappa_e$	dimensionless parameter in Equation 2.187
p_{ji}	stoichiometric coefficients for products in Equation 2.194
P	pressure
Pr	Prandtl number
q	heat flux by conduction and diffusion, Equation 2.111
q_w	wall heat flux
$\vec{q}_r$	radiation heat flux vector
Q	volumetric flow rate
r	radial coordinate in cylindrical coordinates
r_i	chemical reaction rate of species i in mass per unit volume (time)
r_{ji}	stoichiometric coefficients for reactants in Equation 2.43
R_A	chemical reaction rate of species A in mole per unit volume (time)
R	tube radius
R	gas constant
s	$= R-r$ coordinate normal to the wall
s^+	$= v^* s/(\mu/\rho)$ inner law coordinate
S	surface of an arbitrarily shaped control volume
S_{ii}	normal strain
Sc	Schmidt number
T	temperature
T^*	$= q_w/\rho C_p v^*$ characteristic temperature
$\overline{T}$	dimensionless temperature given by Equation 2.227
T^+	$= \rho C_p v^*(\overline{T}-T_R)/q_R$
t	time
U	internal energy; specific internal energy
υ	local fluid velocity
υ_{avg}	averaged velocity for flow in a tube
υ_0	superficial velocity in a porous media

υ_{y0}	velocity at surface of a flat plate
$\upsilon^2/2$	kinetic energy per unit mass
υ_{ds}	diffusion velocity
υ^+	$= v/v^+$ inner law velocity
υ^*	$= \sqrt{\tau_w/\rho_w}$ friction velocity
V	control volume
$\hat{V}$	specific volume
W	mass flow rate or flux
x, y, z	rectangular coordinates
x, θ, z	cylindrical coordinates
y	coordinate normal to surface
y_i	mole fraction of i

2.10.2 GREEK SYMBOLS

α	$= \kappa/\rho C_p$, thermal diffusivity
α	permeability in Darcy's law
α	constant in Blasius boundary layer solution
δ	boundary layer thickness
δ_1^+	wall function variable at edge of laminar sublayer
Δ	difference operator
ε	eddy kinematic viscosity; porosity in Section 2.8.3
ϕ	shearing strain; distorsion angle
κ	second coefficient of viscosity
κ	thermal conductivity in Equation 2.93 and after this equation
κ_e	$= \varepsilon\kappa_g + (1 - \varepsilon)\kappa_s$ effective thermal conductivity for a porous media
κ_v	von Karman's constant in Prandtl's turbulent model, Equation 2.160
σ_{xx}	$= -P + \tau_{xx}$ normal stress given by Equation 2.37a
η	nondimensional stream function given by Equation 2.208
μ	molecular viscosity
ν, υ	kinematic viscosity
ρ	density
ρ_i	concentration mass of species i per unit volume
ρ_A^+	$= v^*(\bar{\rho}_A - \rho_{AR})/n_A$
$\overset{\Rightarrow}{\tau}$	stress tensor
$\overset{\Rightarrow}{\tau}{}_{rz}^l$	time-averaged, laminar shear stress

$\vec{\vec{\tau}}\,^{t}_{rz}$	time-averaged, turbulent shear stress
τ_{xy}	shear stress component
τ_{R}	shear stress at tube radius
Φ_{v}	viscous dissipation function given by Equation 2.114
ψ	property
ψ	stream function; $(\nabla \cdot \vec{v})^{2}$ in Equation 2.115
Ψ	flux of a property
Ψ_{g}	generation of a property
ω	vorticity
Ω	rotation of fluid element

2.10.3 MATHEMATICAL SYMBOLS

∇	Del operator
ΔH_{v}	heat of vaporization
$\vec{\vec{\tau}}$	second-order tensor (τ is any tensor)
$\vec{V}$	vector (V is any vector)
$\bar{m}$	mean value
$\hat{M}$	matrix (M is any matrix)
$\tilde{C}_{v}$	column vector (C_{V} is any column vector))

2.10.4 SUBSCRIPTS

a	apparent value
A	species A in a binary mixture
B	species B in a binary mixture
b	bulk value
c	critical value
g	gas phase property
i	specific property of species i
ℓ	liquid phase
o	centerline or edge condition
o	boundary layer surface value
o	apparent viscosity at zero shear rate
p	constant pressure quantity
r	reduced quantity
s	solid phase property
t	turbulent quantity
R	tube wall property
v	momentum

w	wall property
x, y, z	component of vector or tensor
x, θ, z	component of vector or tensor
∞	boundary layer free stream value; apparent viscosity at infinite shear rate

2.10.5 SUPERSCRIPTS

$*$	dimensionless variable
$*$	friction velocity
$+$	wall function variable
g	external forces
j	concentration
p	pressure
T	temperature, transpose of a matrix

2.10.6 OVERSTRIKES

$\bar{\upsilon}$	turbulent time averaged
$\vec{\upsilon}$	a vector
$\overset{\rightrightarrows}{\tau}$	a tensor

REFERENCES

April, G. C. 1969. Energy transfer in the char zone of a charring ablator. PhD dissertation. Louisiana State University, Baton Rouge, LA.

Bird, R. B., R. C. Armstrong, and O. Hassager. 1987. *Dynamics of Polymer Liquids, Volume I: Fluid Mechanics*, 2nd ed. New York: John Wiley & Sons.

Bird, R. B., W. E. Stewart, and E. N. Lightfoot. 1960. *Transport Phenomena*. New York: John Wiley & Sons.

Bird, R. B., W. E. Stewart, and E. N. Lightfoot. 2002. *Transport Phenomena*, 2nd ed. New York: John Wiley & Sons.

Bennett, C. O. and J. E. Myers. 1962. *Momentum, Heat and Mass Transfer*. New York: McGraw-Hill.

Blasius, H. 1950. *The Boundary Layer in Fluids with Little Friction*. NACA TM 1256, National Advisory Committee for Aeronautics, Washington, D.C. (From Brodkey, R. S. 1967. *The Phenomena of Fluid Motion*. Reading, MA: Addison Wesley).

Brodkey, R. S. 1967. *The Phenomena of Fluid Motion*. Reading, MA: Addison Wesley.

Carslaw, H. S. and J. C. Jaeger. 1959. *Heat Conduction in Solid*. London: Oxford University Press.

Chapman, S. and T. G. Cowling. 1970. *The Mathematical Theory of Non-Uniform Gases; An Account of the Kinetic Theory of Viscosity, Thermal Conduction and Diffusion in Gases*. Cambridge: Cambridge University Press.

Chapman, T. et al. 1969. Ionic Transport and Electrochemical Systems, Chapter 3 in Lectures in Transport Phenomena, AIChE Continuing Education Series 4, American Institute of Chemical Engineers, New York.

Daily, J. W. and D. R. F. Harleman. 1966. *Fluid Mechanics*. Reading, MA: Addison-Wesley.

Del Valle, E. G. 1975. *In Depth Response of Ablative Composites*. PhD dissertation, Louisiana State University, Baton Rouge, LA.

DeSouza-Santos, M. L. 2008. *Analytical and Approximate Methods in Transport Phenomena*. Boca Raton, FL: CRC Press.

Hirschfelder, J. O., C. F. Curtiss, and R. B. Bird. 1954. *Molecular Theory of Gases*. New York: John Wiley & Sons.

Jacob, M. 1949. *Heat Transfer*, Vol. 1. New York: John Wiley & Sons.

Kaufmann, W. 1963. *Fluid Mechanics*. New York: McGraw-Hill.

Knudsen, J. G. and R. L. Katz. 1958. *Fluid Dynamics and Heat Transfer*. New York: McGraw-Hill.

Kuo, K. K. 1986. *Principles of Combustion*, 2nd ed. New York: John Wiley & Sons.

Lamb, H. 1945. *Hydrodyamics*, 6th ed. New York: Dover.

Landau, L. D. and E. M. Liftshitz. 1959. *Fluid Mechanics*. London: Pergamon Press.

Longwell, P. A. 1966. *Mechanics of Fluid Flow*. New York: McGraw-Hill.

Mickley, H. S., R. C. Ross, A. L. Squyers, and W. E. Stewart. 1954. Heat, Mass and Momentum Transfer over a Flat Plate with Blowing or Sucking, NASA-TN 3208, National Aeronautics and Space Administration, Washington, D.C.

Milne-Thompson, L. M. 1968. *Theoretical Hydrodynamics*, 5th ed. New York: MacMillan.

Rouse, H. 1959. *Advanced Mechanics of Fluids*. New York: John Wiley & Sons.

Schlichting, H. 1955. *Boundary Layer Theory*. New York: McGraw-Hill.

Shapiro, A. M. 1953. *The Dynamics and Thermodynamics of Compressible Fluid Flow*, Vol. I. New York: Ronald Press.

Sokolnikoff, I. S. 1964. *Tensor Analysis*, 2nd ed. New York: John Wiley & Sons.

White, F. M. 2006. *Viscous Fluid Flow*, 3rd ed. New York: McGraw-Hill.

Wilkes, J. O. 2006. *Fluid Mechanics for Chemical Engineers*. Upper Saddle River, NJ: Prentice-Hall.

Yuan, S. W. 1967. *Foundations of Fluid Mechanics*. Englewood Cliffs, NJ: Prentice-Hall.

CHAPTER 3

Physical Properties

3.1 OVERVIEW

Much of the literature on transport phenomena describes fluids as being of constant density or as ideal gases, in which case the transport coefficients are the major physical properties of interest. For multicomponent, multiphase flows this is not the case. Rigorous thermodynamic and reaction kinetics properties are of critical importance. Yet the full scope of such information is too vast and complex to be conveniently analyzed. Rather, practical engineering models of these properties must be selected and quantitatively described. The limitations imposed by using such models must be identified and accepted to make use of what computational analysis currently offers for transport analysis. This methodology is both important and nontrivial. Again, this implies an engineering approach and utilization of a production quality computational transport phenomena (CTP) code. The importance of the myriad of journal papers and ongoing research studies are not to be minimized, but somewhere one needs to draw the line and determine the distinction between practical analyses and possible future technological improvements.

For modest temperature and pressure levels, single species fluids may be reasonably analyzed as variable density ideal gases or as constant density liquids. For wide temperature and pressure ranges and for liquid/vapor coexistent conditions, single species fluids must be analyzed with real-fluid thermodynamic properties. Multicomponent fluids may be analyzed the same way, if they do not react. If conditions are such that the multicomponent fluids react very fast, a condition of local chemical equilibrium may exist in the flowfield. If the chemical reactions occur at a rate comparable to the local residence time in the flow, the reactions are controlled by finite-rate chemical reactions. For the purposes of the analyses described herein, multicomponent fluids are always assumed to be intimately mixed at a given location. This means that at a given point, the convection, diffusion, and reactions of the various species are to be predicted, whereas local unmixedness on a molecular scale is not considered. At this time unmixedness can only be described with experimentally measured or arbitrarily postulated probability distribution functions, not with the continuum conservation equations.

Radiative transfer requires integro-partial differential equations for a rigorous analysis. This topic is beyond the current capabilities of the CTP code. However, an introduction to radiation properties is given so that the need for, and a sketch of, such a more comprehensive computational model can be appreciated. Also, radiation analysis is an immensely important experimental tool for investigating transport processes. Thus the radiation property discussion serves to better connect the computational and experimental methodologies.

3.2 REAL-FLUID THERMODYNAMICS

Thermal and caloric equations of state are needed to solve the energy equation. The Gibbs' free energy is needed to describe flows that are in chemical equilibrium and to relate forward and backward reaction rates in reversible reactions. The speed of sound is required to select the algorithms in the difference form of the conservation equations. All of these properties are to describe multicomponent mixtures of real fluids. The properties must cover a wide range of temperatures and pressures. Empirical correlation equations are best suited for this purpose. Data and correlations from the work of Reid et al. (1987) and later editions from Yaws (1999) are best suited for this purpose. NIST data are also acceptable if they are available in other than tabular formats. Handbook data tend to be too sparse with respect to the range of conditions for which they are valid and lack adequate correlation equations. Lastly, the property data should be in a consistent form to expedite numerical computations.

3.2.1 THERMAL EQUATION OF STATE

The thermal equation of state (TEOS) defines the relationship between the pressure (P), molar density (n) or mass density (ρ) specific volume, and absolute temperature (T) for a given single component fluid. For the range of conditions of interest, four regions on a plot of pressure vs. specific volume with temperature as a parameter may be identified. Namely,

Region I—gases at all temperatures and density less than the critical or saturation density

Region II—high density gases at temperatures and densities above their critical values

Region III—liquids at temperatures below critical and pressures above saturation

Region IV—two-phase region where vapors and liquids coexist

Solids are not considered in these categories; they will be described separately as discrete phases.

Most of Region I fluids may adequately be described as ideal gases since the density will be sufficiently low to eliminate molecular interactions. The ideal gas law is appropriate:

$$P = \frac{NRT}{\forall} = nRT = \frac{mRT}{M_w \forall} = \frac{\rho RT}{M_w} \tag{3.1}$$

where

M_w is the molecular weight (i.e., mass)

$\forall$ is volume

m is mass

R is the gas constant

Some authors make a distinction between ideal and perfect gases by defining one to also include an assumption that the constant pressure heat capacity property is also constant. Since there is no consistency between which term is applied to which case, we refer to the aforementioned equation as either the ideal or perfect gas law. If the heat capacity is assumed constant, it will be so stated.

To extend the range of validity of the ideal gas law, a major effort to collect and correlate experimental data was undertaken by Lydersen et al. (1955). Remarkably, the data were found to simply correlate with temperature and pressure expressed as reduced values and with a very weak dependence on a compressibility correction factor (Z). Even this weak dependence could be described with a third parameter, which was Z evaluated at the critical point (Z_c). The result is called the law of corresponding states. It is stated as

$$P = \frac{ZNRT}{\forall} \quad \text{where} \quad Z = Z\left\{P_r, T_r, Z_c\right\} \tag{3.2}$$

Braces are used herein to indicate functionality, not multiplication. Other third parameters have also been suggested, such as Riedel's parameter (α) and Pitzer's acentric factor (ω). These produce similar results. But since they have been modified by various investigators, they will not be defined until they are associated with a specific TEOS. The law of corresponding states is very useful, but it has a major drawback for computational use. The values of the compressibility correction factors are only available as tables.

More general analytical TEOS's have been suggested. Many of them have been based on the van der Waals' equation as

$$P = \frac{RT}{(\forall - b)} - \frac{a}{\forall^2} \tag{3.3}$$

Notice that the equation is written for one mole of a specified fluid. This is a common practice in thermodynamics literature, but care must be taken in applying such equations. The logic for this formulation is that the parameter a is a correction for the attraction between molecules and the parameter b accounts for the volume occupied by the molecules in the gas. When this equation is used to calculate Z, the resulting equation becomes a cubic in Z. The cubic puts a curve in the coexistence region. The curve is a quirk of the equation used that has no physical meaning. Many thermodynamicists like to think otherwise. They have assumed some significance to describe the coexisting phases. They have also modified this curve to fit experimental data to represent multiple phases and species; see Chapter 8 of Sandler (1999). At this point expect no such interpretation; we will simply consider the cubic equation as an empirical fit. The basic cubic equation does not fit pressure–volume–temperature data well, so it has been modified by letting

$$a = a\{T\} \quad \text{and} \quad b = b\{T, \forall\} \tag{3.4}$$

A popular example of the cubic TEOS is the Peng–Robinson (PR) EOS (Sandler, 1999), which is given as:

$$P = \frac{RT}{\forall - b} - \frac{a\{T\}}{\forall(\forall + b) + b(\forall - b)}$$

where

$$a\{T\} = 0.45724 \frac{R^2 T_c^2}{P_c} \alpha\{T\}$$

$$b = 0.07780 \frac{RT}{P_c}$$

$$\sqrt{\alpha} = 1 + \kappa\left(1 - \frac{T}{T_c}\right)$$

$$\kappa = 0.37464 + 1.54226\omega - 0.26992\omega^2$$

(3.5)

Critical properties and the acentric factor are needed to evaluate the parameters in the PR-TEOS.

The acentric factor may be estimated by

$$\omega = -1.0 - \log\left[\frac{P^{vap}}{P_c}\right]$$

(3.6)

where P^{vap} is the vapor pressure at $T_r = 0.7$. Vapor pressure may be estimated by the Lee–Kesler correlation (Reid et al., 1977):

$$\ln P_r^{vap} = f^{(0)}\{T_r\} + \omega f^{(1)}\{T_r\}$$

(3.7)

where

$$f^{(0)} = 5.92714 - \frac{6.09648}{T_r} - 1.28862 \ln T_r + 0.169347 T_r^6$$

$$f^{(1)} = 15.2518 - \frac{15.6875}{T_r} - 13.4721 \ln T_r + 0.43577 T_r^6$$

$$\omega^{app} = \frac{\alpha}{\beta}$$

$$\alpha = -\ln P_c - 5.92714 + \frac{6.09648}{\Theta} + 1.28862 \ln\Theta - 0.169347\Theta^6$$

$$\beta = 15.2518 - \frac{15.6875}{\Theta} - 13.4721 \ln\Theta + 0.43577\Theta^6$$

where $\Theta = T^{nbp}/T_c$ and P_c is in atmosphere

If the reduced normal boiling point temperature (Θ) is equal to 0.7, no iteration is needed to evaluate ω. The superscript "app" indicates that the ω is approximate. The subscripts "r" and "c" represent a reduced value and the critical value, respectively.

Z is computed from the cubic equation:

$$Z^3 - (1-B)Z^2 + (A - 3B^2 - 2B)Z - (AB - B^2 - B^3) = 0 \qquad (3.8)$$

where

$B = pb/RT$
$A = ap/R^2T^2$
$\forall = Z\,RT/P$

The PR-TEOS is very useful, but it has several computational drawbacks. First, if one wished to improve the simulation in any one of the regions, the entire EOS would have to be rewritten and revalidated. Second, the most accurate thermodynamic property data is vapor pressure. This information is not directly utilized in the PR-EOS. Lastly, the boundaries of the coexistence region are determined by calculating the Gibbs' free energy along an isotherm and locating these boundaries by observing when the Gibbs' free energy is equal for the vapor and liquid. A direct determination with a vapor pressure correlation is much more accurate and efficient to evaluate.

Another cubic TEOS was developed by Hirschfelder et al. (1958a,b). In order to analyze the effects of real-fluid properties including phase changes in spray combustion, this, the HBMS TEOS, was selected for use in the CTP code because it treats the gas, dense gas, and liquid regions separately and produces reasonable accuracy over a wide range of pressures and temperatures. The HBMS TEOS is

$$\frac{P}{P_c} = \sum_{j=1}^{4} T_r^{j-2} \sum_{i=1}^{6} B_{ij}\, \rho^{i-2}; \quad T_r = \frac{T}{T_c}; \quad \rho_r = \frac{\rho}{\rho_c} \qquad (3.9)$$

B_{ij} are the coefficients of the thermal property polynomial for a given species for each of three single-phase regions. The HBMS TEOS is of acceptable accuracy for a wide range of conditions, and its component submodels can be easily modified. The vapor pressure curve and the liquid phase density correlations have been improved over the original HBMS formulation.

The original HBMS paper derives the B_{ij}'s as functions of the critical values of pressure, temperature, and specific volume (or Z_c)

for the three single-phase regions. The HBMS TEOS was developed as a generalization of the van der Waals' equation by the following modifications:

$$P = \frac{RT}{\left[\forall - b + b'/\forall\right]} - a\{T\}/\forall^2 - a'\{T\}/\forall^3 \tag{3.10}$$

Note the similarity with the PR-TEOS. The functions a and a' and the constants b and b' were chosen to be consistent with the virial form of the van der Waals' equation, the law of corresponding states, and the requirement of a smooth meeting of the saturated liquid and vapor lines. Riedel's vapor pressure (P^{vap}) equation is used to obtain continuity of the saturated liquid and vapor lines at the critical point:

$$\ln P^{vap} = A + (B/T) + C \ln T + D T^6 \tag{3.11}$$

Two points on the saturated liquid line and the normal boiling point were also required to evaluate the B_{ij}'s. These considerations result in the values of B_{ij} in terms of the critical conditions, which are contained in the CTP code.

To obtain the vapor pressure, the previously mentioned Lee–Kesler correlation equation was used (this equation requires the specification of the normal boiling point). Using the correlations reported in Reid et al. (1987), saturated liquid volumes are calculated with Spencer and Danner's modification of the Racket method:

$$\forall_c = \frac{RT_c}{P_c} Z_{RA}^{\left[1 + (1 - T_r)^{2/7}\right]} \quad \text{where} \quad Z_{RA} = 0.29056 - 0.08775\omega \tag{3.12}$$

Z_{RA} may also be obtained from experiments or from the reduced temperature. Thompson et al.'s model for compressed liquid volumes may be used for Region III:

$$\forall = \forall_{sl} f\{P_c, T_r, P^{vap}, \omega, \text{ and eight constants}\} \tag{3.13}$$

However, only limited validation data are available. It is interesting to note that although the liquid density changes by the least amount (percentage-wise), more empirical constants are used in its estimation than any other thermodynamic property. Carruth and Kobayashi's extension of Pitzer's method is used to calculate the heat of vaporization (ΔH_v) above a reduced temperature of 0.6:

$$\frac{\Delta H_v}{RT_c} = 7.08\left(1 - T_r\right)^{0.354} + 10.95\omega\left(1 - T_r\right)^{0.456} \tag{3.14}$$

Below this limit, Watson's method is used:

$$\Delta H_{v2} = \Delta H_{v1}\left[\frac{1 - T_{r2}}{1 - T_{r1}}\right]^{0.38} \tag{3.15}$$

The saturated vapor density is obtained from the Clapeyron equation:

$$\left(\frac{\partial P^{\text{vap}}}{\partial T}\right)_{s\ell} = \frac{\Delta H_v}{T\left(\forall_{sv} - \forall_{s\ell}\right)} \tag{3.16}$$

All of the models are linearly extrapolated below a reduced temperature of 0.3.

These models are not necessarily the best or most general, but choices are necessary to provide the data needed to estimate sufficient property information to solve the conservation equations. Correlation equations of theoretical or empirical data are preferred over tabular data to expedite calculations and modifications to reflect new information.

3.2.2 CALORIC EQUATION OF STATE

The caloric equation of state (CEOS) defines the relationship between enthalpy (H) (or internal energy U), temperature, molar (or mass) specific volume, and heat capacity (C_p). For a real, single-component fluid:

$$H = H_{\text{REF}} + \int_{T_1,P_1}^{T_2,P_2} C_p \, dT + \int_{T_1,P_1}^{T_2,P_2}\left[\forall - T\left(\frac{\partial \forall}{\partial T}\right)_P\right]dP \tag{3.17}$$

H_{REF} must be chosen. For a single-component fluid, any value is sufficient, even zero. For multicomponent, reacting fluids, the standard heat of formation is used as the reference value. If the substance changes phase, the enthalpy is a weighed average of each phase. T_1 is a convenient low temperature. It is not chosen as zero because entropy is not defined when the temperature is zero degrees absolute. Equation 3.17 has meaning only when applied to an identified quantity of material. Unless stated differently, the quantity will be taken as one mole of pure species. As with all thermodynamic functions, the restrictions placed on a defining equation must be carefully stated. For an ideal gas, the pressure term in the enthalpy is zero. If H_{REF} is zero, the enthalpy is only a function of the heat capacity and temperature. Wilhoit (1975) recognized that gaseous heat capacities are essentially constant ($C_p\{0\}$) at low temperatures (~10–200 K), where only translational and rotational motions are excited, and that at high temperatures the rigid-rotor-harmonic-oscillator model predicts the equipartition of all energy modes, hence the heat capacity becomes a constant, higher value ($C_p\{\infty\}$). For linear molecules, $C_p\{0\}$ is 3.5R for linear molecules and 4R for nonlinear molecules. $C_p\{\infty\}$ is $(3N - 1.5)R$ for linear molecules and $(3N - 2)R$ for nonlinear molecules. N is the number of atoms in the molecule. Wilhoit represented intermediate values of C_p with a polynomial function for 32 compounds. This model implies that the heat capacity is constant at higher temperatures, although for real gases it is known to increase slowly above the high temperature limit.

Over a number of years, NASA scientists at John H. Glenn (formerly Lewis) Research Center developed an ideal gas thermodynamics properties database, with limited liquid and solid species properties included, and computer code (Gordon and McBride, 1976, 1994). This work, now called the CEA code, also includes chemical equilibrium data, transport property data, and some additional thermodynamics data. The CEOS presents and utilizes data correlations as follows:

$$\frac{C_p^o}{R} = a_1 + a_2 T + a_3 T^2 + a_4 T^3 + a_5 T^4 \tag{3.18}$$

$$\frac{H_T^o}{RT} = a_1 + \frac{a_2}{2}T + \frac{a_3}{3}T^2 + \frac{a_4}{4}T^3 + \frac{a_5}{5}T^4 + \frac{a_6}{T} \tag{3.19}$$

$$\frac{S_T^o}{R} = a_1 \ln T + a_2 T + \frac{a_3}{2}T^2 + \frac{a_4}{3}T^3 + \frac{a_5}{4}T^4 + a_7 \tag{3.20}$$

$$\frac{G_T^o}{RT} = a_1 (1 - \ln T) - a_2 T - \frac{a_3}{6}T^2 - \frac{a_4}{12}T^3 - \frac{a_5}{20}T^4 + \frac{a_6}{T} - a_7 \tag{3.21}$$

Notice that C_p^o is fit with a fourth-order polynomial containing five constants. The enthalpy H_T^o fit contains a sixth constant and the entropy a seventh. The Gibbs' free energy G_T^o contains no additional constants as it is calculated from the enthalpy and entropy S_T^o. The superscript "o" indicates ideal gas and the subscript "T" emphasizes that the ideal gas properties are not a function of pressure. The newer versions of the CEA code contain two additional terms in each of these functions and extend the range of applicability to 20,000 K. The old version of the correlations is used in the CTP code. The functions are fit such that the enthalpy and Gibbs' free energy at 298.15 K are the heat and free energy of formation, respectively, of the species. There are other ideal gas databases, but they do not cover the wide range of temperatures and species, they are not as well documented, and the source code for their use is not as generally available. The CEA code has another very useful feature. If only the heat of formation and elemental composition of a reactant are specified and this reactant is not a final product, the equilibrium composition of the mixture can be predicted. For example, if a polymer or gasoline were oxidized (under other than extremely fuel-rich conditions), intermediate-reaction product thermodynamic properties would not need to be known.

Real-fluid thermodynamics are modeled by adding a correction term to the ideal gas property model. For example, if the PR-TEOS is used to calculate real-fluid thermodynamic properties by forming a PR-CEOS, the PR-TEOS is used to calculate "departure functions," which are pressure correction factors to convert the ideal gas enthalpy and entropy to real-fluid values. The ideal gas contribution, including the heat of formation, is obtained from the CEA code data. The enthalpy equation becomes the PR-CEOS:

$$(H - H^\circ)_{T,P} = \int_{T,P=0}^{T,P} \left[\forall - T \left(\frac{\partial \forall}{\partial T} \right)_P \right] dP \qquad (3.22)$$

$$H\{T,P\} - H^\circ\{T,P\} = RT(Z-1) + \int_{\forall=\infty}^{\forall\{T,P\}} \left[T\left(\frac{\partial P}{\partial T}\right)_\forall - P \right] d\forall \qquad (3.23)$$

The entropy form of the PR-CEOS is

$$(S - S^\circ)_{T,P} = - \int_{T,P=0}^{T,P} \left[\left(\frac{\partial \forall}{\partial T}\right)_P - \frac{R}{P} \right] dP \qquad (3.24)$$

$$S\{T,P\} - S^\circ\{T,P\} = R\ln Z + \int_{\forall=\infty}^{\forall\{T,P\}} \left[\left(\frac{\partial P}{\partial T}\right)_\forall - \frac{R}{\forall} \right] d\forall \qquad (3.25)$$

The HBMS TEOS is used in the same fashion to calculate real gas thermodynamic properties. Slightly different variables are used in the HBMS CEOS:

$$H - H^\circ = RT \left[Z_c \int_0^{\rho_r} \left(\frac{P}{T_r} - \left(\frac{\partial P}{\partial T_r} \right)_{\rho_r} \right) \rho_r^{-2} d\rho_r + Z_c \frac{P}{\rho_r T_r} - 1 \right] \quad (3.26)$$

The Gibbs' free energy is calculated from

$$(G - G^\circ) = RT \int_0^\rho (Z - 1) \frac{d\rho}{\rho} + RT \left[(Z - 1) - \ln Z \right] \quad (3.27)$$

where Z is obtained from either the PR-TEOS or HBMS TEOS.

   Note that this standard procedure for calculating departure functions uses no experimental or theoretical information on liquid phase heat capacities. In general, real-fluid EOS studies are weak in the Region III liquid-side analyses and validation. More detailed comparisons with the empirical correlations of liquid properties reported in Reid et al. (1987) would be valuable.

3.2.3 TEOS AND CEOS FOR MULTICOMPONENT FLUIDS

Chemical potentials are defined as

$$\mu_i \equiv \left(\frac{\partial G}{\partial n_i} \right)_{T,P,n_j} = \left(\frac{\partial U}{\partial n_i} \right)_{S,V,n_j} = \left(\frac{\partial H}{\partial n_i} \right)_{S,P,n_j} = \left(\frac{\partial A}{\partial n_i} \right)_{T,V,n_j} \quad (3.28)$$

where A is the Helmholtz free energy. Note that $\mu_i \equiv (\partial G/\partial n_i)_{T,P,n_j} \equiv G_i$, the partial molar Gibbs' free energy. The other chemical potential derivatives are not taken at constant T and P, therefore they are not partial molar quantities. The other partial molar functions are

$$\left(\frac{\partial U}{\partial n_i} \right)_{T,P,n_j} \equiv U_i \quad \left(\frac{\partial H}{\partial n_i} \right)_{T,P,n_j} \equiv H_i$$

$$\left(\frac{\partial A}{\partial n_i} \right)_{T,P,n_j} \equiv A_i \quad \left(\frac{\partial \forall}{\partial n_i} \right)_{T,P,n_j} \equiv \forall_i \quad (3.29)$$

These functions are those commonly used in thermodynamics, but these functions are redefined to put them on a mass basis for most computational analyses:

$$\sigma_i \equiv \left(\frac{\partial G}{\partial m_i} \right)_{T,P,m_j} = \left(\frac{\partial U}{\partial m_i} \right)_{S,V,m_j} = \left(\frac{\partial H}{\partial m_i} \right)_{S,P,m_j} = \left(\frac{\partial A}{\partial m_i} \right)_{T,V,m_j} \quad (3.30)$$

$$\left(\frac{\partial U}{\partial m_i} \right)_{T,P,m_j} = U_i \qquad \left(\frac{\partial H}{\partial m_i} \right)_{T,P,m_j} = H_i$$

$$\left(\frac{\partial A}{\partial m_i} \right)_{T,P,m_j} = A_i \qquad \left(\frac{\partial V}{\partial m_i} \right)_{T,P,m_j} = V_i \qquad (3.31)$$

Let u_i, h_i, g_i, a_i, and v_i be the corresponding thermodynamic functions per unit mass, i.e., the specific properties. These are the quantities that come directly from the TEOS and CEOS when they are divided by the appropriate species molecular weight. The "excess functions" are defined to be the difference between the partial mass (or molar) property and the corresponding pure component specific property. Thus the excess functions account for a property in a mixture being different from its value in the pure state. Several of these functions are

$$V_i^{EX} = \left(\frac{\partial V}{\partial m_i} \right)_{T,P,m_j} - v_i \quad \sum_i^n x_i V_i^{EX} = \Delta v_{MIX} \quad v = \sum_i^n x_i v_i + \Delta v_{MIX} \quad (3.32)$$

$$H_i^{EX} = \left(\frac{\partial H}{\partial m_i} \right)_{T,P,m_j} - h_i \quad \sum_i^n x_i H_i^{EX} = \Delta h_{MIX} \quad h = \sum_i^n x_i h_i + \Delta h_{MIX} \quad (3.33)$$

$$G_i^{EX} = \left(\frac{\partial G}{\partial m_i} \right)_{T,P,m_j} - g_i + RT \ln x_i$$

$$\sum_i^n x_i G_i^{EX} = g^{EX} = \Delta g_{MIX} - RT \ln x_i \quad (3.34)$$

$$g = \sum_i^n x_i g_i + RT x_i \ln x_i + g^{EX}$$

The mass fraction is denoted by x_i in these functions.

For ideal fluid mixtures, the excess functions are all zero. For real fluids, a useful approximation is to assume that the excess functions are zero. Such fluid mixtures are termed "ideal solutions." To investigate the validity of such an assumption, consider the example reported by Balzhiser et al. (1972). A real-fluid mixture of a 3:1 molar ratio of hydrogen to nitrogen is a feed stream to an ammonia synthesis plant. The stream is compressed to 400 atm and heated to 300 K. Measurements are reported in Perry et al. (1963) for these conditions. Several methods were used to estimate mixture compressibility:

1. Experimental values:

$$Z_{MIX} = 1.155 \quad \forall_{MIX} = 2.176 \, ft^3/lb \, mol$$

2. The mixture was assumed to consist of ideal gases:

$$Z_{MIX} = 1.0 \quad \forall_{MIX} = 1.884 \, ft^3/lb \, mol$$

3. Dalton's law of additive pressures. Component compressibilities are based on estimated partial pressures:

$$Z_{MIX} = 1.095 \quad \forall_{MIX} = 2.063 \, ft^3/lb \, mol$$

4. Amagat's law of additive volumes. Component compressibilities are based on system total pressure:

$$Z_{MIX} = 1.16 \quad \forall_{MIX} = 2.186 \, ft^3/lb \, mol$$

5. Pseudo-critical constants were used to compute compressibility:

$$Z_{MIX} = 1.21 \quad \forall_{MIX} = 2.280 \, ft^3/lb \, mol$$

The use of Amagat's law is the same as assuming the mixture is an ideal solution. This is obviously the best model for this example. It is also the simplest, since it just weighs the component volumes by the mass fraction, a dimensionally consistent calculation. The other two real-fluid estimation methods are pure empiricism, which in this case yield poor results.

The PR-EOS uses mixing rules to represent multicomponent systems. These rules use average values of the function a and the constant b:

$$a\{T\} = \sum_{i=1}^{n} \sum_{j=1}^{n} y_i y_j a_{ij}$$

$$b = \sum_{i=1}^{n} y_i b_i \qquad (3.35)$$

$$a_{ij} = \sqrt{a_{ii} a_{jj}} \, (1 - k_{ij}) = a_{ji}$$

where

y_i is mole fraction

n is the number of components

The k_{ij} is the "binary interaction parameter." The k_{ij}'s are determined by experiment. The value of the modeling methodology is very dependent on the chemical systems of interest. Application to mixtures of more than two components is limited. This methodology has several drawbacks when considered for application to CTP. There are only limited data for the binary interaction parameters, and these parameters may influence the property predictions by as much as 20% for gaseous mixtures. For anything other than a binary mixture, the method is very computationally intensive. The method has been applied primarily to mixtures in the coexistence region, where it has been used as a data fit by adjusting k_{ij}. Generally, no consideration has been given to the other three regions, as to how the TEOS accuracy is influenced by the selection of the binary interaction parameter. Using cubic equations of state to represent the coexistence region is generally not as effective as using activity coefficient methods. The use of regular solution models, the UNIFAC model, and the Wong–Sandler mixing rule (Sandler, 1999) are useful for describing mixtures in the coexistence region, but all are too elaborate and system dependent to be practical for consideration in a CTP code. At this time they also appear to be unnecessary, for ideal solution models can only be evaluated and validated for CTP purposes.

The HBMS TEOS and CEOS models along with ideal solution assumptions are currently included in the accompanying CTP code. This approach treats the flow of a multiphase mixture as the flow of a continuum, a multicomponent fluid wherein the local quality determines the relative amounts of liquids and gases. Thus, velocity and temperature equilibrium are assumed to exist between the phases. For many years, dense spray has proved to be too complex to simulate with more elaborate computational models and to measure experimentally. The development of more complete CTP models must be accompanied by an improved experimental methodology. Otherwise, the current modeling techniques are as good as can be expected. There are alternatives; a series of more simplified experiments and accompanying analyses could be conducted to improve CTP methodology. Such an approach would amount to basic research, which is far more difficult to obtain funding for than industrial and military projects. The near term prospect for developing more elaborate mixing models for CTP analyses is dim.

3.2.4 SOUND SPEED IN MULTICOMPONENT FLUIDS

Sound speed is defined as

$$a = \left(\frac{\partial P}{\partial \rho}\right)_S^{0.5} \tag{3.36}$$

but this function is difficult to estimate for multiphase, multicomponent fluids, even in a state of equilibrium. For nonequilibrium conditions, heat and mass exchange and surface tension must also be considered. This function is calculated in the HBMS TEOS by its thermodynamic equivalent as

$$a = \gamma \left(\frac{\partial P}{\partial \rho}\right)_T^{0.5} \quad \text{where} \quad \gamma \equiv \frac{C_p}{C_v} \tag{3.37}$$

The function was calculated for oxygen at 95.58 K and 1.701 atm, i.e., typical equilibrium conditions for two-phase oxygen. NBS data (McCarty and Weber, 1971) for density and sound speed at the saturated vapor and liquid points are $\rho_v = 7.2952\,\text{kg/m}^3$, $a_v = 172.26\,\text{m/s}$, $\rho_L = 1112.874\,\text{kg/m}^3$, and $a_L = 859.45\,\text{m/s}$. Predicted values of these points are 7.255, 181.31, 1119.8, and 896.3, respectively. This agreement is considered to be very acceptable. But how does the sound speed vary in the two-phase region between the saturation points of the vapor and liquid?

Reported analyses and experiments of sound speed in two-phase fluids indicate that the presence of the second phase causes a rapid decrease in sound speed for small additions of the second phase. The real-fluids model simulation for the sound speed of oxygen at the conditions mentioned in the two-phase region are shown in Figure 3.1. The speed of sound for a gaseous/liquid oxygen (GOX/LOX) mixture calculated with Equation 3.37 is shown as the HBMS curve in this figure. Assuming that the equilibrium thermodynamic processes are completely accounted for, the accepted sound speed for the multiphase mixture is given by Brennen (1995) as

$$a = \left\{ \left(y_v \rho_v + y_L \rho_L\right) \left[\frac{y_v}{\rho_v a_v^2} + \frac{y_L}{\rho_L a_L^2}\right] \right\}^{-1/2} \tag{3.38}$$

The sound speeds predicted with the real-fluid model agree quite well with Brennen's equation (Equation 3.38) and should be used for CTP analyses.

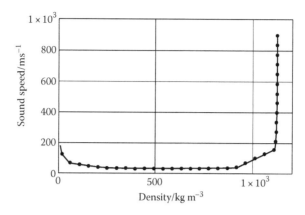

FIGURE 3.1   GOX/LOX sound speed (HBMS ▬▬ Brennen ••••).

3.3 CHEMICAL EQUILIBRIUM AND REACTION KINETICS

At near-ambient temperatures and pressures, most multicomponent flows may be treated as nonreacting. There are exceptions; hypergolic propellants react upon the mixing of their components. In all reacting flows, if the temperature and pressure are sufficiently high, the flow will probably be in local chemical equilibrium. Otherwise, the species continuity equations (SCEs) with appropriate reaction rates must be solved simultaneously with the other conservation equations. The finite-rate solution would require a system of reactions and rate expressions to complete the analysis. This chapter describes the methodology required to determine the local state of a reacting flow in either chemical equilibrium or as governed by finite-rate chemical reactions.

3.3.1 CHEMICAL EQUILIBRIUM

Chemical equilibrium is established when the Gibbs' free energy is minimized while satisfying the local element balances. The local element ratios are determined by the solution of the multicomponent conservation equations. The chemical equilibrium relationships are described as follows.

3.3.1.1 Minimization of Gibbs' Free Energy

Gibbs' free energy per unit volume for a multicomponent, multiphase mixture is

$$G = \sum_{i=1}^{NS} n_i \left[\frac{(h_i - Ts_i)}{RT} + \phi_i \left(\ln\left(P \frac{n_i}{n_g} \right) \right) \right] \equiv \sum_{i=1}^{NS} f_i \qquad (3.39)$$

where

n_g is the molar concentration
ϕ_i is the quality of the fluid
P is pressure in atm

Let

$$C_i = \frac{(h_i - Ts_i)}{RT} + \phi_i \ln P \qquad (3.40)$$

$$b_j = \sum_{i=1}^{NS} n_{oi} e_{ij} \quad \text{for} \quad j = 1, NE \qquad (3.41)$$

where

b_j is the total molar concentration of element j
e_{ij} is the number of atoms of element j in species i
n_{oi} is the initial value of n_i; the "o" denotes initial value
NE is the number of elements

The constraint on minimizing the Gibbs' free energy is that the b_j's are constant. This is accomplished by defining the augmented function:

$$\Phi = G + \sum_{j=1}^{NE} \lambda_j \left[b_j - \sum_{i=1}^{NS} n_i e_{ij} \right] \qquad (3.42)$$

where λ_j are Lagrangian multipliers. Expanding this function as a Taylor series:

$$\Phi = \Phi_o + \sum_{i=1}^{NS} \left[C_i + \phi_i \ln\left(\frac{n_{oi}}{n_{og}} \right) - \sum_{j=1}^{NE} \lambda_j e_{ij} \right] (n_i - n_{oi})$$

$$+ \frac{1}{2} \sum_{i=1}^{NS} (n_i - n_{oi}) \phi_i \sum_{k=1}^{NS} \left(\frac{\delta_{ik}}{n_{ok}} - \frac{\phi_k}{n_{og}} \right) (n_k - n_{ok}) \qquad (3.43)$$

where δ_{ik} is the Kronecker delta.

The augmented function is minimized by driving its derivatives to zero. That is,

$$\frac{\partial \Phi}{\partial n_i} = 0 = C_i + \phi_i \ln\left(\frac{n_{oi}}{n_{og}}\right) - \sum_{k=1}^{NE} \lambda_k e_{ik} + \phi_i\left(\frac{n_i}{n_{oi}} - \frac{n_g}{n_{og}}\right) \qquad (3.44)$$

for i = 1, NE. If the ith species is a gas,

$$n_i = n_{oi}\left[\frac{n_g}{n_{og}} + \sum_{k=1}^{NE} \lambda_k e_{ik}\right] - f_{oi} \qquad (3.45)$$

Summing on all gases,

$$\sum_{k=1}^{NE} \lambda_k \sum_{i=1}^{NS} \phi_i n_{oi} e_{ik} = \sum_{i=1}^{NS} \phi_i f_{oi} \qquad (3.46)$$

If species ℓ is condensed, then $\partial \Phi / \partial n_\ell = 0$ yields

$$\sum_k \lambda_k e_{\ell k} = C_\ell \quad \ell = 1, \; NC \quad \text{or}$$

$$\sum_k \lambda_k n_{o\ell} e_{ik} = f_{o\ell} \qquad (3.47)$$

NC is the number of condensed species.

For condensed species, the constraint equation is

$$\sum_{k=1}^{NE} \lambda_k \sum_{i=1}^{NS} n_{oi} e_{ij} e_{ik} + \frac{n_g}{n_{og}} \sum_{i=1}^{NS} \phi_i n_{oi} e_{ij} + \sum_{\ell=1}^{NC} n_\ell e_{\ell j} = b_j + \sum_{i=1}^{NS} e_{ij} f_{oi} \qquad (3.48)$$

There are NE of these equations. These equations plus Equations 3.46 and 3.47 form a set of $N = NE + 1 + NC$ equations with unknowns

$$\lambda = \left[\lambda_1, \ldots, \lambda_{NE}, \frac{n_g}{n_{og}}, n_\ell\right]^T \quad \text{for} \quad \ell = 1, \; NC \qquad (3.49)$$

This set of N equations is solved to obtain values for the unknowns λ, which are used to obtain new values for species concentrations. Under-relaxation must be used to ensure that species concentrations remain positive and that the Gibbs' free energy is decreasing. The iterative process is repeated until the free energy is minimized. One of the advantages of this method is that only an Nth-order matrix is involved instead of a matrix of the order of the number of species presents (NS), which is typically many.

The previously mentioned CEA code uses a minimization of the free energy approach to obtain equilibrium compositions. The major advantage of using the free energy minimization methods is that the species present at equilibrium do not have to be specified. Such species are a result of the calculation.

3.3.1.2 Equilibrium Constants

If only a small number of species and equilibrium reactions are involved, an alternate approach can also be used. Equilibrium constants can be evaluated from free energy values and a set of equations solved for equilibrium concentrations. For example, the wet-CO mechanism, consisting of eight species, may be modeled using three element balances and five reactions:

N_i is the molar density of species i; E_j is the molar density of element j:

$$0.5O_2 = O \Rightarrow K_O N_{O_2}^{1/2} = N_O \tag{3.50}$$

$$0.5H_2 = H \Rightarrow K_H N_{H_2}^{1/2} = N_H \tag{3.51}$$

$$0.5O_2 + 0.5H_2 = OH \Rightarrow K_{OH} N_{O_2}^{1/2} N_{H_2}^{1/2} = N_{OH} \tag{3.52}$$

$$0.5O_2 + H_2 = H_2O \Rightarrow K_{H_2O} N_{O_2}^{1/2} N_{H_2} = N_{H_2O} \tag{3.53}$$

$$CO + 0.5O_2 = CO_2 \Rightarrow K_{CO_2} N_{O_2}^{1/2} N_{CO} = N_{CO_2} \tag{3.54}$$

where

K_j is the equilibrium constant for reaction j

$N_i = \rho \alpha_i / M_{wi}$ is the molar density of species i, where α_i and M_{wi} are the mass fraction and molecular weight of species i, respectively

The reaction set is not unique, but all other sets can be derived from this set. The above equations are solved to obtain equilibrium species concentrations.

This is an interesting example because it represents the final state of combustion of an oxygen–hydrocarbon system, unless the combustion is so fuel rich that unburned hydrocarbons remain as equilibrium products.

Equilibrium constants for a single reaction (I) are obtained from the Gibbs' free energy. Consider the reaction

$$\nu_A A + \nu_B B = \nu_C C + \nu_D D \tag{3.55}$$

The ν_i's are stoichiometric coefficients and the capital letters represent chemical formulas. Assume that the mixture of species forms an ideal solution. Otherwise there are too many possibilities to make even a partially general analysis. Unfortunately, the Gibbs' free energy is represented by numerous aliases: partial molar Gibbs free energy, chemical potential, fugacity (f), fugacity coefficient, activity (a), and activity coefficient. Using the activity and fugacity, $a_i = y_i f_i/f_i^\circ$. f_i° is the fugacity of i at a standard-state condition. This is usually taken to be 1 atm. This means that it does not appear in any further equations and that all other fugacities and pressures must also be expressed in atmospheres. Otherwise, it must be retained in the following equations. The equilibrium coefficients are defined as

$$K_I = \frac{a_C^{\nu_C} a_D^{\nu_D}}{a_A^{\nu_A} a_B^{\nu_B}} = \frac{y_C^{\nu_C} y_D^{\nu_D}}{y_A^{\nu_A} y_B^{\nu_B}} P^{\nu_C + \nu_D - \nu_A - \nu_B} \left[\frac{(f/P)_C^{\nu_C} \ (f/P)_D^{\nu_D}}{(f/P)_A^{\nu_A} \ (f/P)_B^{\nu_B}} \right] \tag{3.56}$$

The y_i's are mole fractions. If the reaction does not change the number of moles present, the pressure term is eliminated. If the species involved are ideal gases, the fugacity term is eliminated. The equilibrium constant is calculated from the Gibbs' free energy by

$$K_I \equiv \exp \left\{ -\frac{\Delta G_I^\circ}{RT} \right\} \tag{3.57}$$

where the standard-state change in G for this reaction is

$$\Delta G_I^\circ = \nu_C g_C^\circ + \nu_D g_D^\circ - \nu_A g_A^\circ - \nu_B g_B^\circ \tag{3.58}$$

The g's are standard-state Gibbs' free energies for the indicated species per mole:

$$E_H = 2N_{H_2O} + 2N_{H_2} + N_H + N_{OH} \tag{3.59}$$

$$E_O = N_{H_2O} + 2N_{O_2} + N_O + N_{CO} + 2N_{CO_2} + N_{OH} \tag{3.60}$$

$$E_C = N_{CO} + N_{CO_2} \tag{3.61}$$

Given two state variables (P and either T or H), the five equilibrium constants needed in Equations 3.50 through 3.54 can be calculated from the standard-state Gibbs' free energy for a given set of element values. Equations 3.50 through 3.54 and 3.59 through 3.61 can then be solved for the eight species concentrations. This appears to be a simple algebra problem, but for the range of conditions of interest, the solution is sometimes difficult to obtain. The free-energy minimization method, even for a small chemical system, is an attractive backup solution method.

3.3.2 FINITE-RATE CHEMICAL REACTIONS

Multicomponent flows in general involve species that may react with one another at a finite rate. Treating the flow as being in local chemical equilibrium or as being "frozen," i.e., nonreacting, is an approximation that may be entirely appropriate for a given situation. Reactions occur when energetic molecules collide; hence at high temperatures, pressures, and densities, the reactions are fast. Otherwise they are slow. In run-away and combustion situations the reactions are very fast. Simulations of such phenomena require special care. However, these special analyses also describe less severe conditions, so that separate computational tools are not needed.

Specific reaction rates and a reaction mechanism, as well as a description of the convection and diffusion that mix the reactants are needed to simulate, i.e., predict, the local species distributions in a flowfield. This fundamental fact is not appreciated by many who should be in the know. In Fogler's popular chemical engineering textbook on reaction engineering, a satisfactory distinction between mixing, diffusion, and mass transfer is never made, but chemical reactions are discussed for over 950 pages (Fogler, 1999). Chemical reaction rates and mechanisms cannot be accurately predicted; they must be measured. Mechanism, herein, means a set of reactions that start with a set of reactants, and for a given set of conditions yield a realistic set of products. Arguments as to which bonds are made and broken, which type of intermediate unstable complexes may have existed, and which levels of energetic radicals may have taken part are not germane here. Chemists can (occasionally) sort these factors out. This state of affairs means that the reactions of interest for any specific chemical system must be determined by experiment.

The difficulty with determining reaction rates is that transport phenomena get in the way. Because all the important parameters in a laboratory reactor cannot be simultaneously measured, investigators resort to "simplifying" the study to the point where these effects are not important. The result is that a few experiments have emerged which attempt to eliminate convection and diffusion to the point that a mass of reactants at a given temperature and pressure react unaffected by fluid motion until a mass of products at a final time, temperature, and pressure result. Well-stirred reactors, shock tube experiments with gases, and flat flame experiments come close to yielding ideal reactors. Some investigators use a "turbulent" flow reactor to "instantaneously" mix reactants, implying that because the flow is turbulent the mixing is instantaneous. That data obtained thusly are meaningful is wishful thinking. Another example of flawed data comes from flat-flame combustion experiments, which utilize ingenious honeycombed channels to produce a flat flame and make detailed optical species and temperature measurements of the flame. These devices use cooling of the honeycomb to stabilize and position the flame, and often the amount of this cooling is not reported, or even measured, so much for boundary conditions. Nevertheless, if care is exercised in its collection and evaluation, sufficient data exist to warrant analyzing many transport problems.

Much of the worthwhile reaction rate data are in the field of combustion. Because of the importance of this process and the relative ease of making gas phase measurements, this was the first field to truly treat reacting flows as a transport process. The voluminous data from the Combustion Institute Proceedings and such surveys as Gardiner (1984) can be carefully picked over and evaluated to provide useful engineering analyses. Trying to construct kinetics mechanisms for fuels such as kerosene, gasoline, rocket, and jet fuels, which involve 800 reactions, and considering only hydrocarbons up to C8s when the average chain length is C12 is self-defeating. Even industry specialists can only speciate kerosene to the point of identifying 25% of its components. When attempting to construct large complex reaction mechanisms, the data processing techniques described by Frenklach (Gardiner, 1984) should be conducted rather than simply amassing a huge database. Modern computers can integrate large chemical systems for lumped reaction systems, but real geometry and other transport processes cannot be simultaneously accounted for. Or, consider diffusion flames: for 25 years studies have been performed and conferences have been held to measure and develop distribution functions to represent low Reynolds number flames. Professor. R. W. Bilger has measured and statistically modeled such flames for three decades (e.g., see Kent and Bilger, 1972; Bilger et al., 2005). The logic behind studying these "lazy" flames is that a continuum of species is not maintained, meaning that at a given small volume surrounding a point, the molecules are not uniformly mixed, such that a local reaction inefficiency in addition to the kinetic rate of reaction exists. Probability distribution functions derived from experiments, not engineering solutions to the conservation equations, are needed to represent this unmixedness. Where did the engineering models come from that were used to design scramjets, which the country has developed as a scramjet propulsion system for the X-43-A (Wilson, 2007)? Blowout has been measured and the data correlated with one combustion reaction (Longwell and Weiss, 1955). Most important industrial chemical processes can be described with a few reaction expressions. The virtue of doing so is clearly illustrated by Edelman and Fortune (1969) with their development of quasi-global chemical kinetics models to describe mixing and combustion of hydrocarbon fuels. Applying design quality combustion models to the flame data from typical experiments of Bilger's group were studied by Edelman and Harsha (1978). Such comparisons are good approximations, but the models could be improved. Somewhere in between 1 and 800 reactions, there is surely a middle ground in which reacting flows can be practically modeled and analyzed for design purposes. In addition to design analysis, and probably more importantly, when problems arise in complex systems, the comprehensive transport phenomena simulations become invaluable for rectifying the problems.

Even when a relatively simple reaction system is to be modeled, such as constant pressure or the adiabatic combustion of hydrogen/oxygen, serious computational problems arise. The problem is termed stiffness. When going from an unreacted to a reacted condition, thousands of computational steps are required. Moretti (1965) studied and identified the cause of stiffness. He considered the reaction system.

Reaction	Number	Species	Number
$H + O_2 = OH + O$	1	H	1
$O + H_2 = OH + H$	2	O	2
$H_2 + OH = H + H_2O$	3	H_2O	3
$2OH = O + H_2O$	4	OH	4
$H_2 + X = 2H + X$	5	O_2	5
$H_2O + X = OH + H + X$	6	H_2	6
$OH + X = O + H + X$	7		
$O_2 + X = 2O + X$	8		

Let y_i be mole fraction of species i and $\dot{y}_{ij}$ be the rate of change of species i by reaction j. X is a catalyst for third body reactions; here it is the sum of all six species. The change of species 2 by reaction 1 is given by $\dot{y}_{21} = f_1\, y_i\, y_5 + b_1 y_2 y_4$, where f is the forward rate and b is the backward rate.

The total change in $\dot{y}_2$ is the sum of the changes by all eight of the reactions. Repeating this for each species gives six ODEs. They can be combined to make two element balances. The remaining species balances are nonlinear, but they can be linearized (by any of several methods):

$$\dot{y}_i = \sum_{i=1}^{4}\sum_{j=1}^{4} a_{ij}\, y_j + c_i \tag{3.62}$$

These equations can be solved by first finding the four roots of the characteristic equation. This is not a computationally efficient method of solution, but it was used by Moretti (1965) to determine the cause of stiffness. The roots of the characteristic equation were found to change rapidly when the combustion becomes rigorous. Solution stiffness occurs when these roots are not constant but are changing rapidly, which can be the case for locally linearized equations. Such regions must be solved with robust integration methods. The ignition delay region and the postcombustion region, where the roots change slowly, can be solved with large integration steps. This behavior implies that a solution method which uses local time-adjusting computation steps to only employ small steps when they are really needed would be the optimum solution strategy.

Now how is the source term to be evaluated for inclusion in the SCE? It might appear that much of the recommended technology is old, but there is a good reason for this. When these methods were developed, computers were much slower and less readily available. The numerical methodology had to be stronger to make practical calculations. But these techniques are still available for our use, and they allow us to solve meaningful problems on today's PCs.

3.3.3 GENERATION TERM IN THE SPECIES CONTINUITY EQUATION

The conservation equations are perfectly adequate to analyze most reacting flows. If the reacting system is highly exothermic and both the flow and reactions are fast, the possibility of the species equations becoming stiff must be considered. Certainly for combustion and run-away reactions this would be the case. Solution methodology for solving stiff equations will be described, since a computational code that treats such phenomena can equally well be used to analyze a system that does not exhibit stiffness. The converse is not true. Currently, practical turbulence models do not address unmixedness effects, which cannot be treated as continuum phenomena. Lazy flames, for example, would need more consideration. However, most industrial problems would be at a sufficiently high Reynolds number that the continuum approach would be valid. Furthermore, essentially all finite rates are determined by experiment. The measurement of turbulence effects on these experimental data are never determined and reported. This being said, a stiff system of reacting flow equations can be effectively simulated by the following methods.

For reacting flows, the mass, momentum, energy, and species conservation equations must all be solved. The SCE presents special problems due to the behavior of the reaction rate term. The transport equation of the ith species mass fraction (α_j) can be written as

$$\rho \frac{D\alpha_j}{Dt} - \nabla \cdot \left(\frac{\mu + \mu_t}{\sigma_\alpha} \nabla \alpha_j \right) = \omega_j \tag{3.63}$$

where σ_α represents the Schmidt number for turbulent diffusion. For laminar flows, the turbulent viscosity is zero and Sc is the molecular Schmidt number. The generation rate, ω_j, is the production rate of species j, which can be expressed as

$$\omega_j = M_{w_i} \sum_i (v''_{ij} - v'_{ij})(R_{f,i} - R_{b,i}) \tag{3.64}$$

where

$$R_{f,i} = K_{f,i} \prod_j \left(\frac{\rho \alpha_j}{M_{w_j}} \right)^{v'_{ij}} \; ; \; R_{b,i} = K_{b,i} \prod_j \left(\frac{\rho \alpha_j}{M_{w_j}} \right)^{v''_{ij}} \tag{3.65}$$

In the above equation, M_{w_j} is the molecular weight of species j, $K_{f,i}$ and $K_{b,i}$ are the forward and backward rates of reaction i, and v'_{ij} and v'_{ij} are the power dependencies of reactants and products, respectively. The backward reaction rate can be calculated from the forward reaction rate and a chemical equilibrium constant, $K_{e,i}$, i.e., $K_{b,i} = K_{f,i}/K_{e,i}$. Some would argue that measurements have been made to show that this relationship is not true. There is not enough accurate data to prove this point. More importantly, if the forward and backward rates were independently specified, an instability would be set up and remain in the calculations such that a solution could not be obtained. The specific reaction rates are usually represented by

$$K_{f,i} = AT^B \exp(-E/RT) \tag{3.66}$$

where the parameters A, B, and E are determined experimentally. Theoretical attempts to calculate such reaction rates are not practical.

Consider a general system of chemical reactions written in terms of its stoichiometric coefficients (v'_{ij} and v''_{ij}) and the jth chemical species (M_j) of the ith reaction as

$$\sum_j^r v'_{ij} M'_j \leftrightarrow \sum_j^p v''_{ij} M''_j \tag{3.67}$$

Some empirical reaction rate correlations do not use the power dependencies as the stoichiometric coefficients; rather, they are arbitrary parameters.

The stiffness problem is related to the rapidity with which the chemical source term varies in a flowfield. The species production rate equation can also be expressed in terms of the molar density y_j ($=\rho\alpha_j/M_{wj}$). Successful solutions of a stiff system of equations have involved two steps: a linearization and an implicit solution of the resulting system of equations. The source term is evaluated as a first step and then inserted into the species equation for the coupled solution with all of the conservation equations. This two-step process was developed and proven to be efficient by Ferri and Slutsky (1965). Subintervals may be used to evaluate the source term to avoid using very small steps to solve the full system of conservation equations.

The CTP code uses an algorithm called PARASOL to evaluate the chemical source terms. This algorithm is developed as follows. Let y and f be vector valued functions (i.e., arrays), as implied by the following equation:

$$dy/dt = f\{y,T,\rho\} \tag{3.68}$$

Expanding this equation through first-order terms,

$$\left(\frac{dy}{dt}\right) = f\{y_o,T_o,\rho_o\} + \left(\frac{\partial f}{\partial y}\right)_o \Delta y + \left(\frac{\partial f}{\partial T}\right)_o \Delta T + \left(\frac{\partial f}{\partial \rho}\right)_o \Delta\rho + O\{\Delta y^2,\Delta T^2,\Delta\rho^2\} \tag{3.69}$$

where

    $y, f, (\partial f/\partial T)$, and $(\partial f/\partial \rho)$ are column vectors

    $(\partial f/\partial y)$ is a species Jacobian matrix

    Subscript "o" indicates an initial value

    "$O\{\}$" indicates the order of the truncation (or round-off) error caused
        by using only a limited number of terms in the expansion

The concept is that one iteration (or time step) is taken to obtain new values of the species concentrations. Then the process is reinitialized and the next step is taken. The density term can be eliminated with an EOS. The temperature term in the expansion is negligible if the temperature does not change significantly through a single iteration step, but it could be included (Magnus and Schechter, 1966, 1967). The linearized equation becomes

$$\left(\frac{dy}{dt}\right) = \left(\frac{\partial f}{\partial y}\right)_o \Delta y + f\left\{y_o, T_o, \rho_o\right\} + O\{\Delta y^2\} \tag{3.70}$$

After reinitialization and renaming the terms,

$$\tilde{y} = \hat{A}\, y\,\{t\} + \tilde{B} \tag{3.71}$$

where the caret and tilde have been used to emphasize that the terms are column vectors and a matrix $(n \times n)$. This is a system of n-linear ordinary differential equations with constant coefficients. Integrating over $t = h$, the formal solution is

$$\tilde{y}\{h\} = \exp\{h\hat{A}\}(\tilde{y}_o + \hat{A}^{-1}\tilde{B}) - \hat{A}^{-1}\tilde{B} \tag{3.72}$$

The matrix exponential cannot be conveniently calculated; it can be approximated with polynomials. The form of the approximation determines its accuracy, efficiency, and stability.

To obtain a solution compatible with solving the coupled conservation equations, a one-step integration of the generation term is desired. Either Equation 3.70 or 3.72 could be used for this purpose. Also, other finite-difference forms of Equation 3.70 could be used for this purpose.

For reacting gases, the generation term can be implicitly evaluated with a single-step integration scheme by any of the following. The species production/dissipation rate array can be expressed as

$$\frac{dX_j}{dt} = \sum_{i=1}^{nr} \nu_{ij} \left\{ K_{f,i} \prod_{k=1}^{ns} [X_k]^{r'_{ik}} - K_{b,i} \prod_{k=1}^{ns} [X_k]^{r''_{ik}} \right\}; \quad X_j = \frac{\rho \alpha_j}{M_j} \quad (3.73)$$

where

ρ is the density

α_j and M_j are mass fraction and molecular weight of species j, respectively

ν_{ij} is the stoichiometric coefficient of j-species for the ith reaction

r'_{ik} and r''_{ik} are the power dependence of k-species for the ith reaction

$K_{f,i}$ and $K_{b,i}$ are the forward and backward reaction rates for ith reaction, respectively

"ns" and "nr" are the total numbers of involved species and reactions

Equation 3.73 can be rewritten as

$$\frac{dy_j}{dt} = \frac{1}{N} \sum_{i=1}^{nr} \nu_{ij} \left\{ K_{f,i} \prod_{k=1}^{ns} [N y_k]^{r'_{ik}} - K_{b,i} \prod_{k=1}^{ns} [N y_k]^{r''_{ik}} \right\}; \quad N = \frac{N}{\forall} = \frac{P}{R_u T} \quad (3.74)$$

where

P, T, and R_u are the pressure, temperature, and universal gas constant, respectively

y_k is the mole fraction of k-species

If Equation 3.74 is calculated directly based on the current value (i.e., n-level) of species concentrations, then it is called an explicit scheme; whereas, if it is calculated based on the future value (i.e., $n + 1$ level), then it is an implicit method. In order to calculate the species production/dissipation rate using the implicit scheme, Equation 3.74 needs to be linearized based on the Taylor series expansion:

$$f(x_0 + \Delta x) = f(x) + \left(\frac{\partial f}{\partial x}\right)_{x_0} \Delta x + \left(\frac{\partial^2 f}{\partial x^2}\right)_{x_0} \frac{(\Delta x)^2}{2} + \left(\frac{\partial^3 f}{\partial x^3}\right)_{x_0} \frac{(\Delta x)^3}{3!} + \cdots$$

$$(3.75)$$

If at $t = t_0$, $x = x_0 = x^n$, and at $t = t_0 + \Delta t$, $x = x_0 + \Delta x = x^{n+1}$, then Equation 3.75 can be expressed as

$$f(x^{n+1}) = f(x^n) + \left(\frac{\partial f}{\partial x}\right)^n \Delta x + \left(\frac{\partial^2 f}{\partial x^2}\right)^n \frac{(\Delta x)^2}{2} + \left(\frac{\partial^3 f}{\partial x^3}\right)^n \frac{(\Delta x)^3}{3!} + \cdots \quad (3.76)$$

If $f = f(x_1, x_2, x_3, \ldots, x_m)$, then, Equation 3.76 can be approximated as

$$f\left(x_1^{n+1}, x_2^{n+1}, \ldots, x_m^{n+1}\right) \cong f\left(x_1^n, x_2^n, \ldots, x_m^n\right) + \left(\frac{\partial f}{\partial x_1}\right)^n \Delta x_1 + \left(\frac{\partial f}{\partial x_2}\right)^n \Delta x_2 + \cdots + \left(\frac{\partial f}{\partial x_m}\right)^n \Delta x_m + \cdots$$

i.e.,

$$f^{n+1} = f\left(x_1^{n+1}, x_2^{n+1}, \ldots, x_m^{n+1}\right) \cong f\left(x_1^n, x_2^n, \ldots, x_m^n\right) + \sum_{k=1}^{m} \left(\frac{\partial f}{\partial x_k}\right)^n \Delta x_k + \cdots$$

$$\cong f^n + \sum_{k=1}^{m} \left(\frac{\partial f}{\partial x_k}\right)^n \left(x_k^{n+1} - x_k^n\right) + \cdots \tag{3.77}$$

If we let $f_j = \dfrac{dy_j}{dt}$, then the species production/dissipation can be linearized as

$$\left(\frac{dy_j}{dt}\right)^{n+1} = f_j^{n+1} = f_j^n + \sum_{k=1}^{ns} \left(\frac{\partial f_j}{\partial y_k}\right)^n \left(y_k^{n+1} - y_k^n\right) + \cdots$$

$$= \left(\frac{dy_j}{dt}\right)^n + \sum_{k=1}^{ns} \left(\frac{\partial f_j}{\partial y_k}\right)^n \left(y_k^{n+1} - y_k^n\right) + \cdots \tag{3.78}$$

Also, from the Taylor series expansion, we can approximate the species concentration of the j-species as

$$y_j^{n+1} = y_j^n + \left(\frac{dy_j}{dt}\right)^{n+1} \Delta t + \left(\frac{d^2 y_j}{dt^2}\right)^{n+1} \frac{(\Delta t)^2}{2} + \cdots \tag{3.79}$$

Hence, for the first-order implicit scheme

$$y_j^{n+1} = y_j^n + \left(\frac{dy_j}{dt}\right)^{n+1} \Delta t + O\left[(\Delta t)^2\right]$$

$$= y_j^n + \Delta t f_j^n + \Delta t \sum_{k=1}^{ns} \left(\frac{\partial f_j}{\partial y_k}\right)^n \left(y_k^{n+1} - y_k^n\right) + O\left[(\Delta t)^2\right] \tag{3.80}$$

The second-order implicit scheme can be derived using the Crank–Nicolson method:

$$y_j^{n+1} = y_j^n + \left(\frac{dy_j}{dt}\right)^{n+\frac{1}{2}} \Delta t + O\left[(\Delta t)^3\right]$$

$$= y_j^n + \left[f_j^n + \sum_{k=1}^{ns}\left(\frac{\partial f_j}{\partial y_k}\right)^n \left(y_k^{n+\frac{1}{2}} - y_k^n\right)\right]\Delta t + O\left[(\Delta t)^3\right]$$

$$= y_j^n + \Delta t \left(\frac{dy_j}{dt}\right)^n + \frac{\Delta t}{2}\sum_{k=1}^{ns}\left(\frac{\partial f_j}{\partial y_k}\right)^n \left(y_k^{n+1} - y_k^n\right) + O\left[(\Delta t)^3\right] \qquad (3.81)$$

Let the Y array represent y_j, where $j = 1 \rightarrow ns$; then Equation 3.78 can be rewritten as

$$\left(\frac{dY}{dt}\right)^{n+1} = A Y^{n+1} + B \qquad (3.82)$$

where

$$A = \sum_{k=1}^{ns}\left(\frac{\partial f_j}{\partial y_k}\right)^n; \quad B = f_j^n - A Y^n \qquad (3.83)$$

Let $Y = \alpha e^{\lambda t} + \beta$, and substitute it into the above equation. We then can obtain $\lambda = A$, $\beta = -B/A$, and

$$Y^{n+1} = \left(Y^n + \frac{B}{A}\right)e^{A\Delta t} - \frac{B}{A} \qquad (3.84)$$

By using the Padé approximation

$$\exp\{A\Delta t\} = I + A\Delta t + \tfrac{1}{2}(A\Delta t)^2 + \cdots$$
$$= Q^{-1}P + O\left[(A\Delta t)^{p+q}\right] \qquad (3.85)$$

where

$$Q = \sum_{l=0}^{q}\frac{(p+q-l)!q!}{(p+q)!l!(q-l)!}(-A\Delta t)^l$$
$$P = \sum_{l=0}^{p}\frac{(p+q-l)!p!}{(p+q)!l!(p-l)!}(A\Delta t)^l \qquad (3.86)$$

Hence, the species concentration array at the $n + 1$ level can be expressed as

$$Y^{n+1} = \left(Y^n + \frac{B}{A}\right)Q^{-1}P - \frac{B}{A} = Q^{-1}\left[PY + \frac{B}{A}(P - Q)\right] \qquad (3.87)$$

When $p = 0$ and $q = 1$, we will have $Q = I - A\Delta t$ and $P = I$, where I is the unit matrix. Substituting P and Q into Equation 3.87, we can obtain the first-order Padé approximation as

$$Y^{n+1} - Y^n = A\Delta t\,(Y^{n+1} - Y^n) + f_j^n\Delta t \qquad (3.88)$$

i.e.,

$$y_j^{n+1} - y_j^n = \Delta t\left(\frac{dy_j}{dt}\right)^n + \Delta t\sum_{k=1}^{ns}\left(\frac{\partial f_j}{\partial y_k}\right)^n (y_k^{n+1} - y_k^n) \qquad (3.89)$$

This has the same formulation as the one derived from the first-order implicit scheme. When $p = 1$ and $q = 1$, we will have $Q = I - \frac{1}{2}A\Delta t$ and $P = I + \frac{1}{2}A\Delta t$. Substituting P and Q into Equation 3.87, we can obtain the second-order Padé approximation as

$$Y^{n+1} - Y^n = \frac{A\Delta t}{2}(Y^{n+1} - Y^n) + f_j^n\Delta t \qquad (3.90)$$

$$y_j^{n+1} - y_j^n = \Delta t\left(\frac{dy_j}{dt}\right)^n + \frac{\Delta t}{2}\sum_{k=1}^{ns}\left(\frac{\partial f_j}{\partial y_k}\right)^n (y_k^{n+1} - y_k^n) \qquad (3.91)$$

This has the same formulation as the one derived from the second-order implicit (Crank–Nicolson) scheme. When $p = 2$ and $q = 2$, we will have $Q = I - \frac{1}{2}A\Delta t + \frac{1}{12}(A\Delta t)^2$ and $P = I + \frac{1}{2}A\Delta t + \frac{1}{12}(A\Delta t)^2$. Substituting P and Q into Equation 3.87, we can obtain the fourth-order Padé approximation as

$$Y^{n+1} - Y^n = \left(\frac{A\Delta t}{2} - \frac{(A\Delta t)^2}{12}\right)(Y^{n+1} - Y^n) + f_j^n\Delta t \qquad (3.92)$$

$$y_j^{n+1} - y_j^n = \Delta t\left(\frac{dy_j}{dt}\right)^n + \frac{\Delta t}{2}\sum_{k=1}^{ns}\left(\frac{\partial f_j}{\partial y_k}\right)^n (y_k^{n+1} - y_k^n)$$

$$- \frac{(\Delta t)^2}{12}\sum_{k=1}^{ns}\left[\left(\frac{\partial f_j}{\partial y_k}\right)^n\right]^2 (y_k^{n+1} - y_k^n) \qquad (3.93)$$

Either the finite-difference approximates arising from the Taylor expansions or the Padé expansions give the same equations. The Padé methodology gives more information about stability and error propagation. The truncation error means that the Padé approximation agrees with the exponential power series for at least $p + q + 1$ terms. Diagonal Padé approximants are A-stable and have the smallest truncation error. The approximation of the matrix exponential by the diagonal approximants is called the Padé rational approximation. The algorithm for applying the Padé rational approximation to the formal solution of a chemistry equations is herein named the Padé rational solution (PARASOL).

There is an additional error called a propagation error that results from using repeated integration steps. For the Padé integration method this error is represented by

$$e_{k+1} = (Q^{-1}P)e_k \tag{3.94}$$

$(Q^{-1}P)$ is a matrix called the amplification matrix; it determines the stability of the integration. When the eigenvalues of this matrix have negative real parts, e_k decreases as k increases, which means that the method is unconditionally stable. Such is the case when $p = q$ (Varga, 1962).

The PARASOL code has three options: (1) the first-order implicit scheme is used for a fast calculation for a well-behaved chemical system; (2) the second-order Padé scheme is used in general; and (3) the fourth-order Padé scheme is used when convergence with the second-order scheme is difficult to achieve.

Padé integration is not the only single-step implicit method suitable for integrating the linearized source term in the species conservation equation. Frey et al. (1968) developed a one-dimensional kinetics, second-order implicit kinetics code which has since been incorporated into an axisymmetric rocket performance analysis code. Note this rocket analysis is not a fully computational analysis of transport phenomena because it applies to a limited geometry and does not include turbulent transport. Even so, this code has been used successfully for many years,

Several other production quality codes have been developed and used to describe stiff reaction systems. These codes do not treat convective and diffusive aspects of reacting flow problems. Most of them utilize multistep difference algorithms and were designed to handle large systems of reactions and species. In general, they are not expected to be appropriate for including in a CTP analysis with the other multidimensional conservation equations. One of the most popular algorithms for solving stiff systems of ODEs is that of Gear (1971). It was later adapted in GCKP (Bittker and Scullin, 1972) to describe finite-rate chemistry. These schemes employ Adams' explicit methods of variable order to solve nonstiff equations. GCKP84 (Bittker and Scullin, 1984) is a revised version of GCKP that includes an implicit predictor–corrector scheme (Zeleznik and McBride, 1984) to analyze stiff equations. Hindmarsh (1974, 1980, 1982) and Hindmarsh and Byrne (1977) generalized the GEAR algorithms and developed a series of stiff ODE solvers, EPISODE, and LSODE. CREK1D (Pratt and Radhakrishnan, 1984) developed a two part predictor–corrector and an exponentially fitted trapezoidal scheme. CHEMEQ (Young, 1980) applied a predictor–corrector scheme to nonstiff ODEs, but used an asymptotic integration scheme to equations deemed to be stiff. It was previously reported that separating the equations into stiff and nonstiff parts, which were then treated differently, resulted in mass conservation problems (Lomax and Baily, 1967; Kee and Dweyer, 1981). The LENS code (Radhakrishnan and Bittker, 1993) used backward differences to overcome reaction stiffness in analyzing shock tube flow and well-stirred reactors. The work of Hindmarsh (1974) typifies the activity at the Lawrence Livermore National Laboratory, that of Bittker and Radhakrishnan typifies the activity at NASA's Glenn Research Center, and that of Oran and Boris (1987) at the Naval Research Laboratory is described in their comprehensive text.

This brief survey is not meant to be a literature survey; rather, it is an indication of the vast scope of uncoordinated research that has been done on the kinetics stiffness problem. An efficient integration algorithm and a carefully selected set of a reasonable number of reactions for the system under investigation are essential for solving geometrically complex transport problems.

The Padé and the Frey et al. codes were developed specifically to be included in the system of conservation equations that describe multidimensional transport phenomena. They have proven to be efficient and useful. The PARASOL subroutine includes first-order implicit scheme and the two Padé schemes, $(p, q) = (1, 1)$ and $(2, 2)$ as options in the CTP code. Both of these codes have automatically adjusting, variable-step integration features. The variable time step feature is essential to analyze multidimensional reacting flow problems. Modest sized chemical kinetics databases are included in the CTP code, and methods of enlarging these databases will be explained subsequently.

3.4 MOLECULAR TRANSPORT PROPERTIES

3.4.1 BASIC MOLECULAR TRANSPORT COEFFICIENTS

For laminar and time-averaged turbulent flows, momentum, heat, and mass are transferred by convection, turbulent and laminar gradient diffusion, and secondary processes. The transfer by these processes is characterized by fluid mixing, which occurs at a decreasing rate in the order of the processes named. Convection and turbulence are flow properties that must be described by the conservation laws. Diffusion of momentum can be modified by density changes in the flow, which cause additional dissipation. Turbulent diffusion can be modified by nongradient effects resulting from the averaging process; these will be discussed in Chapter 4. Laminar or molecular mixing is controlled mainly by local gradients of the property being transferred. Additional molecular mixing can also be caused by cross-effects, i.e., mass transfer by temperature gradients, etc. Such diffusion is slower than when driven by the primary gradient. The cross-effects are described as nonequilibrium thermodynamic phenomena. Mass diffusion in binary systems is described with diffusion coefficients that relate species 1 motion to that of species 2. When more than two components are present, a method of averaging the binary coefficients to obtain a multicomponent diffusion must be established. To obtain a practical CTP analysis, only gradient diffusion will be utilized herein. The other effects mentioned are real and can be quantified, but they are not usually significant in analyses of transport phenomena.

There is a similarity between the fluxes of momentum, heat and mass, which is why they are lumped together as transport phenomena. For geometrically simple, essentially constant density systems, these fluxes can be written as the first term on the right-hand side and the more general form as the second term:

$$\text{Newton's law: } \tau_{yx} = v\frac{d}{dy}(\rho u) = \mu\frac{du}{dy} \tag{3.95}$$

$$\text{Fourier's law: } q_y = -\alpha\frac{d}{dy}(\rho C_p T) = -\lambda\frac{dT}{dy} \tag{3.96}$$

$$\text{Fick's law: } j_{Ay} = -D_{AB}\frac{d}{dy}(\rho_A) = -\rho D_{AB}\frac{dx_A}{dy} \tag{3.97}$$

Each coefficient of the gradient has a unit of length squared per unit time and they are the kinematic viscosity, v; thermal diffusivity, α; and mass diffusivity, D_{AB}, in the first term. The transport processes are analogous for these simple flows. The analogies cannot be extended to real fluids or to three dimensions since $\vec{\tau}$ is the stress tensor with nine components and $\vec{q}$, the heat flux, and $\vec{j}_A$, the mass flux, are vectors with three components each. The analogies are useful for approximating the transport coefficients and quantitatively evaluating them. The transport coefficients are scalars. The right-hand sides of these equations will change for more complex flows, but the transport coefficients will retain the same values. Note that for a solid, the thermal diffusivity may become a vector by having different values in different directions, e.g., thermal wrapping, where conductivity along the wrapping and across the wrapping will have different values.

The molecular transport properties—viscosity, thermal conductivity, diffusion coefficients, and surface tension—are required to describe laminar flows of real fluids. Empirical correlations of these properties for single and multicomponent fluids are adequate if they cover the temperature and pressure range of interest. In general, these transport coefficients are small and easily measured (Reid et al., 1987). The properties in the following are excerpted from this reference. Although many correlations are available, only one complete set of property data was selected for review herein.

3.4.1.1 Viscosity

Uyehara and Watson (1944) proposed a universal viscosity correlation. The correlation involved defining a critical viscosity at the critical temperature and pressure values to make a chart of reduced viscosity values as a function of reduced temperature and pressure. This chart is reproduced in Bird et al. (2002) and White (2006). The reference viscosity is

$$\mu_c(\text{micropoise}) = \frac{61.6}{(Z_c R)^{2/3}}\left(\frac{M_w^4 P_c^4}{T_c}\right)^{1/6}$$

$$\text{where} \quad T_c(K), P_c(\text{atm}), \forall (cc/g - mole) \tag{3.98}$$

The chart was constructed from fitting experimental data. The rationale for using corresponding states to correlate viscosity data is based on the concept that pressure is composed of two factors, one a kinetic pressure and the other a cohesive pressure (Comings and Egly, 1940). This is reflected in the various gas laws, even as far back as the van der Waals EOS

(Equation 3.3), by the two terms in the equation for pressure. The first term represents the kinetic pressure, the second the cohesive pressure. The correlation chart was developed interpreting the viscosity behavior to be the same as the kinetic pressure when expressed in reduced parameters. For mixtures, Uyehara and Watson (1944) also recommended using pseudocritical properties to represent the mixture. Its accuracy with respect to more recent data and correlations has not been systematically investigated. Obviously, experimental data, such as those reviewed in Reid et al. (1987), would be more accurate.

The most interesting feature of the Uyehara and Watson chart is that it correlated viscosity data with the same reduced thermodynamic variables as were used to represent $P\forall T$ data in the method of corresponding states. This resulted in transport property correlations of this type being termed corresponding states methods. Such methods provide tables and figures depicting the property data but not empirical equations, which would be desirable for CTP analyses.

The alternative to relying on only correlations of experimental data to define transport coefficients is to provide a theoretical model. Kinetic theory has been used to provide such a model. The original application of such a theory was to treat the fluid molecules as hard spheres of constant size all moving at a local mean velocity. Efforts to make such a model more realistic were devised by the Chapman–Enskog theory, which assumed that (1) gases are so dilute that only binary collisions occur, (2) classical mechanics describe the motion during a collision, (3) only elastic collisions occur, and (4) the intermolecular potential function is spherically symmetric. This theory was a major advancement, but it was only applicable to low pressure gases. It did introduce some basic understanding and defined some new variables to predict transport coefficients. Much effort has been devoted to improve this theory, most of which simply introduced correction coefficients based on experimental observations. Nevertheless, the basic theory is worth recounting. Chapman and Enskog (Chapman and Cowling, 1970) determined that

$$\mu = \frac{5}{16}\left[\frac{(\pi MRT)^{0.5}}{\pi\sigma^2\Omega_\upsilon}\right] = \frac{26.69(MT)^{0.5}}{\sigma^2\Omega_\upsilon} \tag{3.99}$$

The molecule is approximated as a hard sphere of diameter σ. The collision integral Ω_υ is determined by a potential function and expressed in terms of a dimensionless temperature $T^* = kT/\varepsilon$. The Lennard-Jones (6-12) potential function is commonly used:

$$\psi\{r\} = 4\varepsilon\left[\left(\frac{\sigma}{r}\right)^{12} - \left(\frac{\sigma}{r}\right)^{6}\right] \tag{3.100}$$

The collision integral was determined by Neufeld et al. (Reid et al., 1987) to be

$$\Omega_v = \left[1.16145(T^*)^{-0.14874}\right] + 0.52487\exp\{-0.77320T^*\}$$
$$+ 2.16178\left[\exp\{-2.43787T^*\}\right] \quad \text{for } 0.3 \le T^* \le 100 \tag{3.101}$$

The distance between molecular centers is r; the minimum in the potential energy curve is ε. Equations 3.99 through 3.101 allow one to calculate the low pressure viscosity of gases within the limitations of the Chapman–Enskog theory. Several new variables are required, but these parameters for many species are in the literature (Reid et al., 1987; Bird et al., 2002). The kinetic theory methodology has been extended by adding more complexity. It is already too complex for efficient use in CTP codes, and it has not yet been accepted for predicting multicomponent liquid viscosities.

Pure component viscosities for gases Lucas corresponding states method for low pressures and his pressure correction for high pressures are suggested for transport analyses. As expected, gaseous viscosity is shown to increases with both pressure and temperature. Liquid pure component viscosities from the polynomial curve fits tabulated in Reid et al. (1987) are suggested. Liquid viscosity decreases with temperature and increases with pressure. Mixture viscosities are determined by Wilke's correlation.

Lucas' low pressure viscosity correlation for gases is

$$\mu_r = \xi\mu° = 0.176\left(\frac{T_c}{M^3 P_c^4}\right)^{1/6} \quad \mu° = F_T°\{T_r\}F_P°\{Z_c T_r d_r\}F_Q°\{T_r, He, H_2, D_2\} \tag{3.102}$$

ξ is an inverse viscosity at the critical point. The reduced dipole moment is represented by d_r and $\mu°$ is the low pressure viscosity:

$$d_r = 52.46\frac{d^2 P_c}{T_c^2} \tag{3.103}$$

The dipole moment, d, is in Debye, and P_c is in bars. Lucas' high pressure viscosity correlation for dense gases is

$$\xi\mu = F_T\{T_r, P_r\}, F_P\{T_r, P_r, F_P^\circ\}, F_Q\{T_r, P_r, F_Q^\circ\} \tag{3.104}$$

The curve fits for low pressure liquids are given as

$$\mu = AT^B \quad \text{or as} \quad \ln\mu = A + B/T + CT + DT^2 \tag{3.105}$$

The viscosity is taken to be that of saturated liquid at its vapor pressure.
For gas mixtures, Wilke's correlation is

$$\mu_m = \sum_{i=1}^{n} \frac{y_i \mu_i}{\sum_{j=1}^{n} y_j \Phi_{ij}} \quad \text{where} \quad \Phi_{ij} = \frac{\left[1 + (\mu_i/\mu_j)^{0.5} (M_j/M_i)^{0.25}\right]^2}{\left[8(1 + M_i/M_j)\right]^{0.5}} \tag{3.106}$$

For liquids, the viscosity is weighed by mole fractions, unless specific data are available.

There is an additional viscosity coefficient that needs considering. Writing the relationships for two of the stress components,

$$\tau_{xx} = -P + 2\mu\frac{\partial u}{\partial x} - \left(\frac{2\mu}{3} - \kappa\right)(\nabla\cdot\vec{U}) \tag{3.107}$$

$$\tau_{xy} = +\mu\left(\frac{\partial u}{\partial y} + \frac{\partial v}{\partial x}\right) \tag{3.108}$$

The bulk viscosity, κ, is associated with the absorption of sound. It is not zero for polyatomic molecules, across shock waves, and in bubbly fluids. It is frequently omitted because (1) the fluid is incompressible, (2) the sum of the two-thirds viscosity and negative bulk is assumed zero, or (3) the fluid is assumed to be a monoatomic gas. The importance of the bulk viscosity is also related to the timescale of the flow phenomena (Rowlinson, 1969). The bulk viscosity will be treated as a constant in the present CTP analyses.

3.4.1.2 Thermal Conductivity

Thermal conductivity for pure component gases for both low and high pressure is determined by the method of Ely and Hanley. Thermal conductivity is calculated from the reference gas methane:

$$\frac{(\lambda - \lambda^*)M}{\mu^* C_v} = 1.32\left(1 - \frac{3/2}{C_v/R}\right) \tag{3.109}$$

where

$$\lambda^* = \lambda_o H, \quad \mu^* = \mu_o H \frac{M}{16.04 \times 10^{-3}}, \quad H = \left(\frac{16.04 \times 10^{-3}}{M}\right)^{0.5} f^{0.5} h^{-0.67}$$

$$(3.110)$$

The functions f and h and the methane values of viscosity and conductivity (denoted by the subscript "o") are calculated from T_c, $\forall_c$, and T_r. High pressure conductivity is calculated from

$$\frac{(\lambda - \lambda^{**})M}{\mu^* C_v} = 1.32\left(1 - \frac{3/2}{C_v/R}\right) \quad \text{where} \quad \lambda^{**} = \lambda\left\{T_o, \rho_o, \mu_o\right\} \quad (3.111)$$

The parameters subscripted with "o" are reference properties for methane. This method is not very accurate, so experimental data for the specific fluids of interest should be sought. Liquid thermal conductivities may be calculated by Missenard's method:

$$\lambda_H = \lambda_L\left(1 + QP_r^{0.7}\right) \quad \text{where} \quad Q\left\{T_r, P_r\right\} \quad (3.112)$$

Mixture thermal conductivities can be evaluated by the method of Mason and Saxena, which parallels the method of Wilke for viscosities. This method is exactly like Wilke's except the viscosity ratio on the right-hand side of Equation 3.112 contains the ratio of thermal conductivities and not viscosities.

3.4.1.3 Diffusion Coefficients

The method of Wilke and Lee was selected to calculate binary diffusion coefficients. This method is a modification of the standard Chapman and Enskog equation:

$$D_{AB} = \frac{\left[3.03 - \left(0.98/M_{AB}^{0.5}\right)\left(10^{-3}\right)T^{1.5}\right]}{PM_{AB}^{0.5}\sigma_{AB}^2\Omega_D} \quad (3.113)$$

where
  Ω and σ are parameters of the Lennard-Jones potential function
  M_{AB} is an average molecular weight

The method uses Neufeld's equation to evaluate the Lennard-Jones potential. Pressure effects are approximated by assuming that the product of pressure and binary diffusion coefficients is constant. Blanc's law may be used to calculate the diffusion coefficient of each component in a multicomponent mixture, resulting in a mixture diffusion coefficient for each component:

$$D_{im} = \left(\sum_{j=1, j \neq i}^{n} \frac{x_j}{D_{ij}} \right)^{-1} \qquad (3.114)$$

This approximation appears reasonable, but lacks substantial validation.

Multicomponent diffusion coefficients may be introduced as described in Bird et al. (2002). These representations are very complex as the binary coefficients are combined in a coupled fashion. Conceptually, this complexity is reduced in a computational analysis that advances a solution by incremental time steps because the coupling may be lagged without introducing numerical difficulty. Such a refinement is not suggested for initially analyzing mass diffusion.

Liquid diffusion coefficients are about a factor of 10^5 smaller than gaseous diffusion coefficients so they can generally be neglected.

3.4.1.4 Surface Tension

Surface tension for a pure liquid is represented by the method of Block and Bird, in terms of the Riedel parameter. The effect of multiple components on surface tension is difficult to predict, since surface effects are often quite different from bulk values. Also, mixture critical points are often quite different from pure component critical points. The multicomponent surface tension may be estimated by the modified Macleod and Sugden method, which involves one-fourth powers of the pure component surface tensions.

The transport coefficient correlations suggested have been compared to oxygen data (McCarty and Weber, 1971), hydrogen data (McCarty and Weber, 1972), and water and steam data (Keenan and Keyes, 1963) by these authors and found to be reasonable.

3.4.2 SECONDARY TRANSPORT

Up to this point we have indicated that laminar mass fluxes of diffusing components are a result of a concentration gradient. Pressure gradients, temperature gradients, and external force differences can also cause movement of species with respect to the mean fluid motion; these are the cross-effects. In a multicomponent system there can be momentum, heat and mass transfer due to velocity, and temperature and concentration gradients. However, there can be a mass flux that is the movement of species relative to the bulk flow, due to pressure gradients, temperature gradients, and external force differences:

$$\vec{j}_i = \vec{j}_i^x + \vec{j}_i^P + \vec{j}_i^T + \vec{j}_i^B \qquad (3.115)$$

Pressure diffusion is used in centrifugal separations where the extremely large gradients can be obtained that are necessary for movement of the species to be separated. Diffusion caused by temperature gradients is referred to as the Soret effect, and thermal diffusion also requires very steep temperature gradients for a significant separation. Forced diffusion caused by external force differences is important in ionic systems, where the force on an ion is the product of the charge of the ion and the strength of the electric field. The equations giving the mass flux of component i due to these various gradients are given by Bird et al. (2002) and are repeated here.

Diffusion due to concentration gradients:

$$\vec{j}_i^x = (c^2/\rho RT)\sum_{j=i}^{n} M_i M_j D_{ij}\left[x_j \sum_{k=1}^{n}(\partial F_j/\partial x_k)_{T,P,s}\nabla x_k \right] \quad \text{for } k \neq j \text{ and } s \neq j, k$$

$$(3.116)$$

In this equation
  F is the partial molal free energy (Gibbs' free energy)
  D_{ij} is the multicomponent mass diffusion coefficient

Diffusion due to pressure gradients:

$$\vec{j}_i^P = (c^2/\rho RT)\sum_{j=1}^{n} M_i M_j D_{ij}\left[x_j M_j (V_j/M_j - 1/\rho)\overline{\nabla}P \right] \quad (3.117)$$

In this equation V_j is the partial molal volume.
  Diffusion due to temperature gradients:

$$\vec{j}_i^T = -D_i^T \overline{\nabla}\ln T \qquad (3.118)$$

In this equation, D_i^T is the multicomponent thermal diffusion coefficient, and the transfer of mass due to a temperature gradient is called the Soret effect.

Diffusion due to external force differences:

$$\bar{j}_i^B = -(c^2/\rho RT)\sum_{j=1}^{n} M_i M_j D_{ij} \left[x_j M_j \left(B_j - \sum_{k=1}^{n} \rho_k \vec{B}_k /\rho \right) \right] \quad (3.119)$$

The multicomponent mass diffusivities D_{ij} are related by the following equation:

$$\sum_{i=1}^{n}(M_i M_j D_{ih} - M_i M_k D_{ik}) = 0 \quad \text{and} \quad D_{ii} = 0 \quad (3.120)$$

For $n < 2$, the diffusivity D_{ij} in general is not equal to D_{ji}. D_{ij} is the diffusivity of the pair i–j in a multicomponent mixture. D_{ij}^* is the diffusivity of the pair i–j in a binary mixture. For diffusion in a binary mixture $D_{ij} = D_{ij}^*$ if i and j form an ideal solution, i.e., activity is proportional to mole fraction.

The Stephan-Maxwell equations can be used to relate the binary mass diffusion coefficients in a multicomponent mixture. It can be shown that these are related by the following equation for an n-component ideal gas mixture (BSL et al., 2002):

$$\left(cD_{im}^*\right)^{-1} = \left(\sum_{j=1}^{n} \left[1/\left(cD_{ij}^* \right) \right]\left(x_j \vec{N}_i - x_i \vec{N}_j \right) \right) \Bigg/ \left(\vec{N}_i - x_i \sum_{j=1}^{n} \vec{N}_j \right) \quad (3.121)$$

The way N is used, diffusion and convection are lumped together, which is not necessary, and it implies that the diffusion is coupled to the mean velocity, which is not true.

This equation can be simplified for several special cases. First, if there is nearly pure ℓ, i.e., only traces of 2, 3,..., n, then

$$D_{im}^* = D_{i\ell}^* \quad (3.122)$$

For systems where all of the diffusivities are equal

$$D_{im}^* = D_{ij}^* \quad (3.123)$$

For systems in which all the components but one move with the same velocity or are stationary

$$(1-x_1)/D_{1m}^* = \sum_{j=2}^{n} x_j /D_{1j}^* \quad (3.124)$$

Unfortunately, there is not sufficient space to delve into the derivations and backgrounds of these equations that describe the transfer of mass due to concentration, pressure, temperature, and external force gradients before energy transfer is discussed. The reader is referred to the works of Bird et al. (2002) and Merk (1959) for further details.

In a multicomponent system, energy is transferred by mechanisms other than conduction. Energy is transferred by interdiffusion and by the Dufour effect. The Dufour effect is the inverse phenomena of the Soret effect and is the formation of a temperature gradient as the result of a concentration gradient. This effect may be described mathematically by adding terms proportional to the concentration gradients to Fourier's law. In general, the energy transferred by this method is not large. In fact, it has only been observed in gas systems, and as a consequence will not be discussed further. However, the energy transferred by interdiffusion can be sizeable and is expressed by the following with respect to a fixed set of coordinates and in terms of molar fluxes:

$$\vec{q}^{d} = \sum_{i=1}^{n} H_{xi} \vec{J}_{i} \tag{3.125}$$

where H_{xi} is the partial molal enthalpy of the ith species. The combined heat flux of conduction and interdiffusion is

$$\vec{q} = -k\nabla T + \sum_{i=1}^{n} H_{xi} \vec{J}_{i} \tag{3.126}$$

These mass and energy fluxes will be combined with the general equations of change for a multicomponent system, to give a set of equations that can be applied to almost any situation. In Chapter 6 these general equations will be developed, and the equations for the fluxes will be incorporated.

If the flow is turbulent, the transport coefficients become larger and are a function of the flow. The laminar coefficients may be further modified by an unmixedness effect. This means that locally the fluid is no longer a continuum because the molecules have not had sufficient time to be uniformly mixed at a given point. These effects are discussed in more detail in Chapter 4.

3.4.3 USE OF DIMENSIONLESS TRANSPORT COEFFICIENTS

The variation of viscosity, thermal conductivity, and diffusion coefficients over a wide range of temperatures and pressures is very similar. Such similarity can be used to advantage if dimensionless Prandtl and Schmidt numbers are used. Thus, only viscosity variation must be modeled and then the dimensionless parameters are used to account for thermal and mass diffusion. This is the same concept that will be subsequently used to describe turbulent transport, hence coding will be simplified to account for all three transport mechanisms in both laminar and turbulent flows.

$$\text{Prandtl number: } Pr = C_p \mu / \lambda \quad \text{Schmidt number} : Sc = \mu / \rho D_{12}$$
$$\text{Lewis number: } Le = Pr/Sc \quad \text{Kinematic viscosity} : \nu = \mu / \rho \tag{3.127}$$

Prandtl number characterization of fluids classifies fluids by type. Liquid metals exhibit low Prandtl numbers like 0.02. Gases have Prandtl numbers in the 0.7–1.0 range. Thin fluids like water have Prandtl numbers from 3 to 10. Heavy oils range roughly from 500 to 10000. Hence, for a given CTP simulation, a single constant value of the Prandtl number might be entirely adequate.

By reviewing the tables of viscosity, and Prandtl and Schmidt numbers found in several references (Fraas and Ozisik, 1965; Sherwood et al., 1975; Bird et al., 2002), it is noted that the dimensionless transport parameters vary less with temperature than the viscosity does. Moreover, correlations of experimental data for specific situations show that the Prandtl and Schmidt number enter into the prediction as parameters are raised to powers less than unity. This further reduces the temperature and pressure sensitivity of the property variation on the prediction. The main factor in causing property variation is the chemical composition of the fluid itself, which would still require definition. Apparently, using correlations of viscosity, Prandtl and Schmidt numbers would be an advantageous method of analyzing transport phenomena where property variation would be important.

Such simplifications are immediately apparent by considering mass transfer. The binary Schmidt number may be estimated for gaseous species by the equation

$$\text{Sc}_{AB} = 1.18 \frac{\Omega_D}{\Omega_i} \left[\frac{M_A}{M_A + M_B} \right]^{0.5} \left(\frac{\sigma_{AB}}{\sigma_B} \right)^2 \approx 1.06 \left[\frac{M_A}{M_A + M_B} \right]^{0.5} \tag{3.128}$$

The omega terms are for the Lennard-Jones potentials for diffusion and viscosity. The sigma terms are for the Lennard-Jones parameters related to the critical molar specific volumes. From the approximation indicated, these ratios are approximately constant at near unity values. This relationship is presented by Sherwood et al. (1975), as well as by many molecular binary diffusion coefficient values. The direct modeling of the dimensionless transport coefficients is a most effective way to make the CTP code applicable to both laminar and turbulent flows.

3.5 THERMAL RADIATION PROPERTIES

Thermal radiation is added to the conservation equations for multicomponent reacting flows by including the radiation heat flux term, $\vec{q}_r$, and a scattering function, Φ, to the energy equation.

$$\frac{\partial \rho h}{\partial t} + \nabla \cdot \rho h \vec{U} = -\nabla \cdot \sum_{i=1}^{n} h_i \vec{j}_i + \nabla \cdot \left(\kappa \nabla T - \vec{q}_r \right) + \frac{DP}{Dt} + \sum_{i=1}^{n} \vec{j}_i \cdot \vec{g}_i + \Phi \quad (3.129)$$

The radiation heat flux term is defined by the equation

$$\nabla \cdot \vec{q}_r = 4 \int_0^\infty \left\{ a_\lambda \{\lambda\} e_{\lambda b} \{\lambda\} - \pi \left[a_\lambda \{\lambda\} + \sigma_{s\lambda} \{\lambda\} \right] \hat{i}_\lambda \{\lambda\} \right\} d\lambda$$
$$+ \int_0^\infty \left\{ \sigma_{s\lambda} \{\lambda\} \int_{\omega_i=0}^{4\pi} i'_\lambda \{\lambda, \omega\} \widehat{\Phi} \{\lambda, \omega_i\} d\omega_i \right\} d\lambda \quad (3.130)$$

The radiation heat flux definition is derived by Siegel and Howell (1992). The symbols used are basically from that definition. The λ is wavelength when used as a subscript or as showing functional dependence. The spectral absorption coefficient is a and the spectral scattering coefficient is σ_s. The blackbody emissive power is $e_{\lambda b}$. The phase function for scattering is Φ. The solid angle is ω; the subscript i indicates direction. The radiation intensity is i; the prime indicates it is a directional quantity. The circumflex indicates the quantity is averaged over all solid angles.

Obtaining sufficient spectral absorption and scattering coefficient data for a fluid mixture and solving the integro-partial differential energy equation is a daunting task. The alternative is to simplify this equation until a solution can be obtained. The general classes of such simplifications will be presented and referenced. Examples will be given to indicate how gaseous and particulate radiation from combustion processes can be estimated.

3.5.1 APPROXIMATE RADIATION TRANSFER ANALYSES

Radiation is an immense and complicated field. Many solution methods are reported, but their accuracy and range of applicability are seldom evaluated. Most disappointing of all is that radiative transfer cannot be scaled. Convective and conductive heat transfers can be evaluated with subscale models and extrapolated to prototype conditions with dimensionless parameters. The integral nature of the radiation precludes this approach to evaluate radiation. Usually, a given problem is analyzed by approximating the transport process until a solution, analytical or numerical, can be obtained. Examples of such simplifying analyses are listed as follows.

Radiation through nonparticipating media within an enclosure is the most simple and most common radiant heating analysis that is used. The enclosure is a control volume wherein the radiation bounces around between the control surfaces. View factors between the surfaces and radiation interaction at the walls must be described. This usually amounts to a geometrically complex problem. Constructing an electrical network analog is a very efficient and graphically simplifying method to address the analysis. Network analyses are well described by Oppenheim (Hartnett, 1961) and Holman (1972). The network methodology does not eliminate the need for determining view factors, but it does provide a systematic way for keeping track of them. The methodology is very useful for studying the effects of the radiation properties of the surfaces.

The most common radiation analysis problem arising in the process industries is combustion gas radiation. The fluid must be hot for radiation to be important. Gaseous radiation is difficult to analyze because of its spiky spectral character, i.e., the photons that exchange radiant energy are emitted and absorbed at discrete frequencies. The classical experiments of Hottel (McAdams, 1954) and Hottel and Sarofim (1967) present these data in a convenient form. The scaling of such data is done by using the experimental heat transfer data without ever attempting to model the spectral character of the radiation.

Geometric simplification is achieved, for example, by considering radiation between parallel plates. The radiating media may have various radiation properties, yet the heat transfer between the plates may be determined. Both absorption and scattering may be considered. Such analyses are described by Sparrow and Cess (1966). Other simple geometries may be considered, but the limitation inherently is that the effects of real geometry are too complex for analysis.

The distance a photon travels before something happens to it is called optical depth. Radiation analysis can be described by the limits of this parameter. Optically thin fluids (long photon travel) or optically thick fluids (short photon travel) are the limiting conditions for radiation. The assumption of nonparticipating media and situations in which energy may be emitted from the fluid but not absorbed are examples of optically thin phenomena. For optically thick cases, the radiation is essentially emitted and/or absorbed at the fluid boundaries. The diffusion approximation for the radiation properties is appropriate for such situations. The limiting radiation analyses methodology is described thoroughly in Siegel and Howell (1992). These limits may be used to produce analyses, but the intermediate optical thicknesses are the real issue.

Gaseous radiation properties may be described in several levels of simplification. The first is to assume that the gas is gray, i.e., the absorption and scattering coefficients are not functions of wavelength. This allows one to simplify the solution of the integral terms in evaluating the radiation heat flux. Unfortunately, no real gases are gray. The major part of radiative heating through gases is accomplished by the vibration–rotation bands in the near infrared. Although the photon transfers are for a distinct transition, these lines are broadened so the radiation is not truly monochromatic. Radiation of diatomic molecules may be calculated, and the results may be represented with band models. However, for polyatomic molecules, there are so many lines closely grouped together that band models must be used to represent the radiative transfer. Two types of band models, narrow and wide, have been used for this purpose. The wideband models represent the entire band parametrically. The narrowband models typically represent the spectrum in $25\ cm^{-1}$ wave number increments. In some instances, $5\ cm^{-1}$ resolution is used. The wideband models have been documented extensively (Edwards, 1981; Siegel and Howell, 1992; Modest, 1993). The expectation is that the narrowband models would average nonisothermal, nonisobaric paths more accurately.

Clean flames like the blue flames used as icons by natural gas companies and hydrogen/oxygen flames only emit and absorb energy. Integrations along lines-of-sight may be used to evaluate radiation from these flames. Fuel-rich hydrocarbon flames generate soot. Although the soot is a particulate, its small size allows it to behave as a gas and not scatter radiation. Larger particulates, like coarsely pulverized coal and the aluminum oxide particulates in solid rocket booster plumes, not only emit and absorb but also scatter radiation. As can be seen in Equation 3.130, the additional scattering term that must be considered makes the solution of the radiation flux much more difficult to evaluate. Radiation along rays cannot be simply evaluated and summed to determine the radiation from a flame. Radiative interchange between the rays must also be accounted for. Such analyses may be made for the radiation field between parallel plates with gradients of temperature and particulates between the plates. Even slightly more complex geometries must be analyzed with statistical Monte Carlo methods. The simple line-of-sight calculations must be replaced by tracking numerous energy bundles through the transport process (Siegel and Howell, 1992). The method is very general such that emitting, absorbing, and scattering media enclosed by arbitrary surfaces can be described. Much ingenuity has gone into reducing the number of energy packets that must be tracked, but the method is still very computationally intensive. Modern computer capability does allow the Monte Carlo method to be used—if the application justifies it.

3.5.2 TRANSPORT PHENOMENA PROBLEM COUPLED WITH RADIATION

Finally, consider a transport process with radiation. Kaplan et al. (1992) reported an analysis of two unsteady, strongly radiating, buoyant, ethylene diffusion-flame. The Reynolds numbers of the two flames were about 3000 and 6000. These were probably the Reynolds numbers at the jet exit, but this was not reported. The appropriate set of conservation equations was solved. This was a very daunting task that was accomplished only by making severely restrictive assumptions on the important submodels necessary to describe the system. The solver used was from the Naval Research Laboratory's group, which was a mature production code. The ethylene jet flow was assumed to be co-current with air. The simulation was for an axisymmetric flow. The gases were assumed to be ideal. Probably the heat capacity in the CEOS was temperature dependent, but this was not specified in the paper. The flow was simulated as laminar with temperature-dependent transport coefficients. The method of simulating multicomponent diffusion from the binary diffusion coefficients was not specified. The ethylene combustion was simulated with a one-step reaction, with an unspecified correction constant supposedly to account for the flame being diffused and not premixed. Although the products of combustion were not specified, the reactions were said to be quasi-global, yielding carbon dioxide and water. Two ODEs were solved to represent the soot volume fraction and number density. Apparently, soot oxidization was not considered. Absorption coefficients for soot and for the mixture of carbon dioxide and water were simulated with two algebraic equations. The discrete-ordinate method was used to evaluate the radiation heat flux term. This method is of variable order; a fourth-order method was used for this analysis. The fourth order required the solution of 12 simultaneous equations. Presumably, all 12 of these equations were solved along with the conservation equations at each time step. The solution obtained with this analysis was discussed at length. However, no evaluation of the limitations imposed by the physical assumptions made was offered. The major conclusion drawn from the work was that a simpler, more computationally efficient method of solving for the radiation term was needed. The authors did evaluate the radiation to conduction ratio suggested by Grosshandler (1993), which indicated that the radiation coupled solution was necessary.

The simulation of the ethylene diffusion flame was an ambitious effort. The purpose of the study was never clearly stated. The effort did consider all of the submodels needed to address the problem being explored. Unfortunately, many of the submodels were too simplistic to result in an acceptable simulation. If one really expected to simulate an unsteady buoyant flame, limiting the geometry to being axisymmetric defeated the purpose. Experiments would show that the buoyant eddies were not axisymmetric. The high Reynolds number test case was probably turbulent. The kinetics model used did not include atomic species and CO, so the predicted temperatures would be too high—of about the same order as the temperature effects attributed to the soot formation. Soot combustion should have been considered. The absorption coefficients used were overly simplistic. If they were to be used, a better justification for their use should have been provided. Since none of the particulates (soot) were considered to be scatters (a good assumption), the effect of leaving the scattering term in the radiation had no meaning. Either it was left out of the discrete-ordinate analysis, or left in and never evaluated. The criticisms offered could not even be addressed with a suitable validation experiment. These comments are not meant to demean the work reported, but to pose questions that should have been addressed to place the work in the proper perspective. These investigators have continued this research, so this is a work in progress. Later papers may have eliminated much of this criticism. The validation of a computational simulation requires experimental data that test the various assumptions made in the computational model. The more elaborate the model, the more definitive the experimental data collected must be. This is a tough requirement that is seldom met.

3.5.3 NARROWBAND MODELS

The narrowband radiation model was mentioned in Section 3.5.1 as an alternative to the wideband model to represent gaseous radiation. The narrowband model accounts for more spectral resolution in predicting the radiation from rocket exhausts, structural and wild fires, furnaces, and boilers. Theoretically, the wideband models could also be used for this purpose, but they generally are not. The increased spectral resolution would probably give a more accurate prediction, and it would definitely be a better tool for interrogating flame character. The initial impetus for developing narrowband radiation models was the NASA conference on molecular radiation (Dahm and Goulard, 1967) to stimulate research on large rocket plumes. The need for improved technology was to provide a better tool for describing base heating to the large Saturn launch vehicles. The primary emitters in the plume were CO_2, H_2O, and soot. At the time the base heat shields weighed about the same as the vehicle payload. Plume analysis was needed because subscale model data could not be scaled to flight conditions. The equally important reentry heating problem required radiation analysis of a different type since the species in the nose cap region primarily emitted line and continuum radiation. More recently, this technology has been applied to structural fires (Grosshandler, 1993) and coal fired combustion facilities (Fiveland, 1987). Military applications of this methodology have been extensive but are not generally available.

The NASA planning conference resulted in an extensive experimental program to obtain radiation property data. These data were collected, modeled, and reported in a handbook (Ludwig et al., 1973). Few, but important, additions have been made to this database in the intervening years. The following development is given to introduce the parameters and terminology used to represent narrowband models. It follows that presented by Reardon and Lee (1979).

Hydrocarbon and hydrogen flames are the primary combustions that require radiation analysis. The vibration–rotation bands of CO_2 and H_2O are the primary gaseous radiators. Fuel-rich hydrocarbon flames also soot. Coal flames contain not only these gaseous species but also pulverized coal particles and ash. The ash and coal particles are large enough that they scatter the radiation. The gaseous species absorb and emit radiation in discrete spectral regions. The soot particles are usually very small, such that they too only emit and absorb; unlike gases they radiate in the continuum. Solid propellant rocket motors produce plumes, which contain fairly large aluminum oxide particles that scatter radiation. Other radiating gaseous species may also be in these plumes. Consider first the narrowband radiation model applied to the description of hot CO_2 and H_2O flames. This is the application that the narrowband model was designed to describe.

The monochromatic absorption coefficient, $k\{\omega, s\}$, is a basic radiation property and is defined in terms of the spectral radiance, $N\{\omega, s\}$, by

$$dN\{\omega,s\}/ds = -k\{\omega,s\}\rho\{s\}N\{\omega,s\} \qquad (3.131)$$

where
ω is the wave number
ρ is the density of a radiating species per unit length
s is the path length

The properties k and ρ depend on the temperature, pressure, and composition of the gas. Integrating this property over path length defines transmissivity:

$$\tau\{\omega,s\}N\{\omega,s\}/N\{\omega,0\} = \exp\left\{-\sum_0^s k\{\omega,s\}\rho\{s\}ds\right\} \qquad (3.132)$$

If the gas is homogeneous,

$$\tau\{\omega,u\} = \exp\left\{-k\{\omega\}u\right\} \qquad \text{where} \qquad u = \rho s \qquad (3.133)$$

The ρs product is dimensionless and is loosely referred to as concentration.

Spectral lines represent energy transitions in a molecule, and they are broadened by collision and the thermal motion of the molecules. The resulting line shapes are termed Lorentz and Doppler, respectively. The line strength (or intensity) is the integrated absorption coefficient over the broadened line:

$$S = \int_0^\infty k\{\omega\}d\omega \qquad (3.134)$$

Do not be alarmed by the wide limits of integration—the absorption coefficient is zero over most of the spectrum. The integrated absorptance is a function of both k and u:

$$W = \int_0^\infty \alpha\{\omega,u\}d\omega = \int_0^\infty \left[1-\exp(-k\{\omega\}u)\right]d\omega \qquad (3.135)$$

W is also called the equivalent width since it represents the width of an equivalent black line. Let γ represent the half-width of the broadened line.

The integration of W with (Su/γ) is termed the curve-of-growth. For both Lorentz and Doppler line shapes, the curve-of-growth is linear at weak line strengths. The shape of the curves diverges at higher values of this parameter. The Lorentz lines approach a square root dependence; the more narrow Doppler lines approach $[\ell n\{Su/\gamma\}]^{0.5}$.

Narrowband models are characterized by the type of lines grouped together to constitute the band. A regular (or Elsasser) model considers absorption by identical, equally spaced lines. Such a model is a good representation of diatomic molecules. A statistical (or random) model assumes randomly spaced lines of a specified intensity distribution. This model represents polyatomic species. A third category used a mixture of these two types of models.

Three parameters are used to characterize the line structure in a small spectral interval—the half-width (γ), the line spacing (d), and the absorption coefficient ($\bar{k}$), which is proportional to the line-strength ratio (S/d). Data for these parameters are given in Ludwig et al. (1973) and Grosshandler (1993). Tables are given for the absorption coefficients and reciprocal line widths. Algebraic equations are given for the half-widths. The half-width expressions are

$$\gamma_{Li} = P\left[c_i\gamma_{ii}\left(273/T\right)^{n_{ii}} + \sum_j c_j\gamma_{ij}\left(273/T\right)^{n_{ij}} \right]$$

$$\gamma_{Di} = 5.94\times10^{-6}\,\omega\left(\frac{T}{273\,M_i}\right)^{0.5} \tag{3.136}$$

Copious other data are in the literature, but these sources are a good basis for radiation studies.

From this point on, the discussion will be limited to describing radiation from CO_2 and H_2O vapor with a statistical band model. To proceed, the intensity distribution must be specified. The probability functions

$$P\{S\} \propto \exp\{-S\} \quad \text{and} \quad P\{S\} \propto S^{-1}\exp\{-S\} \tag{3.137}$$

were investigated by studying the curves-of-growth resulting from their use. These distributions were deemed useful for in-band models. These approximated functions resulted in the exponential line strengths for Lorentz and Doppler lines, respectively:

$$(\overline{W}/d)_L = X_L = \overline{k}\,u\,/\sqrt{1+\overline{k}\,u/(4a_L)} \qquad\qquad (3.138)$$

$$(\overline{W}/d)_D = X_D = 1.70a_D\left(\ln[1+(\overline{k}\,u/1.70a_D)^2]\right)^{0.5} \qquad (3.139)$$

The subscripts L and D refer to Lorentz and Doppler, respectively. If the exponentially tailed-inverse line strength approximation is used,

$$(\overline{W}/d)_L = X_L = 2a_L\left(\sqrt{1+\overline{k}u/(4a_L)}-1\right) \qquad\qquad (3.140)$$

$$(\overline{W}/d)_D = X_D = 0.937a_D\left(\ln[1+(\overline{k}u/0.937a_D)^{2/3}]\right)^{1.5} \qquad (3.141)$$

where
$$a_L = \gamma_L/d$$
$$a_D = \gamma_D/d$$

Either pair of these equations were found to give acceptable results. The Lorentz and Doppler components are combined to give the transmissivity:

$$\ln\overline{\tau} = -\sum_i \overline{k}_i u_i\left(1-y_i^{-0.5}\right)^{0.5} \qquad\qquad (3.142)$$

$$y_i = \left[1-(X_L/\overline{k}u)^2\right]^{-2}+\left[1-(X_D/\overline{k}u)^2\right]^{-2}-1$$

Radiance along a line-of-sight to a point at $s = 0$ is

$$\vec{N}\{\omega,L\} = -\int_0^L N^\circ\{\omega,s\}\left[\frac{d\overline{\tau}\{\omega,s\}}{ds}\right]ds \qquad\qquad (3.143)$$

The Plank function is N° evaluated at T and ω. The radiant heat flux to the point at the terminus of the line-of-sight in the direction of s is

$$\vec{q} = -\int_{\theta_i}^{\theta_f}\int_{\varphi_i}^{\varphi_f}\int_{\omega_i}^{\omega_f}\left\{\int_0^L N^\circ\{\omega,s\}[d\overline{\tau}\{\omega,s\}/ds]ds\right\}\sin\theta\cos\theta\;d\omega\;d\varphi\;d\theta \qquad (3.144)$$

The elevation and azimuth angles measured from the surface normal are θ and φ, respectively.

For moderate inhomogeneity in the hot gas region, the modified Curtis–Godson approximation may be use to represent line-of-sight radiation. This would apply to base heating from a rocket plume or to walls heated by a flame. If the flame is viewed from a long distance away, like a chemical plant flare viewed from afar or a rocket plume viewed by a distant observer, corrections are available for modifying the radiation prediction. The modified Curtis–Godson approximation uses the homogeneous band model formulation with effective parameters (denoted by the subscript "e") defined to account for the inhomogeneity. These parameters are

$$\bar{k}_e\{\omega,i,s\}u\{i,s\} = \int_0^s \bar{k}\{\omega,i,s'\}c\{i,s'\}P\{s'\}ds' \tag{3.145}$$

$$a_{Le}\{\omega,i,s\} = \frac{\int_0^s \bar{k}\{\omega,i,s'\}c\{i,s'\}P\{s'\}a_L\{\omega,i,s'\}ds'}{\bar{k}_e\{\omega,i,s\}u\{i,s\}} \tag{3.146}$$

$$a_{De}\{\omega,i,s\} = \frac{\int_0^s \bar{k}\{\omega,i,s'\}c\{i,s'\}P\{s'\}a_D[\omega,i,s']ds'}{\bar{k}_e\{\omega,i,s\}u\{i,s\}} \tag{3.147}$$

The primes denote a dummy variable of integration. The cP product denotes the ratio of the radiating species density at s' to the density at 1 atm for the same local temperature. Some data refer $\bar{k}$ to reference conditions of 1 atm and a temperature of 273 K. For using such tables, the ρc product must be multiplied by the temperature ratio (273/T).

3.5.3.1 Narrowband Models as a Diagnostic Tool

The spectral radiance for a nonisothermal path is

$$N_\omega = -\int_0^L N_\omega^\circ \left[\frac{d\bar{\tau}\{s,\omega\}}{ds} \right] ds \tag{3.148}$$

N_ω° is the blackbody radiance at the local wave number. Applying the radiance expression to a series of increments along a line-of-sight,

$$N_\omega = \sum_{m=1}^M N_\omega^\circ\{T_m\}(\tau_{\omega,m-1} - \tau_{\omega,m}) \tag{3.149}$$

For $M = 5$ increments of gas and one more, a radiation source impinging on the last increment of gas, we have

$$N_{\omega E} = N_{\omega 1}^{\circ}\left(1-\tau_1\right) + N_{\omega 2}^{\circ}\left(\tau_1 - \tau_2\right) + N_{\omega 3}^{\circ}\left(\tau_2 - \tau_3\right) + N_{\omega 4}^{\circ}\left(\tau_3 - \tau_4\right) + N_{\omega 5}^{\circ}\left(\tau_4 - \tau_5\right)$$

$$(3.150)$$

The subscript "E" denotes radiance emitted from the gases, and "T" represents the radiance impressed on the gas beam:

$$N_{\omega T} = N_{\omega 1}^{\circ}\left(1-\tau_1\right) + N_{\omega 2}^{\circ}\left(\tau_1 - \tau_2\right) + N_{\omega 3}^{\circ}\left(\tau_2 - \tau_3\right) + N_{\omega 4}^{\circ}\left(\tau_3 - \tau_4\right)$$
$$+ N_{\omega 5}^{\circ}\left(\tau_4 - \tau_5\right) + N_{\omega 6}^{\circ}\left(\tau_5 - \tau_6\right) \qquad (3.151)$$

Imagine a single isothermal cell of hot gas with an emission and a transmission measurement for a measured applied radiance. The partial pressure of the radiating species and the temperature of the gas can then be determined. This idea has been applied to measuring the species and temperatures in axisymmetric flames.

Imagine the cross-section of the flame normal to the axis of symmetry. Conceptually, this cross-section would consist of a series of rings, each with a constant partial pressure of a radiating species and temperature. The species can be made unique by choosing a spectral region where it is the only radiator. If a continuum radiator like soot is present, the measurements can be made by correcting the band radiance for the underlying soot radiance. By choosing the spatial resolution desired, a number of paired emission–absorption measurements can be created. This array of measurements can be inverted to produce the desired temperature and species concentration values. The details of such an experiment and instrumentation system required to make the measurements are described by W. Herget (Dahm and Goulard, 1967). A number of such experiments have subsequently been performed. Such measurements were the first quantitative evaluation of the mixing inefficiency in small liquid rocket engines. This inefficiency revealed the error to be expected in using subscale test data to represent prototype engine performance and scale-up.

3.5.3.2 Narrowband Model Applications

To indicate the magnitude and details of the narrowband variables, a simple one-cell emission/absorption analysis is presented. This example is needed because the literature is scattered and the nomenclature is not standardized. The data used in this example were taken from Ludwig et al. (1973). Consider CO_2 radiating at a wavelength (λ) of 4.45 μm or a wave number of 2247 cm^{-1}. Assume a path length (L) of 8 cm. The partial pressure of the CO_2 is 0.23 atm; for the other species $CO/0.27$, $H_2O/0.41$, and $H_2/0.09$. The gas temperature is 2400 K.

The properties given by Ludwig et al. (1973) are for a reference pressure of 1 atm and 273 K. The effective optical depth of the gas is

$$\Delta u = s(P_i/1.0)(273/T) = 8(0.23)(273/2400) = 0.2093/\text{cm} \quad (3.152)$$

The absorption coefficient (k) is 13.66/cm^{-1} and the line spacing ($1/d$) is 502.6/cm from pp. 446 and 457–458, respectively, in Ludwig et al. (1973). The data indexed in the tables are on p. 384. The collision (Lorentz) half-width from p. 223 is given by

$$\gamma_{Li} = \left[\sum_j \gamma_{ij} P_j (273/T)^{n_{ij}}\right] + \gamma_{ii} P_i (273/T)^{n_{ii}} \quad (3.153)$$

where
γ_{ii} for CO_2 is 0.01
γ_{ij}'s for CO_2, H_2O, CO, and H_2 are 0.09, 0.07, 0.06, and 0.08, respectively

The units on these values are reciprocal cm atm. The n_{ii} is 1.0 and all of the n_{ij}'s are 1.0. Thus, $\gamma_{Li} = 2.481 \times 10^{-2}/cm^{-1}$. The Doppler half-width is

$$\gamma_{Di} = 5.94 \times 10^{-6} \omega \left(\frac{T}{273 M_i}\right)^{0.5} = 0.359 \times 10^{-6} 2247 \left(\frac{2400}{44}\right)^{0.5}$$
$$= 5.966 \times 10^{-3}/\text{cm}^{-1} \quad (3.154)$$

The spatially averaged absorption coefficient for one zone is

$$X^* = \int_0^u k\, du' = \sum_{m=1}^M k_m \Delta u_m = \overline{k} u = 13.66 \times 0.2093 = 2.859 \quad (3.155)$$

Also,

$$a_L = 2.481 \times 10^{-2} (502.6) = 2.859 \quad (3.156)$$

$$a_D = 5.966 \times 10^{-3} (502.6) = 2.999 \quad (3.157)$$

The Lorentz and Doppler line strengths become in these new variables

$$X_L = X^*(1+X^*/4a_L)^{-0.5} = 2.859[1+(2.859/(4\times12.47))]^{-0.5} = 2.78 \quad (3.158)$$

$$X_D = 1.7a_D\left[\ln\left\{1+(X^*/1.7a_D)^2\right\}\right]^{0.5}$$

$$= 1.7\times2.999\left[\ln\left\{1+\left(\frac{2.859}{1.7\times2.999}\right)^2\right\}\right]^{0.5} = 2.666 \quad (3.159)$$

Combining these two components and calculating the transmissivity:

$$y_i = \left[1-(X_L/\overline{k}u)^2\right]^{-2} + \left[1-(X_D/\overline{k}u)^2\right]^{-2} - 1$$

$$= \left[1-\left(\frac{2.78}{2.859}\right)^2\right]^{-2} + \left[1-\left(\frac{2.666}{2.859}\right)^2\right]^{-2} - 1$$

$$= 336.7+58.8-1=394.5 \quad (3.160)$$

$$\ln\overline{\tau} = -\sum_i \overline{k}_i u_i (1-y_i^{-0.5})^{0.5} = -2.859(1-394.5^{-0.5})^{0.5}$$

$$= \ln\{-2.786\} \quad \text{or} \quad \overline{\tau} = 0.06166 \quad \text{and} \quad 1-\overline{\tau} = 0.93834 \quad (3.161)$$

The emitted radiation is

$$N_{\omega E} = N_\omega^\circ(\tau_{\omega0}-\tau_{\omega1}) = 4.746\times10^{-3}(0.93834)$$

$$= 4.4534\times10^{-3} \text{ W/cm·sr} \quad (3.162)$$

The impressed (denoted by subscript "s") and transmitted radiation are

$$N_{\omega T} = N_{\omega s}^\circ \times \overline{\tau} = 3.314\times10^{-3}(0.06166) = 0.2043\times10^{-3} \text{ W/cm·sr} \quad (3.163)$$

The radiation detected (denoted by subscript "D") from hot gas and an external source is

$$N_{\omega D} = N_{\omega T} + N_{\omega E} = 4.6577\times10^{-3} \text{ W/cm·sr} \quad (3.164)$$

The concept of using absorption diagnostics has been extended to study three-dimensional nonhomogeneous concentration and temperature fields. The methods are termed optical tomography. The technique employs multiangular scanning of the nonuniform field. Tomography is the same process as used by x-ray resolution to investigate medical conditions. Santoro et al. (1981) demonstrated this technique by studying the mixing regions on a cold flow methane–air jet. The technique was proven feasible and has received further use. In addition to investigating flames, industrial pollutant clouds have been measured. An interesting study was made by W. Herget of the EPA, who mounted a spectrometer on the observation deck atop the 32 story Louisiana state capitol to scan the atmosphere above the two dozen odd chemical complexes in Baton Rouge.

The plumes from hydrocarbon and hydrogen fueled rocket engines and from most structural fires consist of gaseous species and soot. These substances emit and absorb radiation, but do not scatter radiation. To analyze radiant heating from such media, scattering of the radiation does not have to be considered. Base heating of launch vehicles has been predicted for design purposes with an analysis by Reardon and Lee (1979). The analysis evaluated radiation using narrowband models along lines-of-sight from the plume to the vehicle, including radiation from hot structural parts (like the outsides of rocket nozzles). Grosshandler (1993) made a similar analysis to interpret experiments involving fires in mock residences and industrial buildings. These analyses require a solution of the conservation equations, without radiation, to determine the species, temperatures, and pressures along the radiant beams. To date, the calculation has been uncoupled. The flowfield was assumed not to change by losing its radiant energy. This allows the integro-differential energy equation by solving the integral radiation term separately from the PDE for energy. Bhattacharjee and Grosshandler (1989) postulated a radiation/convection interaction parameter (Ψ) to determine the necessity of including coupling to obtain a valid prediction. This parameter is a function of the flame, surrounding wall, and inlet temperatures (T_f, T_w, T_i); the flame inlet mass flux times its heat capacity ($\rho u C_p$); the optical thickness; the absorption coefficient (k); and a streamwise direction (L). The Stefan–Boltzmann constant (σ) also appears in the parameter:

$$\Psi = \frac{\sigma a L}{\rho u C_p} \frac{\left(T_f^4 - T_w^4\right)}{\left(T_f - T_i\right)} \tag{3.165}$$

If Ψ is less than one, the solution of the energy equation would be independent of the radiation loss. Few coupled solutions that are not compromised by oversimplification have been obtained.

3.5.3.3 Radiation Heat Transfer with Narrowband Models and Scattering

If the flames or hot gases contain moderate-size particulates, the media can scatter as well as emit radiation. Scattering complicates the analysis drastically. Rather than simply integrating along lines-of-sight, the later transfer of radiant scattered has to be admitted. Many approximations to solve the integral term have been suggested. Generally, the approximations are severe and even at that the computation is very intensive. Everson and Nelson (1993) have reported on a fairly rigorous, albeit still computationally intensive, reverse Monte Carlo method for accounting for narrowband and soot continuum emission along with particulate emission. The reverse nature of this analysis allows one to specify an area receiving the radiation and trace the computational ray back into the flame. Conventional Monte Carlo schemes discharge rays out of the flame, and only a fraction of these hit the area of interest. This statistical method is sufficiently computationally intensive without additional inefficiency caused by tracing wasted rays.

Reardon and Nelson (1994) used the reversed Monte Carlo method to predict base heating to the advanced solid rocket booster (ASRB) intended for use on the Space Shuttle. Due to the large base area of the shuttle, heating predictions to 1700 locations for six different gimbal positions were made. The increased aluminum loading in the ASRB propellant (16%–19%) exacerbated the heating load due to scattered radiation.

Again, this text is intended to extend the application of computational analyses to real engineering problems. A survey of the multitude of papers in the literature is not a goal of this work.

3.5.4 VALIDATION WITH OPTICAL DATA

Point measurements made with laser scattering, CARS, and fluorescence and laser Doppler velocity measurements are important to the validation process. Such methods for local temperature and concentration measurements have been ably reviewed by Eckbreth (1996). The instrumentation required for such measurements is complex and fragile. It is difficult to provide such measurements outside of the laboratory environment.

3.6 NOMENCLATURE

3.6.1 ENGLISH SYMBOLS

a	absorption coefficient in Section 3.5
a	speed of sound; Equation 3.36
a, a'	TEOS correction factors for molecular attraction
a, b	constants in van der Waals EOS
a_i	empirical constant in Equations 3.18 through 3.21
a_{ij}, b_i	empirical mixing parameters in PR-EOS, Equation 3.35
a_{ij}, c_i	linearized SCE parameters, Equation 3.62
A, a	Helmholtz free energy; specific Helmholtz free energy
A, B	functions in general SCE; Equation 3.84
A, B, C, D, E	parameters in various property correlations
b_j	elemental concentration of j
b, b'	TEOS correction factors for molecular volume
B_{ij}	parameters in HBMS EOS
$\tilde{B}$	column vector in the SCE written in function space
c	molar concentration
C_i	function defined in Equation 3.40
C_p	constant pressure heat capacity
C_v	constant volume heat capacity
d	dipole moment
d	line spacing in Section 3.5
D_{ij}	diffusion coefficient
D_{ij}^*	diffusion coefficient of binary pair in a multicomponent mixture
e_i	elemental concentration of j in species i
e_k	propagation error in the Padé integration scheme; Equation 3.94
$e_{\lambda b}$	blackbody emissive power
E_i	elemental composition
$f^{(i)}$	parameters in vapor pressure correlation
f	linearized SCE function
f, b	forward and backward reaction rates of the H_2/O_2 system
f, h	methane viscosity and conductivity, respectively
f_i	function in Equation 3.39
f_i	partial fugacity
f_i°	standard-state fugacity
$F_T^\circ, F_P^\circ, F_Q^\circ$	viscosity correlation functions

$\vec{g}_i$	gravitational acceleration in "i" direction
G, g	Gibbs free energy; specific Gibbs free energy
$G_i \equiv \mu_i = F$	partial molar Gibbs free energy
H, h	enthalpy; specific enthalpy
H_{xi}	partial molar enthalpy
I	a specific reaction
j	mass diffusion flux
J	molar diffusion flux
k	Boltzmann's constant
$k\{\omega, s\}$	monochromatic absorption coefficient
K	equilibrium constant
K_f, K_b	forward and backward specific rate constants
k_{ij}	binary interaction parameter
Le	Lewis number
m	mass
m_i	mass of i
M	matrix term in the SCE in function space; average solid angle in Section 3.5
M_w, M	molecular weight
M', M''	reactant and product species, respectively
N	number of moles; number of equations in equilibrium calculation
N°	Planck function
$N\{\omega, s\}$	spectral radiance
N_i	molar density of i
$\bar{N}_i$	mass flux vector
n	total molar density
n_i	moles of i
n_{ij}	parameter in half-width correlation
P	pressure
P, p	parameters in Padé approximation; Equation 3.85
Pr	Prandtl number
P^{vap}	vapor pressure
$\vec{q}_r$	radiation heat flux vector
Q	parameter in Equation 3.112
Q, q	parameters in Padé approximation; Equation 3.85
r'_{ij}, r''_{ij}	power dependency of species i in reaction j
R, R_u	gas constant
R_f, R_b	forward and backward reaction rates

s	path length in Section 3.5
S	line strength (or intensity)
S, s	entropy; specific entropy
Sc	Schmidt number
T, T^*	temperature; dimensionless temperature
t	time
u	velocity parallel to surface
u	pseudo-concentration in Section 3.5
U, u	internal energy; specific internal energy
$\vec{U}$	velocity vector
V_j	partial molar volume
υ	velocity normal to surface
W	integrated absorptance
x	coordinate parallel to surface
x	independent variable in linearized SCE
x_A	mass fraction of A
x_i	mass fraction of i
X_i	molar density of i
X_I	parameters in transmissivity
y	coordinate normal to surface
y_i	mole fraction of i
Z	compressibility factor

3.6.2 GREEK SYMBOLS

α	$= \kappa/\rho C_p$, thermal diffusivity
α	Riedel's constant
α_j	mass fraction; Equation 3.63
γ	line half-width in Section 3.5
γ	ratio of specific heats
Δ	incremental operator
ε	characteristic energy
θ	elevation angle in Section 3.5
Θ	reduced normal boiling point temperature
κ	bulk viscosity in Equation 3.107
κ	parameter in function of ω
κ	thermal conductivity in Section 3.5
λ	wave length in Section 3.5
$\lambda, \lambda^*, \lambda^{**}$	thermal conductivities, Equations 3.109 and 3.111

λ_j	Lagrangian multiplier
μ	viscosity
ν	kinematic viscosity
ν_i, ν_i'', ν'	stoichiometric coefficients
ξ	inverse viscosity
ρ	density
σ	hard sphere diameter; scattering coefficient in Section 3.5
σ_i	partial mass quantity
σ_q	Schmidt number
$\tau\{\omega, s\}$	transmissivity
τ_{xy}	shear stress component
ϕ	parameter in Taylor series expansion
φ	azimuthal angle in Section 3.5
Φ	augmented function; scattering phase function in Section 3.5
φ	fluid quality
Φ_{ij}	mixing function for viscosity; Equation 3.106
ψ, Ψ	ratio of radiation to convection; ratio of radiation to convection
ω	acentric factor; solid angle in Section 3.5
ω_i	generation term in SCE
Ω_v	collision integral

3.6.3 MATHEMATICAL SYMBOLS

δ_{ij}	kronecker delta
$\forall, \nu$	volume; specific volume
ΔH_v	heat of vaporization
$\cancel{N}$	molar volume
$\vec{\vec{\tau}}$	second-order tensor (τ is any tensor)
$\vec{V}$	vector (V is any vector)
$\hat{M}$	matrix (M is any matrix)
$\tilde{C}_V$	column vector (C_V is any column vector)
$\bar{k}$	spectral average (k is any variable)

3.6.4 SUBSCRIPTS

c	critical value
D	Doppler line; sum of emitted and transmitted radiation in Equation 3.3
e	average value for inhomogeneous radiation

E	emitted radiation
i	specific property of species *i*
ℓ	liquid phase
L	Lorentz line in Section 3.5
L, *sℓ*	saturated liquid
o	methane value used as a reference
r	reduced quantity
RA	liquid compressibility
REF, B^o, B_o	reference value (*B* is any variable)
sv	saturated vapor
t	turbulent value
T	property is a function of temperature only: transmitted radiation in Section 3.5
vv	saturated vapor

3.6.5 SUPERSCRIPTS

app	approximate value
EX	excess function
nbp	normal boiling point
o	ideal gas value
x, P, T, B	secondary diffusion fluxes in Equation 3.115

3.6.6 ACRONYMS

CEA	thermodynamics code
CEOS	caloric equation of state
CHEMEQ, LENS, CREK1D, PARASOL	stiff ODE solvers
EOS	equation of state
GCKP, GCKP84, GEAR. EPISODE	stiff ODE solvers
HBMS	Hirschfelder, Buehler, McGee, Sutton EOS
NE	number of elements
NS	number of species
ODE	ordinary differential equation
PR	Peng–Robinson
SCE	species continuity equation
TEOS	thermal equation of state
UNIFAC	activity coefficient code

REFERENCES

Balzhiser, R. E., M. R. Samuels, and J. D. Eliassen. 1972. *Chemical Engineering Thermodynamics: The Study of Energy, Entropy, and Equilibrium*. Englewood Cliffs, NJ: Prentice-Hall.

Bhattacharjee, S. and W. L. Grosshandler. 1989. Effect of radiative heat transfer on combustion chamber flows. *Combust. Flame* 24:347–357.

Bilger, R. W., S. B. Pope, K. N. C. Bray, and J. M. Driscoll. 2005. Paradigms in turbulent combustion. *Proc. Combust. Inst.* 30:21–42.

Bird, R. B., W. E. Stewart, and E. N. Lightfoot. 2002. *Transport Phenomena*, 2nd ed. New York: John Wiley and Sons.

Bittker, D. A. and V. J. Scullin. 1972. General Chemical Kinetics Computer Code for Static and Flow Reactions, with Application to Combustion and Shock-Tube Kinetics. NASA TN D-6586. National Aeronautics and Space Administration.

Bittker, D. A. and V. J. Scullin. 1984. General Chemical Kinetics Computer Code for Gas-Phase Flow and Batch Processes Including. Heat Transfer Effects. NASA TP 2320. National Aeronautics and Space Administration.

Brennen, C. E. 1995. *Cavitation and Bubble Dynamics*. New York: Oxford University Press.

Chapman, S. and T. G. Cowling. 1970. *The Mathematical Theory of Non-Uniform Gases*, 3rd ed. London: Cambridge University Press.

Comings, E. W. and R. S. Egly. 1940. Viscosity of gases and vapors at high pressures. *Ind. Eng. Chem.* 32:714–718.

Dahm, W. K. and R. Goulard. 1967. Specialist Conference on Molecular Radiation and Its Application to Diagnostic Techniques. NASA TM X-53711. National Aeronautics and Space Administration.

Eckbreth, A. C. 1996. *Laser Diagnostics for Combustion Temperature and Species*, 2nd ed. Amsterdam: Gordon and Breach.

Edelman, R. B. and O. F. Fortune. 1969. A Quasi-Global Chemical Kinetic Model for the Finite Rate Combustion of Hydrocarbon Fuels with Application to Turbulent Burning and Mixing in Hypersonic Engines. AIAA Paper No. 69-86. Paper presented at the AIAA 7th Aerospace Sciences Meeting, New York.

Edelman, R. B. and P. T. Harsha. 1978. Laminar and turbulent gas dynamics in combustors—current status. *Prog. Energy Combust. Sci.* 4:1–62.

Edwards, D. K. 1981. *Radiation Heat Transfer Notes*. Washington, DC: Hemisphere.

Everson, J. and H. F. Nelson. 1993. Development and Application of a Reverse Monte Carlo Radiative Transfer Code for Rocket Plume Base Heating. AIAA 93-0138. 31st Aerospace Sciences Meeting and Exhibit. Reno.

Ferri, A., G. Moretti, and S. Slutsky. 1965. Mixing processes in supersonic combustion. *J. SIAM* 13:229–258.

Fiveland, W. A. 1987. Discrete ordinate methods for radiative heat transfer in isotropically and anisotropically scattering media. *J. Heat Transfer* 109:809–812.

Folger, F. S. 1999. *Elements of Chemical Reaction Engineering*, 3rd ed. Upper Saddle River, NJ: Prentice-Hall.

Fraas, A. P. and M. N. Ozisik. 1965. *Heat Exchanger Design*. New York: John Wiley & Sons.

Frey, H. M., J. R. Kliegel, G. R. Nickerson, and T. J. Tyson. 1968. ICRPG One-dimensional Kinetic Reference Program. AD841201. Dynamic Science Corp. Monrovia, CA.

Gardiner, W. C., Jr., Ed. 1984. *Combustion Chemistry*. New York: Springer-Verlag.

Gear, G. W. 1971. *Numerical Initial Value Problems in Ordinary Differential Equations*. Englewood Cliffs, NJ: Prentice-Hall.

Gordon, S. and B. J. McBride. 1976. Computer Program for Calculation of Complex Chemical Equilibrium Compositions, Rocket Performance, Incident and Reflected Shocks, and Chapman-Jouget Detonations. NASA SP-273. National Aeronautics and Space Administration.

Gordon, S. and B. J. McBride. 1994. Computer Program for Calculation of Complex Chemical Equilibrium Compositions and Reflected Shocks, and Applications I. Analysis. NASA Reference Publication 1311. National Aeronautics and Space Administration.

Grosshandler, W. L. 1993. RADCAL: A Narrow-Band Model for Radiation Calculations in a Combustion Environment. NIST TN 1402. NIST.

Hartnett, J. P., Ed. 1961. *Recent Advances in Heat and Mass Transfer*. New York: McGraw-Hill.

Hindmarsh, A. C. 1974. GEAR: Ordinary Differential Equation Solver. Rept. UCID-30001. Livermore, CA: Lawrence Livermore Laboratory.

Hindmarsh, A. C. 1980. LSODE and LSODI: Two new initial value ordinary differential equation solvers. ACM *SIGNUM Newsl.* 15:10–11.

Hindmarsh, A. C. 1982. ODEPACK: A Systematized Collection of ODE Solvers. UCRL-88007. Livermore, CA: Lawrence Livermore Laboratory.

Hindmarsh, A. C. and G. D. Byrne. 1977. EPISODE: An Effective Package for the Integration of Ordinary Differential Equations. Rept. UCID-30112. Livermore, CA: Lawrence Livermore Laboratory.

Hindmarsh, A. C. 1980. LSODE and LSODI: Two new initial value ordinary differential equation solvers. *SIGNUM Newsl.* 15:10–11.

Hirschfelder, J. O., R. J. Buehler, H. A. McGee, Jr., and J. R. Sutton. 1958a. Generalized equation of state for gases and liquids. *Ind. Eng. Chem.* 50:375–385.

Hirschfelder, J. O., R. J. Buehler, H. A. McGee, Jr., and J. R. Sutton. 1958b. Generalized thermodynamic excess functions for gases and liquids. *Ind. Eng. Chem.* 50:386–390.

Holman, J. P. 1972. *Heat Transfer*, 3rd ed. New York: McGraw-Hill.

Hottel, H. C. and A. F. Sarofim. 1967. *Radiative Transfer*. New York: McGraw-Hill.

Kaplan, C. R., S. W. Back, E. S. Oran, and J. L. Ellzey. 1992. Dynamics of a strongly radiating unsteady ethylene jet diffusion flame. *Combust. Flame* 96:1–21.

Kee, R. J. and H. A. Dwyer. 1981. Review of stiffness and implicit finite-difference methods in combustion modeling. *Prog. Astronautics Aeronautics* 76:485–500.

Keenan, J. H. and F. G. Keyes. 1963. *Thermodynamic Properties of Steam*. New York: Wiley.

Kent, J. H. and R. W. Bilger. 1972. Measurements of turbulent jet diffusion flames. Tech. Note F-41. Charles Kolling Research Laboratory. University of Sydney.

Lomax, H. and H. Baily. 1967. A Critical Analysis of Various Numerical Integration Methods for Computing the Flow of a Gas in Chemical Nonequilibrium. NASA TN D-4109. National Aeronautics and Space Administration.

Longwell, J. P. and M. A. Weiss. 1955. High temperature reaction rates in hydrocarbon combustion. *Ind. Eng. Chem.* 47:1634–1643.

Ludwig, C. B., W. Malkmus, J. E. Reardon, and J. A. L. Thompson. 1973. Handbook of Infrared Radiation from Combustion Gases. *Combust. Flame* 96: 1–21. NASA SP-3080. National Aeronautics and Space Administration.

Lydersen, A. L., R. A. Greenkorn, and O. A. Hougen. 1955. Generalized Thermodynamic Properties of Pure Substances. Engineering Experiment Station, Report No. 4. University of Wisconsin.

Magnus, D. E. and H. S. Schechter. 1966. Analysis of Error Growth and Stability for the Numerical Integration of the Equations of Chemical Kinetics. GASL TR-607. Westbury, NY: General Applied Sciences Laboratories, Inc.

Magnus, D. E. and H. S. Schechter. 1967. Analysis and Application of the Padé Approximation for the Integration of Chemical Kinetics Equations. GASL TR-642. Westbury, NY: General Applied Sciences Laboratories, Inc.

McAdams, W. H. 1954. *Heat Transmission*, 3rd ed. New York: McGraw-Hill.

McCarty, R. D. and L. A. Weber. 1971. Thermophysical Properties of Oxygen from the Freezing Liquid Line to 600 R for Pressures to 5000 Psia. NBS Technical Note 384. U.S. Department of Commerce.

McCarty, R. D. and L. A. Weber. 1972. Thermophysical Properties of Parahydrogen from the Freezing Liquid Line to 5000 R for Pressures to 10,000 Psia. NBS Technical Note 617. U.S. Department of Commerce.

Merk, H. J. 1959. The macroscopic equations for simultaneous heat and mass transfer in isotropic, continuous and closed systems. *Appl. Sci. Res.* A8:73–99.

Modest, M. F. 1993. *Radiative Heat Transfer*. New York: McGraw-Hill.

Moretti, G. 1965. A new technique for the numerical analysis of nonequilibrium flows. *AIAA J.* 3:223–229.

Oran, E. S. and J. P. Boris. 1987. *Numerical Simulation of Reactive Flow*. New York: Elsevier.

Perry, R. H., C. H. Chilton, and S. D. Kirkpatrick. 1963. *Chemical Engineers' Handbook*, 4th ed. New York: McGraw-Hill.

Pratt, D. T. and K. Radhakrishnan. 1984. CREK1D: A Computer Code for Transient Gas-Phase Combustion Kinetics. NASA TM-83806. National Aeronautics and Space Administration.

Radhakrishnan, K. 1984. Comparison of Numerical Techniques for Integration of Stiff Ordinary Differential Equations Arising in Combustion Chemistry. NASA TP-2372. National Aeronautics and Space Administration.

Radkrishnan, K. and D. A. Bittker. 1993. LSENS, A General Chemical Kinetics and Sensitivity Analysis Code for Gas-Phase Reactions: User's Guide. NASA TM 105851. National Aeronautics and Space Administration.

Reardon, J. E. and Y. C. Lee. 1979. A Computer Program for Thermal Radiation from Gaseous Rocket Exhaust Plumes (GASRAD). RTR 014-9. Huntsville: Remtech, Inc.

Reardon, J. E. and H. F. Nelson. 1994. Rocket plume base heating methodology. *J. Thermodyn. Heat Transfer*. 8:216–222.

Reid, R. C., J. M. Prausnitz, and B. E. Poling. 1987. *The Properties of Gases and Liquids*, 4th ed. Boston, MA: McGraw-Hill.

Reid, R. C., J. M. Prausnitz, and T. K. Sherwood. 1977. *The Properties of Gases and Liquids*, 3rd ed. Boston, MA: McGraw-Hill.

Rowlinson, J. S. 1969. *Liquids and Liquid Mixtures*, 2nd ed. Butterworth: London.

Sandler, S. I. 1999. *Chemical and Engineering Thermodynamics*, 3rd ed. New York: John Wiley and Sons.

Santoro, R. J., H. G. Semerjian, P. J. Emmerman, and R. Goulard. 1981. Optical tomography for flow field diagnostics. *Int. J. Heat Mass Transfer* 24:1139–1150.

Sherwood, T. K., R. L. Pigford, and C. R. Wilke. 1975. *Mass Transfer*. New York: McGraw-Hill.

Siegel, R. and J. R. Howell. 1992. *Thermal Radiation Heat Transfer*, 3rd ed. Washington: Taylor & Francis.

Sparrow, E. M. and R. D. Cess. 1966. *Radiation Heat Transfer*. Belmont: Brooks/Cole.

Uyehara, O. A. and K. M. Watson. 1944. A universal velocity correlation. *National Petroleum News* 36:764–770.

Varga, R. S. 1962. *Matrix Iterative Analysis*. Englewood Cliffs, NJ: Prentice-Hall.

White, F. M. 2006. *Viscous Fluid Flow*, 3rd ed. Boston, MA: McGraw-Hill.

Wilhoit, R. C. 1975. Ideal Gas Thermodynamic Functions. Thermodynamics Research Center. Texas A&M University.

Wilson, J. 2007. NASA's X-43A Scramjet Breaks Speed Record. *News Release 04-59*. National Aeronautics and Space Administration.

Yaws, C. L. 1999. *Chemical Properties Handbook*. New York: McGraw-Hill.

Young, T. R., Jr. 1980. A Subroutine for Solving Stiff Ordinary Equations. NRL-4091, Naval Research Laboratory. Washington.

Zeleznik, F. J. and B. J. McBride. 1984. Modeling the Internal Combustion Engine. NASA RP-1094. National Aeronautics and Space Administration.

CHAPTER 4

Turbulence Modeling Concepts

4.1 REYNOLDS AVERAGING AND EDDY VISCOSITY MODELS

Physically, turbulent flow has some meaning to most of us as some sort of chaotic fluid motion. Three concepts allow us to begin obtaining a quantitative measure of turbulence. In 1883, O. Reynolds observed that a dye jet injected cocurrently into water flowing in a smooth tube retained is jet-like stream until some critical velocity after which the dye stream rapidly broke-up and mixed with the water (McKusick and Wiskind 1959). A larger pressure drop in the pipe was also observed after this critical velocity was reached. Pursuing his studies on turbulence, Reynolds (1895) averaged the Navier–Stokes equations in time to define mean and fluctuating properties, i.e., the "Reynolds stresses." Prandtl in 1904 (Schlichting, 1979) explained the effects of friction as a modification of the mean velocity profile in the near vicinity to the wall, i.e., the boundary layer for laminar and turbulent flow. Thus, turbulent flow mixes more rapidly than laminar flow, has fluctuating fluid properties imposed on an otherwise laminar-like velocity field, and produces more friction when these ragged flows are slowed down by a wall to produce a no-slip condition. Turbulent flow has other characteristics, but these are the most important for influencing transport phenomena.

Early investigators defined turbulence as: "an irregular motion which in general makes its appearance in fluids, gaseous or liquid, when they flow past solid surfaces or even when neighboring streams of the same fluid flow past or over one another." Taylor and Von Karman initiated the application of statistical theories to describe these irregular motions. A collection of their works is compiled in Friedlander and Topper (1961). Hinze (1975) gave a more precise definition as: "Turbulent fluid motion is an irregular condition of flow in which the various quantities show random variation with time and space coordinates, so that statistically distinct average values can be discerned." An addition to these definitions attributed to Bradshaw (Wilcox, 2006) is that "turbulence has a wide range of scales." Bradshaw (1972) also noted: "The problem faced by an engineer, then, is to supply information missing from the time-averaged equations (Reynolds equations) by formulating a model to describe some or all of the six independent Reynolds stresses." Evaluating all of the Reynolds stresses is not possible, but some partial solutions do exist which are useful for analyzing transport phenomena.

Multidimensional convective and diffusive flow must be simulated to address transport problems. Mass and heat transfer problems require describing equilibrium, finite-rate chemical reactions, and real fluid thermodynamics. The character of the momentum transport demands that the interaction of the density and velocity fields be satisfactorily resolved through compressibility and sonic velocity considerations. When appropriate, selections from the multitude of experiments which have already been performed will be cited for providing validation of the various turbulent flow models reviewed. Historically, most research devoted to statistically describing turbulence has been directed toward describing constant density and ideal gas flow. As the statistical turbulence models become more detailed the chemical description of the flow suffers by continuing to be simplified. To obtain a practical simulation of transport phenomena processes, a compromise must be reached to use engineering models of all of the flow and fluid properties such that real processes and fairly accurate flowfields may be simultaneously predicted. Furthermore, to allow a novice to utilize and understand the numerical modeling and the physical restrictions of the model it is desirable that much of the analyses can be accomplished on ones own personal computer. Today this is possible with the eddy viscosity models described in this chapter. More elaborate turbulence models which cannot currently be used to accomplish this goal are being actively researched and are described in Chapter 5.

Even though the importance of the Reynolds stresses has long been known, models to represent them were limited to very simple flows, like boundary layers, until the latter part of the twentieth century. The simplicity of the models precluded them being made universal for application to a large class of flow geometries. With the advent of more computer power, more general models were developed. The first such models to be reasonably accurate and general were the two-equation models which used transport equations to describe the turbulence kinetic energy and energy dissipation. Such models were made even more general and computationally efficient by developing wall-function boundary conditions to expedite their use. These models have proven to be immensely useful to industry and government laboratories for the past quarter century. Many wish to believe that such technology has been superceded by newer turbulence models. This is not the case. The newer more elaborate models are not yet capable of simulating the physics and chemistry of interest, and they come at a higher computational and computer resource cost. These more elaborate models require a much longer learning curve to use and cannot generate sufficient analyses to serve as a learning tool for future analysts.

4.2 TURBULENCE CHARACTERISTICS

Time-averaged turbulent velocity, temperature, and concentration profiles for fully developed pipe flow were shown in Chapter 1. These data were obtained primarily with probe measurements. Such profiles are also those to be expected from boundary layers without wakes and channel flows. The fluids considered were mostly air and water. A few other fluids were studied to investigate the effects of high and low values of Prandtl and Schmidt number fluids. Multicomponent fluids have received very little fundamental study. Compilations of boundary layer and free shear layer data were made for conferences held to compare turbulence model predictions to these data. The boundary layer conference was held at Stanford University (Coles, 1968) and the shear layer conference was held at NASA Langley (Birch et al. 1972). More recent data of this type are reviewed by White (2006) and Wilcox (2006).

Turbulent properties are indicated by fluctuating velocity and pressure measurements by Laufer for air flow in pipes and by Klebanoff for boundary layers. Such fluctuations are always three dimensional. These data and additional measurements have been reviewed by Hinze (1975). These correlations are measured by hot-wire anemometry. The preferred technique observes the heat loss from a small fine wire by maintaining the wire temperature and resistance constant. The current required to maintain this temperature constant is measured, and the instantaneous velocity is deduced from a heat transfer analysis. Hinze (1975) reviews some more recent data, but these experiments are costly to perform because of frequent breakage of the fine wire. In any event, the data show that near the wall the three components of the "turbulent intensity," i.e., the root-mean-square of the velocity fluctuations in each of the three directions are not equal. If the intensity were constant in all directions the flow would be termed isotropic. The intensities become closer to being equal slightly away from the wall. This means that the near-wall turbulence is "anisotropic." This is very inconvenient because it increases the difficulty of producing a very accurate turbulence model. Regardless of the intensity distribution around a point, if such intensities do not vary from point-to-point, the turbulence is said to be "homogeneous."

Turbulence structure consists of coexistent eddies of varying sizes which degrade into smaller and smaller eddies until finally dissipating as heat. This cascade involves the transfer of the (specific) turbulent kinetic energy (k) from the larger eddies to the smaller eddies. Before describing how such eddies can be quantitatively defined, some of their other properties will be considered. Kolmogorov's pioneering work (Hunt et al., 1991) and reviewed by Wilcox (2006) described the flow very near a wall as being locally isotropic and describable by the dimensional analysis to length (ℓ), time (t), and velocity scales (υ) in terms of the energy dissipation $\varepsilon = -dk/dt$ and kinematic viscosity ν.

$$\ell_K \equiv (\nu^3/\varepsilon)^{1/4} \quad t_K \equiv (\nu/\varepsilon)^{1/2} \quad \upsilon_K \equiv (\nu/\varepsilon)^{1/4} \tag{4.1}$$

Taylor (1935) defined an integral length scale (ℓ) by

$$\varepsilon \sim \frac{k^{3/2}}{\ell} \Rightarrow k \sim (\varepsilon\ell)^{2/3}; \quad t_t \sim \frac{k}{\varepsilon}; \quad \ell \gg \ell_K \gg \ell_{mfp} \tag{4.2}$$

where ℓ_{mfp} is the mean free path of the molecules in the fluid. The ratio of large and small length and timescales are

$$\frac{\ell}{\ell_K} = \frac{\ell}{(\mu^3/\rho\varepsilon)^{1/4}} \sim Re_t^{3/4}; \quad \frac{t_t}{t_K} \sim Re_t^{1/2}; \quad \text{where} \quad Re_t \equiv \frac{\rho\sqrt{k}\ell}{\mu} \gg 1 \tag{4.3}$$

By using the local length scale, a wave number (α) is defined to give the energy spectrum function $E\{\alpha\}$ which is related to the Fourier transform of k.

$$k \equiv \sum_i \frac{1}{2}\overline{V_i'V_i'} = \int_0^\infty E(\alpha)\,d\alpha \tag{4.4}$$

Kolmogorov's universal equilibrium theory states the rate that the smallest scale eddies receive energy from the larger scale eddies is approximately equal to the rate that the smallest scale eddied dissipate energy into heat. This theory results in an inertial subrange of eddy sizes and is expressed by his 5/3 power law.

$$E\{\alpha\} = C_K \varepsilon^{2/3} \alpha^{-5/3} \quad \frac{1}{\ell} \ll \alpha = \frac{2\pi}{\lambda} \ll \frac{1}{\ell_K} \tag{4.5}$$

Afzal and Narasimha, as reviewed by Wilcox (2006), extend this to create
Inner region (viscous range):

$$E\{\alpha\} = \varepsilon^{1/4} v^{5/4} f\{\alpha \ell_K\} \quad \ell_K \equiv (v^3/\varepsilon)^{1/4} \tag{4.6}$$

Outer region (large eddies):

$$E\{\alpha\} = k\ell g\{\alpha \ell\} \tag{4.7}$$

Overlapping region (inertial subrange):

$$E\{\alpha\} = C_K \varepsilon^{2/3} \alpha^{-5/3} \tag{4.8}$$

Typical plots of energy spectra are shown in Pope (2000). A specific energy spectrum is given by the following equations from Pope which contain no unspecified functions.

$$E\{\alpha\} = C\varepsilon^{2/3} \alpha^{-5/3} f_L\{\alpha L\} f_\eta\{\alpha \eta\} \tag{4.9}$$

where

$$f_L\{\alpha L\} = \left(\frac{\alpha L}{\left[(\alpha L)^2 + c_L \right]^{0.5}} \right)^{5/3 + p_o}$$

$$f_\eta\{\alpha L\} = \exp\{-\beta \left(\left[(\alpha \eta)^4 + c_\eta^4 \right]^{0.25} - c_\eta \right)\}$$

Assumptions made by Kolmogorov are that the turbulence has a wide separation of scales and that the smaller eddies are in a state where the rate of receiving energy from the large eddies is nearly equal to the rate at which the smallest eddies dissipate the energy to heat. The motion at the smallest scales depends on the rate at which the larger eddies supply energy,

$$\varepsilon = -dk/dt \tag{4.10}$$

Statistical methods must be used to define a length scale more rigorously. Single-point correlations of turbulent eddies depend on turbulent fluctuations at a given point. Unlike the molecular motion of gases, the motion at any point in a turbulent flow affects the motion at other distant points through the pressure field. Meaningful length and timescales can be defined and measured by using at a minimum, two-point correlations. There are two types of such correlations: a temporal separation and a spatial separation. These are the

- Autocorrelation tensor

$$A_{ij}(\vec{r},t;t') \equiv \overline{u_i'(\vec{r},t)u_j'(\vec{r},t+t')} \tag{4.11}$$

- Integral timescale

$$t_1(\vec{r},t) = \int_0^\infty \frac{A_{ii}(\vec{r},t;t')}{2k(\vec{r},t)} \, dt' \quad \text{where} \quad k(\vec{r},t) = \frac{1}{2}A_{ii}(\vec{r},t;0) \tag{4.12}$$

- Two-point correlation tensor in terms of the displacement vector $\vec{s}$

$$V_{ij}(\vec{r},t;\vec{s}) \equiv \overline{u_i'(\vec{r},t)u_j'(\vec{r}+\vec{s},t)} \tag{4.13}$$

- Integral length scale

$$\ell(\vec{r},t) \equiv \frac{3}{16}\int_0^\infty \frac{V_{ii}(\vec{r},t;s)}{2k(\vec{r},t)} \, ds \quad \text{where} \quad s = |\vec{s}| \tag{4.14}$$

The use of time-averaged conservation laws does not consider any spatial correlations of the fluctuating parameters. This is an especially restrictive limitation when one is dealing with environmental transport phenomena. Leonard (1974) introduced spatially averaged conservation equations to help remedy this problem. Such averaging is also referred to as filtering of the conservation equations. In essence, spatial averaging allows two levels of viscous dissipation: one larger level for near-wall flows and another smaller one for large eddies far away from walls. Further discussion of spatial averaging will be presented in Chapter 5 as it pertains to other turbulence models, particularly to DNS and LES models.

Measurements of the parameters just mentioned are sufficient to validate turbulence models needed to simulate transport phenomena processes, but such data are not generally available for real geometric configurations.

Such data are essentially nonexistent for reacting and multiphase flows. The result of this shortage is that most computational simulations are "validated" by comparison to some experiment which is (hopefully) somewhat similar to the desired phenomena. This is not a good situation, but its practice still leads to useful insight to the desired goal of CTP analysis—within acceptable engineering tolerance. Multiyear CFD Working Group conferences at NASA MSFC and NASA GRC's Center for Modeling of Turbulence and Transition have produced numerous studies of this nature to support their major programs. The proceedings of these conferences are reported in NASA publications. Numerous similar conferences sponsored by other agencies/organizations have also been held. A host of more formal publications of similar researches have created an enormous literature.

The remainder of this chapter discusses: Reynolds and Favre averaging, two-equation turbulence models, and the importance of wall-function boundary conditions. Where appropriate, test data to support model validity will be cited. However, the objective of this work is to delineate the concepts of the various turbulence modeling methods. A basic computational tool for the readers' use in making their own evaluation of turbulent modeling methodology is offered and its use is explained. A definitive validation of the methodology with a critique of specific data is not the goal.

4.3 REYNOLDS AND FAVRE AVERAGING

A turbulent velocity measured with an instrument with a slow response time (like a pitot tube) indicates a time-averaged velocity, $\vec{V}$ (or velocity component), while a measurement with a small detector and a rapid response time (like a hot-wire anemometer) measures a highly fluctuating velocity, $\vec{V} + \vec{V}'$. The difference between the two types of measurements indicates the structure of the turbulence. For example, the average of this difference squared is termed the intensity of the turbulence. Similar properties involving pressure, temperature, and concentration may also be measured. Turbulent concentration modeling must be accomplished with caution, since most reaction kinetics data are obtained from experiments (possibility turbulent). This is especially true for gaseous combustion. Not only may meaningful average values of the conserved quantities be defined, but since the turbulent fluctuations are rapid, the more slowly varying mean flowfields may be determined to describe a transient flowfield. The conservation laws must be modified to reflect these two types of flowfield properties in order to analyze turbulent flows.

To define the mean value of some quantity (like a velocity component), time- or space-averaging over some small volume or time increment is most easily measured. Such averaging is defined by Monin and Yaglom (1965):

$$\left\langle f\left\{x_1, x_2, x_3, t\right\}\right\rangle = \int\int\int\limits_{-\infty}^{\infty}\int f\left\{x_1 - \xi_1, x_2 - \xi_2, x_3 - \xi_3, t - \tau\right\}$$
$$\times \omega\left\{\xi_1, \xi_2, \xi_3, \tau\right\} d\xi_1\, d\xi_2\, d\xi_3\, d\tau \tag{4.15}$$

where

x_i's are space coordinates

t is the time

The broken brackets indicate an average quantity

The braces indicate functionality (brackets are not needed when two different symbols (like U and u) are used to indicate mean and fluctuating quantities)

ξ_i's are the small displacements from the corresponding x_i

τ is the small displacement in time.

Weighting function (ω) is chosen to satisfy the normalization condition

$$\int\int\int\limits_{-\infty}^{\infty}\int \omega\left\{\xi_1, \xi_2, \xi_3, \tau\right\} d\xi_1\, d\xi_2\, d\xi_3\, d\tau = 1 \tag{4.16}$$

If the weighting function is constant over some four-dimensional region and zero outside this region, the averaging process is said to be simple. Furthermore, if $\omega = \delta$, the Dirac delta-function, simple space- or time-averaging, respectively, is obtained. Reynolds originated the concept of time-averaging the conservation equations for a constant density fluid and identified important averages of various fluctuating quantities. Certain general rules result from this analysis when applied to products of fluctuating quantities. However, to generalize the analysis a method of evaluating the weighting function must be devised. This is far from a simple requirement. Favre (1961) extended the time-averaging concept to define similar functions for variable density flows.

The three type of averaging techniques most frequently used to describe turbulence are: (1) a time average: appropriate for stationary turbulence (i.e., on the average the turbulent flow does not vary with time),

$$\Phi_t(X_j) = \lim_{\Delta t \to \infty} \frac{1}{\Delta t} \int\limits_{t}^{t+\Delta t} \varphi(X_j, t)\, dt \tag{4.17}$$

(2) the spatial average used for homogeneous turbulence (on the average the turbulent flow is uniform in all directions),

$$\Phi_{V}(t) = \lim_{V \to \infty} \frac{1}{V} \iiint_{V} \varphi(X_j, t) \, dV \tag{4.18}$$

and (3) the ensemble average, the most general type of averaging, is especially useful for representing unsteady flow,

$$\Phi_{E}(X_j, t) = \lim_{N \to \infty} \frac{1}{N} \sum_{i=1}^{N} \varphi_i(X_j, t) \tag{4.19}$$

The "ergodic hypothesis" states that for turbulence which is both stationary and homogeneous, these three averages may be assumed equal.

The turbulent flow properties are random variables (Φ) which can be averaged in a variety of ways. The examples just presented represent averages depending on the definition of the weighting function and its interpretation as a probability.

To expedite further discussion of turbulence models, Appendix 4.A is included to define basic statistical terms. However, for simple time-averaging as used by Reynolds:

$$\overline{\varphi}(X_j, t) = \frac{1}{\Delta t} \int_{t}^{t+\Delta t} \varphi(X_j, t) \, dt \quad t_{\text{turb}} \ll \Delta t \ll t_{\text{flow}}$$

$$\varphi(X_j, t) = \overline{\varphi}(X_j, t) + \varphi'(X_j, t) \tag{4.20}$$

where
  φ is the instantaneous variable
  $\overline{\varphi}$ is the mean variable
  φ' is the fluctuating variable

Rules for time-averaging:

$$\overline{\overline{\varphi}} = \overline{\overline{\varphi} + \varphi'} = \overline{\varphi} + \overline{\varphi'} \Rightarrow \overline{\varphi'} = 0$$

$$\overline{\varphi + \psi} = \overline{\varphi} + \overline{\psi}$$

$$\overline{\varphi \psi} = \overline{\varphi}\,\overline{\psi} + \overline{\varphi' \psi'}$$

$$\overline{\frac{\partial \varphi}{\partial X_i}} = \frac{\partial \overline{\varphi}}{\partial X_i} \quad \overline{\frac{\partial \varphi}{\partial t}} = \frac{\partial \overline{\varphi}}{\partial t}$$

$$\overline{\int \varphi \, dX_i} = \int \overline{\varphi} \, dX_i \tag{4.21}$$

If t_{flow} of mean flow unsteadiness is not much larger than t_{turb}, a time large enough to establish a mean for the turbulent fluctuation, an alternative method needs to be used.

Reynolds decomposition consists of applying these rules to the conservation equations to generate a set of conservation laws for turbulent flow. The time-averaging rules just stated are restricted to flows of constant density fluids. In the literature, such flows are frequently termed incompressible. This is a misnomer, since compressibility really refers to flows with density changes due to compression—not due to temperature or concentrations changes. However, this practice will be used herein. The velocity and pressure terms have mean and fluctuating components as shown in Equation A of Table 4.1. The continuity and three momentum equations from the Reynolds decomposition are shown as Equations B through E in Table 4.1. The averages of the fluctuating velocity component paired-products are the only new terms which appear in these equations. Such terms are called Reynolds stresses and they represent the components of a second-order tensor. Notice, the fluctuating pressure does not appear in these equations. The turbulent kinetic energy is determined by Equation F. Notice that this equation is not closed even if all of the Reynolds stress components are known. This means that additional modeling terms would be needed to solve Equation F. The turbulent shear-stress is comprised of the nine components shown in Equation G. The cross-components are equal so that only six products must be evaluated. The basic definitions of intensity and turbulent kinetic energy are given in Equations H and I, respectively. These definitions for the Favre-averaged equations are also shown in these equations for convenience. An extra continuity equation results involving the fluctuating velocity components. This equation is not used for most turbulent transport process analyses. The equations in Table 4.1 are for a Cartesian coordinate system only. Generalizations to other coordinate systems will be discussed subsequently. The Reynolds-averaged conservation equations are referred to as the (Reynolds-averaged Navier–Stokes turbulence model [RANS]) turbulence model. This is somewhat of a misnomer since species and energy conservation equations and variable density flows are also included in the RANS model.

Favre included density variations and mass-(time-)averaged the conservation equations by using the following definitions and averaging rules.

TABLE 4.1 Reynolds-Averaged Incompressible Conservation Laws in Cartesian Coordinates

$$V_x = \bar{V}_x + V'_x; \quad V_y = \bar{V}_y + V'_y; \quad V_z = \bar{V}_z + V'_z; \quad P = \bar{P} + P' \tag{A}$$

$$\frac{\partial \bar{V}_x}{\partial x} + \frac{\partial \bar{V}_y}{\partial y} + \frac{\partial \bar{V}_z}{\partial z} = 0; \quad \text{and} \quad \frac{\partial V'_x}{\partial x} + \frac{\partial V'_y}{\partial y} + \frac{\partial V'_z}{\partial z} = 0 \tag{B}$$

$$\rho\left[\frac{\partial \bar{V}_x}{\partial t} + \bar{V}_x\frac{\partial \bar{V}_x}{\partial x} + \bar{V}_y\frac{\partial \bar{V}_x}{\partial y} + \bar{V}_z\frac{\partial \bar{V}_x}{\partial z}\right] = -\frac{\partial \bar{P}}{\partial x} + (\nabla\cdot\bar{\tau})_x - \rho\left[\frac{\partial \overline{V'_x V'_x}}{\partial x} + \frac{\partial \overline{V'_y V'_x}}{\partial y} + \frac{\partial \overline{V'_z V'_x}}{\partial z}\right] + \rho g_x + \bar{F}_x \tag{C}$$

$$\rho\left[\frac{\partial \bar{V}_y}{\partial t} + \bar{V}_x\frac{\partial \bar{V}_y}{\partial x} + \bar{V}_y\frac{\partial \bar{V}_y}{\partial y} + \bar{V}_z\frac{\partial \bar{V}_y}{\partial z}\right] = -\frac{\partial \bar{P}}{\partial y} + (\nabla\cdot\bar{\tau})_y - \rho\left[\frac{\partial \overline{V'_x V'_y}}{\partial x} + \frac{\partial \overline{V'_y V'_y}}{\partial y} + \frac{\partial \overline{V'_z V'_y}}{\partial z}\right] + \rho g_y + \bar{F}_y \tag{D}$$

$$\rho\left[\frac{\partial \bar{V}_z}{\partial t} + \bar{V}_x\frac{\partial \bar{V}_z}{\partial x} + \bar{V}_y\frac{\partial \bar{V}_z}{\partial y} + \bar{V}_z\frac{\partial \bar{V}_z}{\partial z}\right] = -\frac{\partial \bar{P}}{\partial z} + (\nabla\cdot\bar{\tau})_z - \rho\left[\frac{\partial \overline{V'_x V'_z}}{\partial x} + \frac{\partial \overline{V'_y V'_z}}{\partial y} + \frac{\partial \overline{V'_z V'_z}}{\partial z}\right] + \rho g_z + \bar{F}_z \tag{E}$$

$$\frac{\partial k}{\partial t} + \rho\sum_i \bar{V}_i\frac{\partial k}{\partial X_i} = -\rho\sum_i\sum_j \overline{V'_i V'_j}\frac{\partial \bar{V}_i}{\partial X_j} + \sum_i \overline{P'\frac{\partial V'_i}{\partial X_i}} - \sum_i\sum_j \mu\overline{\frac{\partial V'_i}{\partial X_j}\frac{\partial V'_i}{\partial X_j}} - \frac{\partial}{\partial X_j}\left[\overline{V'_j\left(P' + \sum_i \tfrac{1}{2}\rho V'_i V'_i\right)} - \mu\frac{\partial k}{\partial X_i}\right] \tag{F}$$

$$\ddot{\bar{\tau}}_t = \begin{pmatrix}(\tau_t)_{xx} & (\tau_t)_{xy} & (\tau_t)_{xz}\\ (\tau_t)_{yx} & (\tau_t)_{yy} & (\tau_t)_{yz}\\ (\tau_t)_{zx} & (\tau_t)_{zy} & (\tau_t)_{zz}\end{pmatrix} = -\rho\begin{pmatrix}\overline{V'_x V'_x} & \overline{V'_x V'_y} & \overline{V'_x V'_z}\\ \overline{V'_y V'_x} & \overline{V'_y V'_y} & \overline{V'_y V'_z}\\ \overline{V'_z V'_x} & \overline{V'_z V'_y} & \overline{V'_z V'_z}\end{pmatrix} \quad \text{or} \quad = -\begin{pmatrix}\overline{\rho V''_x V''_x} & \overline{\rho V''_x V''_y} & \overline{\rho V''_x V''_z}\\ \overline{\rho V''_y V''_x} & \overline{\rho V''_y V''_y} & \overline{\rho V''_y V''_z}\\ \overline{\rho V''_z V''_x} & \overline{\rho V''_z V''_y} & \overline{\rho V''_z V''_z}\end{pmatrix} \tag{G}$$

(continued)

TABLE 4.1 (continued) Reynolds-Averaged Incompressible Conservation Laws in Cartesian Coordinates

$$I_x = \frac{\sqrt{\overline{V'_x V'_x}}}{|\overline{V}|}; \quad I_y = \frac{\sqrt{\overline{V'_y V'_y}}}{|\overline{V}|}; \quad I_z = \frac{\sqrt{\overline{V'_z V'_z}}}{|\overline{V}|}; \quad I_t = \sqrt{\frac{\frac{1}{3}\left(\overline{V'_x V'_x}+\overline{V'_y V'_y}+\overline{V'_z V'_z}\right)}{|\overline{V}|}}$$ (H)

$$k = \frac{\overline{V'_x V'_x}+\overline{V'_y V'_y}+\overline{V'_z V'_z}}{2}=\frac{1}{2}\sum_i \overline{V'_i V'_i} \quad \text{or} \quad \frac{\rho\overline{V''_x V''_x}+\rho\overline{V''_y V''_y}+\rho\overline{V''_z V''_z}}{2\overline{\rho}}=\frac{\sum_i \overline{\rho V''_i V''_i}}{2\overline{\rho}}$$ (I)

$$\tilde{\varphi}(X_j,t) \equiv \frac{1}{\bar{\rho}}\frac{1}{\Delta t}\int_t^{t+\Delta t}\rho(X_j,t)\varphi(X_j,t)\,dt = \frac{\overline{\rho\varphi}}{\bar{\rho}}$$

$$\bar{\rho}(X_j,t) = \frac{1}{\Delta t}\int_t^{t+\Delta t}\rho(X_j,t)\,dt \qquad (4.22)$$

$$\varphi(X_j,t) = \overline{\varphi}(X_j,t) + \varphi'(X_j,t) = \tilde{\varphi}(X_j,t) + \varphi''(X_j,t)$$

φ represent the instantaneous variables
$\tilde{\varphi}$ represent the mass–weight mean variables
φ" represent the superimposed fluctuating variables
$\bar{\rho}$ represent the time-mean density

$$\overline{\rho\varphi} = \overline{\rho(\tilde{\varphi}+\varphi'')} = \overline{\bar{\rho}\tilde{\varphi}} + \overline{\rho\varphi''} \Rightarrow \overline{\rho\varphi''} = 0$$

$$\tilde{\varphi} - \bar{\varphi} = -\overline{\varphi''} = \frac{\overline{\rho'\varphi''}}{\bar{\rho}} = \frac{\overline{\rho'\varphi'}}{\bar{\rho}}$$

$$\frac{\partial\overline{\rho\varphi}}{\partial X_i} = \frac{\partial\overline{\rho\varphi}}{\partial X_i} = \frac{\partial\overline{\bar{\rho}\tilde{\varphi}}}{\partial X_i} \quad \frac{\partial\overline{\rho\varphi}}{\partial t} = \frac{\partial\overline{\rho\varphi}}{\partial t} = \frac{\partial\overline{\bar{\rho}\tilde{\varphi}}}{\partial t} \qquad (4.23)$$

$$\int\overline{\rho\varphi}\,dX_i = \int\overline{\rho\varphi}\,dX_i = \int\overline{\bar{\rho}\tilde{\varphi}}\,dX_i$$

The resulting mass-averaged conservation laws and fluctuating property definitions are shown in Table 4.2, as equations for compressible flow. Equations A through E correspond exactly to the same equations in Table 4.1, except for the definition of the mean fluctuating terms. These terms have been modified to reflect a variation in density. The extra continuity equation does not appear in this formulation of the conservation laws. The Reynolds shear-stress terms now include the variable density, but there are still the nine components, six of which must be determined. Since the density is now a variable, the species continuity equation (Equation F) and the energy variables and the energy conservation law (Equation G) must be utilized to complete description of the transport process. The modifications needed to define the variable density turbulent kinetic energy are shown in Equation I.

The Reynolds stress terms in either the incompressible or compressible form of the conservation are known, hence the equations are not closed. There are more unknowns than equations in the system.

TABLE 4.2 Favre-Averaged Compressible Conservation Equations in Cartesian Coordinates

$$V_x = \tilde{V}_x + V''_x; \quad V_y = \tilde{V}_y + V''_y; \quad V_z = \tilde{V}_z + V''_z; \quad \rho = \bar{\rho} + \rho'; \quad P = \bar{P} + P' \tag{A}$$

$$\frac{\partial \bar{\rho}}{\partial t} + \frac{\partial \bar{\rho}\tilde{V}_x}{\partial x} + \frac{\partial \bar{\rho}\tilde{V}_y}{\partial y} + \frac{\partial \bar{\rho}\tilde{V}_z}{\partial z} = 0 \tag{B}$$

$$\frac{\partial \bar{\rho}\tilde{V}_x}{\partial t} + \frac{\partial \bar{\rho}\tilde{V}_x\tilde{V}_x}{\partial x} + \frac{\partial \bar{\rho}\tilde{V}_x\tilde{V}_y}{\partial y} + \frac{\partial \bar{\rho}\tilde{V}_x\tilde{V}_z}{\partial z} = -\frac{\partial \bar{P}}{\partial x} + (\nabla \cdot \bar{\tau})_x - \frac{\partial \overline{\rho V''_x V''_x}}{\partial x} - \frac{\partial \overline{\rho V''_y V''_x}}{\partial y} - \frac{\partial \overline{\rho V''_z V''_x}}{\partial z} + \bar{\rho}g_x + \bar{F}_x \tag{C}$$

$$\frac{\partial \bar{\rho}\tilde{V}_y}{\partial t} + \frac{\partial \bar{\rho}\tilde{V}_y\tilde{V}_x}{\partial x} + \frac{\partial \bar{\rho}\tilde{V}_y\tilde{V}_y}{\partial y} + \frac{\partial \bar{\rho}\tilde{V}_y\tilde{V}_z}{\partial z} = -\frac{\partial \bar{P}}{\partial y} + (\nabla \cdot \bar{\tau})_y - \frac{\partial \overline{\rho V''_x V''_y}}{\partial x} - \frac{\partial \overline{\rho V''_y V''_y}}{\partial y} - \frac{\partial \overline{\rho V''_z V''_y}}{\partial z} + \bar{\rho}g_y + \bar{F}_y \tag{D}$$

$$\frac{\partial \bar{\rho}\tilde{V}_z}{\partial t} + \frac{\partial \bar{\rho}\tilde{V}_z\tilde{V}_x}{\partial x} + \frac{\partial \bar{\rho}\tilde{V}_z\tilde{V}_y}{\partial y} + \frac{\partial \bar{\rho}\tilde{V}_z\tilde{V}_z}{\partial z} = -\frac{\partial \bar{P}}{\partial z} + (\nabla \cdot \bar{\tau})_z - \frac{\partial \overline{\rho V''_x V''_z}}{\partial x} - \frac{\partial \overline{\rho V''_y V''_z}}{\partial y} - \frac{\partial \overline{\rho V''_z V''_z}}{\partial z} + \bar{\rho}g_z + \bar{F}_z \tag{E}$$

$$\frac{\partial \bar{\rho}\tilde{\alpha}_i}{\partial t} + \sum_j \frac{\partial \bar{\rho}\tilde{V}_j\tilde{\alpha}_i}{\partial X_j} - \sum_j \frac{\partial}{\partial X_j}\left(\frac{\mu}{S_c}\sum_j \frac{\partial \tilde{\alpha}_i}{\partial X_j} - \overline{\rho V''_j \alpha''_i}\right) = \dot{\bar{\omega}}_i; \quad \alpha_i = \tilde{\alpha}_i + \alpha''_i; \quad \tilde{\alpha}_i = \frac{\overline{\rho\alpha_i}}{\bar{\rho}} \tag{F}$$

$$h = \tilde{h} + h''; \quad \tilde{e} + e''; \quad T = \tilde{T} + T''; \quad \tilde{h} = \frac{\overline{\rho h}}{\bar{\rho}}; \quad \tilde{e} = \frac{\overline{\rho e}}{\bar{\rho}}; \quad \tilde{T} = \frac{\overline{\rho T}}{\bar{\rho}} \tag{G}$$

$$vc\;\; \frac{\partial \bar{\rho}\tilde{h}}{\partial t} + \sum_i \frac{\partial \bar{\rho}\tilde{V}_i\tilde{h}}{\partial X_i} = \frac{\partial \bar{P}}{\partial t} + \sum_i \tilde{V}_i\frac{\partial \bar{P}}{\partial X_i} + \sum_i \overline{V''_i\frac{\partial P'}{\partial X_i}} - \sum_i \frac{\partial \bar{q}_i}{\partial X_i}\delta_{ij} - \sum_i \frac{\partial}{\partial X_i}\overline{\rho V''_i h''} + \sum_i\sum_j \overline{\left[\tau_{ij}\frac{\partial V_j}{\partial X_i} + \tau_{ij}\frac{\partial V''_j}{\partial X_i}\right]} \tag{H}$$

$$\frac{\partial \bar{\rho}k}{\partial t} + \sum_i \frac{\partial \bar{\rho}\tilde{V}_i k}{\partial X_i} = -\sum_i\sum_j \overline{\rho V''_i V''_j}\frac{\partial \tilde{V}_i}{\partial X_j} + \sum_i\sum_j \overline{\mu\frac{\partial V''_i}{\partial X_j}\frac{\partial V''_i}{\partial X_j}} - \sum_i \overline{P'\frac{\partial V''_i}{\partial X_i}} - \sum_j \frac{\partial}{\partial X_j}\overline{\left[V''_j\left(P'+\frac{1}{2}\sum_i \rho V''_i V''_i\right) + \tau_{ij}V''_i\right]} - \mu\frac{\partial k}{\partial X_j} \tag{I}$$

Closure problem

- Due to the nonlinearity of the Navier–Stokes equation, successively higher moments generate additional unknowns at each level.
- The Reynolds–Favre averaging process is strictly mathematical in nature, and introduces no additional physical principles.
- The function of turbulence modeling is to devise approximations for the unknown correlations in terms of flow properties that are known so that a sufficient number of equations will be generated.

Although these averaging procedures are very useful, they do not use probability methodology to analyze the conservation equations. Also, much of the experimental hot-wire validation data are averaged electronically using simple analog circuitry, thus precluding their use for validating more elaborate turbulence models. This pragmatic approach to the study of turbulence continues to be widely and successfully used. The advantages offered by the application of probability theory are yet to be realized, but are addressed in Chapter 5 to indicate their potential.

4.4 EDDY VISCOSITY MODELS

Both the Reynolds- and Favre-averaged conservation equations contain correlations (i.e., averaged values) of scalar quantities which are products of fluctuating velocity components, various energy and concentration variables. These correlations cannot be accurately predicted from fundamental principles. Studies over the past century measured and analyzed increasingly complex flows to develop an understanding and modeling methodology for these turbulent flow properties. The flow complexity evolved from boundary layers, pipe flows, and homogeneous wind tunnel airstreams and essentially stopped there. More complex flows and processes were usually studied by performing subscale experiments, in which few fundamental turbulent property measurements were ever attempted. Much esoteric turbulence research is still being enthusiastically, widely, and often cleverly pursued by a host of investigators who are far from producing practical, efficient design tools. The problem they address is difficult.

On the other hand the exoteric practitioner needs to produce engineering design information. For turbulent flow phenomena, this usually means utilizing eddy viscosity flow models and gradient-diffusion heat and mass transfer for process simulations. Such methodology has been much maligned in the literature by those seeking research dollars for a livelihood, but its use has been spectacularly successful.

4.4.1 REYNOLDS STRESSES AND THE STANDARD k–ε MODEL

The correlations of fluctuating velocity component paired-products and of the velocity component products with scalar temperature and mass fraction fluctuations must be evaluated to analyze turbulent transport. The laminar transport coefficients of viscosity, diffusion, and thermal conductivity relate the transport processes to the flowfield in a rigorous manner. In 1877, Boussinesq suggested a turbulence model which simply assumed that laminar-like transport coefficient could be used to describe the turbulence. Before this was done, the velocity fluctuation correlations were recognized as apparent stresses caused by the turbulence. Notice in Tables 4.1 and 4.2, these terms look like the laminar shear-stress terms and employ a minus sign so that they are of the same sign as the laminar shear-stress components, thus: $(\tau_t)_{ij} = -\rho\langle V_i' V_j'\rangle$ or $-\langle \overline{\rho} V_i'' V_j''\rangle$. These are the Reynolds stresses for incompressible and compressible flow, respectively. The Reynolds stresses are elements of a second-order tensor. This tensor is symmetric, so there are only six unknown stresses. However, they are unknown and must be modeled, or an alternative found. Utilizing the Boussinesq analogy

$$(\tau_t)_{ij} = -\rho\overline{V_i'V_j'} = \mu_t\left(\frac{\partial \overline{V}_i}{\partial X_j} + \frac{\partial \overline{V}_j}{\partial X_i} - \frac{2}{3}\sum_k \frac{\partial \overline{V}_k}{\partial X_k}\delta_{ij}\right) - \frac{2}{3}\rho k\delta_{ij}$$

$$(\tau_t)_{ij} = -\rho\overline{V_i''V_j''} = \mu_t\left(\frac{\partial \tilde{V}_i}{\partial X_j} + \frac{\partial \tilde{V}_j}{\partial X_i} - \frac{2}{3}\sum_k \frac{\partial \tilde{V}_k}{\partial X_k}\delta_{ij}\right) - \frac{2}{3}\rho k\delta_{ij}$$

$$\vec{q}_t = (q_t)_i\delta_i = \overline{\rho V_i''h''}\delta_i = -\lambda_t\left(\frac{\partial \tilde{T}}{\partial X_i}\right)\delta_i = -\frac{\mu_t}{Pr_t}C_p\left(\frac{\partial \tilde{T}}{\partial X_i}\right)\delta_i = -\frac{\mu_t}{Pr_t}\left(\frac{\partial \tilde{h}}{\partial X_i}\right)\delta_i$$

$$\overline{\rho V_j''\alpha_i''} = -\rho D_t\frac{\partial \tilde{\alpha}_i}{\partial X_j}\delta_j = -\frac{\mu_t}{Sc_t}\frac{\partial \tilde{\alpha}_i}{\partial X_j}\delta_j$$

$$(4.24)$$

where

μ_t is the eddy viscosity
λ_t is the eddy conductivity
D_t is the eddy mass diffusivity
Pr_t is the turbulent Prandtl number
Sc_t is the eddy Schmidt number

The laminar kinematic viscosity as determined by elementary kinetic theory is proportional to the molecular velocity times the mean free path of the molecules. In 1925, Prandtl took this model to represent the eddy viscosity by replacing the molecular velocity with the mean velocity of the flow and the mean free path with a mixing length. This model correlated data, but different mixing lengths had to be specified for different flow geometries and in boundary layers three such lengths had to be used to model the entire boundary layer. This does not provide any generality in the turbulence model. Kolmogorov in 1941 and Prandtl in 1945 replaced the average fluid velocity with the square root of the turbulent kinetic energy (Hunt, Phillips, and Williams 1991). The length parameter in the two-equation type models has been replaced by the parameters: turbulent kinetic energy dissipation (ε) or the dissipation rate per unit of turbulent kinetic energy (ω). Other second parameters have also been suggested. In any event, two-transport equations are solved for the two parameters. The use of the eddy viscosity to utilize calculated values of k and ε means that the model simulates only isotropic turbulence. The "standard k–ε model" is that developed by Jones and Launder (1972). The greatest variation in modified formulations of two-equation models is in the choice of the second parameter to be used. The choice of which two parameters are to be used and the form of the transport equations to be solved for the best or most general solution depends mostly on the experience and level of effort which have been devoted to their study. The relationships between these parameters are

$$\nu_t \sim \frac{k^2}{\varepsilon}; \quad \ell \sim \frac{k^{3/2}}{\varepsilon} \Rightarrow \mu_t = \bar{\rho} C_\mu \frac{k^2}{\varepsilon}; \quad \ell = C_\mu \frac{k^{3/2}}{\varepsilon}$$

$$\nu_t \sim \frac{k}{\omega}; \quad \ell \sim \frac{\sqrt{k}}{\omega}; \quad \varepsilon \sim \omega k \Rightarrow \omega = \frac{\varepsilon}{C_\mu k} = \frac{\sqrt{k}}{\ell}; \quad \mu_t = \bar{\rho} \frac{k}{\omega}$$

(4.25)

The Jones and Launder k–ε turbulence model is

$$\mu_t = \bar{\rho} C_\mu k^2 / \varepsilon$$

$$\bar{\rho}\frac{\partial k}{\partial t} + \sum_j \bar{\rho}\bar{V}_j \frac{\partial k}{\partial X_j} = \sum_j \frac{\partial}{\partial X_j}\left[\left(\mu + \frac{\mu_t}{\sigma_k}\right)\frac{\partial k}{\partial X_j}\right] + P_k - \bar{\rho}\varepsilon$$

$$\bar{\rho}\frac{\partial \varepsilon}{\partial t} + \sum_j \bar{\rho}\bar{V}_j \frac{\partial \varepsilon}{\partial X_j} = \sum_j \frac{\partial}{\partial X_j}\left[\left(\mu + \frac{\mu_t}{\sigma_\varepsilon}\right)\frac{\partial \varepsilon}{\partial X_j}\right] + C_{\varepsilon 1}\frac{\varepsilon}{k}P_k - C_{\varepsilon 2}\frac{\bar{\rho}\varepsilon^2}{k}$$

$$\sigma_k = 1.0; \quad \sigma_\varepsilon = 1.3; \quad C_{\varepsilon 1} = 1.44; \quad C_{\varepsilon 2} = 1.92; \quad C_\mu = 0.09$$

(4.26)

This model is good for high Reynolds number flows except very near walls where special treatments are needed. It has been argued that two-equation models should be modified to include molecular viscosity terms and the steep velocity-gradients near the wall calculated with many grid points to create a low-Reynolds number option of this model. This is not practical. A better procedure is to use wall-function boundary conditions. For example, local equilibrium of the production rate and dissipation rate could be assumed and modeled as follows.

$$\tau_t \frac{\partial \bar{V}_i}{\partial X_j} = \bar{\rho}\varepsilon = \frac{\bar{\rho}^2 C_\mu k^2}{\mu_t}; \quad \text{and} \quad \tau_t = \tau_w \Rightarrow u_\tau = \sqrt{\tau_w / \bar{\rho}} = k^{1/2} C_\mu^{1/4}$$

$$\bar{u} = \frac{\tau_w}{\bar{\rho}\kappa C_\mu^{1/4} k^{1/2}} \ln y + C; \quad u^+ = \frac{1}{\kappa}\ln y + C' = \frac{1}{\kappa}\ln(Ey^+)$$

(4.27)

$$\tau_w = \frac{\kappa \bar{u}\bar{\rho} C_\mu^{1/4} k^{1/2}}{\ln(Ey^+)}$$

where
  y is the distance normal to the wall
  $\bar{u}$ is the mean velocity at the wall-function point
  C, C', E, κ, and C_μ are empirical constants

Notice that one of these equations is for the wall shear-stress. The wall-function approach is not only computationally efficient, but lends itself easily to being modified to include other boundary conditions. In this case (as is the case for most of the published turbulence modeling simulations), the wall is assumed to be smooth. For rough walls and for walls which are transferring heat, the appropriate wall shear-stress equation can be modified without changing any of the other modeling functions. This allows not only more efficient but more general solutions.

The standard k–ε model does not have to be used. When improvements can be identified they can be included in the model. For example, a production rate timescale (k/P_k) was added to the ε-transport equation to give the "extended k–ε model."

$$\mu_t = \frac{\bar{\rho} C_\mu k^2}{\varepsilon}$$

$$\bar{\rho}\frac{\partial k}{\partial t} + \sum_j \bar{\rho}\bar{V}_j \frac{\partial k}{\partial X_j} = \sum_j \frac{\partial}{\partial X_j}\left[\left(\mu + \frac{\mu_t}{\sigma_k}\right)\frac{\partial k}{\partial X_j}\right] + \bar{\rho}P_k - \bar{\rho}\varepsilon$$

$$\bar{\rho}\frac{\partial \varepsilon}{\partial t} + \sum_j \bar{\rho}\bar{V}_j \frac{\partial \varepsilon}{\partial X_j} = \sum_j \frac{\partial}{\partial X_j}\left[\left(\mu + \frac{\mu_t}{\sigma_\varepsilon}\right)\frac{\partial \varepsilon}{\partial X_j}\right] + C_{\varepsilon1}\frac{\bar{\rho}\varepsilon P_k}{k} - C_{\varepsilon2}\frac{\bar{\rho}\varepsilon^2}{k} + C_{\varepsilon3}\frac{\bar{\rho}P_k^2}{k}$$

$$P_k = \sum_i \sum_j \frac{(\tau_t)_{ij}}{\bar{\rho}}\left(\frac{\partial \bar{V}_i}{\partial X_j} + \frac{\partial \bar{V}_j}{\partial X_i}\right); \quad (\tau_t)_{ij} = \mu_t\left(\frac{\partial \tilde{V}_i}{\partial X_j} + \frac{\partial \tilde{V}_j}{\partial X_i}\right)$$

$$\sigma_k = 0.75; \quad \sigma_\varepsilon = 1.15; \quad C_{\varepsilon1} = 1.15; \quad C_{\varepsilon2} = 1.9; \quad C_{\varepsilon3} = 0.25; \quad C_\mu = 0.09$$

$$(4.28)$$

Using the extended k–ε model, improved predictions for flow over a back-step (Chen, 1988). Other such modifications and improvements will be discussed subsequently.

Initial and boundary conditions must be supplied for k and ε to utilize this turbulence model. It is difficult to provide general values for these parameters; each problem should be evaluated to determine the best values to specify. If no specific values can be determined, the following values are suggested.

$$k = \sum_i \frac{1}{2}\overline{V_i'^2} \approx \frac{3}{2}\overline{u'^2}; \quad u' = \alpha U_\infty; \quad \alpha < 0.2 \text{ (typically 0.1)}$$

$$\varepsilon = \frac{k^{1.5}}{\ell}; \quad \omega = \frac{k^{0.5}}{C_\mu \ell}$$

$$\ell = \begin{cases} 0.41\delta, & \text{for B.L. flow} \\ 0.53\delta, & \text{for Plane jet} \\ 0.44\delta, & \text{for Round jet} \\ 0.95\delta, & \text{for Plane wake} \\ 0.25 \sim 0.3 D_p, & \text{for Pipe flow} \end{cases}$$

$$(4.29)$$

4.4.2 k–ω MODEL

Wilcox (1998) proposed an updated "k–ω turbulence model" as follows. The model was designed to maintain the same predictions for boundary layers and to reduce the jet spreading rate for free-shear layers.

$$\bar{\rho}\frac{\partial k}{\partial t}+\sum_j \bar{\rho}\bar{V}_j \frac{\partial k}{\partial X_j}=\sum_j \frac{\partial}{\partial X_j}\left[\left(\mu+\frac{\mu_t}{\sigma_k}\right)\frac{\partial k}{\partial X_j}\right]+\sum_i\sum_j (\tau_t)_{ij}\frac{\partial \bar{V}_i}{\partial X_j}-\bar{\rho}C_\mu k\omega$$

$$\bar{\rho}\frac{\partial \omega}{\partial t}+\sum_j \bar{\rho}\bar{V}_j \frac{\partial \omega}{\partial X_j}=\sum_j \frac{\partial}{\partial X_j}\left[\left(\mu+\frac{\mu_t}{\sigma_\omega}\right)\frac{\partial \omega}{\partial X_j}\right]+C_{\omega 1}\frac{\omega}{k}\sum_i\sum_j (\tau_t)_{ij}\frac{\partial \bar{V}_i}{\partial X_j}-C_{\omega 2}\bar{\rho}\omega^2$$

$$\mu_t=\bar{\rho}k/\omega;\quad \sigma_k=2.0;\quad \sigma_\omega=2.0;\quad C_{\omega 1}=0.556;\quad C_{\omega 2}=0.075;\quad C_\mu=0.09$$

$$(4.30)$$

Wilcox (2006) has been the primary developer of the k–ω turbulence model. His continued improvements and generalizations of this model are well documented in his text. The model with various modifications has been utilized to describe constant density flows involving wall roughness, mass injection at the wall, separated flows, low-Reynolds number flows, and compressible flows of ideal gases. He concludes that these k–ω models are superior to the k–ε models because they are more accurate near the wall, and also that the k–ε models are more accurate further from the wall. Notice, that to predict the near-wall effects very tight grid spacing is required to resolve the near-wall effects—this is computationally expensive when wall-function methodology is not used.

4.4.3 SST MODEL AND ITS IMPLICATIONS

Menter (1994) made a logical extension to the ε/ω models by blending them together to use the k–ε model away from walls and the k–ω model near the wall. Most turbulence model developers, including Wilcox and Menter, have used test data from low speed (incompressible) and high speed (compressible) air flow experiments to validate their model development. This is necessary to have meaningful velocity and velocity fluctuation data. Such limitations are in the same vein as those of heat and mass transfer researchers who use small temperature differences and small mass transfer rates to simplify the process so that constant fluid properties and unperturbed free stream velocities can be assumed in analyzing the test data. This does not mean that such models are not valuable; it means only that the validation is incomplete and that the model generality is undetermined.

4.4.4 FURTHER EXTENSIONS TO THE k–ε TURBULENCE MODEL

The authors of this text have made many analyses of aerospace-related transport processes over the years. Such applications are mainly of high Reynolds number and pressure, high-temperature reacting fluids and cryogenic fluids, speed ranging from subsonic to high supersonic, and very complex geometries. These studies were generally performed for NASA, the Air Force, and private aerospace companies; all of which were the pioneers in developing and using computational fluid dynamics simulations for engineering design. These authors and most other contemporary investigators used k–ε models to describe the turbulent flows which they studied. Such models were modified as necessary and appropriate boundary constructed as needed.

High-speed flow analyses required Mach number corrections to the standard k–ε model. The k-transport equation was modified to be (Sarkar et al., 1989):

$$\bar{\rho}\frac{\partial k}{\partial t}+\sum_{j}\bar{\rho}\bar{V}_{j}\frac{\partial k}{\partial X_{j}}=\sum_{j}\frac{\partial}{\partial X_{j}}\left[\left(\mu+\frac{\mu_{t}}{\sigma_{k}}\right)\frac{\partial k}{\partial X_{j}}\right]+P_{k}-\bar{\rho}\varepsilon(1+M_{t}^{2});\quad M_{t}=\sqrt{\frac{k}{\gamma RT}}$$

(4.31)

The ε-transport equation was modified for high-Mach numbers to be (Smith et al., 1989):

$$\bar{\rho}\frac{\partial \varepsilon}{\partial t}+\sum_{j}\bar{\rho}\bar{V}_{j}\frac{\partial \varepsilon}{\partial X_{j}}=\sum_{j}\frac{\partial}{\partial X_{j}}\left[\left(\mu+\frac{\mu_{t}}{\sigma_{\varepsilon}}\right)\frac{\partial \varepsilon}{\partial X_{j}}\right]+C_{\varepsilon1}^{*}\frac{\varepsilon}{k}P_{k}-C_{\varepsilon2}\frac{\bar{\rho}\varepsilon^{2}}{k}+C_{\varepsilon3}\frac{\bar{\rho}P_{k}^{2}}{k}$$

$$C_{\varepsilon1}^{*}=C_{\varepsilon1}(1+0.08M^{0.25});\quad M=\frac{V}{\sqrt{\gamma RT}}$$

(4.32)

The ε-transport equation was modified for high temperatures to be (Cheng et al., 1994):

$$\bar{\rho}\frac{\partial \varepsilon}{\partial t}+\sum_{j}\bar{\rho}\bar{V}_{j}\frac{\partial \varepsilon}{\partial X_{j}}=\sum_{j}\frac{\partial}{\partial X_{j}}\left[\left(\mu+\frac{\mu_{t}}{\sigma_{\varepsilon}}\right)\frac{\partial \varepsilon}{\partial X_{j}}\right]+C_{\varepsilon1}\frac{\varepsilon}{k}P_{k}-C_{\varepsilon2}\frac{\bar{\rho}\varepsilon^{2}}{k}+C_{\varepsilon3}^{*}\frac{\bar{\rho}P_{k}^{2}}{k}$$

$$C_{\varepsilon3}^{*}=C_{\varepsilon3}(T/T_{\text{ref}})^{\alpha};\quad \alpha=0.4 \text{ or } 0.6$$

(4.33)

Combustion generated turbulence has been reported (Ballal, 1988), primarily for situations where the initial flow was low speed. In high-speed, high-pressure flows, the temperature rise associated with combustion drops the density sufficiently to suppress the turbulence, as indicated by this modified ε-equation.

As previously mentioned, wall-function boundary conditions can be used to avoid using an excessive number of grid points to simulate near-wall flow behavior. This is a common means of improving computational efficiency of CFD codes (Craft et al., 2002; Patel et al., 1985; Viegas and Rubesin, 1983). Surface roughness and heat transfer are easily handled in this manner. This also reduces the advantage of using the k–ω model or the blended SST model for near-wall flows and allows a more computationally efficient simulation. Depending on what information is available or can be best estimated, conjugate heat transfer can be used to replace wall/fluid boundary conditions for simulations. This process simply sets the velocities of grid points within a solid wall as zero and changes the fluid properties within the wall to be those of the solid structure.

The need of special boundary conditions to represent high blowing rates can be eliminated by treating the blowing (or sucking) surface as an inlet (or exit) for simulation. This practice has been used successfully to analyze hybrid rocket motor combustion chambers (Cheng et al., 1998) and is commonly used to analyze ablative protection systems for spacecraft reentry heating.

Another example of using inventive boundary condition specification and turbulent modeling is the experiment and simulation of flow in a stirred-tank reactor (Ju et al., 1990). The turbulence was modeled by using the standard k–ε model which gives an isotropic eddy viscosity: $\mu_t = C_\mu \rho k^2/\varepsilon$. The C_μ constant was then given different values in the three coordinate direction used to specify the tank geometry. The modified constant values were evaluated from the experimental data. Similar to the hybrid motor analysis, the turbine-impeller flow was simulated with inflow and outflow boundary conditions.

The value of the turbulence model extensions and the nonstandard boundary specifications just mentioned is illustrated by the following examples. First, the slot injection and subsequent combustion of hydrogen into an air stream was simulated and compared to experiments. The reduced flame turbulence was necessary in order to correctly simulate the wall heating by keeping the cooling jet intact longer. Second, the high-speed hydrogen/oxygen flame was simulated and good agreement was obtained when compared to total temperature and concentration measurements. The analyses were conducted with the CTP code and were reported by Chen et al. (1992). The simulations obtained for these two experiments are quite sufficient for engineering design purposes and for demonstrating the validity of the computational methodology.

4.4.5 SUMMARY OF TWO-EQUATION TURBULENCE MODELS

Advantages of using two-equation models:

- Using the parameters k and ε (or ω) to model turbulence eliminates the need for specifying a length scale which varies with flow geometry.
- The model applies to a wide range of transport phenomena.
- Solving two-transport equations for the two parameters is computationally efficient.

Disadvantages of these models:

- Further tuning and final selection of the two modeling parameters requires constant updating and validation.
- The prediction of the turbulence as being isotropic is not always acceptable.
- Additional wall-function boundary conditions are needed.

Applications posing concerns which have not been fully evaluated:

- Flow with sudden changes in the mean strain rate
- Flow over curved surfaces
- Secondary flow in curved ducts
- Flow in rotating fluids
- Flow with boundary layer separation
- Complex three-dimensional flows

4.5 NOMENCLATURE

4.5.1 ENGLISH SYMBOLS

A_{ij} autocorrelation tensor
B event B

$C, C', E,$ C_μ, κ	empirical constants in wall functions
$C_{\varepsilon1}, C_{\varepsilon2}, C_\mu$	empirical constants in the k–ε model
$C_{\varepsilon2}, C_{\varepsilon3}, C_{\varepsilon1}^*$	Mach number correction parameters, Equation 4.32
D, D_t	laminar and turbulent diffusion coefficients
e	specific internal energy
$E\{\alpha\}$	energy spectrum function
c, f, g	empirical functions to fit the energy spectrum
$f\{\vec{r}, t\}$	arbitrary function
$f\{V\}$	probability distribution function; PDF, f
f_{12}	joint probability distribution function (JPDF), Equation 4.A.19
F	cumulative probability; cumulative probability function, CDF, F
h	specific enthalpy
H	dummy symbol to indicate mathematical operations
I_i	turbulent intensity
k	turbulent kinetic energy
ℓ	length scale
ℓ_K	Kolmogorov's length scale
ℓ_{mfp}	mean free path
L	integral length scale
M_t	Mach number correction parameter, Equation 4.31
p_o	constant in the energy spectrum model
$P = P\{B\}$	probability of event B occurring; pressure in Tables 4.1 and 4.2
P_k	production rate of k
Pr, Pr_t	laminar and turbulent Prandtl numbers
Q	function of U
Re_t	Reynolds number bases on timescale, t_K
s	eddy size
Sc, Sc_t	laminar and turbulent Schmidt numbers
t	time
t_i	time variable of type i
t_K	Kolmogorov's timescale
T	temperature
$u = U - \langle U \rangle$	variance of U: velocity in wall function
U	function; standardized (normalized) function

u^+	dimensionless distance from a wall
u_i'	fluctuating component about the mean
u_τ	friction velocity
V	sample space variable; velocity otherwise
V_{ij}	two-point correlation tensor
υ_K	Kolmogorov's velocity scale
x_i, X_i	spatial coordinates
y	spatial coordinate; distance from a wall
z	spatial coordinate

4.5.2 GREEK SYMBOLS

α	wave number
α_i	mass fraction of i
α, T_{REF}	high temperature correction parameters, Equation 4.33
ε	energy dissipation
ε	turbulent kinetic energy dissipation
λ, λ_t	laminar and turbulent thermal conductivity; wavelength in Equation 45.5
μ, μ_t	laminar and turbulent viscosity
μ_n	the nth central moment
ν	kinematic viscosity
ν_t	turbulent kinematic viscosity
ξ_i	small displacement in x_i
ρ	density
ρ_{12}	correlation coefficient
$\sigma_k, \sigma_\varepsilon, a$	empirical constants in the k–ε model
$\sigma_u = \sqrt{U}$	standard deviation
τ	small displacement in time; shear-stress otherwise
τ_w	wall shear-stress
φ, ψ	property to be averaged
Φ_E	general ensemble average
Φ_t	general time average
Φ_∇	general spatial average
Ψ	characteristic function, Equation 4.A.19
ω	specific energy dissipation rate per unit k in Equation 4.25; per unit volume Equation 4.38
$\omega\{\}$	weighting function

4.5.3 MATHEMATICAL SYMBOLS

$\vec{e}$	unit vector
$f_{2|1}\{V_2|V_1\}$	conditional probability density function
H'	fluctuating component
$\overline{H}$	time average value
$\check{H}$	standardized parameter
$\tilde{H}$	Favre average
H''	fluctuating component in Favre averaging
H_i	denotes a component
$\vec{\overline{L}}_{ii}$	length scale
$\overline{q}_t$	mean specific energy fluctuation
$\vec{r}, \vec{s}$	displacement vectors
$\langle Q\{U_1, U_2 | U_1 = V_1\}\rangle$	conditional mean of Q on V_1
$\vec{\overline{R}}_{ij}$	two-point, one-time autocorrelation
$\overline{u}$	mean velocity at the wall-function point
$\langle U \rangle$	mean or expectation of U
$\{U_1, U_2\}_{cov}$	covariance
$\vec{\overline{U}}$	mean velocity vector
$\overline{V}_i$	time mean of a velocity component
$\tilde{V}_i$	Favre mean of a velocity component
$\overline{V_i' V_i'}$	component of the Reynolds stress
$-\rho\langle V_i' V_j' \rangle$	Reynolds stress
$-\langle \rho V_i'' V_j'' \rangle$	Reynolds stress in Favre averaging system
δ	Dirac delta function
$(\tau_t)_{ij}$	component of the Reynolds stress tensor

4.5.4 ACRONYMS

CDF	cumulative distribution function
JPDF	joint probability distribution function
PDF	probability density function
RANS	Reynolds-averaged Navier–Stokes turbulence model

APPENDIX 4.A   BASIC PROBABILITY PARAMETERS

The chaotic motion of turbulent flows is most fundamentally described in terms of probability theory. Probability theory is complex, voluminous, and computationally inefficient to analyze. Texts of Monin and Yaglom and of Hinze provide an introduction to the terms used, but the definitions of these terms are obscured in the details of the discussion. Pope (2000) and Fox (2003) provide a more lucid discussion, but the many terms with specific technical definitions also make this material difficult to read and understand. A very brief introduction to this field is summarized below. Pope's nomenclature is used whenever possible to provide an easier read of his more complete work.

To apply probability theory, the turbulent flow is considered to be a statistical ensemble of similar flows created by a fixed set of external boundary conditions. The need for such a consideration is illustrated by a fundamental difference in laminar and turbulent flow. For example, consider an experiment involving a cylinder in cross-flow. If the flow is laminar, a measurement of a flow property-like velocity ($\vec{u}\{\vec{x},t\}$) at a given point is the same every time the experiment is repeated (assuming that there is negligible error in making the measurement). For turbulent flow, the measurement would not be the same (again assuming no error in the measurement). Due to the randomly fluctuating field, the velocity would be oscillating rather rapidly, such that no two measurement records would be identical. These fluctuations are due to small uncontrollable disturbances in both the flow-field and the boundary conditions creating the flow. If the experiment were repeated several times in the same or other laboratories, a database could be established. The entire data set would constitute an ensemble of all values of $\vec{u}\{\vec{x},t\}$ measured in these experiments. Each data set would be one realization of the these values. An arithmetic average of the data would be a probability mean of the velocity and is denoted by broken brackets, $\langle \vec{u}\{\vec{x},t\} \rangle$. This meaning of the broken brackets is applied only to this section or where otherwise specified. The same type of measurements and averaging procedure could be applied to any other of the fluid dynamic variables. It is to be expected that the probability mean determined from several subsets of data would not vary substantially, if the data truly represents a statistical ensemble. For example, consider that the placement of the cylinder slips once such that the cylinder becomes yawed so part of the data would actually be collected for one set of boundary conditions and the remainder of the data collected for another set of boundary conditions. Probability means calculated from such data could vary substantially, indicating that the data no longer created a statistical ensemble.

Before defining additional statistical parameters, consider an experiment in which one screens a mixture of particles. The upper screen has openings of $10\,\mu$m and a lower one $1\,\mu$m openings. Upon screening a sample would remain between the screens which certainly contained particles smaller than $10\,\mu m$ and larger than $1\,\mu m$. A distribution of particle sizes in between these sizes would remain on the bottom screen. If the number of particles between 1 and $2\,\mu m$ were counted, and the procedure repeated for each cut of $1\,\mu m$. The distribution would be by number. Knowing the particle density, this distribution could be converted to a mass distribution. If the cuts were progressively summed, a cumulative distribution would be obtained. Various mean sizes could be calculated, depending on the definition of the mean value used. If we picked out a particle at random, we could assign a probability to what that size would be. Now we shall formally define what these statistical properties are termed.

Random variables (U) (for example, a velocity component) are characterized by measuring or estimating a series of (sample-space) variables (V) in a sample space. The sample space is bounded by the largest and smallest value of the variables it contains. The value of a sample-space variable is a realization. Probability is a number between 0 and 1 which expresses the likelihood that an event (B) occurs. Expressing this as Equation 4.A.1.

$$p = P\{B\} = P\{V_{min} \leq U \leq V_{max}\} \tag{4.A.1}$$

For an impossible event, p is zero; for a certain event p is unity. In the example, p is the likelihood that a particle of size which lies between, say 6 and $7\,\mu m$, can be selected from the cut of particles. The cumulative distribution function (CFD, F) is characterized by

$$F\{V_{min}\} = 0 \quad F\{V_{max}\} = 1$$
$$F\{V_b\} - F\{V_a\} = P\{V_a \leq U \leq V_b\} \geq 0 \tag{4.A.2}$$

The index a could be $-\infty$ and b could be $+\infty$. The CDF states that all the particles below, say $7\,\mu m$, is greater than all the particles below, say $3\,\mu m$, for example. The probability density function (PDF, f) is the derivative of the CDF.

$$f\{V\} = dF\{V\}/dV \tag{4.A.3}$$

Since the CFD is always increasing, f is always greater than zero. Also

$$\int_{-\infty}^{+\infty} f\{V\}dV = 1 \quad \text{and} \quad f\{-\infty\} = f\{+\infty\} = 0 \tag{4.A.4}$$

Furthermore

$$\int_{V_a}^{V_b} f\{V\}dV = F\{V_b\} - F\{V_a\} \quad \text{and} \quad f\{V\}dV = F\{V + dV\} - F\{V\} \tag{4.A.5}$$

The PDF is the probability per unit distance in the sample space V, therefore it is said to be a density. (OK, this a mathematician's definition of density.) The units on f are the inverse of the units on U. The CDF and the product $f\{V\}dV$ are dimensionless.

The "mean" or "expectation" of the random variable U is

$$\langle U \rangle = \sum_{-\infty}^{+\infty} V f\{V\} dV \qquad (4.A.6)$$

If Q is a function of U, its mean is

$$\langle Q\{U\} \rangle = \sum_{-\infty}^{+\infty} Q\{V\} f\{V\} dV \qquad (4.A.7)$$

The fluctuation of U is

$$u = U - \langle U \rangle \qquad (4.A.8)$$

The variance of U is the mean-square fluctuation:

$$U_{var} = \langle u^2 \rangle = \sum_{-\infty}^{+\infty} (V - \langle U \rangle)^2 f\{V\} dV \qquad (4.A.9)$$

The square root of the variance is the standard deviation:

$$\sigma_u = \sqrt{U_{var}} = \langle u^2 \rangle^{1/2} \qquad (4.A.10)$$

The nth central moment is

$$\mu_n = \langle u^n \rangle = \sum_{-\infty}^{+\infty} (V - \langle U \rangle)^n f\{V\} dV \qquad (4.A.11)$$

"Standardized random variables" (as denoted by the breve) are defined to have zero mean and zero variance.

$$\breve{U} = (U - \langle U \rangle)/\sigma_u \qquad (4.A.12)$$

The standardized PDF is

$$\breve{f}\{\breve{V}\} = \sigma_u f\{\langle U \rangle + \sigma_u \breve{V}\} \qquad (4.A.13)$$

The standardized moments are

$$\breve{\mu}_n = \frac{\langle u^n \rangle}{\sigma_u^n} = \frac{\mu_n}{\sigma_u^n} = \int_{-\infty}^{+\infty} \breve{V}^n \breve{f}\{\breve{V}\} d\breve{V} \qquad (4.A.14)$$

A summary of the moments and their meaning is

$$\mu_0 = \bar{\mu}_0 = 1 \quad \mu_1 = \bar{\mu}_1 = 0 \quad \mu_2 = \sigma_u^2 \quad \bar{\mu}_2 = 1 \tag{4.A.15}$$

The moments μ_3 and μ_4 describe skewness and flatness, respectively.

The characteristic function of U is

$$\Psi\{s\} = \langle e^{iU_s} \rangle = \int_{-\infty}^{+\infty} f\{V\} e^{iV_s} \, dV \tag{4.A.16}$$

The integral in Equation 4.A.16 is an inverse Fourier transform. The variables Ψ and f contain the same information.

Joint random variables are collections of random variables that have some common meaning. For example, a two-dimensional turbulent flowfield. Each point in space and time would represent each of the two velocity components. Say a number of points were measured to create a sample space at a given time. The CDF would be represented by

$$F_{12}\{V_1, V_2\} = P\{U_1 < V_1, U_2 < V_2\} \tag{4.A.17}$$

The joint probability distribution function (JPDF) is

$$f_{12}\{V_1, V_2\} = \frac{\partial^2 F_{12}\{V_1, V_2\}}{\partial V_1 \partial V_2} \tag{4.A.18}$$

The fundamental property of the JPDF is

$$P\{V_{1a} \le U_1 < V_{1b}, \quad V_{2a} \le U_2 < V_{2b}\} = \int_{V_{1a}}^{V_{1b}} \int_{V_{2a}}^{V_{2b}} f_{12}\{V_1, V_2\} \, dV_1 \, dV_2 \tag{4.A.19}$$

The means $\langle U_1 \rangle$ and $\langle U_2 \rangle$ and the variances $\langle u_1^2 \rangle$ and $\langle u_2^2 \rangle$ can be determined. As can the covariance of U_1 and U_2:

$$\{U_1, U_2\}_{cov} = \langle u_1 u_2 \rangle = \int_{-\infty}^{+\infty} \int_{-\infty}^{+\infty} (V_1 - \langle U_1 \rangle)(V_2 - \langle U_2 \rangle) f_{12}\{V_1, V_2\} \, dV_1 \, dV_2$$

$$\tag{4.A.20}$$

The correlation coefficient is defined as

$$\rho_{12} = \langle u_1 u_2 \rangle / [u_1^2 u_2^2]^{1/2} \tag{4.A.21}$$

A correlation coefficient of zero means the U_1 and U_2 are not correlated; a value of +1 means they are perfectly correlated; and a value of −1 means perfectly negatively correlated. A conditional PDF is defined by

$$f_{2|1}\{V_2|V_1\} = f_{12}\{V_1, V_2\}/f_1\{V_1\} \qquad (4.A.22)$$

This is read as the PDF of U_2 conditional on $U_1 = V_1$. For a function Q of U_1 and U_2, the conditional mean of Q on V_1 is:

$$\left\langle Q\{U_1, U_2 | U_1 = V_1\} \right\rangle = \int_{-\infty}^{+\infty} Q\{V_1, V_2\}\, f_{2|1}\{V_2|V_1\}\, dV_1 \qquad (4.A.23)$$

Consider a three dimensional, unsteady turbulent flowfield. The one-point, one-time joint CDF may be defined as

$$F\{\vec{V}, \vec{x}, t\} = P\{U_i\{\vec{x}, t < V_i\}\} \quad \text{for } i = 1, 2, 3 \qquad (4.A.24)$$

The JPDF is

$$f\{\vec{V}; \vec{x}, t\} = \frac{\partial^3 F\{\vec{V}; \vec{x}, t\}}{\partial V_1 \partial V_2 \partial V_3} \qquad (4.A.25)$$

The semicolon is a shorthand to indicate that the variables preceding it are in sample-space and those after it are the independent variables of the random field. This JPDF fully characterizes the turbulent velocity field at a point, but it provides no information at other times or positions. The mean velocity is indicated by

$$\left\langle \vec{U}\{\vec{x}, t\} \right\rangle = \iint_{-\infty}^{+\infty}\!\!\int \vec{V} f\{\vec{V}; \vec{x}, t\}\, d\vec{V} \qquad (4.A.26)$$

The fluctuating velocity is described by

$$\vec{u}\{\vec{x}, t\} = \vec{U}\{\vec{x}, t\} - \left\langle \vec{U}\{\vec{x}, t\} \right\rangle \qquad (4.A.27)$$

The one-point, one-time covariance of the velocity is

$$\left\langle u_i\{\vec{x}, t\} u_j\{\vec{x}, t\} \right\rangle = \left\langle u_i u_j \right\rangle \qquad (4.A.28)$$

This quantity is also obtained from the Reynolds time-averaging of the momentum equations where it is called the Reynolds stress.

The two-point, one-time autocorrelation is defined by

$$\vec{R}_{ij}\{\vec{r},\vec{x},t\} = \langle u_i\{\vec{x},t\}\, u_j\{\vec{x}+\vec{r},t\}\rangle \tag{4.A.29}$$

This correlation may be used to give information about the spatial structure of the random field by defining various length scales, for example:

$$\vec{L}_{11}\{\vec{x},t\} = \frac{1}{\vec{R}_{11}\{0,\vec{x},t\}}\int_0^\infty \vec{R}_{11}\{\vec{e}_1 r,\vec{x},t\}\,\mathrm{d}\vec{r} \tag{4.A.30}$$

where $\vec{e}_1$ is a unit vector in the x_1-coordinate direction. Thus, an infinite number of length scales may be defined depending on how $\vec{x}$ and $\vec{r}$ are related.

Evaluating the length scales and the distribution functions to describe the structure of the random turbulent field is the goal of applying probability theory to the description of turbulence.

REFERENCES

Ballal, D. R. 1988. Combustion-generated turbulence in practical combustors. *Journal of Propulsion* 4:385–390.

Birch, S. F., D. H. Rudy, and D. M. Bushnell. 1972. *Free Turbulent Shear Flows*, Vol. I—Conference Proceedings. NASA SP-321. National Aeronautics and Space Administration, Washington D.C.

Bradshaw, P. 1972. The understanding and prediction of turbulent flow. *The Aeronautical Journal.* 76:403–418.

Chen, Y. S. 1988. Viscous Flow Computations Using a Second-Order Upwind Differencing Scheme. AIAA 88-0417. AIAA 26th Aerospace Sciences Meeting, Reno, NV.

Chen, Y. S., G. C. Cheng, and R. C. Farmer. 1992. Reacting and Non-Reacting Flow Simulation for Film Cooling in a 2-D Supersonic Flows. AIAA 93-3602. AIAA/SAE/ASME/ASEE 28th Joint Propulsion Conference and Exhibit, Nashville, TN.

Cheng, G. C., R. C. Farmer, and Y. S. Chen. 1994. Numerical Study of Turbulent Flows with Compressibility Effects and Chemical Reactions. AIAA 94-2026. 6th Joint Thermophysics and Heat Transfer Conference, Colorado, Springs, CO.

Cheng, G. C., R. C. Farmer, S. H. Jones, and J. P. Alrves. 1998. A Practical CFD Model for Simulating Hybrid Motors. JANNAF 35th Combustion, Airbreathing Propulsion and Propulsion Systems Hazards Subcommittee Joint Meeting. Tucson, AZ.

Coles, D. E. and E. A. Hirst. 1968. Proceedings of the Computation of Turbulent Boundary Layers—1968 AFOSR-IFP-Stanford Conference, Vol. II. Compiled data. Stanford, CA.

Craft, T. J., A. V. Gerasimov, H. Iacovides, and B. E. Launder. 2002. Progress in the generalization of wall-function treatments. *International Journal of Heat and Fluid Flow* 23:148–160.

Favre, A. 1961. Statistical Equations of Turbulent Gases. AFOSR translation of summary of Favre's papers. AFOSR European Office, London.

Fox, R. O. 2003. *Computational Models for Turbulent Reacting Flows.* Cambridge: Cambridge University Press.

Friedlander, S. K. and L. Topper. 1961. *Turbulence, Classical Papers on Statistical Theory.* New York: Interscience.

Hinze, J. O. 1975. *Turbulence,* 2nd ed. New York: McGraw-Hill.

Hunt, J. C. R., O. M. Phillips, and D. Williams, Ed. 1991.*Turbulence and Stochastic Processes: Kolmogorov's Ideas 50 Years On.* London: The Royal Society.

Jones, S. A. and B. E. Launder. 1972. The prediction of laminarization with a two-equation model of turbulence. *International Journal of Heat and Mass Transfer* 15:301–315.

Ju, S. -Y., T. M. Mulvahill, and R. W. Pike. 1990. Three-dimensional, turbulent flow in agitated vessels using a nonisotropic turbulence model. *Canadian Journal of Chemical Engineering* 68:3–16.

Leonard, A. 1974. Energy Cascade in Large-Eddy Simulations of Turbulent Flows. In *Advances in Geophysics,* Vol. 18A, Ed. F. N. Frenkiel and R. E. Munn. New York: Academic Press, pp. 237–248.

McKusick, V. A. and H. K. Wiskind. 1959. Osborne Reynolds of Manchester. *Bulletin of the History of Medicine* 33:116–136.

Menter, F. R. 1994. Two-equation eddy-viscosity turbulence models for engineering applications. *AIAA Journal* 32:1598–1605.

Monin, A. S. and A. M. Yaglom. 1965. *Statistical Fluid Mechanics: Mechanics of Turbulence,* Vols. I and II. Cambridge, MA: MIT Press.

Patel, V. C., W. Rodi, and G. Scheuerer. 1985. Turbulence models for near wall and low Reynolds number flow: A review. *AIAA Journal* 23:1308–1319.

Pope, S. B. 2000. *Turbulent Flows.* Cambridge: Cambridge University Press.

Reynolds, O. 1895. On the dynamical theory of incompressible viscous fluids and the determination of the criterion. *Philosophical Transactions of the Royal Society of London Series A* 186:123–164.

Sarkar, S., G. Erlebacher, M. Y. Hussaini, and H. O. Kreiss. 1989. The Analysis and Modeling of Dilatational Terms in Compressible Turbulence. ICASE Report No. 89-79. NASA Langley Research Center, Hampton, VA.

Schlichting, H. 1979. *Boundary Layer Theory,* 7th ed. New York: McGraw-Hill.

Smith, S. D., J. A. Freeman, and R. C. Farmer. 1989. Model Development for Exhaust Plume Effects on Launch Stand Design. Report No. SECA-89-20. SECA. Huntsville, AL.

Viegas, J. R. and M. W. Rubesin. 1983. Wall-Function Boundary Conditions in the Solution of the Navier-Stokes Equations for Complex Compressible Flows. AIAA 83-1694. AIAA 16th Fluid and Plasma Dynamics Conference. Danvers, MA.

White, F. M. 2006. *Viscous Fluid Flow*, 3rd ed. Boston, MA: McGraw-Hill.

Wilcox, D. C. 1998. *Turbulence Modeling for CFD*, 2nd ed. La Canada: DCW.

Wilcox, D. C. 2006. *Turbulence Modeling for CFD*, 3rd ed. La Canada: DCW.

CHAPTER 5

Other Turbulence Models

5.1 MORE COMPREHENSIVE TURBULENCE MODELS

The eddy viscosity models (EVM) which have become the workhorses of engineering turbulence analyses do not describe the eddy structure of the turbulent fields, anisotropic turbulence, turbulence/chemistry interactions, or transition from laminar to turbulent flow. Much research to address such issues has been and is being pursued to understand and model these effects. Solving more complete equations for the Reynolds stresses by differential second-moment (DSM) closure is expected to remove some of the shortcomings of the EVM analyses. Using probability density functions (PDF) to address more of the statistical properties of turbulent fields, including turbulence/chemistry interactions, is being actively researched. The Navier–Stokes equations are being integrated directly to provide direct numerical simulations (DNS) of turbulence to represent all spatial scales of the turbulent field. Large eddy simulations (LES) treat large and small eddy scales separately to provide a model intermediate between DNS and PDF models in computational efficiency. Models to describe transitional flow from laminar to turbulent are being researched. All of these models are being studied with the goal of providing a better description of turbulent flows. Most have provided some interesting insight. None have been developed to the point of being used by nonspecialists to analyze transport phenomena. All are extremely computationally inefficient and have no demonstrated performance on other than very simple flows. These models are not expected to be practical for simulations on personal computer systems. Nevertheless, they are tools which will be utilized in future engineering work, as indicated by the active research surveyed by Launder and Sandham (2002).

5.2 DIFFERENTIAL SECOND-MOMENT CLOSURE METHODS

EVMs approximate the Reynolds stresses with a single effective viscosity. If transport equations are written for the DSM of the fluctuating velocity products, the resulting equations are not closed. However, if the unknown terms are approximated closure can be obtained. Solving these equations should provide a more complete and hopefully more accurate description of the turbulent field. The second-moment transport equations are shown in Table 5.1—Equation A is the Reynolds averaged transport equation for the Reynolds stresses and Equation B is the Favre averaged transport equation for the Reynolds stresses. Six new equations for either of the Reynolds stresses are generated by this operation, but 22 new unknowns result. Models for terms needed to close these equations have been developed by Launder et al. (1975) and are shown below. This is the best known and most widely tested of the Reynolds-stress models (RSM).

$$C_{ij} = \overline{\rho}\frac{D\overline{V_i'V_j'}}{Dt} = P_{ij} + \Pi_{ij} - \overline{\rho}\varepsilon_{ij} + D_{ij}; \quad D_{ij} = D_{ij}^t + D_{ij}^v$$

$$\overline{\rho}\frac{\partial\varepsilon}{\partial t} + \sum_j \overline{\rho}\overline{V}_j\frac{\partial\varepsilon}{\partial X_j} = \sum_j\frac{\partial}{\partial X_j}\left[\mu\frac{\partial\varepsilon}{\partial X_j}\right] + C_{\varepsilon 1}\frac{\varepsilon}{k}P_k - C_{\varepsilon 2}\frac{\overline{\rho}\varepsilon^2}{k}$$

$$+ C_\varepsilon\sum_j\frac{\partial}{\partial X_j}\left[\sum_m\frac{\overline{\rho}k}{\varepsilon}\overline{V_j'V_m'}\frac{\partial\varepsilon}{\partial X_m}\right] \tag{5.1}$$

$$P_{ij} = -\overline{\rho}\sum_m\left[\overline{V_i'V_m'}\frac{\partial\overline{V}_j}{\partial X_m} + \overline{V_j'V_m'}\frac{\partial\overline{V}_i}{\partial X_m}\right]$$

$$D_{ij}^v = \sum_m\frac{\partial}{\partial X_m}\left[\mu\frac{\partial\overline{V_i'V_j'}}{\partial X_m}\right]$$

The dissipation for isotropic flow is defined by

$$\varepsilon_{ij} = 2\frac{\mu}{\rho}\sum_k\overline{\frac{\partial V_i'}{\partial X_k}\frac{\partial V_j'}{\partial X_k}} = \frac{2}{3}\delta_{ij}\varepsilon; \quad \varepsilon = \frac{\mu}{\rho}\sum_k\overline{\frac{\partial V_i'}{\partial X_k}\frac{\partial V_i'}{\partial X_k}} \tag{5.2}$$

TABLE 5.1 Transport Equation for Reynolds Stresses

$$\rho\frac{\partial \overline{V_i'V_j'}}{\partial t} + \rho\overline{V}\sum_k \frac{\partial \overline{V_i'V_j'}}{\partial X_k} = -\rho\sum_k\left[\overline{V_i'V_k}\frac{\partial \overline{V_j}}{\partial X_k} + \overline{V_j'V_k'}\frac{\partial \overline{V_i}}{\partial X_k}\right] + \overline{P'\left(\frac{\partial V_i'}{\partial X_j} + \frac{\partial V_j'}{\partial X_i}\right)} - 2\sum_k \mu\overline{\frac{\partial V_i'}{\partial X_k}\frac{\partial V_j'}{\partial X_k}}$$

$$- \sum_k \frac{\partial}{\partial X_k}\left(\rho\overline{V_i'V_j'V_k'} + \overline{V_i'P'}\delta_{jk} + \overline{V_j'P'}\delta_{ik} - \mu\frac{\partial \overline{V_i'V_j'}}{\partial X_k}\right)$$

(A)

$$\frac{\partial \rho\overline{V_i''V_j''}}{\partial t} + \sum_k \tilde{V}_k\frac{\partial \rho\overline{V_i''V_j''}}{\partial X_k} = -\sum_k\left[\rho\overline{V_i''V_k''}\frac{\partial \tilde{V}_j}{\partial X_k} + \rho\overline{V_j''V_k''}\frac{\partial \tilde{V}_i}{\partial X_k}\right] + \overline{P'\left(\frac{\partial V_i''}{\partial X_j} + \frac{\partial V_j''}{\partial X_i}\right)} - 2\sum_k \mu\overline{\frac{\partial V_i''}{\partial X_k}\frac{\partial V_j''}{\partial X_k}}$$

$$- \left[\sum_k \frac{\partial}{\partial X_k}\left(\rho\overline{V_i''V_j''V_k''} + \overline{V_i''P'}\delta_{jk} + \overline{V_j''P'}\delta_{ik} - \nu\frac{\partial \rho\overline{V_i''V_j''}}{\partial X_k}\right)\right]$$

(B)

The corresponding term for anisotropic flow is

$$
\varepsilon_{ij} = 2\frac{\mu}{\rho}\sum_k \overline{\frac{\partial V_i'}{\partial X_k}\frac{\partial V_j'}{\partial X_k}} = \frac{2}{3}\delta_{ij}\varepsilon + 2f_s\varepsilon b_{ij};
$$

$$
b_{ij} = \frac{\overline{V_i'V_j'}}{2k} - \frac{1}{3}\delta_{ij}; \quad f_s = (1+0.1Re_t)^{-1}
$$

(5.3)

Hanjalic and Launder (1976) modeled the diffusion term as

$$
D_{ij}^t = C_s\sum_m \frac{\partial}{\partial X_m}\left[\overline{\rho}\frac{k}{\varepsilon}\sum_n \left(\overline{V_i'V_n'}\frac{\partial \overline{V_j'V_m'}}{\partial X_n} + \overline{V_j'V_n'}\frac{\partial \overline{V_i'V_m'}}{\partial X_n} + \overline{V_m'V_n'}\frac{\partial \overline{V_i'V_j'}}{\partial X_n}\right)\right]
$$

(5.4)

Daly and Harlow (1970) and Mellor and Herring (1973) also offered models of this term.

Launder et al. modeled the pressure strain term as

$$
\Pi_{ij} = -C_1\overline{\rho}\frac{\varepsilon}{k}\left(\overline{V_i'V_j'} - \frac{2}{3}k\delta_{ij}\right) - C_2\left(P_{ij} - \frac{2}{3}P_k\delta_{ij}\right)
$$

$$
-C_3\left(D_{ij} - \frac{2}{3}P_k\delta_{ij}\right) - C_4\overline{\rho}k\left(S_{ij} - \frac{1}{3}\sum_m S_{mm}\delta_{ij}\right)
$$

$$
D_{ij} = -\overline{\rho}\sum_m\left[\overline{V_i'V_m'}\frac{\partial \overline{V}_m}{\partial X_j} + \overline{V_j'V_m'}\frac{\partial \overline{V}_m}{\partial X_i}\right]; \quad S_{ij} = \frac{1}{2}\left(\frac{\partial \overline{V}_i}{\partial X_j} + \frac{\partial \overline{V}_j}{\partial X_i}\right)
$$

(5.5)

Speziale et al. (1991) also offered pressure-strain models. The principal difficulty in these models is in how the pressure-strain term is modeled.

Wilcox (1998) developed a high and a low Reynolds number stress-ω model and a multiscale version He demonstrated that these models represented flat-plate boundary layers and pipe flow well. The multiscale version represented supersonic flow over a compression corner marginally better than the more simple single scale model.

Hanjalic and Jakirlic (2002) surveyed various RSMs. They presented many examples of relatively simple incompressible flow simulations in which anisotropy was important. Many of these examples were from their own work. Simulations of oscillating and secondary flow circulations were particularly interesting. However, the authors own assessment was that

- The RSMs offer potential advantages for describing anisotropy, the evolution of the stress field, and streamline curvature effects.
- Numerical problems exist in applying the model and its complexity requires extensive computational resources.
- The use of wall function boundary conditions is necessary for reasonable flow analyses.
- The potential of the RSMs cannot be achieved without additional research.

The conservation equations shown in Table 4.2 are for compressible fluids. The equation for scalar transport of energy and mass (as indicated by the generic symbol φ) are written as

$$\frac{\partial \langle \varphi \rangle}{\partial t} + \nabla \cdot \left(\langle \vec{V} \rangle \langle \varphi \rangle \right) = \nabla \cdot \left(\Gamma \nabla \langle \varphi \rangle - \langle \vec{V}' \varphi' \rangle \right) \tag{5.6}$$

The laminar transport coefficient is Γ and $\langle \vec{V}' \varphi' \rangle$ is the turbulent transport of the scalar.

Very few comparison analyses have been reported for determining the level of improvement that can be expected versus the additional cost incurred for DSM simulations compared to EVM analyses. None of the simulations reviewed account for complex fluid properties or system geometries.

5.3 PROBABILITY DENSITY FUNCTION MODELS

The simple averaging methods just discussed allow analysts to address many practical turbulent transport processes. However, even with current computational power, these are not trivial or definitive simulations. Even so, not all of the issues defining turbulence can be addressed. Further statistical methods must be employed to represent other important turbulence properties. The simple averaging methodology must be replaced with probability theory.

Turbulent flow is ably described with probability theory by Pope (2000), Fox (2003), Monin and Yaglom (1965), and Hinze (1975). The nomenclature in these references is not the same; therefore, the nomenclature and development used by Pope will be used herein. When vectors or tensors are used by these authors, they use it for only Cartesian coordinate systems, unless specifically stated otherwise. The summary of probability parameters shown as Appendix A to Chapter 4 mainly follows that of Pope.

5.3.1 PDF DESCRIPTION OF TURBULENCE

The mean velocity and the Reynolds stresses are the first and second moments of the Eulerian PDF of velocity. An additional transport equation or an arbitrary specification is needed to define the PDF. However, such an analysis is not closed; a value of the turbulent energy dissipation or turbulent frequency must also be supplied to provide information on the turbulent timescale. An alternative is to utilize a joint PDF for velocity and turbulent frequency. If reacting, diffusive flows are to be described, additional correlations of velocity fluctuations and scalar variables are needed.

Pope derived the transport equation for the Eulerian PDF of velocity for incompressible flow to be

$$\frac{\partial f}{\partial t} + V_i \frac{\partial f}{\partial x_i} - \frac{1}{\rho} \frac{\partial \langle P \rangle}{\partial x_i} \frac{\partial f}{\partial V_i} = -\frac{\partial}{\partial V_i} \left[f \left\langle \nu \nabla^2 U_i - \frac{1}{\rho} \frac{\partial P'}{\partial x_i} \middle| \vec{V} \right\rangle \right] \quad (5.7)$$

The Navier–Stokes equation was used in this derivation. The RHS of this equation is not closed. Several closures were discussed by Pope. For example, the generalized Langevin model (GLM) results in the following:

$$\frac{\partial f}{\partial t} + V_i \frac{\partial f}{\partial x_i} - \frac{1}{\rho} \frac{\partial \langle P \rangle}{\partial x_i} \frac{\partial f}{\partial V_i} = -\frac{\partial}{\partial V_i} \left[f G_{ij} (V_j - \langle U_j \rangle) \right] + \frac{1}{2} C_o \varepsilon \frac{\partial^2 f}{\partial V_i \partial V_i} \quad (5.8)$$

G_{ij} and C_o are coefficients which are functions of $\langle U_i U_j \rangle$, ε and $\partial \langle U_i \rangle / \partial x_j$. Terms like ε or ω are also needed to close the equations. The conditional mean turbulence frequency and the normalized turbulent frequency are the generalizations needed to define the general velocity–frequency joint PDF. The first two of these quantities are

$$\Omega = C_\Omega \int_{\langle \omega \rangle}^{\infty} \theta f_\omega \{\theta\} d\theta \middle/ \int_{\langle \omega \rangle}^{\infty} f_\omega \{\theta\} d\theta \quad (5.9)$$

$$d\omega^* = -C_3 (\omega^* - \langle \omega \rangle) \Omega dt - \Omega \omega^* S_\omega dt + (2C_3 \sigma^2 \langle \omega \rangle \Omega \omega^*)^{0.5} dW \quad (5.10)$$

W is the Wiener process variable which is defined by diffusion and drift coefficients.

For the velocity–frequency PDF, the GLM model is used to yield ($\bar{f}$)

$$\frac{\partial \bar{f}}{\partial t} + V_i \frac{\partial \bar{f}}{\partial x_i} - \frac{1}{\rho} \frac{\partial \langle p \rangle}{\partial x_i} \frac{\partial \bar{f}}{\partial V_i} = -G_{ij} \frac{\partial}{\partial V_i} \left[\bar{f}(V_j - \langle U_j \rangle) \right] + \frac{1}{2} C_o \Omega k \frac{\partial^2 \bar{f}}{\partial V_i \partial V_i}$$

$$+ \frac{\partial}{\partial \theta} \left[\bar{f} \left(C_3 \left(\theta - \langle \omega \rangle \right) \Omega + \Omega \theta S_\omega \right) \right] + C_3 \delta^2 \langle \omega \rangle \Omega \frac{\partial^2 (\bar{f} \theta)}{\partial \theta^2} \quad (5.11)$$

This PDF along with the inclusion of a scalar reaction parameter was used to model a swirling combustion flow (Anand et al. 1997). The simulation was compared to experimental test data and fair agreement was demonstrated. The simplistic chemistry model alone could have accounted for several hundred degrees Kelvin discrepancy between the test data and the simulation.

For this methodology to realistically be useful for simulating three-dimensional combusting flows, drastic improvement in computational time is needed. For a three-dimensional flow three position variables, three velocity components, and time are the independent variables, this makes the methodology computationally prohibitive.

A major step in shortening run time is to utilize notional particles and Lagrangian formulations of the PDFs involved. An ensemble of "N" fluid particles is used to represent the flow. The location and velocity of the particles, denoted by $\vec{X}$; and $\vec{U}$, respectively, define the particle properties. One-point, one-time Eulerian PDFs for the fluid (f) and for the particles (f^*) are used to describe the statistics of the flow. Since the particles are the fluid, they move with the local fluid velocity. Particle properties are determined by solving the Equations 5.12 through 5.14 to represent the evolution of these properties in time.

$$d\vec{X}^*\{t\}/dt = \vec{U}^*\{t\} \quad (5.12)$$

$$dU^*\{t\} = \vec{a}\{\vec{U}^*\{t\}, \vec{X}^*\{t\}, t\} \, dt + b\{\vec{X}^*\{t\}, t\} \, d\vec{W} \quad (5.13)$$

This is a stochastic model equation for velocity where $\vec{a}$ and b are the drift and diffusion coefficients which will be described subsequently. In this case, $\langle \vec{a} | \vec{x} \rangle = -\nabla P/\rho$

$$dU_i^*\{t\} = -\frac{1}{\rho} \frac{\partial P}{\partial x_i} + G_{ij} \left(U_i^*\{t\} - \langle U_i^*\{t\} | \vec{X}^*\{t\} \rangle \right) \, dt + \left[C_o \varepsilon \left(\vec{X}^*\{t\}, t \right) \right]^{0.5} dW_i \quad (5.14)$$

The conditional means must be estimated and this is usually done with kernel estimation. For example,

$$\langle Q|\vec{x}\rangle_{N,h} = \frac{(\forall/N)\sum_{n=1}^{N} K\{\vec{x}-X^{(n)}, h\}\, Q^{(n)}}{(\forall/N)\sum_{n=1}^{N} K\{\vec{x}-X^{(n)}, h\}} \quad \text{where} \ \ \vec{x}\geq h \quad (5.15)$$

The kernel function in Equation 5.15 is defined by

$$K\{\vec{r},h\} = \alpha_D h^{-D}\left(1-\frac{r}{h}\right)$$
$$\text{for} \ \frac{r}{h}\leq 1 \ \text{and} = 0, \quad \text{for} \ \frac{r}{h}>1 \tag{5.16}$$

The error in this approximation slowly approaches zero, as the number of particles approach infinity.

To analyze reacting flows, the conservation equation for species (i.e., a reactive scalar) must be satisfied.

$$\frac{D\varphi}{Dt} = \Gamma\nabla^2\varphi + S\{\varphi\{\vec{x}, t\}\} \tag{5.17}$$

The solution of this equation with various accompanying submodels has been addressed exhaustively (not by the investigators, but by the readers). The most informative works are by Pope (1985, 2000), Fox (1996, 2003), Corrsin (1974), and Roekaerts (2002).

To apply probabilistic methods to the species equation, a one-point, one-time Eulerian PDF denoted by $f_\phi\{\psi;\vec{x},t\}$ is obtained from the evolution equation:

$$\frac{\partial f_\phi}{\partial t} + \frac{\partial}{\partial x_i}\left[f_\phi(\langle U_i\rangle + \langle u_i|\psi\rangle)\right] - \frac{\partial}{\partial\psi}\left(f_\phi\left\langle\frac{D\phi}{Dt}\Big|\psi\right\rangle\right)$$
$$= -\frac{\partial^2}{\partial\psi^2}(f_\phi[\langle\Gamma\nabla^2\phi|\psi\rangle + S\{\psi\}]) \tag{5.18}$$

The source term does not appear as an averaged value:

$$\langle S\{\phi\}|\phi = \psi\rangle = S\{\psi\} \tag{5.19}$$

A "turbulent convection flux" arises from the fluctuating velocity field, which is evaluated with a gradient-diffusion model and a turbulent diffusivity (Γ_T):

$$-f_\phi\langle\vec{u}|\psi\rangle = \Gamma_T\nabla f_\phi \tag{5.20}$$

A micro- or molecular-mixing model is also employed; one early example of which is

$$\left\langle \Gamma \nabla^2 \phi \middle| \psi \right\rangle = -\frac{1}{2} C_\phi \frac{\varepsilon}{k} \left(\psi - \langle \phi \rangle \right) \tag{5.21}$$

Several micro-mixing models are discussed by Jones (2002).

Micro-mixing models characteristically reduce the reaction rates in very lean or rich regions of non-premixed flames. Sometimes this predicts a better fit of laboratory test data; sometimes the fit is worse.

Combining these submodels the composition-PDF equation becomes

$$\frac{\overline{D}f_\phi}{Dt} = \frac{\partial}{\partial x_i} \left(\Gamma_T \frac{\partial f_\phi}{\partial x_i} \right) + \frac{\partial}{\partial \psi} \left[f_\phi \left(\frac{1}{2} C_\phi \frac{\varepsilon}{k} \left(\psi - \langle \varphi \rangle \right) - S \{ \psi \} \right) \right] \tag{5.22}$$

A turbulence model calculation must be made to predict the mean value of U, k, ε, and Γ_T before this equation may be solved for f_ϕ.

These notional particle models lead to requiring that a large number of particles must be tracked. Monte Carlo methods are used to accomplish this tracking. Such methods have been developed and used to simulate laboratory axisymmetric-burner flows (Tang et al., 2000). This is but one of the many simulations which Pope and his colleagues have pioneered and published. This work has utilized a wide variety of submodels and PDFs which the authors developed during the past decade. However, superiority over Reynolds averaged Navier–Stokes (RANS) turbulence model methods is claimed. Such claims are not substantiated by one-to-one simulation comparisons of the same experiments. All numerical analyses of combustor flows offer many opportunities for errors and intentional simplifications for computational convenience to affect the predictions. For example, the PDF methodology effectively modifies the reaction rates of the chemical system involved. These rates were determined by experiment, often involving turbulent flow. In effect, corrections for effects which are already imbedded in the component submodels are being duplicated. The PDF models are so much more complex, computationally intensive, and expensive that they better serve as computational experiments for further testing RANS models than for practical, day-to-day use. Highly three-dimensional flows can only be modeled with RANS methodology. Exotic PDF simulations are interesting, but they should be an adjunct to the more practical RANS models not a replacement.

5.3.2 COMMENTS ON STATISTICAL ANALYSIS OF DIFFUSION

The statistical analysis of diffusion from a point source into a homogeneous turbulent field was first addressed by Taylor (1921). A homogeneous turbulent field, to a good approximation, is generated by air flow through screens in a wind tunnel. The analysis described diffusion as a Fickian process (gradient diffusion) which at longer times took on the character of a random walk process. Corrsin (1974) showed that the gradient diffusion was obtained as a first-order approximation to a more complete statistical analysis of particle (fluid lump) motion.

Pasquill (1962) gave a similar analysis which he illustrated by describing the flow of a plume of smoke in the atmosphere (the atmosphere is approximately a homogeneous turbulent field). The plume is first observed to spread as the square of the distance from the source and at longer distances (or times) the spread rate decreases and becomes linear with the distance. This behavior is predicted by the statistical analysis. For analyzing transport phenomena in the atmosphere or in river plumes, the effect of nongradient diffusion could be significant. However, all of the discussion to this point has assumed that the turbulent field is of homogeneous turbulence. For the large distances required to change the character of fluid dispersion, there is a high probability that stronger convective currents might have an overriding effect on the dispersed flow. For most process analyses, the gradient diffusion models should be adequate since the length scales of the flow are rather small.

5.4 DIRECT NUMERICAL SIMULATION

All attempts to analyze turbulent flow begin with the laminar conservation laws. Rather than averaging the conservation equations to create a turbulent flow model, the unsteady, three-dimensional, constant-density, laminar momentum, and continuity equations are solved to represent turbulent flow. Analytical solutions are not possible due to the nonlinearity of the momentum equation. This means that numerical solutions using very small temporal and spatial steps are required to produce a DNS of the flow. The idea is to resolve the eddies for unbounded flows by using scalers of the order of Kolmogorov's micro-scale (η) and micro-timescale (τ).

$$\eta = \left(v^3/\varepsilon\right)^{0.25} \quad \text{and} \quad \tau = \left(v/\varepsilon\right)^{0.5} \tag{5.23}$$

The energy dissipation (ε) is the rate that the kinetic energy of the turbulent fluctuations (k) is dissipated as heat by the smallest eddies in the flow (per unit mass). Or, $\varepsilon = -dk/dt$. Actually, smaller timescales are needed for stable solutions, and about five times the spatial scales have been found acceptable. For wall-bounded flows, 10 grid points below y^+ of 10 are desired. The largest spatial scale of interest is dictated by the size of the region being analyzed. The solution is time dependent so statistical properties are generated to describe the flowfield. Methods to manage the copious amounts of computed data must then be devised to visualize and utilize turbulent flow properties.

The resolution requirements impose a very severe computational burden on DNS analyses. Only small computational domains, low Reynolds number flows, and simple geometries have been analyzed to date. It is considered to provide high-accuracy solutions for the flowfields which it predicts. However, the errors introduced by the boundary conditions imposed on the simulations and by the numerical methods used to obtain solutions have not been thoroughly evaluated. Simply getting a solution seems to be the criteria for getting a good answer. The major virtue of this methodology is that it provides numerical experiments which generate turbulence properties which cannot be experimentally measured. Besides cost of the simulations, the virtue is also a major drawback. How can the simulations be validated? The meager comparisons available are encouraging, but much more research is required to make use of this methodology.

To solve the laminar flow conservation equations initial and boundary conditions are needed. The pertinent equations require a highly accurate numerical solution over some finite domain. It is not a question as to whether or not the laminar flow equations apply, all of the turbulence models assume that this is true even when these equations are averaged. Rather, are the solution and the applied boundary conditions sufficiently accurate to represent the physics of the turbulent flow? The expectation is that all important turbulent eddy scales are to be obtained from the DNS solution. So the solution domain size cannot be extremely small to expedite the numerical solution, without compromising the evaluation of the large-scale eddies. And, since the computational requirements become impractical at high Reynolds number, only low Reynolds number flow may currently be analyzed.

At best, a DNS solution is not a solution of the laminar conservation laws; rather it is a solution of the difference analog which is used to numerically solve the equations for whatever boundary conditions are imposed on the flow. This means that high-order numerical methods must be used for the solver. Usually, the second- and fourth-order schemes are used. Spectral methods are also used to expedite the solutions, even though such methods are difficult to apply to nonuniform grids.

DNS methods are described in Rogallo and Monin (1984), Pope (2000), and Launder and Sandham (2002). Boundary layer simulations by Spalart (1988) for a Reynolds number based on momentum thickness up to 1410 show good agreement with Klebanhof's experimental data and correctly show the anisotropic nature of the flow near the wall. Spalart's simulations utilized up to about 10^7 grid points and a spectral method to solve the flow equations. Periodic spanwise and streamwise boundary conditions were applied. A multiscale procedure was used to approximate the local effects of streamwise growth of the flow. This was a careful and thoughtful study, but due to its difficulty many approximations to the idealized DNS theory were required. The study was concluded by analyzing the spectra deduced from the data and comparing the results to experimental measurements.

Further implications from such boundary layer simulations are discussed by Monin and Mahesh (1998). A major deduction from these computed data was to elucidate the important role played by mean shear rather than bursting phenomena in boundary layers. Incompressible channel flow and the trailing-edge wake of flow over a plate have been simulated with DNS and the results summarized by Launder and Sandham (2002). Other DNS simulations are referenced by these same authors. Such computational experiments are expensive and difficult to conduct with acceptable accuracy; however, once successfully completed they offer a statistical database for testing more computationally efficient turbulence models.

In summary, DNS is a useful research tool for investigating fundamental turbulence characteristics which is not expected to provide engineering design information anytime in the foreseeable future.

5.5 LARGE EDDY SIMULATION

The need for simulating large-scale meteorological (Smagorinsky, 1963; Deardorff, 1973) and hydrological (Bedford and Dakhoul, 1982) flow phenomena suggested the desirability of averaging the Navier–Stokes equations over space instead of or in addition to over time. Leonard (1974) formally performed such averaging by introducing a filter function. Subsequently, this methodology has been recognized as a bridge to

construct a hybrid turbulence model to DNS and RANS models to produce a more computationally efficient model. The resulting LES models
have served to represent the large-scale flows. They still leave much to be
desired as practical turbulent flow simulators.

5.5.1 LES METHODOLOGY

The large eddies are to be resolved by computation. The smallest eddies
are to be modeled with a subgrid scale model (SGS). A spatial filtering
operation is defined to decompose an instantaneous quantity into the sum
of a resolved scale quantity and a subgrid scale quantity. The large-scale
motion is computed explicitly by solving the Navier–Stokes equation, and
is called resolved scale. The SGS terminology is a misnomer because it
is not necessarily the small-grid spacing, but the spatial-filter width that
dictates when the small eddy models need to be modeled. The finer the
grids and filters the more of the energy is resolved in the calculations.
The energy is usually produced in the large-scale eddies and dissipated in
small eddies near a wall.

The smallest grid size needed in the LES method is much larger than
the Kolmogorov scale needed in the DNS methods. The difficulty in posing
boundary and initial conditions and in computing the smallest resolved
eddies incurred in using DNS methods is the same in using LES methods.
The concept is that stress-bearing and dissipative eddies are separated, in
fact they overlap.

Methodology:

- A filtering operation is defined to decompose an instantaneous
  quantity ϕ into the sum of a resolved scale quantity and a subgrid
  scale quantity ϕ.
- The large-scale motion is computed explicitly by solving the
  Navier–Stokes equation, and is called the resolved scale.
- The small-scale motion, assumed to be nearly universal, is simulated by solving the modeled turbulence transport equation with
  the length scale given by the filter, and is called the SGS.
- The finer the filter and grids are the higher percentage of the
  energy is resolved.
- The energy transfer is usually from resolved to SGSs, but may be
  reversed near solid boundaries, where small productive eddies are not
  resolved and the SGS model must account for the lost production.

The filtering operation is performed using a generalized filter (Leonard, 1974)

$$\overline{\varphi} = \int_{-\infty}^{\infty} G(\vec{x} - \vec{\xi}; \Delta)\varphi(\vec{\xi}, t)d\xi; \quad \varphi = \overline{\varphi} + \varphi'; \quad \overline{\frac{\partial \varphi}{\partial x}} = \frac{\partial \overline{\varphi}}{\partial x} \tag{5.24}$$

The overbar denotes a spatial average and the prime the spatial variation of the arbitrary property φ. Unlike time averaging, the mean value of the filtered residue φ' is not zero. G is the generalized filter and Δ the filter width. For a three-dimensional velocity field

$$\overline{V}_i(\vec{x}, t) = \iiint G(\vec{x} - \vec{\xi}; \Delta)V_i(\vec{\xi}, t)d^3\xi \quad \text{and}$$

$$\iiint G(\vec{x} - \vec{\xi}; \Delta)d^3\vec{\xi} = 1 \quad V_i' = V_i - \overline{V}_i; \quad \Delta = (\Delta x \Delta y \Delta z)^{1/3} \tag{5.25}$$

In this example the filter width is the cube root of the grid volume. For Favre filtering for compressible flow

$$\overline{\rho\varphi} = \int_{-\infty}^{\infty} G(\vec{x} - \vec{\xi}; \Delta)\rho\varphi(\vec{\xi}, t)d\xi;$$

$$\overline{\rho} = \int_{-\infty}^{\infty} G(\vec{x} - \vec{\xi}; \Delta)\rho(\vec{\xi}, t)d\xi \quad \tilde{\varphi} = \frac{\overline{\rho\varphi}}{\overline{\rho}}; \quad \varphi = \tilde{\varphi} + \varphi'' \tag{5.26}$$

The filter function must satisfy the following conditions:

$$\int_{-\infty}^{\infty} G(\xi) = 1 \tag{5.27}$$

$$G(\xi) \to 0 \text{ as } |\xi| \to 0 \text{ such that } \int_{-\infty}^{\infty} G(\xi)\xi^n d\xi (n \geq 0) \text{ exist} \tag{5.28}$$

$$G(\xi) \approx 0 \quad \text{if} \quad \xi > \frac{1}{2}\Delta, \quad \text{or} \quad \xi < -\frac{1}{2}\Delta \tag{5.29}$$

Common filter functions are the volume-averaged box filter (Deardorff, 1970),

$$G(\vec{x} - \vec{\xi}; \Delta) = \begin{cases} 1/\Delta^3, & |x_i - \xi_i| < \Delta x_i/2 \\ 0, & \text{otherwise} \end{cases} \tag{5.30}$$

the Fourier (spectral) filter (Orsag et al., see Ferziger, 1976),

$$G(\vec{x} - \vec{\xi}; \Delta) = \frac{1}{\Delta^3} \prod_{i=1}^{3} \frac{\sin(x_i - \xi_i)/\Delta}{(x_i - \xi_i)/\Delta} \tag{5.31}$$

and the Gaussian filter (Ferziger, 1976)

$$G(\vec{x} - \vec{\xi}; \Delta) = \left(\frac{C}{\pi \Delta^2}\right)^{1.5} \exp\left(\frac{-C|\vec{x} - \vec{\xi}|^2}{\Delta^2}\right); \quad 2 \leq C \leq 6 \qquad (5.32)$$

The filtered incompressible momentum equation is

$$\rho \frac{\partial \overline{V}_i}{\partial t} + \sum_j \rho \frac{\partial \overline{V_i V_j}}{\partial X_j} = \rho \frac{\partial \overline{V}_i}{\partial t} + \sum_j \rho \frac{\partial \overline{V}_i \overline{V}_j}{\partial X_j} - \frac{\partial \tau^r_{ij}}{\partial X_i} = -\frac{\partial \overline{P}}{\partial X_i} + \mu \sum_j \frac{\partial^2 \overline{V}_i}{\partial X_j^2}$$

where $\quad \tau^R_{ij} = \overline{V_i V_j} - \overline{V}_i \overline{V}_j; \quad \tau^r_{ij} = \tau^R_{ij} - (2/3) k_r \delta_{ij} \quad$ and $\quad k_r = 0.5 \tau^R_{ij}$

$$(5.33)$$

The anisotropic residual-stress tensor is τ^r_{ij} and k_r is the residual kinetic energy.

The residual stress $\tau^R_{ij} = \overline{V_i V_j} - \overline{V}_i \overline{V}_j$. Terms called stresses which do not include the density are stresses per unit mass. If the residual stress is defined and the filter used is spatially uniform, the momentum equation will be closed and may be solved. If the isotropic residual stress is included in the pressure $\overline{P} = \overline{P} + (2/3)k_r$, the Poisson equation may be solved for the modified pressure.

One evaluation of the residual stress is the Galilean-invariant decomposition suggested by Germano (Pope, 2000):

$$\rho \frac{\partial \overline{V}_i}{\partial t} + \sum_j \rho \frac{\partial \overline{V_i V_j}}{\partial X_j} = -\frac{\partial \overline{P}}{\partial X_i} + \mu \sum_j \frac{\partial^2 \overline{V}_i}{\partial X_j^2}$$

and

$$\overline{V_i V_j} = \overline{V}_i \overline{V}_j + L_{ij} + C_{ij} + R_{ij} \qquad (5.34)$$

$$\text{Leonard stress: } L_{ij} = \overline{\overline{V}_i \overline{V}_j} - \overline{V}_i \overline{V}_j \qquad (5.35)$$

$$\text{Cross-term stress: } C_{ij} = \overline{\overline{V}_i V'_j} + \overline{\overline{V}_j V'_i} - \overline{\overline{V}_i V'_j} - \overline{V'_i \overline{V}_j} \qquad (5.36)$$

$$\text{SGS Reynolds stress: } R_{ij} = \overline{V'_i V'_j} - \overline{V}'_i \overline{V}'_j \qquad (5.37)$$

The residual stress is analogous to the Reynolds stress. This term removes significant energy from the resolvable scales and can be computed directly. SGS stress term will be modeled. The cross stress tensor also drains significant energy from the resolvable scales, and can be treated as L_{ij}, or modeled as R_{ij}. Alternatively, spectral filters can be defined and utilized for LES analyses.

Smagorinsky in 1963 was the first to propose a residual-stress model (SGS).

$$\left(\tau_t^{SGS}\right)_{ij} = 2\mu_t S_{ij}; \quad S_{ij} = \frac{1}{2}\left(\frac{\partial \overline{V}_i}{\partial X_j} + \frac{\partial \overline{V}_j}{\partial X_i}\right); \quad \mu_t = \rho(C_S \Delta)^2 \sqrt{2S_{ij}S_{ij}} \quad (5.38)$$

This resembles a mixing length EVM with the length being $C_S\Delta$. Using the grid spacing as Δ is crude; frequently C_S is set at 0.1–0.24 to improve this approximation. The dynamic SGS model is a modification which reevaluates the filter during the course of the simulation.

Advantages of the LES model are first that it allows energy transfer between the resolved and SGS in roughly the correct magnitude. Compared to RANS models, LES models can describe large-scale turbulent structure, which allows approximations of unsteady aerodynamic loads and estimates of sound generation. It is particularly useful for simulating very large-scale environmental phenomena.

Disadvantages are that it is computationally inefficient. Run times are approximately twice that of using RSMs. Application to high-speed compressible and reacting flows will probably require additional development and increased complexity.

5.5.2 LES APPLICATIONS

Developing LES methodology is a popular research area. The simulations are very computationally intensive and boundary condition specification and implementation is difficult. Several major centers for computational research are actively pursuing LES development. Sandia National Laboratories is conducting advanced engine combustion R&D based on LES techniques. The laboratory computational efforts are supported by two "Beowulf" clusters which were funded by two major DOE programs. Progress reports on this effort (Oefelein, 2006) indicate that the codes being developed and utilized are all inclusive for describing spray combustors. The simulations are being validated by accompanying experiments. The submodels included to represent spray combustion processes and grid resolution reported are impressive. Specific numerical issues, for example, the use of staggered grids and preconditioning, are not generally good solution methodology, but with the extreme computer power they are utilizing it cannot be said that such practices impact their work. Hard comparisons to critical combustion properties do not yet justify the confidence placed in or cost of this major research program. Justification of LES methods over RANS methods has not been justified by one-to-one simulation comparisons. The major success of the Sandia work is their ability to secure and sustain major research funding for computational research.

The Building and Fire Research Laboratory of NIST is utilizing LES methodology to simulate fire-driven flows in rooms (Mell et al. 1995). In such flows, the fire itself is small in comparison to the room size. Like applications to meteorological and hydrological problems, the disparity in size of the domain to be analyzed to the turbulent interior flows is good justification for using LES methodology. Also like the Sandia work, a very large initial research effort was required to develop and validate (with nonreacting flow simulations) the LES codes used for the desired application. Again, continued support by government agencies has been required to develop the LES technology to a useful status.

Laurence (2002) addresses the question is "Large Eddy Simulation of (for) Industrial Flows?" and concludes "…at present and in the near future, industry is not likely to use LES for actual engineering applications, despite, or rather because of, the daily use of (RANS) CFD." He cites examples of LES technology in describing acoustics and fluid–structure interaction phenomena. Even in these examples, first the small pressures predicted with LES have to be supplemented by using an acoustics program to predict noise generation. For analyzing a tube bundle in cross-flow, a subset of the tubes was considered in a two-dimensional simulation. One tube surrounded by four-quarter tubes constituted the computational domain. LES results for the wake-to-impingement axis were much better than those obtained from a RANS analysis. Finally, Laurence concludes that the utility of LES is to support, not replace, RANS methodology by analyzing subregions of complex flows where the advantages of LES could provide improved predictions. The small-scale region simulations he reviews are examples of the subregions which could be analyzed with LES methodology.

5.6 LAMINAR-TO-TURBULENT TRANSITION MODELS

Since Reynolds observed laminar and turbulent flow in a pipe by noticing the different rate of spread of injected dye, analytical attempts have been made to describe these flows. Further experiments showed that the two types of flow did not instantly turn from one to the other. Initially, laminar flow began to fluctuate and eventually become completely chaotic. The dye filament was noticed to retrain its initial thinness for some distance downstream from the injection point. The higher the Reynolds number the shorter the region of retained thinness. The transition region, where an unsteadiness in changing from laminar to turbulent flow was observed, was difficult to analyze. Numerous attempts to describe flow transition have not yet yielded a universal solution. Because of the analytical difficulty of this phenomenon, boundary layers with simple geometry have received the most attention. Despite the many years of experimental and analytical effort, current modeling methods still depend almost completely on the flow geometry and thus require modification to simulate specific geometries.

Four classes of transition modeling methodology have been reported. The numerous variants of these methods have been reviewed by Cheng et al. (2009) and Launder and Sandham (2002). The classes are (1) linear stability theory, (2) models with specified transition onset, (3) practical models which predict onset and transition, and (4) models which are still under development. The classes of models will be described in the following. Some of the class 3 models will be shown to be natural extensions of the computational transport phenomenon (CTP) code methodology.

5.6.1 LINEAR STABILITY THEORY

Natural transition occurs when a laminar flow reacts to small disturbances by fluctuating to a point which does not recover, but becomes increasingly chaotic. Parallel flows of boundary layers, free shear layers, and wakes can be effectively analyzed with linear stability theory. When this method is applied to incompressible flow, the conservation laws are linearized and expressed in terms of mean and fluctuating variables. For the simple geometry involved, the equations are reduced to three variables, the fluctuating velocity components parallel and normal to the main flow and the fluctuating pressure. These are further simplified using complex variables to produce one fourth-order linear homogeneous equation; the Orr–Sommerfeld equation (White, 2006). The resulting solution is referred to as the Tollmien–Schlichting waves.

Subsequent analyses have removed some of the highly restrictive assumptions made in describing stability in this manner. For example, consideration of temperature effects on viscosity and pressure gradient effects on stability have been estimated. Favorable pressure gradients and fluid cooling increase stability. Converse effects reduce stability. However, the over riding restriction is that the onset of transition is predicted, the transition process to fully turbulent flow is not.

An application of the linear stability theory to the prediction of the onset of transition is given by Smith and Gamberoni (1956). Such predictions and experiments show that the natural transition of a laminar to a turbulent boundary layer is a slow process involving a long induction length. More likely, the main flow will not be quiet enough and the wall not smooth enough, for such natural transitions to occur. Skipping the initial steps of the natural process results in what is termed bypass transition for which surface vortices are assumed to be formed immediately. The impact of such an analysis shall be described subsequently.

A third cause of the onset of transition is the reattachment of separated flow. When the flow reattaches it immediately forms a turbulent boundary layer. This behavior is not only useful for analytical purposes, but is the reason that boundary layer trips are used to expedite experimental investigations of turbulent boundary layers. Experimentally one does not wish to depend on the long run required for natural transition onset to occur, followed by the transition region to develop before creating a fully turbulent boundary layer for study.

5.6.2 TRANSITION MODELS BASED ON A SPECIFIED ONSET VALUE

A transition-onset model (usually that of Smith and Gamberoni, 1956), a weighting function (usually an empirical intermittency), and a conventional EVM of any of several correlations is used to predict laminar to turbulent transition. Various degrees of success have been reported with this methodology. Also, some predictions have been made using modifications to two-equation turbulence models. The intermittency correlations frequently used are those of Dhawan and Narasimha (1958).

The mechanics of this method are awkward. The laminar boundary layer is calculated, and the onset of transition points estimated. The turbulent boundary layer with the turbulence model of choice is made from the onset point onward. Then a third calculation is made from the onset point onward using the blended laminar/turbulent viscosity model. Thus, this methodology is not practical for complex geometries, involving other than parallel flows.

5.6.3 TRANSITION MODELS WITH PREDICTED ONSET

The literature review of Cheng et al. (2009) covered several methods of modeling transition including its onset. Two of the models deemed to be the most promising for coupling to a RANS solver with a two-equation turbulence model, for example, the CTP code, were selected for further study. These models were one that described both a laminar and a turbulent kinetic energy contribution to fluctuating velocities (Walters and Leylek, 2004) and the second that calculates a vorticity Reynolds number to determine the onset and transition of the boundary layer (Langtry and Menter, 2005; Menter et al., 2006).

The Walters–Leylek model is based on the concept that bypass transitions are caused by very high amplitude laminar streamwise fluctuations. Starting with a low Reynolds number k–ε turbulence model, the source term in the k-equation was modified, a second transport equation for laminar kinetic energy, and a dissipation rate equation which included both the laminar and turbulent kinetic energies were solved to obtain Walters–Leylek transition model simulations. For energy transfer, laminar Prandtl numbers and turbulent Prandtl numbers based on the turbulent kinetic energy are used.

The local correlation-based transition model (LCTM) uses the vorticity Reynolds number as the measure which controls transition. The model solves transport equations for the intermittency and for the momentum thickness Reynolds number. The vorticity Reynolds number is calculated from the momentum thickness Reynolds number. If separation occurs, the intermittency is increased such that correct reattachment is achieved. The calculated intermittency is used to modify the production and destruction terms in the shear-stress transport (SST) turbulence model to complete the model. Critical values of the momentum thickness Reynolds number were reexamined to improve the model's capability for predicting hypersonic flows (Cheng et al., 2009).

These solution methodologies do not require multiple solutions over the computation domain. For instance, a laminar flow solution and a turbulent flow solution followed by a blended solution of the two is not required. This produces a somewhat more efficient solution, but, more importantly, the methods can be used for more geometrically complex process simulations.

5.6.4 VALIDATION CASES

In general, a wealth of information exists describing transitional flows. However, effective simulations with reasonably general, transition models have only been successful recently. Few comprehensive benchmark experimental test cases are yet available. Since so many different investigators and so many different computational approaches have been reported, a definitive state of the art has not been established. Though successes have been reported by various turbulence transition model developers, those transition models were only tuned and validated for a limited number of test cases, and thus the generality of these transition models are questionable. Furthermore, essentially all of the models and their validation is still more of a research nature than that of a mature technology. The work reported by Cheng et al. (2009) is a step in remedying this situation.

Cheng et al. (2009) has conducted third-party evaluation of the Walters–Lelek and the LCTM transition models by comparing simulations to test data from three experiments: (1) three subsonic flows over a flat plate, (2) hypersonic flow over a cylindrical cone, and (3) subsonic flow over a turbine stator cascade. The general performance of the models was encouraging, but additional model development and validation testing were recommended. In regard to the latter, the establishment of a database for benchmark quality transition experiments was also recommended.

5.6.5 OTHER MODELING APPROACHES

Several other transitional flow modeling approaches have been reported, mostly by the European community (Launder and Sandham, 2002). The aforementioned subsonic boundary layer experiments are compiled in the ERCOFTAC database (Coupland, 1993). These data were modeled in a pragmatic fashion by predicting an onset point followed by an arbitrary blending into a fully turbulent boundary layer. The data were also analyzed by directly using a low Reynolds number k–ε turbulence model and by using a Reynolds stress turbulence model. These results are summarized by Savill (2002).

DNS has been performed for some boundary layer flows (Durbin et al. 2002). Due to the large computational domain required of analyzing natural transition, only bypass simulations have been conducted. These authors estimated that free-stream turbulence levels of 1% were sufficient to bypass the Tollmien–Schlichting route to transition. As previously mentioned, DNS methodology is so computationally intense that only numerical experiments for benchmark data are expected to be produced. The same can be said of possible LES turbulence models.

Additional transitional modeling research has been directed toward increasing the applicability of low Reynolds number turbulence models. These efforts have been reviewed by Savill (2002). The current conclusion is that the near wall region must be described with such a fine grid that practical simulations are not possible.

The research being conducted to describe transitional flows is essentially studies of relatively simple geometries with complex turbulent models. Recommendations for establishment of databases for representative are being made. The recommendations are generally in line with establishing benchmark test cases for computational studies. It appears reasonable to extend this idea to also include experimental data of complex flows, which could be used for establishing empirical correlations for practical applications.

5.7 NOMENCLATURE

b	diffusion coefficient in stochastic diffusion equation
b_{ij}, f_s	functions in the anisotropic term for energy dissipation in the RSM model
C_o	constant in the Langevin equation
C_Ω	constant in definition of Ω
C_3	constant in definition of ω^*
Cn	constants in the RSM model. $n = 1, 2, 3, 4, \varepsilon, \varepsilon_1, \varepsilon_2$
C_{ij}	substantial derivative of the Reynolds stress in the RSM
C_s	constant in SGS model
C_{ij}	cross-stress term
f, f^*	Eulerian PDF for incompressible flow; for notional particles
f_φ	one-time, one-point Eulerian PDF
G	generalized filter
G_{ij}	function ij; the GLM

h	grid spacing
H	dummy symbol to indicate mathematical operations
H^*	notional particle
k, k_r	turbulent kinetic energy; residual kinetic energy
K, D, α_D	parameters in kernel function
L_{ij}	Leonard stress
P	pressure
$P\{B\}$	probability of event B occurring
P_{ij}	production term in RSM
R_{ij}	SGS Reynolds stress
$S\{\psi\}$	source term in species continuity equation (SCE)
Q	kernel function
S_{ij}	component of the pressure-strain term in the RSM
S_{ij}	stress component in SGS model
$S_\omega, S\{\varphi\}$	mean source of turbulent frequency; source of property φ
t	time
t_K	Kolmogorov's timescale
U	function; standardized (normalized) function
$\vec{U}, \vec{U}^*$	velocity vector; for notional particles
V_i	velocity component; sample space variable in Section 5.3
W	Wiener process variable
x_i, X_i, X_i^*	spatial coordinates; *for notional particles
Y	spatial coordinate; distance from a wall
y^+	dimensionless distance from the wall
Z	spatial coordinate

5.7.1 GREEK SYMBOLS

E	energy dissipation
ε_{ij}	dissipation term in the RSM
μ, μ_t	laminar and turbulent viscosity
Γ, Γ_t	laminar, turbulent diffusion coefficients
ν	laminar kinematic viscosity
Δ	filter function
Ξ	dummy length parameter
H	Kolmogorov micro-scale
T	Kolmogorov timescale
P	density
φ, ψ	property to be averaged
ϕ	subgrid property in LES model

Π_{ij}	pressure-strain term in the RSM
Ω	conditional mean turbulent frequency
ω	turbulent frequency or specific dissipation rate
ω^*	model for turbulent frequency with sample space θ

5.7.2 MATHEMATICAL SYMBOLS

$\vec{a}$	drift coefficient in stochastic diffusion equation
D_{ij}	component of the pressure-strain in the RSM
D_{ij}	diffusion term in he RSM
D_{ij}^t	turbulent diffusion term in the RSM
D_{ij}^v	laminar diffusion term in the RSM
$\tilde{f}$	velocity–frequency PDF with the GLM
H', H''	fluctuating component
$\vec{H}, \overline{H}, \tilde{H}$	vector, time or spatial average, Favre average
τ_{ij}^r	anisotropic residual stress
τ_{ij}^R	residual stress
τ_{ij}^{SGS}	subgrid stress

5.7.3 ACRONYMS

DNS	direct numerical simulation
DSM	differential second-moment model
EVM	eddy viscosity model
GLM	generalized Langevin model
LES	large eddy simulation
PDF	probability density function
RANS	Reynolds averaged Navier–Stokes turbulence model
RSM	Reynolds-stress model
SCE	species continuity equation
SGS	subgrid scale

REFERENCES

Anand, M. S., A. T. Hsu, and S. B. Pope. 1997. Calculations of swirl combustors using joint velocity-scalar probability density function method. *AIAA J.* 35:1143–1150.

Bedford, K. W. and Y. M. Dakhoul. 1982. Applying LES turbulence modeling to open channel flow. In P. E. Smith (Ed.), *Proc. Conf. Applying Res Hydraulic Practice*. Jackson, MS.

Cheng, G. C., R. Nichols, K. D. Neroorkar, and P. G. Radhamony. 2009. Validation and assessment of turbulence transition models. Submitted to *47th AIAA Aerospace Sciences Meeting*. Orlando, FL.

Corrsin, S. 1974. Limitations of gradient transport models in random walks and in turbulence. In *Advances in Geophysics, Vol. 18A*, F. N. Frenkiel and R. E. Munn (Eds.), pp. 237–248. New York: Academic Press.

Coupland, J. 1993. ERCOFTAC classic database. http://cfd.mc.umist.ac.uk/ercoftac/

Daly, B. and F. Harlow. 1970. Transport equations in turbulence. *Phys. Fluids*. 13:2634–2649.

Deardorff, J. W. 1970. A numerical study of three-dimensional turbulent channel flow at large Reynolds numbers. *J. Fluid Mech*. 41:453–480.

Deardorff, J. W. 1973. The use of subgrid transport equations in a three-dimensional model of atmospheric turbulence. *ASME J. Fluids Engr*. 95:429–438.

Dhawan, S. and R. Narasimha. 1958. Some properties of boundary layer flow during transition from layinar to turbulent motion. *J. Fluid Mech*. 3:418–436.

Durbin, P. A., R. G. Jacobs, and X. Wu. 2002. DNS of bypass transition. In *Closure Strategies for Turbulent and Transitional Flows*. Launder, B. E. and N. D. Sandham (Eds.), pp. 449–463. Cambridge, U.K.: Cambridge University Press.

Ferziger, J. H. 1976. Large eddy numerical simulations of turbulent flows. AIAA Paper 76–347. AIAA.

Fox, R. O. 1996. Computational methods for turbulent reacting flows in the chemical process industry. *Revue de L'Institut Francais du Petrole*. 51:215–243.

Fox, R. O. 2003. *Computational Models for Turbulent Reacting Flows*. Cambridge, U.K.: Cambridge University Press.

Hanjalic, K. and S. Jakirlic. 2002. Second-moment turbulence closure modelling. In *Closure Strategies for Turbulent and Transitional Flows*. Launder, B. E. and N. D. Sandham (Eds.), pp. 47–101, Cambridge, U.K.: Cambridge University Press.

Hanjalic, K. and B. E. Launder. 1976. Contribution towards a Reynolds-stress closure for low-Reynolds-number turbulence. *J. Fluid Mech*. 74:593–610.

Jones, W. P. 2002. The joint scalar probability density function method. In *Closure Strategies for Turbulent and Transitional Flows*. Launder, B. E. and N. D. Sandham (Eds.), pp. 626–655. Cambridge, U.K.: Cambridge University Press.

Langtry, R. B. and F. R. Menter. 2005. Transition modeling for general CFD applications in aerodynamics. AIAA Paper 2005–522.

Launder, B. E. and N. D. Sandham, Eds. 2002. *Closure Strategies for Turbulent and Transitional Flows*. Cambridge: Cambridge University Press.

Launder, B. E., G. J. Reece, and W. Rodi. 1975. Progress in the development of a Reynolds-stress turbulence closure. *J. Fluid Mech*. 68:537–566.

Laurence, D. 2002. Large eddy simulation of industrial flows? In *Closure Strategies for Turbulent and Transitional Flows*. Launder, B. E. and N. D. Sandham (Eds.), pp. 392–406. Cambridge, U.K.: Cambridge University Press.

Leonard, A. 1974. Energy cascade in large-eddy simulations of turbulent flows. In *Advances in Geophysics, Vol. 18A*, F. N. Frenkiel and R. E. Munn (Eds.), pp. 237–248. New York: Academic Press.

Mell, W., K. B. McGrattan, and H. R. Baum. 1995. Large eddy simulations of fire-driven flows. *HTD-ASME*. 2:73–77.

Mellor, G. L. and H. J. Herring. 1973. A survey of the mean turbulent field closure models. *AIAA J.* 11:590–599.

Menter, F. R., R. B. Langtry, S. R. Likki, Y. B. Suzen, P. G. Huang, and S. Volker. 2006. A correlation-based transition model using local variables, Part 1 – Model formulation. *J. Turbomach.* 128:413–422.

Monin, P. and K. Mahesh. 1998. Direct numerical simulation: A tool for turbulence research. *Ann. Rev. Fluid Mech.* 30:539–578.

Monin, A. S. and A. M. Yaglom. 1965. *Statistical Fluid Mechanics: Mechanics of Turbulence, Vols. I and II.* Cambridge: MIT Press.

Oefelein, J. C. 2006. Large eddy simulation applied to low-temperature and hydrogen engine combustion. FY 2006 progress Report. Sandia National Laboratories. Livermore.

Pasquill, F. 1962. *Atmospheric Diffusion.* New York: D. Van Nostrand.

Pope, S. B. 1985. PDF methods for turbulent reactive flows. *Prog. Energy combust. Sci.* 11:119–192.

Pope, S. B. 2000. *Turbulent Flows.* Cambridge: Cambridge University Press.

Roekaerts, D. 2002. Reacting flows and probability density function methods. In *Closure Strategies for Turbulent and Transitional Flows.* Launder, B. E. and N. D. Sandham (Eds.), pp. 328–337. Cambridge, U.K.: Cambridge University Press.

Rogallo, R. S. and P. Monin. 1984. Numerical simulation of turbulent flows. *Ann. Rev. Fluid Mech.* 16:99–137.

Savill, A. M. 2002. New strategies in modelling by-pass. In *Closure Strategies for Turbulent and Transitional Flows.* Launder, B. E. and N. D. Sandham, (Eds.), pp. 493–521, Cambridge, U.K.: Cambridge University Press.

Smagorinsky, J. 1963. General circulation experiments with the primitive equations. 1. The basic experiment. *Mon. Weather Rev.* 91:99–164.

Smith, A. M. O. and N. Gamberoni. 1956. Transition, pressure gradient and stability theory. Douglas Aircraft Co. Dept. ES26388, El Segundo, CA.

Spalart, P. R. 1988. Direct simulation of a turbulent boundary layer up to $R_\theta = 1410$. *J. Fluid Mech.* 187:61–98.

Speziale, C. G., R. Abid, and T. B. Gatski. 1991. Modeling the pressure-strain correlation of turbulence. *Ann. Rev. Fluid Mech.* 227:245–272.

Tang, Q., J. Xu, and S. B. Pope. 2000. Probability density function calculations of local extinction and NO production in piloted-jet turbulent methane/air flames. *Proc. Combust. Inst.* 28:133–139.

Taylor, G. I. 1921. Diffusion by continuous movements. Reprinted in 1961 *Turbulence: Classical Papers on Statistical Theory.* S. K. Frielander and L. Topper (Eds.), pp. 1–16. New York: Interscience Publishers.

Walters, D. K. and J. H. Leylek. 2004. A new model for boundary layer transition using a single-point RANS approach. *J. Turbomach.* 126:193–202.

Wilcox, D. C. 2006. *Turbulence Modelling for CFD*, 3rd edn. La Canada: DCW Industries.

Wilcox, D. C. 1998. *Turbulence Modelling for CFD*, 3rd edn. La Canada: DCW Industries.

White, F. M. 2006. *Viscous Fluid Flow*, 3rd edn. New York: McGraw-Hill.

CHAPTER 6

Computational Coordinates and Conservation Laws

6.1 OVERVIEW

The conservation laws, fluid properties, and turbulence models presented in Chapters 2 through 4 represent a wide class of transport phenomena. In general, these equations and models are too complex to solve analytically. Simple geometries consisting of pseudo-one-dimensional flows and flows which are contained within orthogonal Cartesian or cylindrical walls may be solved for a variety of interesting problems by a number of approximate methods. However, the most complete solutions to date are numerical solutions of the nonlinear partial differential form of the governing laws. This is the purview of the computational transport phenomenon (CTP). To produce a practical, general-purpose computational tool, turbulence will be modeled with a two-equation model and radiation will be neglected. These restrictions leave an immense number of challenging transport problems which may be addressed with the code described herein.

Since numerical solutions require a defined coordinate system of uniquely identified grid points, the geometry issue will be addressed in this chapter. Pseudo-one-dimensional flows utilize control volumes in which convection along one direction is to be evaluated. Rectangular Cartesian coordinates and cylindrical coordinates constitute the remaining practical control volumes which may be defined with algebraic coordinate systems. More general geometries require curvilinear orthogonal or nonorthogonal coordinates. Vector and tensor analysis is the mathematical language invented to describe curvilinear geometry. Tensor analysis describes both coordinate systems and components of vector and tensor quantities. It also serves as a type of shorthand, but this is a trivial consequence. Such formalism looks concise on paper, but it introduces considerable unnecessary complexity. The definition of coordinate lines is a necessity, but base vectors which change direction within the computational domain cannot be conveniently integrated throughout the flowfield. As an alternative, one may define curvilinear coordinates as a mathematical transformation of independent variables into what is called function space or vector space. The scalar-dependent variables from a coordinate system which utilizes base vectors which are invariant in direction are used without change. This is a straightforward, although complex, process, which is described in this chapter. Be advised! Mathematicians like to generalize, engineers do not. This causes some confusing literature which, hopefully, will be elucidated in this chapter. The essentials of these issues are presented in this chapter in order to arrive at the discretized equations which are to be solved by the CTP code.

The objective of this text is to describe and explain the use of the mature CTP computer code, and then to illustrate its application to basic transport phenomena analyses. The intent is not to teach one how to write such a code; there are an abundance of books and thousands of papers available for such a purpose. Detailed explanations and derivations will be avoided whenever possible to maintain the focus on understanding and using the computational tools already available. Appropriate references will be given to locate existing back-up material, but where such material is unclear or obtuse further explanation will be provided in Appendix B on the basics of tensor analysis.

6.2 COORDINATES

Consider a few simple flow systems, such as flow through a converging diverging nozzle, an orifice, or an elbow. These flows cannot be accurately described with simple orthogonal Cartesian or cylindrical coordinates. They can be described using body-fitted coordinates. In fact, very complex flows such as the flow about an entire aircraft can now be described with such coordinates. Since friction and heat and mass transfer cause severe gradients near solid walls, coordinates which increase resolution near such walls are also useful. Body-fitted coordinates allow boundary layer resolution. On occasion, one wishes to reposition the coordinates used during the course of a calculation. This too is possible using adaptive grid generation. The grid system used in the CTP code is a structured, multizone body-fitted coordinate system, it is not adaptive. There are other useful grid systems with the same, and in some cases even more general, discretization capability, but in no case are more than three spatial and one temporal coordinates needed.

The conservation equations contain both dependent and independent variables. The independent variables are the grid system being used. If the grid is adaptive, that is, it is allowed to vary as the computation progresses, time (or the iteration number) would also be a variable in defining the coordinates. For general flow domains, body-fitted curvilinear coordinates would be used to describe the geometry. The dependent velocity components could be transformed to curvilinear values, or remain the scalars used in an orthogonal Cartesian system. Scalar variables like pressure, temperature, and density would be unaffected by coordinate transformations. The CTP code utilizes the scalar-dependent variables derived in Chapter 2 for orthogonal Cartesian or cylindrical coordinates.

The objective of this section is to derive and explain the form of the conservation laws which will then be discretized and numerically solved. To utilize general nonorthogonal coordinates, two methods of transforming the orthogonal Cartesian coordinates and the conservation equations expressed in these coordinates are available. First, the coordinates may be considered multivalued functions and linearly transformed by matrix operations. Since curvilinear coordinates are a desired result, the transformation must be applied to differential coordinates. The relationship between the Cartesian coordinates (x, y, z) and the general curvilinear coordinates (η, ξ, ζ) is not specified until after the transformation is accomplished. This is the method described by Anderson et al. (1997) and Anderson (1995) and is the method which is used in this chapter. Their nomenclature will also be used wherever possible in order to make comparisons to their work easier. Unfortunately, the straightforward matrix methodology is contaminated by using terms, specifically vectors, to describe an array which has no directional property associated with it. This contamination is due to mathematicians "generalizing" terms for their use. For example, they wish to consider a vector to be n-dimensional. Courant and Hilbert (1953, p. 1) define n-dimensional vectors, then state "For $n > 3$ geometrical visualization is no longer possible but geometrical terminology remains suitable." The mechanics of this matrix transformation will contain no unit or base vectors, yet the method is still adequate for our purposes. Herein, when a nongeometrical *vector* is implied, the term *vector* will be italicized. Such vectors are really one-dimensional arrays or column vectors. When a symbol is needed, a tilde overbar will be used instead of the usual arrow to represent vectors. Matrices will be indicated with a caret overbar.

The second method for coordinate transformation is to use the methodology of tensor analysis. This method is not used here because its nomenclature is too awkward and no single text on the subject can be recommended. The methodology has been "generalized" by mathematicians to the point that its application to three-dimensional geometry problems has been obscured. Another problem not encountered with the matrix methodology is that base vectors are required which change direction from point to point in the field and may even change magnitude. This means that the integration of vector terms in PDEs would have to be considered. Note cylindrical coordinates are orthogonal and utilize unit vectors, but these unit vectors change direction throughout the control volume. Thus, this coordinate system is of intermediate complexity. Nevertheless, in order for the equations presented in this chapter to be placed in proper perspective with respect to existing literature, curvilinear coordinate formulations of the conservation equations will also be derived and presented in Appendix B.

6.2.1 COORDINATE TRANSFORMATIONS

Symbolically coordinate transformations and the inverse coordinate transformations are represented by

$$\begin{aligned}
\xi &= \xi\{x, y, z\} & x &= x\{\xi, \eta, \zeta\} \\
\eta &= \eta\{x, y, z\} \quad \text{and} \quad & y &= y\{\xi, \eta, \zeta\} \\
\zeta &= \zeta\{x, y, z\} & z &= z\{\xi, \eta, \zeta\}
\end{aligned} \tag{6.1}$$

This inversion is possible because there is a one-to-one correspondence of grid points between the coordinate systems, i.e., the transformation is admissible. However, there are several major problems for performing this transformation if the curvilinear coordinate system is not analytic. Since it is desirable to treat body-fitted geometries of general shapes, the transformations must be made in a convenient form. But even in simple analytical boundaries the curvilinear systems are difficult to describe. The base (or unit) vectors point in different directions at each point in the field. Curvilinear coordinates are not linear; hence the curved coordinates must be defined incrementally. To fit general shapes, the transformations must be performed numerically. Even then, additional considerations must be given to the numerical solution of accompanying transport phenomena problems. The nonlinear transport equations must be solved numerically also. For this numerical solution to be performed efficiently, scaling of grid lines and both grid metrics and inverse grid metrics are used to form the difference equations. This additional information is needed to produce a computational grid. These issues are discussed in the following.

The local, differential form of the transformation relations are

$$d\xi = \xi_x dx + \xi_y dy + \xi_z dz$$
$$d\eta = \eta_x dx + \eta_y dy + \eta_z dz \tag{6.2}$$
$$d\zeta = \zeta_x dx + \zeta_y dy + \zeta_z dz$$

or

$$\begin{bmatrix} d\xi \\ d\eta \\ d\zeta \end{bmatrix} = \begin{bmatrix} \xi_x & \xi_y & \xi_z \\ \eta_x & \eta_y & \eta_z \\ \zeta_x & \zeta_y & \zeta_z \end{bmatrix} \begin{bmatrix} dx \\ dy \\ dz \end{bmatrix} \tag{6.3}$$

where the subscripts indicate partial differentiation. The inverse transformations are

$$dx = x_\xi d\xi + x_\eta d\eta + x_\zeta d\zeta$$
$$dy = y_\xi d\xi + y_\eta d\xi + y_\zeta d\zeta \tag{6.4}$$
$$dz = z_\eta d\eta + z_\xi d\xi + z_\zeta d\zeta$$

or

$$\begin{bmatrix} dx \\ dy \\ dz \end{bmatrix} = \begin{bmatrix} x_\xi & x_\eta & x_\zeta \\ y_\xi & y_\eta & y_\zeta \\ z_\xi & z_\eta & z_\zeta \end{bmatrix} \begin{bmatrix} d\xi \\ d\eta \\ d\zeta \end{bmatrix} \tag{6.5}$$

The derivative transformations are

$$\frac{\partial}{\partial x} = \xi_x \frac{\partial}{\partial \xi} + \eta_x \frac{\partial}{\partial \eta} + \zeta_x \frac{\partial}{\partial \zeta}$$
$$\frac{\partial}{\partial y} = \xi_y \frac{\partial}{\partial \xi} + \eta_y \frac{\partial}{\partial \eta} + \zeta_y \frac{\partial}{\partial \zeta} \tag{6.6}$$
$$\frac{\partial}{\partial z} = \xi_z \frac{\partial}{\partial \xi} + \eta_z \frac{\partial}{\partial \eta} + \zeta_z \frac{\partial}{\partial \zeta}$$

or

$$
\begin{bmatrix} \xi_x & \xi_y & \xi_z \\ \eta_x & \eta_y & \eta_z \\ \zeta_x & \zeta_y & \zeta_z \end{bmatrix} = \begin{bmatrix} x_\xi & x_\eta & x_\zeta \\ y_\xi & y_\eta & y_\zeta \\ z_\xi & z_\eta & z_\zeta \end{bmatrix}^{-1}
$$

$$
= J \begin{bmatrix} y_\eta z_\zeta - y_\zeta z_\eta & -(x_\eta z_\zeta - x_\zeta z_\eta) & x_\eta y_\zeta - x_\eta y_\xi \\ -(y_\xi z_\zeta - y_\zeta z_\xi) & x_\xi z_\zeta - x_\zeta z_\xi & -(x_\xi y_\zeta - x_\zeta y_\xi) \\ y_\xi z_\eta - y_\eta z_\xi & -(x_\xi z_\eta - x_\eta z_\xi) & x_\xi y_\eta - x_\eta y_\xi \end{bmatrix}
\tag{6.7}
$$

where J is the Jacobian of the transformation and

$$
\begin{aligned}
\xi_x &= J(y_\eta z_\zeta - y_\zeta z_\eta) \\
\xi_y &= -J(x_\eta z_\zeta - x_\zeta z_\eta) \\
\xi_z &= J(x_\eta y_\zeta - x_\zeta y_\eta) \\
\eta_x &= -J(y_\xi z_\zeta - y_\zeta z_\xi) \\
\eta_y &= J(x_\xi z_\zeta - x_\zeta z_\xi) \\
\eta_z &= -J(x_\xi y_\zeta - x_\zeta y_\xi) \\
\zeta_x &= J(y_\xi z_\eta - y_\eta z_\xi) \\
\zeta_y &= -J(x_\xi z_\eta - x_\eta z_\xi) \\
\zeta_z &= J(x_\xi y_\eta - x_\eta y_\xi)
\end{aligned}
\tag{6.8}
$$

$$
J = \frac{\partial(\xi, \eta, \zeta)}{\partial(x, y, z)} = \begin{vmatrix} \xi_x & \xi_y & \xi_z \\ \eta_x & \eta_y & \eta_z \\ \zeta_x & \zeta_y & \zeta_z \end{vmatrix}
\tag{6.9}
$$

$$
J = 1/J^{-1} = 1 \Big/ \frac{\partial(x, y, z)}{\partial(\xi, \eta, \zeta)} = 1 \Big/ \begin{vmatrix} x_\xi & x_\eta & x_\zeta \\ y_\xi & y_\mu & y_\zeta \\ z_\xi & z_\eta & z_\zeta \end{vmatrix}
$$

$$
= 1 \Big/ \left[x_\xi (y_\eta z_\zeta - y_\zeta z_\eta) - x_\eta (y_\xi z_\zeta - y_\zeta z_\xi) - x_\zeta (y_\xi z_\eta - y_\eta z_\xi) \right]
\tag{6.10}
$$

Notice that these equations do not contain base or unit vectors. The effect of such vectors must be contained in the coordinate lengths only. Since there is not a one-to-one correspondence between curvilinear coordinate increments and the rectilinear incremental lengths, another measure must be preserved in the transformation. This measure is the incremental distance between two position (or grid) points. This measure is defined with the metric tensor, as explained in Appendix B. The relationships between the (x, y, z) and the (ξ, η, ζ) coordinates called "metrics" in the above equations are related to but not identical with the elements of the metric tensor.

To solve the transformation relationship between the curvilinear and rectilinear coordinate systems, a definition of the curvilinear system must be made. Analytical relationships for orthogonal coordinates have been developed to yield cylindrical coordinates, spherical coordinates, etc. For nonorthogonal curvilinear coordinates, coordinate lines may be chosen as the intersection of two coordinate surfaces (contravariant coordinates) or as normals to coordinate surfaces (covariant coordinates). A coordinate surface is one for which two of the three coordinates are constants. These various coordinate systems are given specific names and characterization by the formalism by tensor analysis as described in Appendix B. However, analytical relationships are of limited value; therefore, numerical relationships based on the use of boundary-fitted coordinates have been developed.

6.2.2 BODY-FITTED COMPUTATIONAL COORDINATES

To define body-fitted and computational coordinates, the length along a curvilinear boundary must be evaluated. Ideally this could be accomplished by using analytical curves to represent the boundaries, but this does not provide a general solution to constructing computational grids. Complex grid generation is usually established by numerical methods, for which both simple and complex geometries can be described with one computer code. This is readily accomplished by calculating incremental lengths in orthogonal Cartesian coordinates and numerically converting them into the general curvilinear coordinates that fit body surfaces and are nonlinearly dispersed to give higher resolution near body (i.e., wall) surfaces. The grids so generated are frequently structured so that arrays spanning the three coordinate directions may be utilized for computational convenience. This is the methodology used in the CTP code.

Other grid generation methodology such as unstructured, adaptive, finite-element, and very fine rectangular Cartesian grids are also used (Anderson, 1995). Modern computer speed and storage has largely supplanted the use of more elaborate grid generation methods. However, to work CTP problems on personal computers (PCs), the methodology used in the CTP code is efficient and accurate for many practical applications.

6.2.3 ARC LENGTH AND COORDINATE LINES

The control volume within which the transport equations are to be solved must be filled with a grid system, the nodes of which will be points where the equations are numerically solved. For arbitrary curvilinear walls, curvilinear coordinates are generated by locating a coordinate line along the surface. This line may be defined by an analytical function, a set of specified points which are curve fit, or results from a computer-aided design (CAD) calculation. The length of the coordinate line must be determined. The following describes how this is accomplished.

An increment of arc length (ds) in a plane in an orthogonal Cartesian coordinate system is given by (Korn and Korn, 1968)

$$ds = \left(\left(dx \right)^2 + \left(dy \right)^2 \right)^{0.5} \tag{6.11}$$

or by

$$ds = \left(\frac{ds}{dt} \right) dt = \left[\left(\frac{dx}{dt} \right)^2 + \left(\frac{dy}{dt} \right)^2 \right]^{0.5} dt \tag{6.12}$$

where t is a parameter. Arc length along a line in a three-dimensional surface described with an orthogonal Cartesian coordinate system is

$$ds = \left[(dx)^2 + (dy)^2 + (dz)^2 \right]^{0.5} \tag{6.13}$$

If x, y, and z are functions of ξ along that line

$$ds = \left(\frac{ds}{d\xi} \right) d\xi = \left[\left(\frac{dx}{d\xi} \right)^2 + \left(\frac{dy}{d\xi} \right)^2 + \left(\frac{dz}{d\xi} \right)^2 \right]^{0.5} d\xi \tag{6.14}$$

Therefore, given $\xi\{x, y, z\}$, the length along this coordinate line can be calculated. Likewise, the other two curvilinear coordinates (η and ζ) could be evaluated.

Had we wished to use this line as the ξ coordinate and have it vary from 0 to 1.0, a choice would still have to be made as to how the line was to be incremented. Regardless of the method of discretization (linear, exponential, arbitrarily nonuniform), a length scale for the coordinate would have to be defined. The obvious choice would be the physical length of the coordinate. This means that the length of the line in the (x, y, z) coordinate system would have to be calculated before the ξ coordinate could be defined. For example, say 10 increments of length were to be used. "s" would be calculated first then $\Delta s = s/10$. Now the (x, y, z) coordinate positions along ξ from 0 to 1.0 can be located. This is precisely the manner in which a boundary-fitted coordinate system is generated.

An alternative to relating the general curvilinear coordinate system to the orthogonal Cartesian system would be to relate it to another type of orthogonal system. For example, a cylindrical coordinate system for pipe flow or for a circular duct of varying cross-section analyses. Such a system must allow all of the coordinates to produce linear measures along the coordinate lines. The transformation required would then be from (ξ, η, ζ) to (r, θ, z) or to (x, y, z). If the centerline of the conduit is curved (such as for an elbow), the grid generation is more complex. However, analyses of such configurations are in the literature. An example is described in Patankar (1980).

6.2.4 BODY-FITTED COORDINATE SYSTEMS

Consider the two-dimensional region bounded by the $x = 0$, $x = 20$, $y = 0$ and a quadrant of an elliptic curve:

$$y = 10 + \left[1600 - 4(x^2 - 40x + 400)\right]^{0.5} \tag{6.15}$$

Curvilinear coordinates are to be introduced as $\xi = 0$ along $x = 0$ and $\xi = 1$ along $x = 20$ and $\eta = 0$ along $y = 0$ and $\eta = 1$ along the elliptic segment. Twenty-one grid lines are defined along the ξ direction and six along the η direction. The grid lines are to be equally spaced. This means that the distance along the elliptic segment must be calculated so that it can be divided into the equal segments. Since distance has not yet been defined in the ξ direction along the $\eta = 1$ coordinate, the required distance is calculated in the x–y coordinate system.

The interior grid lines and node points are calculated by transfinite interpolation (TFI) between the two-boundary coordinates $\eta = 0$ and $\eta = 1$. This method of interpolation is described in detail by Shih et al. (1991). These investigators described the application of this methodology to parallelepipeds with two-, four-, and six-sides being curvilinear and unsteady. The extension to adaptive gridding, the inclusion of orthogonal coordinates near walls, and the treatment of stretching was also described. Finally, these investigators provided a grid generation code (the GRID2D/3D code) for its implementation. Their work is an application of the basic boundary fitting methodology first introduced by Thompson et al. (1985).

The nonadaptive, two-boundary TFI version of this methodology was used to construct the curvilinear grid for this example. The X, Y, and Z functions are the fits of the boundaries. The ℓ functions are the interpolation functions. These functions are given by

$$
\begin{aligned}
x\{\xi,\eta,\zeta\} &= X_1\{\xi,\zeta\}\ell_1\{\eta\} + X_2\{\xi,\zeta\}\ell_2\{\eta\} \\
y\{\xi,\eta,\zeta\} &= Y_1\{\xi,\zeta\}\ell_1\{\eta\} + Y_2\{\xi,\zeta\}\ell_2\{\eta\} \\
z\{\xi,\eta,\zeta\} &= Z_1\{\xi,\zeta\}\ell_1\{\eta\} + Z_2\{\xi,\zeta\}\ell_2\{\eta\} \\
\text{where} \quad &\ell_1\{\eta\} = 1-\eta \quad \text{and} \quad \ell_2\{\eta\} = \eta
\end{aligned}
\tag{6.16}
$$

The grid for this illustrative problem is shown in Figure 6.1. The nodal numbers I are associated with the η-grid values, and the nodal numbers J correspond to the ξ-grid values. Thus, a structured grid that spans the I–J domain was created.

Notice, no unit or base vectors have been defined or used in this grid generation procedure. Neither has orthogonally between the arbitrarily

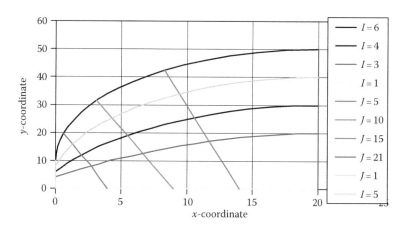

FIGURE 6.1 (See color insert following page 294.) Curvilinear coordinate system.

scaled curvilinear coordinates been assumed or plotted. Many papers on grid generation indicate that such orthogonality results from the transformation, including the two just referenced. The relationship between the curvilinear coordinates and the orthogonal Cartesian coordinates is simply their respective values at the common I-J nodal values of the grid line intersections. This correspondence is of immense value in establishing structured grid differencing analyses. It has no other purpose. The literature continually refers to these arrays as "rectangular." This description is misleading and unnecessary. Specifically, no base vectors which indicate orthogonally of the curvilinear coordinate system are specified or implied.

6.3 CONSERVATION LAWS IN COMPUTATIONAL COORDINATES

6.3.1 FORMULATION OF THE CONSERVATION LAWS FOR THE CTP CODE

The chemical and physical fluid properties and the laminar and turbulent conservation laws have been discussed in Chapters 3 through 5. The CTP code was designed as a practical tool for analyzing the transport processes. All of the known transport processes could not be included and have the code remain practical, much less understandable. In our judgment, the basic code should treat multicomponent real, reacting fluids in laminar and turbulent flow. In the course of the CTP code development, some extensions to these processes were also found to be conveniently included. Simple multiphase flow and fluid structural interaction such as conjugate heat transfer were included in the code capability. These additions are described only as specific examples in this work. Other inclusions such as dense multiphase flow and radiation transfer are not treated as they would involve major coding modifications. The basic code is still a complicated model consisting of a discretized set of nonlinear partial differential equations (PDEs) and their numerical solution. Not only must such equations be solved, but a computational grid which accurately represents complex walls must be used to avoid the need of an excessively dense grid to resolve boundary layer effects.

The evolution of the CTP code development resulted in the HBMS thermal and caloric equations of state being used to describe real multicomponent fluid properties. The Pade' integration scheme was selected to integrate finite-rate chemical reactions. A set of laminar transport property correlations was recommended for use. Turbulent flows were modeled with time and mass-averaged conservation equations. A two equation k–ε correlation was used to close these conservation equations. Effective eddy transport coefficients were determined from the k–ε values to produce an eddy viscosity model (EVM) for the turbulence effects. Since the k–ε model was developed for incompressible flow, extensions to the original model to account for more realistic fluid and flow properties were included in the CTP code. A generalized force term is included in the momentum and energy equations to account for such effects as buoyancy, Coriolis forces, and centrifugal forces.

The appropriate conservation equations stated in orthogonal Cartesian coordinates are shown in Table 6.1. The orthogonal adjective may be dropped for convenience, but it is always to be implied herein. The solution involves seven plus i PDEs, a thermal and a caloric equation of state. This provides one more variable than necessary. The extra variable may be used for checking the solution to the species continuity equation. The restrictions implied by these equations have been described in previous chapters. However, this set of equations serves to represent an extremely wide range of transport phenomena. To utilize these equations, a body-fitted curvilinear grid and a numerical algorithm for solving a discretized analog of these conservation laws must be available. The remainder of this and further chapters and the appendices of this work describe the complete computational methodology.

A single effective diffusion coefficient was used to represent multicomponent mixtures. This is a very good approximation for turbulent flow. For laminar flow, it represents diffusion of a single species in a mixture of other species, which is strictly correct for diffusion in a binary mixture. For more species involved, a correction to the single value could be made using the generalized Maxwell–Stefan equation by using an explicit calculation. The explicit calculation would involve evaluating the diffusion fluxes of all of the species present, which is feasible since the solution strategy involves time stepping until all of the conservation equations are solved. This would be a simple calculation, but it is not in the present code, nor has its importance been investigated by these authors. The thermal energy equation in Table 6.1 is cast in terms of enthalpy. The diffusion flux for enthalpy is given by

$$\vec{j}_h = \vec{q} + \sum_k \vec{j}_k h_k = -\kappa \nabla T - \rho D^m \sum_k h_k \nabla \omega_k \qquad (6.17)$$

TABLE 6.1 Conservation Form of the Transport Equations in Orthogonal Cartesian Coordinates

$$\frac{\partial \rho}{\partial t} + \frac{\partial \rho u}{\partial x} + \frac{\partial \rho v}{\partial y} + \frac{\partial \rho w}{\partial z} = 0 \tag{A}$$

$$\frac{\partial \rho \omega_a}{\partial t} + \left[\frac{\partial \rho u \omega_a}{\partial x} + \frac{\partial \rho v \omega_a}{\partial y} + \frac{\partial \rho w \omega_a}{\partial z}\right] - \left[\frac{\partial}{\partial x}\left(\phi_D \frac{\partial \omega_a}{\partial x}\right) + \frac{\partial}{\partial y}\left(\phi_D \frac{\partial \omega_a}{\partial y}\right) + \frac{\partial}{\partial z}\left(\phi_D \frac{\partial \omega_a}{\partial z}\right)\right] = K_a \tag{B}$$

$$\frac{\partial \rho u}{\partial t} + \frac{\partial \rho u u}{\partial x} + \frac{\partial \rho u v}{\partial y} + \frac{\partial \rho u w}{\partial z} - \left[\frac{\partial}{\partial x}\left(\mu_e \frac{\partial u}{\partial x}\right) + \frac{\partial}{\partial y}\left(\mu_e \frac{\partial u}{\partial y}\right) + \frac{\partial}{\partial z}\left(\mu_e \frac{\partial u}{\partial z}\right)\right] \tag{C}$$
$$= -\frac{\partial p}{\partial x} + \frac{\partial}{\partial x}\left[\left(\frac{2\lambda_c}{3}\mu_e\right)\left(\frac{\partial u}{\partial x} + \frac{\partial v}{\partial y} + \frac{\partial w}{\partial z}\right)\right] + \frac{\partial}{\partial x}\left(\mu_e \frac{\partial u}{\partial x}\right) + \frac{\partial}{\partial y}\left(\mu_e \frac{\partial v}{\partial x}\right) + \frac{\partial}{\partial z}\left(\mu_e \frac{\partial w}{\partial x}\right) + \rho g_x$$

$$\frac{\partial \rho v}{\partial t} + \frac{\partial \rho u v}{\partial x} + \frac{\partial \rho v^2}{\partial y} + \frac{\partial \rho v w}{\partial z} - \left[\frac{\partial}{\partial x}\left(\mu_e \frac{\partial v}{\partial x}\right) + \frac{\partial}{\partial y}\left(\mu_e \frac{\partial v}{\partial y}\right) + \frac{\partial}{\partial z}\left(\mu_e \frac{\partial v}{\partial z}\right)\right] \tag{D}$$
$$= -\frac{\partial p}{\partial y} + \frac{\partial}{\partial y}\left[\left(\frac{2\lambda_c}{3}\mu_e\right)\left(\frac{\partial v}{\partial y} + \frac{\partial u}{\partial x} + \frac{\partial w}{\partial z}\right)\right] + \frac{\partial}{\partial x}\left(\mu_e \frac{\partial u}{\partial y}\right) + \frac{\partial}{\partial y}\left(\mu_e \frac{\partial v}{\partial y}\right) + \frac{\partial}{\partial z}\left(\mu_e \frac{\partial w}{\partial y}\right) + \rho g_y$$

$$\frac{\partial \rho w}{\partial t} + \frac{\partial \rho u w}{\partial x} + \frac{\partial \rho v w}{\partial y} + \frac{\partial \rho w^2}{\partial z} - \left[\frac{\partial}{\partial x}\left(\mu_e \frac{\partial w}{\partial x}\right) + \frac{\partial}{\partial y}\left(\mu_e \frac{\partial w}{\partial y}\right) + \frac{\partial}{\partial z}\left(\mu_e \frac{\partial w}{\partial z}\right)\right] \tag{E}$$
$$= -\frac{\partial p}{\partial z} + \frac{\partial}{\partial z}\left[\left(\frac{2\lambda_c}{3}\mu_e\right)\left(\frac{\partial w}{\partial z} + \frac{\partial v}{\partial y} + \frac{\partial u}{\partial x}\right)\right] + \frac{\partial}{\partial x}\left(\mu_e \frac{\partial u}{\partial z}\right) + \frac{\partial}{\partial y}\left(\mu_e \frac{\partial v}{\partial z}\right) + \frac{\partial}{\partial z}\left(\mu_e \frac{\partial u}{\partial z}\right) + \rho g_z$$

(continued)

TABLE 6.1 (continued)   Conservation Form of the Transport Equations in Orthogonal Cartesian Coordinates

$$\frac{\partial \rho h}{\partial t} + \left[\frac{\partial \rho u h}{\partial x} + \frac{\partial \rho \upsilon h}{\partial y} + \frac{\partial \rho w h}{\partial z}\right] - \left[\frac{\partial}{\partial x}\left(\phi_h \frac{\partial h}{\partial x}\right) + \frac{\partial}{\partial y}\left(\phi_h \frac{\partial h}{\partial y}\right) + \frac{\partial}{\partial z}\left(\phi_h \frac{\partial h}{\partial z}\right)\right] = \frac{Dp}{Dt} + \bar{\bar{\tau}} : \nabla \bullet \vec{U} = \frac{Dp}{Dt} + \Phi \qquad \text{(F)}$$

$$\frac{\partial \rho k}{\partial t} + \left[\frac{\partial \rho u k}{\partial x} + \frac{\partial \rho \upsilon k}{\partial y} + \frac{\partial \rho w k}{\partial z}\right] - \left[\frac{\partial}{\partial x}\left(\phi_k \frac{\partial k}{\partial x}\right) + \frac{\partial}{\partial y}\left(\phi_k \frac{\partial k}{\partial y}\right) + \frac{\partial}{\partial z}\left(\phi_k \frac{\partial k}{\partial z}\right)\right] = \rho(P_r - \varepsilon) \qquad \text{(G)}$$

$$\frac{\partial \rho \varepsilon}{\partial t} + \left[\frac{\partial \rho u \varepsilon}{\partial x} + \frac{\partial \rho \upsilon \varepsilon}{\partial y} + \frac{\partial \rho w \varepsilon}{\partial z}\right] - \left[\frac{\partial}{\partial x}\left(\phi_\varepsilon \frac{\partial \varepsilon}{\partial x}\right) + \frac{\partial}{\partial y}\left(\phi_\varepsilon \frac{\partial \varepsilon}{\partial y}\right) + \frac{\partial}{\partial z}\left(\phi_\varepsilon \frac{\partial \varepsilon}{\partial z}\right)\right] = \frac{\rho k}{\varepsilon}\left[\left(\frac{C_1 + C_3 P_r}{\varepsilon}\right)P_r - C_2 \varepsilon\right] \qquad \text{(H)}$$

$$P_r = \frac{\mu_t}{\rho}\left[2\left(u_x^2 + \upsilon_y^2 + w_z^2\right) + \left(\upsilon_x + u_y\right)^2 + \left(w_y + \upsilon_z\right)^2 + \left(u_z + w_x\right)^2 - \frac{2\lambda_c}{3}\left(u_x + \upsilon_y + w_z\right)^2\right] \qquad \text{(I)}$$

If the Lewis number $Le = \rho C_p D^m/\kappa$ is unity, $\rho D^m = \kappa/C_p$. Also, $h = \Sigma_k h_k \omega_k$.

It follows that $\vec{j}_h = -(\kappa/C_p)\nabla h = -(\mu/Pr)\nabla h$. This relationship is interesting (and a similar one involving the viscosity and Schmidt number) because the Prandtl and Schmidt numbers are primarily a function of composition; whereas, the viscosity is a strong function of temperature and pressure. Practically, a correlation for viscosity in terms of temperature and pressure for an average composition and the use of constant values of Prandtl and Schmidt numbers would provide a good approximation for the laminar transport properties. Such an approximation is even better for turbulent flow. This explains the logic in using the effective viscosity model in the transport equations. A further benefit of using the enthalpy form of the energy equation is that real fluid properties for the enthalpy of a multicomponent mixture are easily obtained.

It bears repeating, the intent of this text is to provide the readers a tool with which one can use their own PCs to become familiar with computational solutions of the conservation equations representing transport phenomena. In this vain, the CTP code is chosen as the tool. The CTP code is serial, that is, it runs one computer sequentially until a solution is obtained. This implies that only modest size grid systems should be analyzed. Current computational methodology utilizes an ensemble of PCs running in parallel to analyze larger problems. The CTP code would be typical of an element in a parallel computing environment. However, much can be learned and practiced by analyzing more modest simulations. If more ambitious simulations are desired, the parallel processing methodology could be employed or analyses could be obtained from leased codes such as Fluent. This work is designed to educate one in the use of computational methodology for transport phenomena. It is time consuming to write or even use one's own code, but the learning curve for using comprehensive, leased software is even more excessive.

6.3.2 VECTOR FORM OF THE CTP CONSERVATION EQUATIONS IN CARTESIAN COORDINATES

The multicomponent conservation equations from Table 6.1 are expressed in vector form as follows for a Cartesian coordinate system. These vectors are column vectors or one-dimensional arrays. This formalism is introduced to expedite discussing the equations and to place their development in the proper historical perspective.

$$\frac{\partial \tilde{U}}{\partial t} + \frac{\partial \tilde{E}}{\partial x} + \frac{\partial \tilde{F}}{\partial y} + \frac{\partial \tilde{G}}{\partial z} = \tilde{S} \quad \text{or} \quad \tilde{U}_t + \tilde{E}_x + \tilde{F}_y + \tilde{G}_z = \tilde{S} \tag{6.18}$$

$$\tilde{U} = \begin{bmatrix} \rho \\ \rho u \\ \rho \upsilon \\ \rho w \\ \rho h \\ \rho \omega_i \\ \rho k \\ \rho \varepsilon \end{bmatrix} \tag{6.19}$$

$$\tilde{E} = \begin{bmatrix} \rho u \\ \rho u^2 + p - \tau_{xx} \\ \rho u \upsilon - \tau_{xy} \\ \rho u w - \tau_{xz} \\ \rho u h + j_{hx} \\ \rho u k + j_{kx} \\ \rho u \varepsilon + j_{\varepsilon x} \end{bmatrix} \tag{6.20}$$

$$\tilde{F} = \begin{bmatrix} \rho \upsilon \\ \rho u \upsilon - \tau_{yx} \\ \rho \upsilon^2 + p - \tau_{yy} \\ \rho \upsilon w - \tau_{yz} \\ \rho \upsilon h + j_{hy} \\ \rho \upsilon \omega_i + j_{iy} \\ \rho \upsilon k + j_{ky} \\ \rho \upsilon \varepsilon + j_{\varepsilon y} \end{bmatrix} \tag{6.21}$$

$$\tilde{G} = \begin{bmatrix} \rho w \\ \rho u w - \tau_{zx} \\ \rho \upsilon w - \tau_{zy} \\ \rho w^2 + p - \tau_{zz} \\ \rho w h + j_{hz} \\ \rho w \omega_i + j_{iz} \\ \rho w k + j_{kz} \\ \rho w \varepsilon + j_{\varepsilon z} \end{bmatrix} \tag{6.22}$$

$$\tilde{S} = \begin{bmatrix} 0 \\ F_x \\ F_y \\ F_z \\ \dfrac{DP}{Dt} + \Phi + \vec{U} \cdot \vec{F} \\ K_i \\ \rho(P_r - \varepsilon) \\ \dfrac{\rho k}{\varepsilon}\left[\left(\dfrac{C_1 + C_3 P_r}{\varepsilon}\right) P_r - C_2 \varepsilon\right] \end{bmatrix} \tag{6.23}$$

The diffusive terms in the conservation equations are now expressed in terms of the gradients which drive the processes. For momentum transport

$$\begin{aligned} \tau_{xx} &= 2\mu \frac{\partial u}{\partial x} - \left(\frac{2}{3}\mu - \lambda\right)\left(\frac{\partial u}{\partial x} + \frac{\partial \upsilon}{\partial y} + \frac{\partial w}{\partial z}\right) \\ \tau_{yy} &= 2\mu \frac{\partial \upsilon}{\partial y} - \left(\frac{2}{3}\mu - \lambda\right)\left(\frac{\partial \upsilon}{\partial y} + \frac{\partial u}{\partial x} + \frac{\partial w}{\partial z}\right) \\ \tau_{zz} &= 2\mu \frac{\partial w}{\partial z} - \left(\frac{2}{3}\mu - \lambda\right)\left(\frac{\partial w}{\partial z} + \frac{\partial u}{\partial x} + \frac{\partial \upsilon}{\partial y}\right) \end{aligned} \tag{6.24}$$

The second coefficient of viscosity (λ) is assumed negligible for the CTP equations. However, a parameter, λ_c, is included to expedite distinguishing between compressible and incompressible flows. Herein, compressible is meant to be variable density flow and incompressible, constant density flow.

$$\tau_{xx} = 2\mu \frac{\partial u}{\partial x} - \left(\frac{2\lambda_c}{3}\mu\right)\left(\frac{\partial u}{\partial x} + \frac{\partial v}{\partial y} + \frac{\partial w}{\partial z}\right)$$

$$\tau_{yy} = 2\mu \frac{\partial v}{\partial y} - \left(\frac{2\lambda_c}{3}\mu\right)\left(\frac{\partial v}{\partial y} + \frac{\partial u}{\partial x} + \frac{\partial w}{\partial z}\right) \qquad (6.25)$$

$$\tau_{zz} = 2\mu \frac{\partial w}{\partial z} - \left(\frac{2\lambda_c}{3}\mu\right)\left(\frac{\partial w}{\partial z} + \frac{\partial u}{\partial x} + \frac{\partial v}{\partial y}\right)$$

$$\tau_{xy} = \tau_{yx} = \mu\left(\frac{\partial u}{\partial y} + \frac{\partial v}{\partial x}\right)$$

$$\tau_{xz} = \tau_{zx} = \mu\left(\frac{\partial w}{\partial x} + \frac{\partial u}{\partial z}\right) \qquad (6.26)$$

$$\tau_{yz} = \tau_{zy} = \mu\left(\frac{\partial v}{\partial z} + \frac{\partial w}{\partial y}\right)$$

$$
\begin{array}{llll}
j_{hx} = -\dfrac{\mu}{Pr}\dfrac{\partial h}{\partial x} & j_{ix} = -\dfrac{\mu}{Sc}\dfrac{\partial \omega_i}{\partial x} & j_{kx} = -\dfrac{\mu}{\sigma_k}\dfrac{\partial k}{\partial x} & j_{\varepsilon x} = -\dfrac{\mu}{\sigma_\varepsilon}\dfrac{\partial \varepsilon}{\partial x} \\[2ex]
j_{hy} = -\dfrac{\mu}{Pr}\dfrac{\partial h}{\partial y} & j_{iy} = -\dfrac{\mu}{Sc}\dfrac{\partial \omega_i}{\partial y} & j_{ky} = -\dfrac{\mu}{\sigma_k}\dfrac{\partial k}{\partial y} & j_{\varepsilon y} = -\dfrac{\mu}{\sigma_\varepsilon}\dfrac{\partial \varepsilon}{\partial y} \\[2ex]
j_{hz} = -\dfrac{\mu}{Pr}\dfrac{\partial h}{\partial y} & j_{iz} = -\dfrac{\mu}{Sc}\dfrac{\partial \omega_i}{\partial z} & j_{kz} = -\dfrac{\mu}{\sigma_k}\dfrac{\partial k}{\partial z} & j_{\varepsilon z} = -\dfrac{\mu}{\sigma_\varepsilon}\dfrac{\partial \varepsilon}{\partial z}
\end{array}
$$

$$(6.27)$$

For turbulent flow replace all μ's with μ_e's and let

$$\mu_e = \mu + \mu_t; \quad Pr = \sigma_q; \quad Pr_t = \sigma_{qt}; \quad Sc = \sigma_q;$$
$$Sc_t = \sigma_{qt}; \quad \sigma_k = \sigma_q; \quad \sigma_\varepsilon = \sigma_q \qquad (6.28)$$

$$\mu_t = \rho C_\mu k^2/\varepsilon \quad \text{and} \quad \mu_e = \left(\mu + \mu_t\right)/\sigma_\mu \qquad (6.29)$$

$$\phi_D = \frac{\mu}{Sc} + \frac{\mu_t}{Sc_t} \quad \text{and} \quad \phi_h = \frac{\mu}{Pr} + \frac{\mu_t}{Pr_t} \qquad (6.30)$$

$$\phi_k = \frac{\mu_t}{\sigma_k} \quad \text{and} \quad \phi_\varepsilon = \frac{\mu_t}{\sigma_\varepsilon} \qquad (6.31)$$

Values for the σ_i's are given in Table 6.2. The σ_i's for k and ε can be considered infinitely large for laminar flow because the k–ε turbulence parameters are negligible. The $\tilde{U}$ and $\tilde{S}$ vectors are unchanged. The others become

$$
\tilde{E} = \begin{bmatrix}
\rho u \\[2mm]
\rho u^2 + p - 2\mu_e \dfrac{\partial u}{\partial x} + \left(\dfrac{2\lambda_c}{3} \mu_e \right) \left(\dfrac{\partial u}{\partial x} + \dfrac{\partial v}{\partial y} + \dfrac{\partial w}{\partial z} \right) \\[2mm]
\rho u v - \mu_e \left(\dfrac{\partial u}{\partial y} + \dfrac{\partial v}{\partial x} \right) \\[2mm]
\rho u w - \mu_e \left(\dfrac{\partial w}{\partial x} + \dfrac{\partial u}{\partial z} \right) \\[2mm]
\rho u h - \left(\phi_h \dfrac{\partial h}{\partial x} \right) \\[2mm]
\rho u \omega_i - \left(\phi_D \dfrac{\partial \omega_i}{\partial x} \right) \\[2mm]
\rho u k - \left(\phi_k \dfrac{\partial k}{\partial x} \right) \\[2mm]
\rho u \varepsilon - \left(\phi_\varepsilon \dfrac{\partial \varepsilon}{\partial x} \right)
\end{bmatrix}
\tag{6.32}
$$

$$
\tilde{F} = \begin{bmatrix}
\rho v \\[2mm]
\rho u v - \mu_e \left(\dfrac{\partial u}{\partial y} + \dfrac{\partial v}{\partial x} \right) \\[2mm]
\rho v^2 + p - 2\mu_e \dfrac{\partial v}{\partial y} + \left(\dfrac{2\lambda_c}{3} \mu_e \right) \left(\dfrac{\partial v}{\partial y} + \dfrac{\partial u}{\partial x} + \dfrac{\partial w}{\partial z} \right) \\[2mm]
\rho v w - \mu_e \left(\dfrac{\partial v}{\partial z} + \dfrac{\partial w}{\partial y} \right) \\[2mm]
\rho v h - \left(\phi_h \dfrac{\partial h}{\partial y} \right) \\[2mm]
\rho v \omega_i - \left(\phi_D \dfrac{\partial \omega_i}{\partial y} \right) \\[2mm]
\rho v k - \left(\phi_k \dfrac{\partial k}{\partial y} \right) \\[2mm]
\rho v \varepsilon - \left(\phi_\varepsilon \dfrac{\partial \varepsilon}{\partial y} \right)
\end{bmatrix}
\tag{6.33}
$$

TABLE 6.2 Values for σ_i

σ	σ_ω	σ_μ	σ_h	σ_k	σ_ε
Laminar	$1.0 = Sc$	1.0	$0.72 = Pr$	—	—
Turbulent	$0.9 = Sc_t$	1.0	$0.90 = Pr_t$	0.8927	1.15

$$
\tilde{G} = \begin{bmatrix}
\rho w \\[4pt]
\rho u w - \mu_e\left(\dfrac{\partial w}{\partial x} + \dfrac{\partial u}{\partial z}\right) \\[4pt]
\rho \upsilon w - \mu_e\left(\dfrac{\partial \upsilon}{\partial z} + \dfrac{\partial w}{\partial y}\right) \\[4pt]
\rho w^2 + p - 2\mu_e\dfrac{\partial w}{\partial z} + \left(\dfrac{2\lambda_c}{3}\mu_e\right)\left(\dfrac{\partial w}{\partial z} + \dfrac{\partial u}{\partial x} + \dfrac{\partial \upsilon}{\partial y}\right) \\[4pt]
\rho w h - \left(\phi_h\dfrac{\partial h}{\partial z}\right) \\[4pt]
\rho w \omega_i - \left(\phi_D\dfrac{\partial \omega_i}{\partial z}\right) \\[4pt]
\rho w k - \left(\phi_k\dfrac{\partial k}{\partial z}\right) \\[4pt]
\rho w \varepsilon - \left(\phi_\varepsilon\dfrac{\partial \varepsilon}{\partial z}\right)
\end{bmatrix}
\tag{6.34}
$$

In general, the continuity (conservation of mass) equation may be combined to simplify the appearance of all the other equations, as discussed in Chapter 2. When such a formulation is used, it is referred to as the nonconservative form of the conservation equations. Such a formulation introduces errors in numerical calculations and is, therefore, not generally used. The form of the equations presented is termed the "strong conservation" form of the transport equations.

If the flow is of a single component fluid, the species continuity equation is eliminated from these arrays as it is not needed. If the flow is laminar, the turbulence model equations are not needed. Some terms of the source arrays may be placed in the other arrays and others may be assumed negligible, in order to minimize and eliminate the source array. When all of the source terms are eliminated, the set of equations becomes the strong form of the conservation laws; otherwise they are referred to as the weak form of the conservation laws. When wave phenomena, such as shock waves, are imbedded in the flowfield, the strong conservative form of the conservation laws gives more accurate resolution across the shock discontinuity. When finite-rate chemical reactions or nongradient forces like buoyancy, the fictitious Coriolis, or centrifugal forces are to be simulated, eliminating all of the source terms is not possible. This has not proved to be a major impediment to simulating transport problems. Usually, shock waves are simply smeared over a few grid points. For example, see Wang and Chen's (1993) analysis of the shock structure in rocket plumes. Such smearing causes concern only to aerodynamicists interested in very accurate simulations. Even so, other approximations in the modeling process may introduce comparable errors, if they are introduced solely to expedite the removal of source terms. Specifically, assuming the fluid is an ideal gas or the fluid is not turbulent are such approximations. As will be shown, additional terms may be included in the source arrays to expedite the numerical solution.

The mathematical model represented by the PDEs in Table 6.1 and by the vector equations just discussed provides a reasonably accurate and general description of many of the physical and chemical processes governing transport phenomena. Conservation laws formulated in orthogonal (rectangular) Cartesian coordinates are inadequate to efficiently simulate the complex geometry of many transport processes. Introduction of a general curvilinear grid which coincides with the geometric boundaries of the flow to be analyzed is a major advancement in computational technology. Such grid considerations are developed in the Section 6.3.3. The curvilinear coordinate system is used to define the independent variables of the simulation. The dependent variables will be kept as those utilized in the Cartesian formulation. Otherwise, more complex tensor analysis must be used to transform the dependent variables to the curvilinear system. Geometries of intermediate complexity, such as the orthogonal curvilinear coordinates, could be developed. Cylindrical coordinates are the most useful of such grids. However, it will be shown that the orthogonal curvilinear coordinates can be described with a code written for the more general case. Thus, separate computer codes are not needed for intermediately complex systems. The concept of thinking of the conservation laws in terms of vectors and matrices as arrays will be retained, since this is the same methodology as that employed in computer instructions for solving the conservation equations.

6.3.3 TRANSFORMING THE VECTOR FORM OF THE CTP EQUATIONS

The coordinate transformation equations for converting the independent variables to curvilinear coordinates given in Section 6.2.1 will be applied first to the compressible Navier–Stokes (CNS) equations and then to the CTP equations. Not only will the coordinates be transformed, but the equations will be rearranged to place them in strong conservation form. This development is similar to that for the CNS equations derived and represented in general curvilinear coordinates by Anderson et al. (1984, 1997).

6.3.3.1 Transformed CNS Equations

The CNS equations will be discussed first because they are somewhat simpler, but still contain the basic transformation mechanics. The CNS equations differ from the CTP equations by (1) being restricted to a one component ideal gas; (2) including the pressure term in the flux vectors, i.e., the $\tilde{E}$, $\tilde{F}$, and $\tilde{G}$ column vectors; (3) replacing the energy equation with one appropriate for the fluid involved and neglecting external forces. The total energy (E_T) equation becomes

$$E_5 = (E_T + p)u - u\tau_{xx} - \upsilon\tau_{xy} - w\tau_{xz} + q_x \tag{6.35}$$

$$F_5 = (E_T + p)\upsilon - \upsilon\tau_{yy} - w\tau_{yz} - u\tau_{yx} + q_y \tag{6.36}$$

$$G_5 = (E_T + p)w - w\tau_{zz} - u\tau_{zx} - \upsilon\tau_{zy} + q_z \tag{6.37}$$

Although indicial notation is not used, the three equations reflect cyclic permutation of the indices. The vector form of the conservation laws becomes

$$\frac{\partial \tilde{U}}{\partial t} + \frac{\partial \tilde{E}}{\partial x} + \frac{\partial \tilde{F}}{\partial y} + \frac{\partial \tilde{G}}{\partial z} = 0 \tag{6.38}$$

Notice that the source term is zero for this form of the CNS equations. The transformation equation is

$$\tilde{U}_t + \xi_x \tilde{E}_\xi + \eta_x \tilde{E}_\eta + \zeta_x \tilde{E}_\zeta + \xi_y \tilde{F}_\xi + \eta_y \tilde{F}_\eta + \zeta_y \tilde{F}_\zeta$$
$$+ \xi_z \tilde{G}_\xi + \eta_z \tilde{G}_\eta + \zeta_z \tilde{G}_\zeta = 0 \tag{6.39}$$

The vector form of the conservation equations for ideal gas flows is particularly accurate for calculating flows with shock waves. It is termed the strong-conservation form of the equations. This form is not as useful for flows with gravity effects or for flows with nondifferentiated source terms like chemically reacting and multiphase flows. The second-derivatives associated with viscous, diffusive, and thermally conductive flow also pose additional problems for realizing the advantages of using the strongly conservative form of the transport equations.

When this transformation law is applied source terms are created. Vinokur (1974) showed that this is the case even when the flow is inviscid. Viviand (1974) and Vinokur (1974) rearranged the transformed equation and recognized that terms like

$$\left[\left(\frac{\xi_x}{J}\right)_\xi + \left(\frac{\eta_x}{J}\right)_\eta + \left(\frac{\zeta_x}{J}\right)_\zeta\right] = 0 \tag{6.40}$$

to maintain the strong-conservation form of the equations for ideal gases. The modified vectors in the conservation equations are expanded as

$$\frac{\partial \tilde{U}^*}{\partial t} + \frac{\partial \tilde{E}^*}{\partial \xi} + \frac{\partial \tilde{F}^*}{\partial \eta} + \frac{\partial \tilde{G}^*}{\partial \zeta} = 0 \tag{6.41}$$

$$\tilde{U}^* = \frac{1}{J} \begin{bmatrix} \rho \\ \rho u \\ \rho \upsilon \\ \rho w \\ E_t \end{bmatrix} \tag{6.42}$$

$$\tilde{E}^* = \frac{1}{J} \begin{bmatrix} \rho U^c \\ \rho u U^c + \xi_x(p - \tau_{xx}) \\ \rho \upsilon U^c - \xi_y \tau_{xy} \\ \rho w U^c - \xi_z \tau_{xz} \\ (E_t + p)U^c - \xi_x(u\tau_{xx} + \upsilon\tau_{xy} + w\tau_{xz} - q_x) \end{bmatrix} \tag{6.43}$$

$$\tilde{F}^* = \frac{1}{J} \begin{bmatrix} \rho V^c \\ \rho u V^c - \eta_x \tau_{xy} \\ \rho \upsilon V^c + \eta_y(p - \tau_{yy}) \\ \rho w V^c - \eta_z \tau_{yz} \\ (E_t + p)V^c - \eta_y(u\tau_{xy} + \upsilon\tau_{yy} + w\tau_{yz} - q_y) \end{bmatrix} \tag{6.44}$$

$$\tilde{G}^* = \frac{1}{J} \begin{bmatrix} \rho W^c \\ \rho u W^c - \zeta_x \tau_{xy} \\ \rho \upsilon W^c - \zeta_z \tau_{yz} \\ \rho w W^c + \zeta_z (p - \tau_{zz}) \\ (E_t + p)W^c - \zeta_z (u\tau_{xz} + \upsilon\tau_{yz} + w\tau_{zz} - q_z) \end{bmatrix} \qquad (6.45)$$

where J is the Jacobian and the contravariant velocities are

$$U^c = \xi_x u + \xi_y \upsilon + \xi_z w$$
$$V^c = \eta_x u + \eta_y \upsilon + \eta_z w \qquad (6.46)$$
$$W^c = \zeta_x u + \zeta_y \upsilon + \zeta_z w$$

This is an unfortunate choice of symbols for the contravariant velocity components, because they are not associated specifically with individual Cartesian velocity components. They are the velocity components tangent to the curvilinear coordinate directions. They are termed contravariant components of velocity (Hawkins, 1963). Notice some literature erroneously terms these vectors normal to the coordinate surfaces. The U's and u's, etc., bear no relation to each other; this is what makes this choice of symbol confusing. But what is shown is more prevalent in the literature.

The inclusion of the Jacobian does convert the transformed CNS equations to a strong conservation form. For the inviscid case considered by Vinokur, the strong conservation feature is obtained; but what about the stress terms? These are evaluated with a viscosity and velocity gradients. Anderson (1995) discusses this issue, but does not clearly state that the source terms must be zero for strong conservation. The implication is that the stress terms would cause a loss of the strong conservation feature.

6.3.3.2 Transformed CTP Equations

The CTP conservation equations are transformed into general curvilinear coordinates for their solution in the CTP code. The continuity equation contains terms for accumulation and convection. The species continuity equations, the energy equation, turbulent kinetic energy, and energy dissipation equation, contain as expected accumulation, convection, diffusion, and generation terms. The momentum equations contain not only such terms, but additional terms arising from the more complex stress/rate-of-strain relationships. The stress terms are expressed in terms of velocity gradients and in both laminar and turbulent eddy viscosity. If the diffusion like terms from the momentum equations are identified, they may be treated like the scalar diffusion terms of species and energy. This leaves only extra velocity gradient terms for further consideration.

Table 6.3 shows the continuity and species continuity equation transformed to general curvilinear coordinates. The accumulation and convection terms in the continuity equation are the same as for all of the conserved quantities, once the specific conserved quantity (q) is chosen. The diffusion term is the same in the species equation as in all the remaining equations, once the specific conserved property is identified. The transformed diffusion term contains several second-derivative terms and several other mixed second-derivative terms. The mixed second-derivative terms are moved to the right-hand side (RHS) of the equation along with the generation terms. The strategy is to evaluate the RHS of the conservation laws explicitly.

The species source term is to be evaluated separately due to the very strong nonlinear nature of possibility fast reactions in the fluid. However, the species equation contains no velocity derives in the generation term. The source terms of the momentum, enthalpy, turbulent kinetic energy, and turbulent dissipation equations contain first and second and mixed second-derivative velocity terms. These terms are excessively lengthy, so they are described in Appendix 6.A. The derivation of the transformed diffusion terms is also given in Appendix 6.A. The transformed source terms in energy equation, exclusive of the diffusion terms are in Appendix 6.A. The transformed pressure gradient term from the momentum equations are in this appendix also. The viscous terms in the momentum equations which are not part of the diffusion of momentum are all described in terms of transformed velocity gradients in Appendix 6.A. Writing the entire momentum equations in terms of transformed coordinates is unwieldy, so the momentum equations and the energy equation are summarized with many of the untransformed terms on the RHS. In the CTP code, all of the transformed terms are included in the solution. Leaving part of these terms untransformed is simply a shorthand scheme for summarizing the CTP conservation equations.

To take full advantage of the similarity of the conservation equations, the coordinates and velocities will be renamed to fit an indicial format. Namely,

TABLE 6.3 Mass and Species Conservation Equations in Curvilinear Coordinates

$$\frac{\partial}{\partial t}\left(\frac{\rho}{J}\right)+\frac{\partial}{\partial \xi}\left(\frac{\rho(\xi_x u+\xi_y \upsilon+\xi_z w)}{J}\right)+\frac{\partial}{\partial \eta}\left(\frac{\rho(\eta_x u+\eta_y \upsilon+\eta_z w)}{J}\right)+\frac{\partial}{\partial \zeta}\left(\frac{\rho(\zeta_x u+\zeta_y \upsilon+\zeta_z w)}{J}\right)=0 \quad \text{(A)}$$

$$\frac{\partial}{\partial t}\left(\frac{\rho}{J}\right)+\frac{\partial}{\partial \xi}\left(\frac{\rho W^{\xi}}{J}\right)+\frac{\partial}{\partial \eta}\left(\frac{\rho W^{\eta}}{J}\right)+\frac{\partial}{\partial \zeta}\left(\frac{\rho W^{\zeta}}{J}\right)=0 \quad \text{(B)}$$

$$W^{\xi}=\xi_x u+\xi_y \upsilon+\xi_z w$$
$$W^{\eta}=\eta_x u+\eta_y \upsilon+\eta_z w \quad \text{(C)}$$
$$W^{\zeta}=\zeta_x u+\zeta_y \upsilon+\zeta_z w$$

$$\frac{1}{J}\frac{\partial \rho\omega_a}{\partial t}+\left[\frac{\partial}{\partial \xi}\left(\frac{\rho\omega_a}{J}(u\xi_x+\upsilon\xi_y+w\xi_z)\right)+\frac{\partial}{\partial \eta}\left(\frac{\rho\omega_a}{J}(u\eta_x+\upsilon\eta_y+w\eta_z)\right)+\frac{\partial}{\partial \zeta}\left(\frac{\rho\omega_a}{J}(u\zeta_x+\upsilon\zeta_y+w\zeta_z)\right)\right]$$

$$\frac{1}{J}\frac{\partial \rho\omega_a}{\partial t}+\left[\frac{\partial}{\partial \xi}\left(\frac{\rho\omega_a}{J}(u\xi_x+\upsilon\xi_y+w\xi_z)\right)+\frac{\partial}{\partial \eta}\left(\frac{\rho\omega_a}{J}(u\eta_x+\upsilon\eta_y+w\eta_z)\right)+\frac{\partial}{\partial \zeta}\left(\frac{\rho\omega_a}{J}(u\zeta_x+\upsilon\zeta_y+w\zeta_z)\right)\right]$$
$$-\left\{\frac{\partial}{\partial \xi}\left[\frac{\xi_x^2+\xi_y^2+\xi_z^2}{J}\phi_D\frac{\partial \omega_\alpha}{\partial \xi}\right]+\frac{\partial}{\partial \eta}\left[\frac{\eta_x^2+\eta_y^2+\eta_z^2}{J}\phi_D\frac{\partial \omega_\alpha}{\partial \eta}\right]+\frac{\partial}{\partial \zeta}\left[\frac{\zeta_x^2+\zeta_y^2+\zeta_z^2}{J}\phi_D\frac{\partial \omega_\alpha}{\partial \zeta}\right]\right\}$$
$$=+\frac{\partial}{\partial \xi}\left[\frac{\xi_x\eta_x+\xi_y\eta_y+\xi_z\eta_z}{J}\phi_D\frac{\partial \omega_\alpha}{\partial \eta}\right]+\frac{\partial}{\partial \xi}\left[\frac{\xi_x\zeta_x+\xi_y\zeta_y+\xi_z\zeta_z}{J}\phi_D\frac{\partial \omega_\alpha}{\partial \zeta}\right]$$
$$+\frac{\partial}{\partial \eta}\left[\frac{\xi_x\eta_x+\xi_y\eta_y+\xi_z\eta_z}{J}\phi_D\frac{\partial \omega_\alpha}{\partial \xi}\right]+\frac{\partial}{\partial \eta}\left[\frac{\eta_x\zeta_x+\eta_y\zeta_y+\eta_z\zeta_z}{J}\phi_D\frac{\partial \omega_\alpha}{\partial \zeta}\right]$$
$$+\frac{\partial}{\partial \zeta}\left[\frac{\xi_x\zeta_x+\xi_y\zeta_y+\xi_z\zeta_z}{J}\phi_D\frac{\partial \omega_\alpha}{\partial \xi}\right]+\frac{\partial}{\partial \zeta}\left[\frac{\eta_x\zeta_x+\eta_y\zeta_y+\eta_z\zeta_z}{J}\phi_D\frac{\partial \omega_\alpha}{\partial \eta}\right]+K_\alpha/J \quad \text{(D)}$$

$$(u, \upsilon, w) \Rightarrow (U_1, U_2, U_3) \quad (x, y, z) \Rightarrow (X_1, X_2, X_3)$$
$$\text{and} \quad (\xi, \eta, \zeta) \Rightarrow (Z^1, Z^2, Z^3) \tag{6.47}$$

The contravariant velocities will also be renamed to emphasize that they are associated with specific transformed coordinate directions.

$$(U^c, V^c, W^c) \Rightarrow (W^\xi, W^\eta, W^\zeta) \tag{6.48}$$

The individual terms have been defined such that a general form of the conservation laws may be stated as

$$\frac{1}{J}\frac{\partial(\rho q)}{\partial t} + \frac{\partial(\rho W^i q)}{\partial Z^i} - \frac{\partial}{\partial Z^i}\left(\phi_q\, G_{ii}\, \frac{\partial q}{\partial Z^i}\right) = \sum_m \frac{\partial}{\partial Z^i}\left(\Gamma_q G_{ij}\, \frac{\partial q}{\partial Z^j}\right) + S_{qq} \equiv S_q \tag{6.49}$$

where $q = \rho,\ \rho U_1,\ \rho U_2,\ \rho U_3,\ \rho h,\ \rho k,\ \rho \varepsilon,\ \rho \omega_i$. S_{qq} is the part of the source term that is not due to the mixed-derivative diffusion terms. For

$$W^i \equiv \frac{U_j}{J}\frac{\partial Z^i}{\partial X_j} \quad \text{and} \quad G_{ij} = \sum_m \sum_k \frac{1}{J}\frac{\partial Z^i}{\partial X_k}\frac{\partial Z^j}{\partial X_k} \tag{6.50}$$

The generalized conservation law and the corresponding source terms are listed in Table 6.4. Since the CTP code does not utilize adaptive gridding, all of the metric coefficients and the Jacobian calculated from them are constants. Even though the code might have such terms under partial signs, their exact location is not important for correctly stating the generalized form of the equations.

Part of the reason that the diffusion terms of all of the conserved quantities are complex is that when taking the second derivative of the two required, the general diffusion coefficient is placed under this derivative operator. If it is placed outside, the equations become much simpler. The generalized diffusion coefficient would still be evaluated locally. It must be varying very rapidly and over a course grid for this not to be a good approximation. While the turbulent transport coefficients vary rapidly near a wall, both using a fine grid and using wall functions suggest that the local evaluation of the transport coefficients without including them under the second partial operator may be acceptable. Such a procedure should be evaluated to simplify the solution algorithm.

TABLE 6.4 Generalized Conservation Equations

$$\frac{1}{J}\frac{\partial(\rho q)}{\partial t}+\frac{\partial\left(\rho W^{i}q\right)}{\partial Z^{i}}-\frac{\partial}{\partial Z^{i}}\left(\phi_{q}G_{ii}\frac{\partial q}{\partial Z^{i}}\right)=S_{q} \tag{A}$$

S_q denotes the source vector.

$$S_1 = 0 \tag{B}$$

$$S_{\omega_\alpha}=\frac{1}{J}\left[\sum_m\left[\frac{\partial}{\partial Z^i}\left(\phi_D JG_{ij}\frac{\partial\omega_\alpha}{\partial Z^j}\right)\right]+K_\alpha\right] \tag{C}$$

$$S_{X_1}=\frac{1}{J}\left[\begin{array}{l}\sum_m\left[\frac{\partial}{\partial Z^i}\left(\mu_e JG_{ij}\frac{\partial U_1}{\partial Z^j}\right)\right]-\frac{\partial p}{\partial X_1}\\+\left[\frac{\partial}{\partial X_a}\left(\mu_e\frac{\partial U_a}{\partial X_1}\right)-\frac{\partial}{\partial X_1}\left(\frac{2\lambda_c}{3}\mu_e\right)\left(\frac{\partial U_a}{\partial X_a}\right)\right]+\rho g_{X_1}\end{array}\right] \tag{D}$$

$$S_{X_2}=\frac{1}{J}\left[\begin{array}{l}\sum_m\left[\frac{\partial}{\partial Z^i}\left(\mu_e JG_{ij}\frac{\partial U_2}{\partial Z^j}\right)\right]-\frac{\partial p}{\partial X_2}\\+\left[\frac{\partial}{\partial X_a}\left(\mu_e\frac{\partial U_a}{\partial X_2}\right)-\frac{\partial}{\partial X_2}\left(\frac{2\lambda_c}{3}\mu_e\right)\left(\frac{\partial U_a}{\partial X_a}\right)\right]+\rho g_{X_2}\end{array}\right] \tag{E}$$

$$S_{X_3}=\frac{1}{J}\left[\begin{array}{l}\sum_m\left[\frac{\partial}{\partial Z^i}\left(\mu_e JG_{ij}\frac{\partial U_3}{\partial Z^j}\right)\right]-\frac{\partial p}{\partial X_3}\\+\left[\frac{\partial}{\partial X_a}\left(\mu_e\frac{\partial U_a}{\partial X_3}\right)-\frac{\partial}{\partial X_3}\left(\frac{2\lambda_c}{3}\mu_e\right)\left(\frac{\partial U_a}{\partial X_a}\right)\right]+\rho g_{X_3}\end{array}\right] \tag{F}$$

$$S_h=\frac{1}{J}\left[\sum_m\left[\frac{\partial}{\partial Z^i}\left(\phi_h JG_{ij}\frac{\partial h}{\partial Z^j}\right)\right]+\frac{Dp}{Dt}+\vec{\vec{\tau}}:\nabla\cdot\vec{U}\right] \tag{G}$$

$$S_k=\frac{1}{J}\left[\sum_m\left[\frac{\partial}{\partial Z^i}\left(\phi_k JG_{ij}\frac{\partial k}{\partial Z^j}\right)\right]+\rho(P_r-\varepsilon)\right] \tag{H}$$

$$S_\varepsilon=\frac{1}{J}\left[\sum_m\left[\frac{\partial}{\partial Z^i}\left(\phi_\varepsilon JG_{ij}\frac{\partial\varepsilon}{\partial Z^j}\right)\right]+\frac{\rho k}{\varepsilon}\left[\left(\frac{C_1+C_3 P_r}{\varepsilon}\right)P_r-C_2\varepsilon\right]\right] \tag{I}$$

For $W^i\equiv\dfrac{U_j}{J}\dfrac{\partial Z^i}{\partial X_j}$ and $G_{ij}=\displaystyle\sum_m\sum_k\frac{1}{J}\frac{\partial Z^i}{\partial X_k}\frac{\partial Z^j}{\partial X_k}$. The summation m is for ij, if $i\neq j$.

In general, repeated indices indicate summations.

6.4 NOMENCLATURE

6.4.1 ENGLISH SYMBOLS

A, D, Ψ	damping parameters
C_p	constant pressure heat capacity
D, h, k, ε	used as subscripts to indicate type of diffusion: mass, enthalpy, turbulent kinetic energy, dissipation of turbulent kinetic energy, respectively
g_n	gravitational acceleration in the direction n
G_{ij}	coordinate transformation function
H	dummy symbol to indicate mathematical operations
h, h_i	specific enthalpy, specific enthalpy of species i
i, j, k	nodal indices
I, J, K	nodal indices
j_{mn}	diffusion flux for variable m in direction n
K	turbulent kinetic energy
K_i	reaction rate
k, ε	turbulent kinetic energy and energy dissipation
Le, Le_t	laminar and turbulent lewis numbers
ℓ_i	interpolation function
p	pressure
P_k	production rate of k
Pr, Pr_t	laminar and turbulent Prandtl numbers
q	any conserved quantity
$\vec{q}$	heat flux vector
Q	any conserved quantity
r, θ, z	cylindrical coordinates
S	distance along a coordinate
Sc, Sc_t	laminar and turbulent Schmidt numbers
s	arc length
t	coordinate variable which may be time
T	temperature
U	function; standardized (normalized) function
U^c, V^c, W^c	the contravariant velocity components
u, v, w	velocity components in Cartesian coordinate system
U_1, U_2, U_3	alternate form of the velocity components in Cartesian coordinates
W^ξ, W^η, W^ζ	alternate form of the contravariant velocity components
x, y, z	Cartesian coordinates

X_1, X_2, X_3	alternate form of the Cartesian coordinates
X, Y, Z	boundary-fitted functions
Z^1, Z^2, Z^3	alternate form of the curvilinear contravariant coordinates

6.4.2 GREEK SYMBOLS

ε	turbulent kinetic energy dissipation
ϕ	generalized diffusion coefficient
κ	thermal conductivity
λ	second coefficient of viscosity
λ_c	parameter to indicate compressible flow
μ, μ_t, μ_e	laminar, turbulent viscosity, and effective viscosity
ξ, η, ζ	nonorthogonal curvilinear coordinates
ρ	density
σ_q	parameter to indicate type of diffusive quantity
τ_{ij}	shear–stress component
ω_i	mass fraction of i

6.4.3 MATHEMATICAL SYMBOLS

$\tilde{H}$	generalized column vector ($\tilde{U}, \tilde{E}, \tilde{F}, \tilde{G},$ or $\tilde{S}$) used to write transport equations in a vector form
$\hat{H}$	matrix
$H*$	modified version of $\tilde{H}$
H_i	denotes a component
H_t	turbulent quantity
$\vec{U}$	velocity vector
$(\tau_t)_{ij}$	component of the turbulent shear stress

6.4.4 ACRONYMS

CAD	computer-aided design
CNS	compressible Navier–Stokes
EVM	eddy viscosity model
FLUENT	commercial computer code
GRID2D/3D	computer grid code
HBMS	Hirschfelder, Buehler, McGee, Sutton equation of state
PC	personal computer

PDE	partial differential equation
RHS	right-hand side
TFI	transfinite interpolation

APPENDIX 6.A TRANSFORMED TERMS WHICH COMPLETE THE SYSTEM OF CONSERVATION LAWS

6.A.1 TRANSFORMATION OF THE DIFFUSION TERMS FOR U-MOMENTUM EQUATION

$$\frac{1}{J}\frac{\partial}{\partial x}\left(\mu_e\frac{\partial u}{\partial x}\right)+\frac{1}{J}\frac{\partial}{\partial y}\left(\mu_e\frac{\partial u}{\partial y}\right)+\frac{1}{J}\frac{\partial}{\partial z}\left(\mu_e\frac{\partial u}{\partial z}\right)=\frac{1}{J}\left[\frac{\partial A}{\partial x}+\frac{\partial B}{\partial y}+\frac{\partial C}{\partial z}\right]$$

$$=\frac{1}{J}\left[\begin{array}{l}\dfrac{\partial A}{\partial \xi}\dfrac{\partial \xi}{\partial x}+\dfrac{\partial A}{\partial \eta}\dfrac{\partial \eta}{\partial x}+\dfrac{\partial A}{\partial \zeta}\dfrac{\partial \zeta}{\partial x}+\dfrac{\partial B}{\partial \xi}\dfrac{\partial \xi}{\partial y}+\dfrac{\partial B}{\partial \eta}\dfrac{\partial \eta}{\partial y}\\[2mm]+\dfrac{\partial B}{\partial \zeta}\dfrac{\partial \zeta}{\partial y}+\dfrac{\partial C}{\partial \xi}\dfrac{\partial \xi}{\partial z}+\dfrac{\partial C}{\partial \eta}\dfrac{\partial \eta}{\partial z}+\dfrac{\partial C}{\partial \zeta}\dfrac{\partial \zeta}{\partial z}\end{array}\right]$$

$$=\underbrace{\frac{\partial}{\partial \xi}\left[\frac{1}{J}\left(A\frac{\partial \xi}{\partial x}+B\frac{\partial \xi}{\partial y}+C\frac{\partial \xi}{\partial z}\right)\right]}_{\text{I}}+\underbrace{\frac{\partial}{\partial \eta}\left[\frac{1}{J}\left(A\frac{\partial \eta}{\partial x}+B\frac{\partial \eta}{\partial y}+C\frac{\partial \eta}{\partial z}\right)\right]}_{\text{II}}$$

$$+\underbrace{\frac{\partial}{\partial \zeta}\left[\frac{1}{J}\left(A\frac{\partial \zeta}{\partial x}+B\frac{\partial \zeta}{\partial y}+C\frac{\partial \zeta}{\partial z}\right)\right]}_{\text{III}} \qquad (6.A.1)$$

where

$$A=\mu_e\frac{\partial u}{\partial x}=\mu_e\left(\frac{\partial u}{\partial \xi}\frac{\partial \xi}{\partial x}+\frac{\partial u}{\partial \eta}\frac{\partial \eta}{\partial x}+\frac{\partial u}{\partial \zeta}\frac{\partial \zeta}{\partial x}\right)=\mu_e\left(u_\xi\xi_x+u_\eta\eta_x+u_\zeta\zeta_x\right)$$

$$B=\mu_e\frac{\partial u}{\partial y}=\mu_e\left(\frac{\partial u}{\partial \xi}\frac{\partial \xi}{\partial y}+\frac{\partial u}{\partial \eta}\frac{\partial \eta}{\partial y}+\frac{\partial u}{\partial \zeta}\frac{\partial \zeta}{\partial y}\right)=\mu_e\left(u_\xi\xi_y+u_\eta\eta_y+u_\zeta\zeta_y\right)$$

$$C=\mu_e\frac{\partial u}{\partial z}=\mu_e\left(\frac{\partial u}{\partial \xi}\frac{\partial \xi}{\partial z}+\frac{\partial u}{\partial \eta}\frac{\partial \eta}{\partial z}+\frac{\partial u}{\partial \zeta}\frac{\partial \zeta}{\partial z}\right)=\mu_e\left(u_\xi\xi_z+u_\eta\eta_z+u_\zeta\zeta_z\right)$$

$$(6.A.2)$$

Substitute the above equation into the Equation 6.A.1, we can get

$$
\text{Term I} \quad = \frac{\partial}{\partial \xi}\left[\frac{\xi_x^2 + \xi_y^2 + \xi_z^2}{J}\mu_e \frac{\partial u}{\partial \xi}\right] + \frac{\partial}{\partial \xi}\left[\frac{\xi_x \eta_x + \xi_y \eta_y + \xi_z \eta_z}{J}\mu_e \frac{\partial u}{\partial \eta}\right]
$$

$$
+ \frac{\partial}{\partial \xi}\left[\frac{\xi_x \zeta_x + \xi_y \zeta_y + \xi_z \zeta_z}{J}\mu_e \frac{\partial u}{\partial \zeta}\right]
$$

$$
\text{Term II} \quad = \frac{\partial}{\partial \eta}\left[\frac{\xi_x \eta_x + \xi_y \eta_y + \xi_z \eta_z}{J}\mu_e \frac{\partial u}{\partial \xi}\right] + \frac{\partial}{\partial \eta}\left[\frac{\eta_x^2 + \eta_y^2 + \eta_z^2}{J}\mu_e \frac{\partial u}{\partial \eta}\right]
$$

$$
+ \frac{\partial}{\partial \eta}\left[\frac{\eta_x \zeta_x + \eta_y \zeta_y + \eta_z \zeta_z}{J}\mu_e \frac{\partial u}{\partial \zeta}\right]
$$

$$
\text{Term III} = \frac{\partial}{\partial \zeta}\left[\frac{\xi_x \zeta_x + \xi_y \zeta_y + \xi_z \zeta_z}{J}\mu_e \frac{\partial u}{\partial \xi}\right] + \frac{\partial}{\partial \zeta}\left[\frac{\eta_x \zeta_x + \eta_y \zeta_y + \eta_z \zeta_z}{J}\mu_e \frac{\partial u}{\partial \eta}\right]
$$

$$
+ \frac{\partial}{\partial \zeta}\left[\frac{\zeta_x^2 + \zeta_y^2 + \zeta_z^2}{J}\mu_e \frac{\partial u}{\partial \zeta}\right]
\tag{6.A.3}
$$

Hence, the transformed diffusion terms are expressed as

$$
\frac{\partial}{\partial \xi}\left[\frac{\xi_x^2 + \xi_y^2 + \xi_z^2}{J}\mu_e \frac{\partial u}{\partial \xi}\right] + \frac{\partial}{\partial \eta}\left[\frac{\eta_x^2 + \eta_y^2 + \eta_z^2}{J}\mu_e \frac{\partial u}{\partial \eta}\right] + \frac{\partial}{\partial \zeta}\left[\frac{\zeta_x^2 + \zeta_y^2 + \zeta_z^2}{J}\mu_e \frac{\partial u}{\partial \zeta}\right]
$$

$$
+ \frac{\partial}{\partial \xi}\left[\frac{\xi_x \eta_x + \xi_y \eta_y + \xi_z \eta_z}{J}\mu_e \frac{\partial u}{\partial \eta}\right] + \frac{\partial}{\partial \xi}\left[\frac{\xi_x \zeta_x + \xi_y \zeta_y + \xi_z \zeta_z}{J}\mu_e \frac{\partial u}{\partial \zeta}\right]
$$

$$
+ \frac{\partial}{\partial \eta}\left[\frac{\xi_x \eta_x + \xi_y \eta_y + \xi_z \eta_z}{J}\mu_e \frac{\partial u}{\partial \xi}\right] + \frac{\partial}{\partial \eta}\left[\frac{\eta_x \zeta_x + \eta_y \zeta_y + \eta_z \zeta_z}{J}\mu_e \frac{\partial u}{\partial \zeta}\right]
$$

$$
+ \frac{\partial}{\partial \zeta}\left[\frac{\xi_x \zeta_x + \xi_y \zeta_y + \xi_z \zeta_z}{J}\mu_e \frac{\partial u}{\partial \xi}\right] + \frac{\partial}{\partial \zeta}\left[\frac{\eta_x \zeta_x + \eta_y \zeta_y + \eta_z \zeta_z}{J}\mu_e \frac{\partial u}{\partial \eta}\right]
\tag{6.A.4}
$$

To solve the governing equation using a tri-diagonal matrix solver, only those terms in the first line of the above equation will stay on the LHS of the discretized equation, and the remaining terms will be moved to the RHS of the discretized equation as the source term. To represent the diffusion terms for the dependent variable (q) in a compact form, the above expression can be written as

$$\underbrace{\frac{\partial}{\partial \xi}\left[G_{11}\phi_q \frac{\partial q}{\partial \xi}\right] + \frac{\partial}{\partial \eta}\left[G_{22}\phi_q \frac{\partial q}{\partial \eta}\right] + \frac{\partial}{\partial \zeta}\left[G_{33}\phi_q \frac{\partial q}{\partial \zeta}\right]}_{\text{LHS}}$$

$$+ \frac{\partial}{\partial \xi}\left[G_{12}\phi_q \frac{\partial q}{\partial \eta}\right] + \frac{\partial}{\partial \xi}\left[G_{13}\phi_q \frac{\partial q}{\partial \zeta}\right] + \frac{\partial}{\partial \eta}\left[G_{21}\phi_q \frac{\partial q}{\partial \xi}\right] + \frac{\partial}{\partial \eta}\left[G_{23}\phi_q \frac{\partial q}{\partial \zeta}\right]$$

$$+ \frac{\partial}{\partial \zeta}\left[G_{31}\phi_q \frac{\partial q}{\partial \xi}\right] + \frac{\partial}{\partial \zeta}\left[G_{32}\phi_q \frac{\partial q}{\partial \eta}\right] \tag{6.A.5}$$

or

$$\underbrace{\sum_n \frac{\partial}{\partial Z^i}\left[G_{ii}\,\phi_q \frac{\partial q}{\partial Z^i}\right]}_{\text{LHS}} + \underbrace{\sum_m \frac{\partial}{\partial Z^i}\left[G_{ij}\phi_q \frac{\partial q}{\partial Z^j}\right]}_{\text{RHS}} \tag{6.A.6}$$

where indicial nomenclature is introduced for the independent variables. The position of the index on the variables is explained in Appendix B.

6.A.2 TRANSFORMATION OF SOURCE TERMS IN THE MOMENTUM AND ENERGY EQUATIONS

For the momentum equations

$$-\frac{\partial p}{\partial x} = -\left[\frac{\partial}{\partial \xi}\left(\frac{1}{J}(\xi_x p)\right) + \frac{\partial}{\partial \eta}\left(\frac{1}{J}(\eta_x p)\right) + \frac{\partial}{\partial \zeta}\left(\frac{1}{J}(\zeta_x p)\right)\right]$$

$$-\frac{\partial p}{\partial y} = -\left[\frac{\partial}{\partial \xi}\left(\frac{1}{J}(\xi_y p)\right) + \frac{\partial}{\partial \eta}\left(\frac{1}{J}(\eta_y p)\right) + \frac{\partial}{\partial \zeta}\left(\frac{1}{J}(\zeta_y p)\right)\right] \tag{6.A.7}$$

$$-\frac{\partial p}{\partial z} = -\left[\frac{\partial}{\partial \xi}\left(\frac{1}{J}(\xi_z p)\right) + \frac{\partial}{\partial \eta}\left(\frac{1}{J}(\eta_z p)\right) + \frac{\partial}{\partial \zeta}\left(\frac{1}{J}(\zeta_z p)\right)\right]$$

For the energy equation

$$\frac{Dp}{Dt} = \frac{\partial p}{\partial t} + \vec{U} \cdot \nabla p = \frac{\partial p}{\partial t} + u\frac{\partial p}{\partial x} + \upsilon\frac{\partial p}{\partial y} + w\frac{\partial p}{\partial z} = \frac{\partial p}{\partial t}$$

$$+ u\left[\frac{\partial}{\partial \xi}\left(\frac{1}{J}(\xi_x p)\right) + \frac{\partial}{\partial \eta}\left(\frac{1}{J}(\eta_x p)\right) + \frac{\partial}{\partial \zeta}\left(\frac{1}{J}(\zeta_x p)\right)\right]$$

$$+ \upsilon\left[\frac{\partial}{\partial \xi}\left(\frac{1}{J}(\xi_y p)\right) + \frac{\partial}{\partial \eta}\left(\frac{1}{J}(\eta_y p)\right) + \frac{\partial}{\partial \zeta}\left(\frac{1}{J}(\zeta_y p)\right)\right]$$

$$+ w\left[\frac{\partial}{\partial \xi}\left(\frac{1}{J}(\xi_z p)\right) + \frac{\partial}{\partial \eta}\left(\frac{1}{J}(\eta_z p)\right) + \frac{\partial}{\partial \zeta}\left(\frac{1}{J}(\zeta_z p)\right)\right] \qquad (6.A.8)$$

$$\Phi = \bar{\bar{\tau}} : \nabla \bar{U} = \mu_e\left(2\frac{\partial u}{\partial x} - \frac{2}{3}\left[\frac{\partial u}{\partial x} + \frac{\partial \upsilon}{\partial y} + \frac{\partial w}{\partial z}\right]\right)\left(\frac{\partial u}{\partial x}\right)$$

$$+ \mu_e\left(2\frac{\partial \upsilon}{\partial y} - \frac{2}{3}\left[\frac{\partial \upsilon}{\partial y} + \frac{\partial w}{\partial z} + \frac{\partial u}{\partial x}\right]\right)\left(\frac{\partial \upsilon}{\partial y}\right)$$

$$+ \mu_e\left(2\frac{\partial w}{\partial z} - \frac{2}{3}\left[\frac{\partial w}{\partial z} + \frac{\partial u}{\partial x} + \frac{\partial \upsilon}{\partial y}\right]\right)\left(\frac{\partial w}{\partial z}\right)$$

$$+ \mu_e\left(\frac{\partial u}{\partial y} + \frac{\partial \upsilon}{\partial x}\right)\left(\frac{\partial u}{\partial y} + \frac{\partial \upsilon}{\partial x}\right)$$

$$+ \mu_e\left(\frac{\partial \upsilon}{\partial z} + \frac{\partial w}{\partial y}\right)\left(\frac{\partial \upsilon}{\partial z} + \frac{\partial w}{\partial y}\right) + \mu_e\left(\frac{\partial w}{\partial x} + \frac{\partial u}{\partial z}\right)\left(\frac{\partial w}{\partial x} + \frac{\partial u}{\partial z}\right)$$

$$(6.A.9)$$

6.A.3 TRANSFORMATION OF REMAINING VELOCITY DERIVATIVES

Other velocity derivatives necessary to transform the momentum, enthalpy, turbulent kinetic energy, and turbulent dissipation equations are given as follows. The use of a diffusion-type metric is not convenient for these terms, so it is not used.

$$\frac{\partial A}{\partial x} = \frac{\partial}{\partial x}\left[\mu_e\left(\frac{\partial u}{\partial x}\right)\right] = \xi_x \frac{\partial}{\partial \xi}\left[\mu_e\left(u_\xi \xi_x + u_\eta \eta_x + u_\zeta \zeta_x\right)\right]$$

$$+\eta_x \frac{\partial}{\partial \eta}\left[\mu_e\left(u_\xi \xi_x + u_\eta \eta_x + u_\zeta \zeta_x\right)\right]$$

$$+\zeta_x \frac{\partial}{\partial \zeta}\left[\mu_e\left(u_\xi \xi_x + u_\eta \eta_x + u_\zeta \zeta_x\right)\right]$$

$$\frac{\partial B}{\partial x} = \frac{\partial}{\partial x}\left[\mu_e\left(\frac{\partial u}{\partial y}\right)\right] = \xi_x \frac{\partial}{\partial \xi}\left[\mu_e\left(u_\xi \xi_y + u_\eta \eta_y + u_\zeta \zeta_y\right)\right]$$

$$+\eta_x \frac{\partial}{\partial \eta}\left[\mu_e\left(u_\xi \xi_y + u_\eta \eta_y + u_\zeta \zeta_y\right)\right] \qquad (6.A.10)$$

$$+\zeta_x \frac{\partial}{\partial \zeta}\left[\mu_e\left(u_\xi \xi_y + u_\eta \eta_y + u_\zeta \zeta_y\right)\right]$$

$$\frac{\partial C}{\partial x} = \frac{\partial}{\partial x}\left[\mu_e\left(\frac{\partial u}{\partial z}\right)\right] = \xi_x \frac{\partial}{\partial \xi}\left[\mu_e\left(u_\xi \xi_z + u_\eta \eta_z + u_\zeta \zeta_z\right)\right]$$

$$+\eta_x \frac{\partial}{\partial \eta}\left[\mu_e\left(u_\xi \xi_z + u_\eta \eta_z + u_\zeta \zeta_z\right)\right]$$

$$+\zeta_x \frac{\partial}{\partial \zeta}\left[\mu_e\left(u_\xi \xi_z + u_\eta \eta_z + u_\zeta \zeta_z\right)\right]$$

$$\frac{\partial A}{\partial y} = \frac{\partial}{\partial y}\left[\mu_e\left(\frac{\partial u}{\partial x}\right)\right] = \xi_y \frac{\partial}{\partial \xi}\left[\mu_e\left(u_\xi \xi_y + u_\eta \eta_y + u_\zeta \zeta_y\right)\right]$$

$$+\eta_y \frac{\partial}{\partial \eta}\left[\mu_e\left(u_\xi \xi_y + u_\eta \eta_y + u_\zeta \zeta_y\right)\right]$$

$$+\zeta_y \frac{\partial}{\partial \zeta}\left[\mu_e\left(u_\xi \xi_y + u_\eta \eta_y + u_\zeta \zeta_y\right)\right]$$

$$\qquad (6.A.11)$$

$$\frac{\partial A}{\partial z} = \frac{\partial}{\partial z}\left[\mu_e\left(\frac{\partial u}{\partial x}\right)\right] = \xi_z \frac{\partial}{\partial \xi}\left[\mu_e\left(u_\xi \xi_z + u_\eta \eta_z + u_\zeta \zeta_z\right)\right]$$

$$+\eta_z \frac{\partial}{\partial \eta}\left[\mu_e\left(u_\xi \xi_z + u_\eta \eta_z + u_\zeta \zeta_z\right)\right]$$

$$+\zeta_z \frac{\partial}{\partial \zeta}\left[\mu_e\left(u_\xi \xi_z + u_\eta \eta_z + u_\zeta \zeta_z\right)\right]$$

$$\frac{\partial E}{\partial z} = \frac{\partial}{\partial z}\left[\mu_e \frac{\partial \upsilon}{\partial y}\right] = \xi_z \frac{\partial}{\partial \xi}\left[\mu_e\left(\upsilon_\xi \xi_y + \upsilon_\eta \eta_y + \upsilon_\zeta \zeta_y\right)\right]$$

$$+\eta_z \frac{\partial}{\partial \eta}\left[\mu_e\left(\upsilon_\xi \xi_y + \upsilon_\eta \eta_y + \upsilon_\zeta \zeta_y\right)\right]$$

$$+\zeta_z \frac{\partial}{\partial \zeta}\left[\mu_e\left(\upsilon_\xi \xi_y + \upsilon_\eta \eta_y + \upsilon_\zeta \zeta_y\right)\right]$$

$$\frac{\partial E}{\partial x} = \frac{\partial}{\partial x}\left[\mu_e \frac{\partial \upsilon}{\partial y}\right] = \xi_x \frac{\partial}{\partial \xi}\left[\mu_e\left(\upsilon_\xi \xi_y + \upsilon_\eta \eta_y + \upsilon_\zeta \zeta_y\right)\right]$$

$$+\eta_x \frac{\partial}{\partial \eta}\left[\mu_e\left(\upsilon_\xi \xi_y + \upsilon_\eta \eta_y + \upsilon_\zeta \zeta_y\right)\right]$$

$$+\zeta_x \frac{\partial}{\partial \zeta}\left[\mu_e\left(\upsilon_\xi \xi_y + \upsilon_\eta \eta_y + \upsilon_\zeta \zeta_y\right)\right]$$

(6.A.12)

$$\frac{\partial I}{\partial x} = \frac{\partial}{\partial x}\left[\mu_e\left(\frac{\partial w}{\partial z}\right)\right] = \xi_x \frac{\partial}{\partial \xi}\left[\mu_e\left(w_\xi \xi_z + w_\eta \eta_z + w_\zeta \zeta_z\right)\right]$$

$$+\eta_x \frac{\partial}{\partial \eta}\left[\mu_e\left(w_\xi \xi_z + w_\eta \eta_z + w_\zeta \zeta_z\right)\right]$$

$$+\zeta_x \frac{\partial}{\partial \zeta}\left[\mu_e\left(w_\xi \xi_z + w_\eta \eta_z + w_\zeta \zeta_z\right)\right]$$

$$\frac{\partial I}{\partial y} = \frac{\partial}{\partial y}\left[\mu_e\left(\frac{\partial w}{\partial z}\right)\right] = \xi_y \frac{\partial}{\partial \xi}\left[\mu_e\left(w_\xi \xi_z + w_\eta \eta_z + w_\zeta \zeta_z\right)\right]$$

$$+\eta_y \frac{\partial}{\partial \eta}\left[\mu_e\left(w_\xi \xi_z + w_\eta \eta_z + w_\zeta \zeta_z\right)\right]$$

$$+\zeta_y \frac{\partial}{\partial \zeta}\left[\mu_e\left(w_\xi \xi_z + w_\eta \eta_z + w_\zeta \zeta_z\right)\right]$$

(6.A.13)

$$D = \mu_e \frac{\partial \upsilon}{\partial x} = \mu_e \left(\upsilon_\xi \xi_x + \upsilon_\eta \eta_x + \upsilon_\zeta \zeta_x \right)$$

$$\frac{\partial D}{\partial y} = \frac{\partial}{\partial y} \left[\mu_e \left(\frac{\partial \upsilon}{\partial x} \right) \right] = \xi_y \frac{\partial}{\partial \xi} \left[\mu_e \left(\upsilon_\xi \xi_x + \upsilon_\eta \eta_x + \upsilon_\zeta \zeta_x \right) \right]$$

$$+ \eta_y \frac{\partial}{\partial \eta} \left[\mu_e \left(\upsilon_\xi \xi_x + \upsilon_\eta \eta_x + \upsilon_\zeta \zeta_x \right) \right]$$

$$+ \zeta_y \frac{\partial}{\partial \zeta} \mu_e \left(\upsilon_\xi \xi_x + \upsilon_\eta \eta_x + \upsilon_\zeta \zeta_x \right)$$

$$E = \mu_e \frac{\partial \upsilon}{\partial y} = \mu_e \left(\upsilon_\xi \xi_y + \upsilon_\eta \eta_y + \upsilon_\zeta \zeta_y \right)$$

$$\frac{\partial E}{\partial y} = \frac{\partial}{\partial y} \left[\mu_e \frac{\partial \upsilon}{\partial y} \right] = \xi_y \frac{\partial}{\partial \xi} \left[\mu_e \left(\upsilon_\xi \xi_y + \upsilon_\eta \eta_y + \upsilon_\zeta \zeta_y \right) \right]$$

$$+ \eta_y \frac{\partial}{\partial \eta} \left[\mu_e \left(\upsilon_\xi \xi_y + \upsilon_\eta \eta_y + \upsilon_\zeta \zeta_y \right) \right] \qquad \text{(6.A.14)}$$

$$+ \zeta_y \frac{\partial}{\partial \zeta} \left[\mu_e \left(\upsilon_\xi \xi_y + \upsilon_\eta \eta_y + \upsilon_\zeta \zeta_y \right) \right]$$

$$F = \mu_e \frac{\partial \upsilon}{\partial z} = \mu_e \left(\upsilon_\xi \xi_z + \upsilon_\eta \eta_z + \upsilon_\zeta \zeta_z \right)$$

$$\frac{\partial F}{\partial y} = \frac{\partial}{\partial y} \left[\mu_e \frac{\partial \upsilon}{\partial z} \right] = \xi_y \frac{\partial}{\partial \xi} \left[\mu_e \left(\upsilon_\xi \xi_z + \upsilon_\eta \eta_z + \upsilon_\zeta \zeta_z \right) \right]$$

$$+ \eta_y \frac{\partial}{\partial \eta} \left[\mu_e \left(\upsilon_\xi \xi_z + \upsilon_\eta \eta_z + \upsilon_\zeta \zeta_z \right) \right]$$

$$+ \zeta_y \frac{\partial}{\partial \zeta} \left[\mu_e \left(\upsilon_\xi \xi_z + \upsilon_\eta \eta_z + \upsilon_\zeta \zeta_z \right) \right]$$

$$G = \mu_e \frac{\partial w}{\partial x} = \mu_e \left(w_\xi \xi_x + w_\eta \eta_x + w_\zeta \zeta_x \right)$$

$$\frac{\partial}{\partial z}\left[\mu_e \left(\frac{\partial w}{\partial x} \right) \right] = \xi_z \frac{\partial}{\partial \xi}\left[\mu_e \left(w_\xi \xi_x + w_\eta \eta_x + w_\zeta \zeta_x \right) \right]$$

$$+ \eta_z \frac{\partial}{\partial \eta}\left[\mu_e \left(w_\xi \xi_x + w_\eta \eta_x + w_\zeta \zeta_x \right) \right]$$

$$+ \zeta_z \frac{\partial}{\partial \zeta}\left[\mu_e \left(w_\xi \xi_x + w_\eta \eta_x + w_\zeta \zeta_x \right) \right]$$

$$H = \mu_e \frac{\partial w}{\partial y} = \mu_e \left(w_\xi \xi_y + w_\eta \eta_y + w_\zeta \zeta_y \right)$$

$$\frac{\partial}{\partial z}\left[\mu_e \left(\frac{\partial w}{\partial y} \right) \right] = \xi_z \frac{\partial}{\partial \xi}\left[\mu_e \left(w_\xi \xi_y + w_\eta \eta_y + w_\zeta \zeta_y \right) \right]$$

(6.A.15)

$$+ \eta_z \frac{\partial}{\partial \eta}\left[\mu_e \left(w_\xi \xi_y + w_\eta \eta_y + w_\zeta \zeta_y \right) \right]$$

$$+ \zeta_z \frac{\partial}{\partial \zeta}\left[\mu_e \left(w_\xi \xi_y + w_\eta \eta_y + w_\zeta \zeta_y \right) \right]$$

$$I = \mu_e \frac{\partial w}{\partial z} = \mu_e \left(w_\xi \xi_z + w_\eta \eta_z + w_\zeta \zeta_z \right)$$

$$\frac{\partial}{\partial z}\left[\mu_e \left(\frac{\partial w}{\partial z} \right) \right] = \xi_z \frac{\partial}{\partial \xi}\left[\mu_e \left(w_\xi \xi_z + w_\eta \eta_z + w_\zeta \zeta_z \right) \right]$$

$$+ \eta_z \frac{\partial}{\partial \eta}\left[\mu_e \left(w_\xi \xi_z + w_\eta \eta_z + w_\zeta \zeta_z \right) \right]$$

$$+ \zeta_z \frac{\partial}{\partial \zeta}\left[\mu_e \left(w_\xi \xi_z + w_\eta \eta_z + w_\zeta \zeta_z \right) \right]$$

REFERENCES

Anderson, J. D. Jr. 1995. *Computational Fluid Dynamics: The Basic Applications.* New York: McGraw-Hill.

Anderson, D. A., J. C. Tannehill, and R. H. Pletcher. 1984. *Computational Fluid Mechanics and Heat Transfer.* New York: Hemisphere Publishing.

Anderson, D. A., J. C. Tannehill, and R. H. Pletcher. 1997. *Computational Fluid Mechanics and Heat Transfer.* 2nd ed. Philadelphia, PA: Taylor & Francis.

Courant, R. and D. Hilbert. 1953. *Methods of Mathematical Physics.* Vol. 1. New York: Interscience Publishers.

Hawkins, G. A. 1963. *Multilinear Analysis for Students in Engineering and Science.* New York: John Wiley & Sons.

Korn, G. A. and T. M. Korn. 1968. *Mathematical Handbook for Scientists and Engineers*. New York: McGraw-Hill.

Patanker, S. V. 1980. *Numerical Heat Transfer and Fluid Flow*. Philadelphia, PA: Taylor & Francis.

Shih, T. I.-P., R. T. Bailey, H. L. Nguyen, and R. J. Roelke. 1991. Algebraic grid generation for complex geometries. *Int. J. Num. Methods Fluids*. 13:1–31.

Thompson, J. F., Z. U. A. Warsi, and C. W. Mastin. 1985. *Numerical Grid Generation: Foundations and Applications*. New York: North-Holland.

Vinokur, M. 1974. Conservation equations of gasdynamics in curvilinear coordinate systems. *J. Comp. Physics*. 14:105–125.

Viviand, H. 1974. Conservative forms of gas dynamics equations. *La Reserche Aerospatiale*. No. 1974–1:65–68.

Wang, T.-S. and Y. S. Chen. 1993. Unified Navier-Stokes flowfield and performance analysis of liquid rocket engines. *J. Propul. Power*. 5:678–685.

CHAPTER 7

Numerical Methods for Solving Governing Equations

7.1 OVERVIEW

This chapter introduces the basic concepts of various numerical methods which have been developed to solve the governing conservation equations of transport phenomena. These numerical methods when used to solve the governing equations (including the nonlinear Navier–Stokes equations) of fluid motion are often called computational fluid dynamics (CFD). Readers who are interested in learning an in-depth description of these numerical methods should refer to the references listed at the end of this chapter.

CFD methodology and computers have improved dramatically since the late 1980s. As a result, the development of numerical tools and physical models has become popular not only in academic research but also in industry, as CFD tools are increasingly being employed in routine design practices. In addition, CFD technology has advanced to a state that flows with both complicated geometry and complex physics can be analyzed with reasonable accuracy. There are numerous methodologies employed by CFD analysts. Among these, the two most widely used methods are the density- and pressure-based approaches, and each approach has its own strengths and weaknesses. The basic concepts and differences between these two methods will be described later.

For each numerical approach, three popular numerical methods are used to discretize and solve the set of governing equations. These three numerical methods are finite difference, finite volume, and finite element. Within each numerical method, the set of discretized governing equations can be solved in a fully coupled fashion, or sequentially, or a combination of both. For the fully coupled solution scheme, all the dependent variables are calculated simultaneously. This procedure requires a huge computer memory to store and solve the coupled coefficients of the system of discretized equations. If the system of discretized equation is solved sequentially (also known as the segregated solution scheme), then only one dependent variable is calculated at a time. Thus, a smaller computer memory is required, but more iterations must be performed to get a converged solution. Some CFD solvers use a combination of both, i.e., solving the critical dependent variables in coupled fashion, while calculating the others sequentially.

The discretized equations are solved in a series of small regions within the domain of interest (the so-called computational domain). These numerous small regions are called numerical meshes (or grids or cells). Three commonly seen grid topologies are used to discretize the computational domain: structured, unstructured, and hybrid meshes. These methods are described herein. Recently, two new grid topologies have been proposed to reduce the difficulty of generating good-quality numerical meshes for problems with complex geometries. These two grid topologies are the meshless method (Löhner and Onate, 1998; Sridar and Balakrishnan, 2003) and the Cartesian grid method (Russell et al., 1995; Pember and Wang, 2003; Marella et al., 2005). Details may be obtained from these references.

Despite the success claimed for different numerical methods, different forms of numerical damping still exist in each method. The use of excessive numerical damping can affect the accuracy of numerical results, especially in simulating unsteady flow. Recently, a space–time conservation-element/solution-element (CE/SE) framework has been developed to address the issue of numerical damping and improve the numerical accuracy in solving various partial differential equations. This numerical method will also be introduced at the end of this chapter. It should be noted that even though not included here, there are other numerical methods for solving the governing equations of fluid dynamics, such as the boundary element method (Brebbia et al., 1984), the direct simulation Monte-Carlo (DSMC) method (Bird, 1994), the lattice Boltzmann equation (LBE), or direct Boltzmann equation (DBE) solver. They are more useful for specific groups of problems, and thus are not covered here. For example, most of the CFD methods are employed to solve the Navier–Stokes equations to model fluid motion, which is valid for continuum flows only where the Knudsen number (the ratio of the mean free path of the fluid to the characteristic length of the flow) is very small (<0.01). However, when the flow approaches the slip or free-molecular flow regime (Knudsen number >0.01) or thermal nonequilibrium regime (where rotational, vibrational, and electron energies are not negligible), the use of the Navier–Stokes equations is inappropriate. Though DSMC, LBE, and DBE are valid for simulating noncontinuum and thermally nonequilibrium flows, they are very computationally expensive, especially for problems involving continuum flow. Lately, a hybrid numerical approach has been employed to couple the DSMC and Navier–Stokes solvers to account for the proper physics with good computational efficiency (Roveda et al., 1998; Sun et al., 2004; Wu et al., 2006). References for these numerical methods are readily available.

7.2 DENSITY-BASED AND PRESSURE-BASED METHODS

As discussed in previous chapters, a set of governing equations will be solved for a set of dependent variables for each problem of interest. For example, for a three-dimensional laminar, compressible, non-reacting flow problem, seven dependent variables (density, three velocity components, pressure, temperature, and energy) need to be calculated from five transport equations (continuity equation, three momentum equations, and energy equation). Since there are more unknowns than the number of transport equations, two equations of state are also required to close the system of equations. The difference between the density-based and the pressure-based methods is determined by which dependent variables are calculated from the transport equations and which are calculated from the equations of state. Details of these two methods are presented in the following.

7.2.1 DENSITY-BASED METHOD

The density-based method calculates density directly from the continuity equation and pressure from an equation of state. The density, three velocity components, and energy are calculated as the primary dependent variables by solving five transport equations, and the pressure and temperature, treated as the secondary dependent variables, are calculated from the equations of state based on the density and energy. Hence, in solving the momentum and energy equations, the values of pressure and energy calculated from the previous time step are used. In high-speed flows such as transonic, supersonic, and hypersonic flows, density is a primary variable and is a function of time and space. The governing equations for such flows without viscosity are hyperbolic in nature. A large number of numerical schemes have been developed to solve such problems and are available in the literature (Yee, 1989; Toro, 1999). The wave propagation properties of the hyperbolic equations have been taken into consideration for the development of these types of schemes. As a rule of thumb, the flow of air at Mach numbers less than 0.3 can be considered incompressible, and the compressible schemes without modifications become unsuitable for such flows (Turkel et al., 1997; Roller and Munz, 2000).

Within the low subsonic region, the magnitude of the flow velocity becomes very small compared to the acoustic speed. This makes the eigenvalues of the system vary in magnitude. Also, the condition number of the system becomes very large resulting in stiffness of the linear system and numerical instability. The condition number is the ratio of the largest to the smallest eigenvalue matrices being solved. One of the methods used to overcome this problem is the application of preconditioning in which the time derivative is premultiplied by an appropriate matrix to rescale the condition number (Turkel, 1987; van Leer et al., 1991). One of the drawbacks of preconditioning is the difficulty in predicting the transient solution because of the introduction of artificial transient terms during the preconditioning process. Another drawback is the lack of numerical stability at stagnation points. The second approach for the application of hyperbolic schemes developed for compressible flows to incompressible flows is based on the artificial compressibility method, which was developed by Chorin (1967). In this approach, an artificial time derivative term based on the pressure is added to the continuity equation to cast the equation in hyperbolic form. This method has been successfully extended to various applications and has been presented in the literature (Taylor, 1991; Sheng et al., 1999; Koomullil and Soni, 2001).

7.2.2 PRESSURE-BASED METHOD

In contrast, for pressure-based methods, which were originally developed for incompressible flows, the pressure is treated as one of the primary dependent

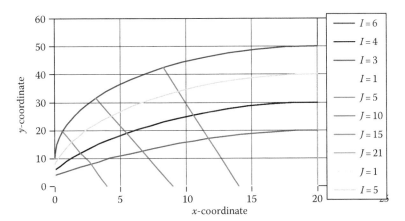

FIGURE 6.1 Curvilinear coordinate system.

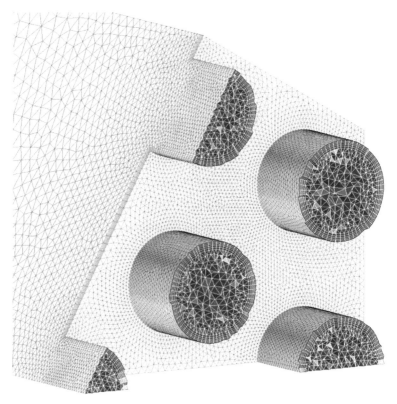

FIGURE 7.4 Example of a hybrid grid system. (Courtesy of Dr. Yasushi Ito, University of Alabama at Birmingham.)

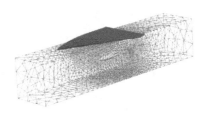

(a) Overall view

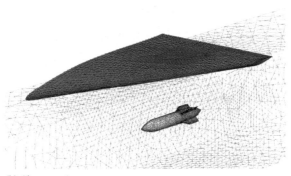

(b) Close-up view

FIGURE 7.5 An overset grid system for a wing-store configuration. (Courtesy of Dr. Roy P. Koomullil, University of Alabama at Birmingham.)

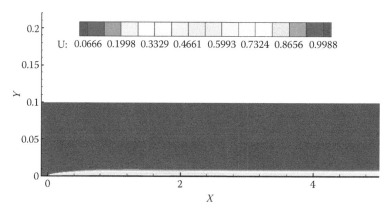

FIGURE A.1 Boundary layer flow over a flat plate (working directory: z01-BL).

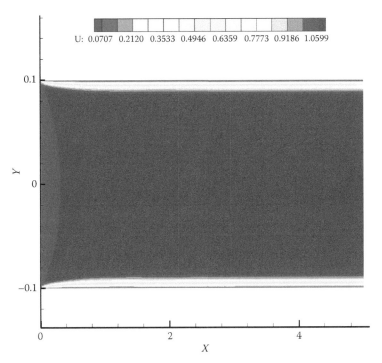

FIGURE A.2 Developing and fully developed pipe flow (working directory: z02-DPF).

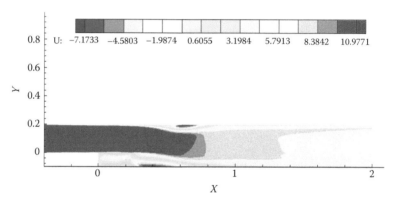

FIGURE A.3 Flow over a backstep.

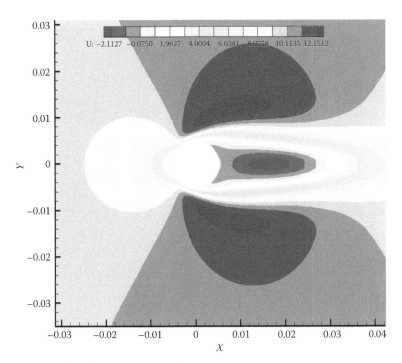

FIGURE A.4 A cylinder in cross-flow.

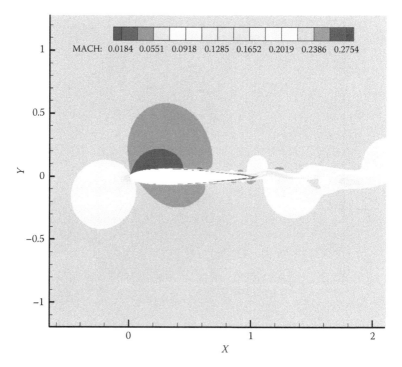

FIGURE A.5 Flow over an airfoil.

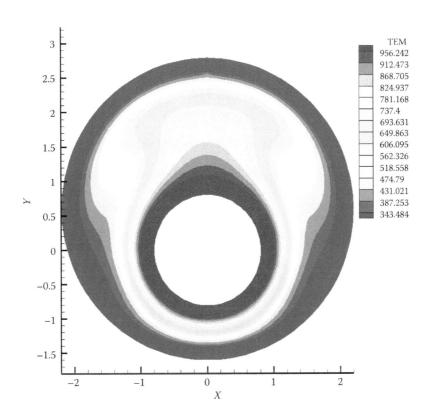

FIGURE A.6 Shell and tube heat exchanger.

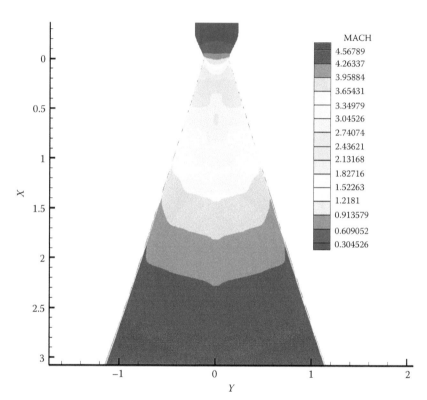

FIGURE A.7 Converging–diverging nozzle flow.

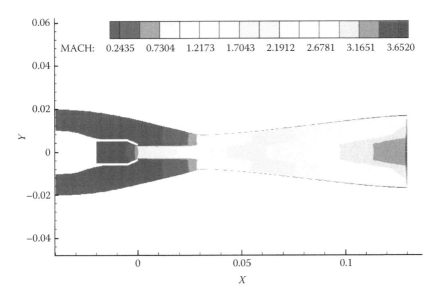

FIGURE A.8 Orifice flow and an ejector pump.

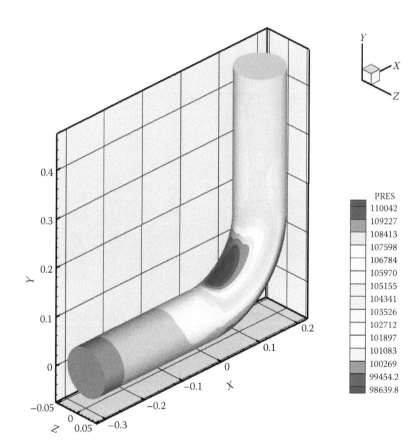

FIGURE A.9 Flow through a pipe elbow.

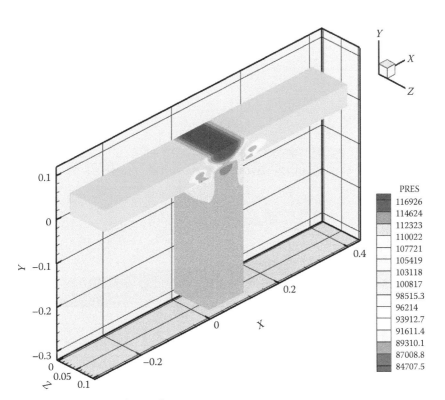

FIGURE A.10 Flow through a pipe tee.

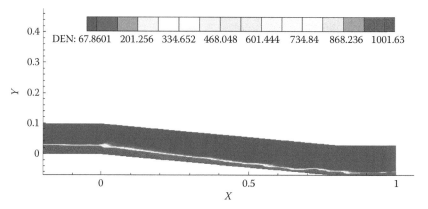

FIGURE A.11 Free-surface flow in an open duct.

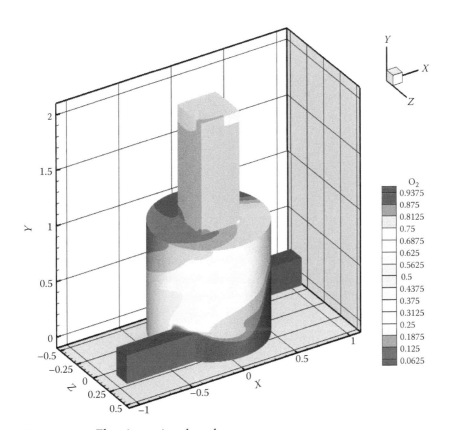

FIGURE A.12 Flow in a stirred-tank.

variables and is dependent on space and time. Hence, this method calculates the pressure, three velocity components, and energy by solving five transport equations. Since the continuity equation involves only the density and velocity components as dependent variables, it needs to be reformulated and cast in terms of the pressure and velocity components. The transport equation for pressure is obtained by manipulating the momentum and continuity equations to produce an equation for the second derivative of pressure correction (or pressure change). This equation is termed the Poisson equation. Once the primary variables can be calculated from the equations of state. This numerical approach is called the pressure correction method. Various pressure correction methods were developed to recast the continuity equation and solve the system of equations in an iterative fashion. Among them, a Semi-Implicit Method for Pressure Linked Equations (SIMPLE) scheme was first proposed by Patankar and Spalding (1972), and then revised by Patankar (1980) (SIMPLER) for the pressure–velocity calculation procedure on the staggered grid topology (this grid will be explained in Section 7.4). Various numerical schemes, such as SIMPLEC by Van Doormal and Raithby (1984), and PISO by Issa (1986), were also developed for the pressure-based method. The basic ideals of the pressure correction method are

1. To let the velocity vector and density be $\vec{V} = \vec{V}^* + \vec{V}'$ and $\rho = \rho^* + \rho'$, respectively, where the superscripts * and $'$ denote the guessed and corrected values
2. To solve the momentum equations in a predictor step to obtain $\vec{V}^*$ where the value of pressure at the previous time step is used
3. To solve the reformulated continuity equation to obtain the value of pressure correction (p')
4. To perform the corrector step to calculate velocity correction based on pressure correction

A correlation between the pressure change and density change based on either the isothermal or isentropic process is then used to replace the density correction with pressure correction in the continuity equation so that the pressure-based method can be used to simulate the compressible flows. With this substitution and mathematical manipulation (details will be shown in Chapter 8), the reformulated continuity equation, also known as the pressure-correction equation, can be expressed as

$$\beta_p \frac{\partial p'}{\partial t} + \nabla \cdot (\vec{V}^* \beta_p p') - \nabla \cdot (\rho^* D_u \nabla p') = -\left[\frac{\partial \rho^*}{\partial t} + \nabla \cdot (\rho^* \vec{V}^*) \right] \qquad (7.1)$$

where

$$\beta_p = \gamma/a^2; \quad \vec{V}' \approx -D_u \nabla p'; \quad p' = p^{n+1} - p^n$$

where

the superscripts n and $n+1$ denote the value at the previous and current time steps

D_u is the inverse of the matrix of the coefficients of the convective terms in the finite difference form of the inviscid equations of motion

a and γ are the speed of sound and the ratio of specific heats, respectively

This method is valid for both perfect-gas and real-fluid flows (Farmer et al., 2005; Cheng and Farmer, 2006). With the pressure-based approach, pressure variation remains finite regardless of the Mach number, and thus avoids the shortcomings of density-based methods. Since the pressure-based solution method is an iterative predictor–corrector solution procedure, it does not need a large amount of computer memory to resolve a huge matrix of the system of equations, thus it is more computer friendly.

7.3 NUMERICAL METHODS

As mentioned earlier, the most commonly reported numerical methods used to discretize and solve the governing equations are finite difference (Anderson et al., 1984; Hirsch, 1990; Anderson Jr., 1995), finite volume (Hirsch, 1990; Toro, 1999; Chung, 2002; Leveque, 2002; Versteeg and Malalasekera, 2007), and finite element (Chung, 1978, 2002; Baker, 1983; Hirsch, 1990). The finite difference method discretizes and solves partial differential in the form

$$\frac{\partial \rho \phi}{\partial t} + \nabla \cdot (\rho \vec{V} \phi) + \nabla \cdot (\nabla \Gamma_\phi \phi) = S_\phi \tag{7.2}$$

where ϕ is the dependent variable. However, the finite volume method discretizes and solves the system of governing equation in the integral form,

$$\iiint \frac{\partial \rho \phi}{\partial t} d\forall + \oiint \rho \phi \vec{V} \cdot \vec{n} dA + \oiint (\nabla \Gamma_\phi \phi) \cdot \vec{n} dA = \iiint S_\phi d\forall \tag{7.3}$$

The finite element method, originated in the field of structural analysis, has been used to solve various linear and nonlinear continuous field problems with good success. This method has also been used to solve fluid flow problems. Similar to the finite volume method, the finite element method solves the integral form of the transport equations and discretizes the computational domain into numerous elements (cells) of arbitrary shape and size. The value of the dependent variables is approximated as a sum of their values at the vertices of each element, as modified by an interpolation function corresponding to the vertex of each element. Various interpolation functions have been used, such as spline, Legendre, Chebyshev, or trigonometric functions. The finite element method formulates the integral form of the discretized transport equations using either a variational principle (where the dependent variable is expressed as the extremum of a functional) or the weighted residual method through a weak formulation. The latter one is more widely used because it can define an equivalent integral formulation for various cases including discontinuities such as shock waves.

7.4 GRID TOPOLOGIES

In CFD simulations, a computational domain is divided into numerous small subdomains, which are called meshes (or grids or cells or elements). The vertices of each cell are called nodes (or grid points). A control volume is the volume of a finite region within a computational domain, where the physical conservation laws are imposed and solved to obtain the value of dependent variables. The control volume is typically the same as a cell, but could be different for some numerical methods. The dependent variables can be calculated at either the grid point (denoted as the node-centered method), or the center of a cell (denoted as the cell-centered method). For some cell-centered CFD solvers, once the dependent variables are calculated, they are interpolated or averaged to get their values at each node point so that the coordinate and dependent variables are stored at the same location. Figure 7.1 illustrates the difference between these two methods. The shaded area is the control volume, the filled circle denotes the location where the dependent variable is calculated, and the cross symbol indicates the grid point.

The control volume is typically set to be that of the cell. In the staggered grid method, the control volume is defined differently for different governing equations. As shown in Figure 7.2a, the scalar variable (such as pressure) is calculated at the grid point (denoted as filled circle), while the vector variable (such as velocity components) is calculated at the midpoint between two grid nodes (denoted as open circle). Selection of the control volume is different for different conservation equations, such that the unknown dependent variable for the corresponding equation is located at the center of a control volume and the derivative of scalar variable can

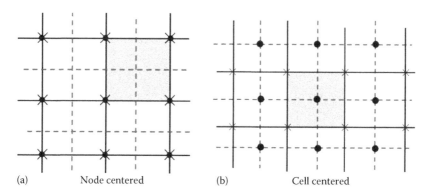

FIGURE 7.1 Control volumes for the node-centered and cell-centered topologies.

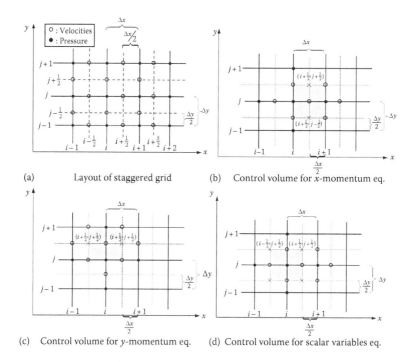

FIGURE 7.2 Storage locations and control volumes of a staggered grid system.

be evaluated directly. Control volumes for x- and y-momentum equations and for the transport equation for a scalar variable (such as pressure) are shown in Figure 7.2b through d, respectively.

There are three basic types of mesh systems: structured, unstructured, and hybrid (a combination of structured and unstructured meshes). For the structured mesh, the nodes are indexed in a certain order, and the shape of a cell has to be quadrilateral (2-D), and hexahedral (3-D). As a result, the grid connectivity and interface between neighboring cells are implicitly defined, and there is no need to search and store such information. The nodes of the unstructured mesh are indexed randomly, and thus are not in any order. Hence, the grid connectivity and interface between neighboring cells have to be explicitly defined (Weatherill, 1988; Marcum and Weatherill, 1995) in a database and provided to the CFD code. In finite element parlance, this is termed the assembly step. Comparisons between the structured and unstructured mesh topologies are illustrated in Figure 7.3. The use of structured meshes is computationally more efficient and requires less computer memory than the use of unstructured meshes. However, structured meshes are difficult to construct for complex geometries even when using multiblock grid topology (Thompson, 1987). The unstructured mesh topology provides flexibility, simplicity, and automation for modeling complex geometries. It is also easier to conduct parallel computing with domain decomposition and to achieve computer load balance using the unstructured mesh. Grid refinement for obtaining more accurate numerical solutions can also be easily accomplished with the unstructured grid. The use of unstructured meshes not only requires more computer memory and computational time but also has relatively poor numerical accuracy due to the presence of skewed triangular (2-D) or tetrahedral (3-D) elements in sensitive regions like boundary layers.

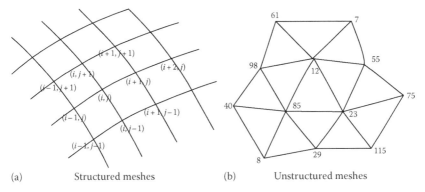

(a) Structured meshes (b) Unstructured meshes

FIGURE 7.3 Illustration of structured and unstructured mesh systems.

In an attempt to combine the advantages of both structured and unstructured grids, a hybrid grid topology was proposed (Shaw, 1998). In the hybrid mesh system, pyramid or prismatic or hexahedral cells are used in the boundary layer region, while the rest of the domain is filled with tetrahedral cells (Kallinderis, 1998; Shaw, 1998). Figure 7.4 shows an example of the hybrid grid system. It has been observed that a hybrid grid in viscous regions creates a fewer number of elements than a completely unstructured grid for the same resolution. Recently, the idea of using grids of different element-topologies has been extended further to give rise to what is called a generalized grid topology (Koomullil and Soni, 1999; Thompson et al., 2000; Cheng et al., 2007). The generalized grid topology employs polyhedral cells and has no restrictions on the number of edges or faces for a cell, which makes it extremely flexible for adapting the mesh to complex geometry and maintaining good grid quality.

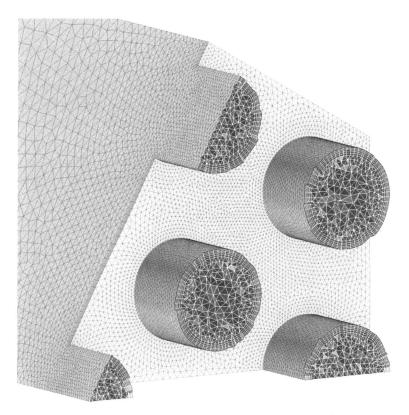

FIGURE 7.4 (See color insert following page 294.) Example of a hybrid grid system. (Courtesy of Dr. Yasushi Ito, University of Alabama at Birmingham.)

The aforementioned fixed grid topologies are typically employed to model objects which are stationary relative to fluid flows. Transport phenomena of fluid flow involving multiple bodies in relative motion are commonly seen in the store/stage separation process, control surface deflections for air vehicles, debris transport, turbomachinery, etc. For the simulation of moving body problems, the mesh associated with each body needs to be relocated to preserve a body conforming grid; thus, a conventional fixed grid approach cannot be used to simulate such problems. Currently, there are two major approaches used to tackle this type of problem. The first approach utilizes a grid deformation and remeshing strategy to handle the relative motion among multiple bodies. In this approach, as different bodies move relative to each other, the grid between them is deformed using different numerical strategies such as a tension or torsion spring analogy (Sing et al., 1995; Farhat et al., 1998), a Laplacian based approach (Burg, 2005), or a linear elasticity data interpolation (Gao et al., 2002). The mesh must be locally or globally regenerated once the mesh quality degrades. If the body movement is sufficiently small, the grid deformation can be smoothly transferred to the interior points in such a way that it will be dampened at the fixed outer boundary. The advantages of these methods provide ease of implementation into a CFD flow solver without data interpolation, as long as no regeneration of the grid is required. It is very difficult to maintain the grid quality (which has a strong effect on the numerical accuracy) of the deforming grid and is very time consuming to check the grid quality during the mesh deformation process. Another drawback of this approach is that the mesh quality must be checked at each time-iteration. This approach is best suited for bodies involving small relative motion between different components.

The second approach is the use of an overset grid topology (Benek et al., 1985; Noack, 2002, 2003; Buning et al., 2004; Fasta and Shelley, 2004; Cheng et al., 2005, 2007; Koomullil et al., 2008). In this approach, individual grids are generated for the objects involved in the problem of interest, and they are overlaid with each other to form a complete mesh system. An example of overset grid for a wing-store configuration is demonstrated in Figure 7.5. This makes the grid generation less labor intensive, and the grid movements associated with each body can be independently modeled for the individual mesh systems. Since an overset composite grid consists of several overlapping grids, procedures have to be performed to interpolate and exchange information between the flow variables from different mesh systems. Appropriate algorithms are used to remove the mesh

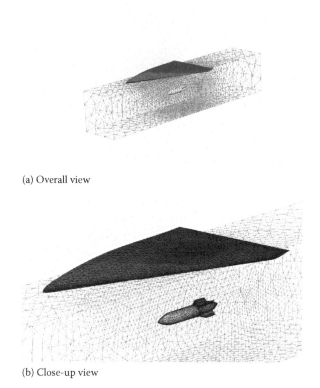

(a) Overall view

(b) Close-up view

FIGURE 7.5 (See color insert following page 294.) An overset grid system for a wing-store configuration. (Courtesy of Dr. Roy P. Koomullil, University of Alabama at Birmingham.)

points outside the domain of interest (such as mesh points inside a solid object) and to find the regions where the flow variables need to be interpolated from other meshes. During the time integration of the governing equations, appropriate information is transferred between different meshes for an accurate simulation. As one or more of the objects move relative to another, the meshes associated with them also move rigidly along with the body. The use of overset meshes makes the flow simulation process more efficient and allows mesh motion without deformation or remeshing. In addition, the overset grid topology allows the structured mesh to easily model complex geometries. However, implementation of this approach into a flow solver is extremely complicated because (1) the grid connectivity between zonal meshes needs to be tracked and updated at each time iteration, (2) the grid points that fall into the solid object need to be identified and excluded from flow calculations, and (3) interpolations between overlapping grids need to be performed, which can decrease numerical accuracy. The numerical accuracy in calculating heat transfer in a thin boundary layer such as high-speed flows is of great concern because of the disparity among different mesh systems in that region.

The accuracy of numerical solutions is highly dependent on the grid spacing employed. However, it is not feasible to use very fine grid throughout the entire computational domain due to the limitation of computer power and time. Hence, the common practice of CFD is to use fine grid only for the region where flow variables change rapidly, such as boundary layer, flow separation, and shock waves. Unfortunately, the exact location and the size of the high-gradient regions are unknown until the numerical solution is obtained. Trial-and-error can be used to determine the regions which require fine mesh. Another alternative is to automatically adapt the mesh with refinement or redistribution based on the solution, which is called mesh refinement or mesh adaptation/redistribution. The basic structure of grid refinement procedure consists of (a) solving the governing equations on the current grid; (b) identifying cells for refinement or coarsening; (c) subdividing the cells identified for refinement; (d) coalescing the cells identified for coarsening; and (e) refining additional cells to maintain a smooth grid density variation which is required to guarantee the stability and accuracy of the solver. Grid adaptation/redistribution has the same procedure as grid refinement, except that step (c)-(e) will be replaced with relocation of grid points. For the structured mesh, grid adaptation (Soni et al., 1998, 2000) can lead to deterioration of grid quality (skewed mesh), and grid refinement can greatly increase the number of grid points. Compared to the structured mesh, unstructured meshes offer more flexibility in both grid adaptation and refinement (Khawaja et al., 2000; Zhang et al., 2001; Suerich-Gulick et al., 2004; Shephard et al., 2005).

7.5 SPACE–TIME CONSERVATION-ELEMENT/ SOLUTION-ELEMENT METHODS

In recent years, a growing trend in the field of CFD has been the demand for increased accuracy and capability to handle complex flows. In particular, the focus of CFD research has shifted away from steady-state inviscid problems, toward the more complicated regime of unsteady viscous flows that may involve shocks, combustion, etc. In such regimes, the traditional finite difference, finite volume, and finite element methods, such as the

Lax-Wendroff scheme or the two-step MacCormack's scheme, generally seem to suffer from numerical wiggles near a discontinuity (Hirsch, 1990). Motivated by the idea that this difficulty may be overcome by addition of artificial dissipation, as suggested by von Neumann and Richtmeyer (von Neumann and Richtmyer, 1950), many effective wiggle-suppressing, shock-capturing schemes such as Harten's (1983) high-resolution scheme, Yee's total variation diminishing (TVD) scheme (Yee et al., 1985), and the essentially nonoscillatory (ENO) method (Harten et al., 1987) were developed during the 1980s. However, these schemes are burdened with additional effects such as monotonicity, entropy conditions, and TVD properties, which may not be consistent with flow physics (Chang et al., 1998). Moreover, as rightly pointed out by Shu et al. (1992), these methods have difficulties in capturing small-scale flow features due to the excessive damping.

In contrast, there are popular schemes such as spectral methods (Gottlieb and Orszag, 1977) and the compact finite difference scheme with spectral-like resolution (Lele, 1992), which on one hand have high accuracy and low numerical dissipation and thus can resolve small-scale flow features. On the other hand, these schemes are handicapped by their limited capability to handle practical problems, e.g., those involving complex geometries and (or) shock waves (Shu et al., 1992). In fact, these schemes are not flux-conserving, hence making them unsuitable for resolving shocks. In general, while solving nonlinear partial differential equations, stability cannot be maintained without the presence of adequate dissipation. Hence, schemes such as these, with low numerical dissipation are susceptible to numerical instability and may require some ad hoc treatment to maintain stability.

The inadequacy of the popular schemes mentioned above can be illustrated vividly by their general inability to resolve both strong shocks and small disturbances (e.g., acoustic waves) simultaneously. Note that, while numerical dissipation is needed for shock capturing, too much of it will result in annihilation of small disturbances. By design, these schemes are not equipped to overcome this difficulty. Another challenge with respect to the development of flux-conserving flow solvers has been the procedure for evaluating the convective flux. For most established schemes, this procedure is constructed using a characteristics-based approximate solution to the Riemann problem, drawing motivation from Godunov's work (Godunov, 1969). As the Riemann problem is fundamentally one dimensional, this approach does not allow for a natural extension into multiple dimensions. In practice, the extensions are implemented using artificial dimensional-splitting techniques (Batten et al., 1997; Chang et al., 1998). Not only are these techniques difficult to apply over non-Cartesian meshes, they also make it much more complicated to enforce flux conservation over both space and time (a property required by physics), if it can be done at all. Batten et al. (1997) has a good discussion on this issue and explores some of the newer approaches in regards to flux evaluation procedures. In addition, because of numerical dissipation being introduced in those aforementioned methods through upwind biasing, its strength in a stagnation region often becomes too low to sustain computational stability and even results in spurious solutions such as the so-called carbuncle phenomenon (Peery and Imlay, 1988). Another thorny issue that is overlooked in the established methods is the conflict between stability and accuracy in time-accurate computations, i.e., too much numerical dissipation would degenerate accuracy, while too little of it may cause numerical instability. In fact, to meet both accuracy and stability requirements, computation must be performed slightly away from the limit of instability, without going too far away from it.

The space–time conservation element and solution element (CESE) method (Chang and To, 1991; Chang, 1995; Chang et al., 1998, 1999) is an emerging numerical framework and was developed as an attempt to overcome the aforementioned numerical difficulties. This method treats time and spatial coordinates in exactly the same manner. It is substantially different, in both concept and approach, from those well-established methods, such as finite difference, finite volume, and finite element methods. This high-resolution, multidimensional, numerical framework has been built from scratch with extensive consideration on physics and rigorous mathematical proof, thereby doing away with some of the limitations of traditional numerical simulation methods. In addition to being mathematically simple, it has many other attractive features for solving problems involving flow instability or unsteadiness. These features include the following

1. Unified treatment of both space and time
2. Enforcement of both local and global space–time flux conservation
3. Use of a space–time staggered mesh that allows for evaluation of fluxes at the cell interfaces without solving the Riemann problem
4. Schemes built from a neutrally stable (i.e., nondissipative) core scheme, allowing for control of numerical dissipation (if needed) effectively and with mathematical justification (not ad hoc)

5. Treating mesh values of the flow variables and their spatial derivatives as independent unknowns
6. For flows in multiple spatial dimensions, no directional splitting is employed, leading to a truly multidimensional scheme
7. Avoids using ad hoc numerical damping as much as possible

For inviscid flow problems (without flow discontinuity), the nondissipative core scheme is a natural fit, because the flow is reversible in time. However, the characteristics of viscous flow and inviscid flow problems with shocks are irreversible in time, and thus the addition of numerical dissipation may be required. Since its inception in 1991 (Chang and To, 1991), the unstructured-mesh compatible CESE method has been successfully adapted to model several different applications (Chang et al., 2000, 2005; Qin et al., 2000; Loh et al., 2001; Loh and Zaman, 2002; Kim et al., 2004; Zhang et al., 2004; He et al., 2005; Yen and Wagner, 2005) in unsteady Euler flows, acoustic waves, traveling and interacting shocks, detonation waves, cavitation, etc. Batten et al. (1997) in their paper that explored several new implicit schemes for the solution of the compressible Navier–Stokes equations, have high praise for the CESE framework among the conventional second-order accurate schemes and consider it a worthy candidate to model the Navier–Stokes equations. Nevertheless, as the CESE framework generates an additional convective flux of the flow variables near the wall, that can diffuse the boundary layer, Batten et al. (1997) suggest the need for developing appropriate boundary conditions that can remedy this situation. Zhang et al. (2000) and Venkatachari et al. (2008) have proposed unified wall boundary treatment, and show good success for both steady and unsteady viscous flow simulations. Unsteady flow problems such as combustion instability and aeroacoustics have gained a lot of attention in recent times. The CESE methodology, with its unique feature of enforcing conservation laws in both space and time, appears to be an ideal fit for simulating such problems. Hopefully, this methodology will be more fully developed.

7.6 NOMENCLATURE

See Section 6.5, exceptions are the replacement of the symbols: g with ϕ, ϕ with Γ, and p with P.

REFERENCES

Anderson, D. A., J. C. Tannehill, and R. H. Pletcher. 1984. *Computational Fluid Mechanics and Heat Transfer*. New York: McGraw-Hill.

Anderson Jr., J. D. 1995. *Computational Fluid Dynamics—The Basics with Applications*. New York: McGraw-Hill.

Baker, A. J. 1983. *Finite Element Computational Fluid Mechanics*. New York: Hemisphere/McGraw-Hill.

Batten, P., M. A. Leschziner, and U. C. Goldberg. 1997. Average-state Jacobians and implicit methods for compressible viscous and turbulent flows. *Journal of Computational Physics* 137:38–78.

Benek. J. A., P. G. Bunning, and J. L. Steger. 1985. A 3D CHIMERA grid embedding Technique. AIAA Paper 1985–1523.

Bird, G. A. 1994. *Molecular Gas Dynamics and the Direct Simulation of Gas Flows*. New York: Oxford University Press.

Brebbia, C. A., J. C. F. Telles, and L. C. Wrobel. 1984. *Boundary Element Techniques–Theory and Applications in Engineering*. Berlin: Springer-Verlag.

Buning. P. G., R. J. Gomez, and W. I. Scallion. 2004. CFD approaches for simulation of wing-body stage separation. AIAA paper 2004-4838.

Burg, C. O. E. 2005. Analytic study of 2D and 3D grid motion using modified Laplacian. *International Journal of Numerical Methods in Fluids* 52:163–197.

Chang, S. C. 1995. The method of space-time conservation element and solution element—A new approach for solving the Navier-Stokes and Euler equations. *Journal of Computational Physics* 119:295–324.

Chang, S. C. and W. M. To. 1991. A New Numerical Framework for Solving Conservation Laws—The Method of Space-Time Conservation Element and Solution Element. NASA TM 104495.

Chang, I. S., C. L. Chang, and S. C. Chang. 2005. Unsteady Navier–Stokes Rocket Nozzle Flows. AIAA Paper 2005-4353.

Chang, S. C., X. Y. Wang, and C. Y. Chow. 1998. The space-time conservation element and solution element method—A new high resolution and genuinely multidimensional paradigm for solving conservation laws I. The Two Dimensional Time Marching Schemes, NASA TM 1998-208843.

Chang, S. C., X. Y. Wang, and C. Y. Chow. 1999. The space-time conservation element and solution element method—A new high-resolution and genuinely multidimensional paradigm for solving conservation laws. *Journal of Computational Physics* 156(1):89–136.

Chang, S. C., X. Y. Wang, and W. M. To. 2000. Application of the space-time conservation element and solution element method to one-dimensional convection-diffusion problems. *Journal of Computational Physics* 165:189–215.

Chang, S. C., S. T. Yu, A. Himansu, X. Y. Wang, C. Y. Chow, and C. Y. Loh. 1998. The method of space-time conservation element and solution element—A new paradigm for numerical solution of conservation laws. *Computational Fluid Dynamics Review 1998*, Eds. M. M. Hafez and K. Oshima, 1:206–240. Singapore: World Scientific.

Cheng, G. C. and R. C. Farmer. 2006. Validation of a practical dense-spray combustion model for liquid rocket engine injector analyses. *AIAA Journal of Propulsion & Power* 22(6):1373–1381.

Cheng, G. C., R. P. Koomullil, and R. W. Noack. 2005. A Library Based Overset Grid Development for Density- and Pressure-Based Flow Solvers. AIAA Paper 2005-5119.

Cheng, G. C., R. P. Koomullil, and B. K. Soni. 2007. Multidisciplinary & multi-scale computational field simulations—algorithms and applications. *Mathematics and Computers in Simulation* 75(5–6):161–170.

Chorin, A. J. 1967. A numerical method for solving incompressible viscous flow problems. *Journal Computational Physics* 2:12–26.

Chung, T. J. 1978. *Finite Element Analysis in Fluid Dynamics*. New York: McGraw-Hill.

Chung, T. J. 2002. *Computational Fluid Dynamics*. Cambridge: Cambridge University Press.

Farhat, C., C. Degand, B. Koobus, and M. Lesoinne. 1998. Torsional springs for two-dimensional dynamic unstructured fluid meshes. *Computer Methods in Applied Mechanics and Engineering* 163:231–245.

Farmer, R. C., R. Pike, and G. C. Cheng. 2005. CFD analyses of complex flows. *Computers & Chemical Engineering* 29(11–12):2386–2403.

Fasta, P. and M. J. Shelley. 2004. A moving overset grid method for interface dynamics applied to non-Newtonian Hele-Shaw flow. *Journal of Computational Physics* 195:117–142.

Gao, X. W., P. C. Chen, and L. Tang. 2002. Deforming mesh for computational aeroelasticity using a nonlinear elastic boundary element method. *AIAA Journal* 40:1512–1517.

Godunov, S. K. 1969. A difference scheme for numerical computation of discontinuous solution of hydrodynamic equations. *Matematicheskii Sbornik* 47:271–306 (Russian), Translated U.S. Joint Public Research Service, JPRS 7226.

Gottlieb, D. and S. Orszag. 1977. *Numerical Analysis of Spectral Methods: Theory and Applications*. Philadelphia, PA: SIAM-CBMS.

Harten, A. 1983. High resolution schemes for hyperbolic conservation laws. *Journal of Computational Physics* 49:357–393.

Harten, A., B. Engquist, S. Osher, and S. R. Chakravarthy. 1987. Uniformly high order accurate essentially non-oscillatory schemes. *Journal of Computational Physics* 71:231–303.

He, H., S. T. Yu, and Z. C. Zhang. 2005. Direct Calculations of One-, Two-, and Three-Dimensional Detonations by the CESE Method. AIAA 2005-0229.

Hirsch, C. 1990. *Numerical Computation of Internal and External Flows*. Vols. 1 and 2, New York: Wiley & Sons.

Issa, R. I. 1986. Solution of the implicitly discretised fluid flow equations by operator-splitting. *Journal of Computational Physics* 62:40–65.

Kallinderis, Y. 1998. Hybrid grids and their applications. In *Handbook of Grid Generation*, Eds. J. F. Thompson, B. K. Soni, and N. P. Weatherill, pp. 25-1–25-18. Boca Raton, FL: CRC Press.

Khawaja, A., T. Minyard, and Y. Kallinderis. 2000. Adaptive hybrid grid methods, computer methods in applied mechanics and engineering. *Computer Methods in Applied Mechanics and Engineering* 189:1231–1245.

Kim, C. K., S. T. Yu, and Z. C. Zhang. 2004. Cavity flow in scramjet engine by the space-time conservation element and solution element method. *AIAA Journal* 42(5):912–919.

Koomullil, R. P. and B. K. Soni. 1999. Flow simulation using generalized static and dynamics grids. *AIAA Journal* 37(12):1551–1557.

Koomullil, R. P. and B. K. Soni. 2001. Wind Field Simulations in Urban Area. AIAA Paper 2001-2621.

Koomullil, R. P., G. C. Cheng., B. K. Soni, R. W. Noack, and N. Prewitt. 2008. Moving-body simulation using overset framework with rigid body dynamics. *Mathematics and Computers in Simulation* 78:618–626.

Lele, S. K. 1992. Compact finite-difference schemes with spectral-like resolution. *Journal of Computational Physics* 103:16–42.

Leveque, R. J. 2002. *Finite Volume Methods for Hyperbolic Problems*. Cambridge: Cambridge University Press.

Loh, C. Y., L. S. Hultgren, and S. C. Chang. 2001. Wave computation in compressible flow using the space-time conservation element and solution element method. *AIAA Journal* 39(5):794–801.

Loh, C. Y. and K. B. M. Q. Zaman. 2002. Numerical investigation of transonic resonance with a convergent–divergent nozzle. *AIAA Journal* 40(12):2393–2401.

Löhner, R. and E. Onate. 1998. An advancing front point generation technique. *Communications in Numerical Methods in Engineering* 14:1097–1108.

Marcum, D. L. and N. P. Weatherill. 1995. Unstructured grid generation using iterative point insertions and local reconnection. *AIAA Journal* 33: 1619–1625.

Marella, S., S. Krishnan, H. Liu, and H. S. Udaykumar. 2005. Sharp interface Cartesian grid method I: An easily implemented technique for 3D moving boundary computations. *Journal of Computational Physics* 210:1–31.

Noack, R. W. 2002. DiRTlib: A library to add overset capability to flow solvers. *6th Symposium on Overset Composite Grid and Solution Technology*, Fort Walton Beach, FL.

Noack, R. W. 2003. Resolution Appropriate Overset Grid Assembly for Structured and Unstructured Grids. AIAA Paper 2003-4123.

Patankar, S. V. 1980. *Numerical Heat Transfer and Fluid Flow*. Washington D.C.: Hemisphere Publishing Co., Taylor & Francis Group.

Patankar, S. V. and D. B. Spalding. 1972. A calculation procedure for heat, mass and momentum transfer in three-dimensional parabolic flows. *International Journal of Heat Mass Transfer* 15:1787–1806.

Peery, K. M. and S. T. Imlay. 1988. Blunt-Body Flow Simulation. AIAA Paper 88-2904.

Pember, D. and Z. J. Wang. 2003. A Cartesian grid method for modeling multiple moving objects in 2D incompressible viscous flow. *Journal Computational Physics* 191:177–205.

Qin, J., S. T. Yu, Z. C. Zhang, and M. C. Lai. 2000. Direct calculations of cavitating flows in fuel delivery pipe by the space-time CESE method. *Journal of Fuels & Lubricants, SAE Transactions* 108(4):1720–1725.

Roller, S. and C. D. Munz. 2000. A low Mach number scheme based on multi-scale asymptotics. *Computing and Visualization in Science* 3:85–91.

Roveda, R., D. B. Goldstein, and P. L. Varghese. 1998. Hybrid Euler/particle approach for continuum/rarefied flows. *Journal of Spacecraft and Rockets* 35:258–265.

Russell, R. B., J. B. Bell, P. Colella, W. Y. Crutchfield, and M. L. Welcome. 1995. An adaptive Cartesian grid method for unsteady compressible flow in irregular regions. *Journal of Computational Physics* 120:278–304.

Shaw, J. A. 1998. Hybrid grids. In *Handbook of Grid Generation*, Eds. J. F. Thompson, B. K. Soni, and N. P. Weatherill, pp. 23-1–23-18. Boca Raton, FL: CRC Press.

Sheng, C., D. Hyams, K. Sreenivas, A. Gaither, D. Marcum, and D. Whitfield. 1999. Three-Dimensional Incompressible Navier-Stokes Flow Computations about Complete Configurations Using a Multiblock Unstructured Grid Approach. AIAA Paper 99-0778.

Shephard, M. S., J. E. Flaherty, K. E. Jansen, X. Li, X. Luo, N. Chevaugeon, J. F. Remacle, M. W. Beall, and R. M. O'Bara. 2005. Adaptive mesh generation for curved domains. *Applied Numerical Mathematics* 52(2–3):251–271.

Shu, C. W., T. A. Zang, G. Erlebacher, D. Whitaker, and S. Osher. 1992. High-order ENO schemes applied to two- and three-dimensional compressible flow. *Journal of Applied Numerical Mathematics* 1(9):45–71.

Sing, K. P., J. C. Newman, and O. Basyal. 1995. Dynamic unstructured methods for flows past multiple objects in relative motion. *AIAA Journal* 33(4):641–649.

Soni, B. K., R. P. Koomullil, D. S. Thompson, and H. Thornburg. 2000. Solution adaptive grid strategies based on point redistribution. *Computer Methods in Applied Mechanics and Engineering* 189(4):1183–1204.

Soni, B. K., H. J. Thornburg, R. P. Koomullil, M. Apte, and A. Madhavan. 1998. PMAG: Parallel multiblock adaptive grid system. In *Proceedings of the 6th International Conference on Numerical Grid Generation in Computational Field Simulation*, London, England, pp. 769–779.

Sridar, M. and N. Balakrishnan. 2003. An upwind finite difference scheme for meshless solvers. *Journal of Computational Physics* 189:1–29.

Suerich-Gulick, F., C. Lepage, and W. Habashi. 2004. Anisotropic 3-D Mesh Adaptation for Turbulent Flows. AIAA Paper 2004-2533.

Sun, Q., I. D. Boyd, and G. V. Candler. 2004. A hybrid continuum/particle approach for modeling subsonic, rarefied gas flows. *Journal of Computational Physics* 194:256–277.

Taylor, L. K. 1991. Unsteady three-dimensional incompressible algorithm based on artificial compressibility. PhD Dissertation, Mississippi State University, starkville, Mississippi.

Thompson, J. F. 1987. A general three-dimensional elliptic grid generation system on a composite block structure. *Computer Methods in Applied Mechanics and Engineering* 64:377–411.

Thompson, D. S., S. Chalasani, and B. K. Soni. 2000. Generations of generalized grids by extrusion from surface meshes of arbitrary topology. In *Proceedings of the 7th International Conference on Numerical Grid Generation in Computational Field Simulations*, Eds. M. Cross, P. R. Eiseman, J. Hauser, and B. K. Soni, British Columbia, Canada, pp. 1061–1070.

Toro, E. F. 1999. *Riemann Solvers and Numerical Methods for Fluid Dynamics: A Practical Introduction*. 2nd ed. Berlin: Springer-Verlag.

Turkel, E. 1987. Preconditioned methods for solving the incompressible and low speed compressible equations. *Journal of Computational Physics* 72:277–298.

Turkel, E., R. Radespiel, and N. Kroll. 1997. Assessment of preconditioning methods for multi-dimensional aerodynamics. *Journal of Computers and Fluids* 26(6):613–634.

Van Doormal, J. P. and G. D. Raithby. 1984. Enhancement of the SIMPLE method for predicting incompressible fluid flows. *Numerical Heat Transfer* 7:147–163.

van Leer, B., W. T. Lee, and P. L. Roe. 1991. Characteristic Time-Stepping or Local Preconditioning of the Euler Equations. AIAA Paper 91-1552.

Venkatachari, B., G. C. Cheng, B. K. Soni, and S. C. Chang. 2008. Validation and verification of courant number insensitive CE/SE method for transient viscous flow simulations. *Mathematics and Computers in Simulation* 78:653–670.

Versteeg, H. K. and W. Malalasekera. 2007. *An Introduction to Computational Fluid Dynamics–The Finite Volume Method*. 2nd ed., New York: Prentice Hall.

von Neumann, J. and R. D. Richtmyer. 1950. A method for the numerical calculation of hydrodynamic shocks. *Journal of Applied Physics* 21:232–237.

Weatherill, N. P. 1988. A method for generating irregular computational grids in multiply connected planar domains. *International Journal for Numerical Methods in Fluids* 8:181–197.

Wu, J. S., Y. -Y. Lian, G. C. Cheng, R. P. Koomullil, and K. -C. Tseng. 2006. Development and verification of a coupled DSMC-NS scheme using unstructured mesh. *Journal of Computational Physics* 219(1):579–607.

Yee, H. C. 1989. A Class of High Resolution Explicit and Implicit Shock Capturing Methods. NASA TM-101088.

Yee, H. C., R. F. Warming, and A. Harten. 1985. Implicit total variation diminishing schemes for steady state calculations. *Journal of Computational Physics* 57:327–360.

Yen, J. C. and D. A. Wagner. 2005. Computational Aeroacoustics Using a Simplified Courant Number Insensitive CESE Method. AIAA Paper 2005-2820.

Zhang, S. J., J. Liu, Y. -S. Chen, and T. -S. Wang. 2001. Adaptation for Hybrid Unstructured Grid with Hanging Node Method. AIAA Paper 2001-2657.

Zhang, M., S. T. Yu, S. C. Lin, S. C. Chang, and I. Blankson. 2004. Solving magnetohydrodynamic equations without special treatment for divergence-free magnetic field. *AIAA Journal* 42(12):2605–2608.

Zhang, Z. C., S. T. Yu, X. Y. Wang, S. C. Chang, A. Himansu, and P. C. E. Jorgenson. 2000. The CESE Method for Navier-Stokes Equations Using Unstructured Meshes for Flows at All Speeds. AIAA Paper 2000-0393.

CHAPTER 8

The CTP Code

8.1 GRIDS

The CTP code solves the conservation form of the fluid transport equations by dividing the flow domain into a network of small control volumes using a system of grids. Based on this network of cells, the conservation equations are integrated for each control volume. Discretization in the temporal direction is applied separately based on a specified time-step size. The result is a system of discretized algebraic equations organized in matrix form that can be solved through iterative methods. Solutions for each time step are obtained the same way until a required length of time is reached for transient solutions or when a converged steady-state solution is achieved.

To discretize a computational domain that may be encompassed by boundaries with arbitrarily irregular shapes, a system of body-fitted grids can be incorporated to describe the geometry. The CTP code utilizes a structured grid system where the grid lines run in orderly curvilinear directions and can be identified by indices in each direction. A body-fitted grid, or body-conforming grid, as illustrated in Figure 8.1, gives a close representation of the boundary shapes of the flow domain under investigation. This grid is called a single-block structured grid. Clearly, the exact boundary shapes is approached, as the mesh size of this grid is continuously refined. Therefore, it is important to note that for boundaries with strong curvature or discontinuity in boundary slope, more grid points along the boundaries may be needed to predict good solutions of the flowfield. That is, aside from physical model effects, keeping smoothness

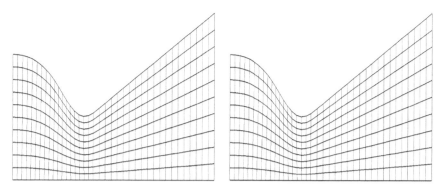

FIGURE 8.1 Typical body-fitted grid.

or minimizing the changes in mesh cell slopes and volumes is a good practice in getting good numerical solutions using the CTP code.

For complicated flow domains, a system of body-fitted grids (or multiblock grids) may be employed to model the geometry in order to give smooth representation of the boundaries. Using multiblock grid option also allows generation of better grid quality that is measured if the grid lines are intersected as close to be orthogonal as possible. Figure 8.2 illustrates an example of multiblock grid system. Here, interfaces between grid blocks (called block interfaces) are identified such that the discretization of the conservation equations at the interface cells, or control volumes, can be formulated the same way as for the interior cells of each block.

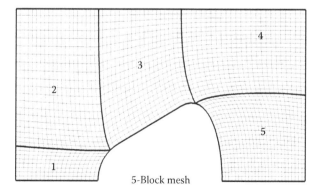

FIGURE 8.2 Multiblock grid system.

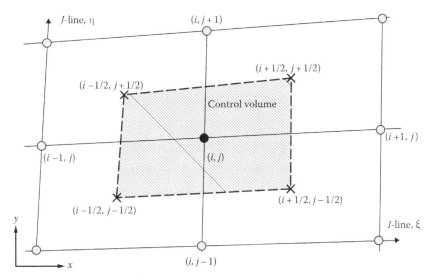

FIGURE 8.3 Control volumes and the grid index system.

To keep track of the grid system numerically, the grid points on each grid line are numbered in an orderly fashion. For a two-dimensional (2D) grid, grid lines in two running directions are named as *I*-line and *J*-line, respectively. The right-hand rule is observed in organizing the running directions of the grid indices. Since a control volume method is adopted to integrate the transport equations, cell-centered locations, defined as geometric average of the coordinates of the surrounding grid points, of grid cells are identified to construct control volumes. Figure 8.3 shows an example of the grid indices and the definition of control volumes. Here, cell-centered locations surrounding grid point (i, j) are represented as $(i - 1/2, j - 1/2)$, $(i + 1/2, j - 1/2)$, $(i - 1/2, j + 1/2)$, and $(i + 1/2, j + 1/2)$. A control volume of grid point (i, j) is then constructed by connecting the points of the cell-centered locations. This control volume definition can be directly extended to 3D grid systems. For boundary cells, the cell-centered points are pushed to the midpoint locations of the boundary surface segments. This system of control volumes forms a watertight domain representing the flow domain geometry. Here, for good numerical representation of the geometry and solution accuracy, it is assumed that the grid lines are smooth and the grid density is tailored to the local curvature of the grid lines.

8.2 DISCRETIZED CONSERVATION EQUATIONS

As described in Chapter 6 and Table 6.1, the transport equations in 2-D general curvilinear coordinates can be represented in a general form as

$$
\frac{\partial}{\partial t}\left(\frac{\rho\phi}{J}\right) + \frac{\partial}{\partial\xi}\left(\frac{\rho U^{c}\phi}{J}\right) + \frac{\partial}{\partial\eta}\left(\frac{\rho V^{c}\phi}{J}\right) - \frac{\partial}{\partial\xi}\left(\frac{\xi_{x}^{2}+\xi_{y}^{2}}{J}\Gamma_{\phi}\frac{\partial\phi}{\partial\xi}\right) - \frac{\partial}{\partial\eta}\left(\frac{\eta_{x}^{2}+\eta_{y}^{2}}{J}\Gamma_{\phi}\frac{\partial\phi}{\partial\eta}\right)
$$
$$
= \frac{S_{\phi}}{J} + \frac{\partial}{\partial\xi}\left(\frac{\xi_{x}\eta_{x}+\xi_{y}\eta_{y}}{J}\Gamma_{\phi}\frac{\partial\phi}{\partial\eta}\right) + \frac{\partial}{\partial\eta}\left(\frac{\eta_{x}\xi_{x}+\eta_{y}\xi_{y}}{J}\Gamma_{\phi}\frac{\partial\phi}{\partial\xi}\right) \qquad (8.1)
$$

where

ϕ represents all flow variables of the governing equations, i.e., 1, u, v, w, h, k, ε, and $(S_{\omega})_{k}$ for continuity, momentum, energy, turbulence kinetic energy, turbulence energy dissipation rate, and species equations, respectively

Γ_{ϕ} represents the diffusion coefficient, and can be expressed as

$$
\Gamma_{\phi} = \begin{cases} \dfrac{\mu}{\sigma_{\phi}} \text{ for laminar flows} \\[2ex] \dfrac{\mu}{\sigma_{\phi}} + \dfrac{\mu_{t}}{\sigma_{\phi,t}} \text{ for turbulent flows} \end{cases} ; \quad \text{where } \mu_{t} = \rho C_{\mu}\frac{k^{2}}{\varepsilon}
$$

Values of σ_{ϕ} and $\sigma_{\phi,t}$ for various governing equations are listed in Table 8.1. The curvilinear coordinate transformation Jacobian and contravariant velocities are defined as

TABLE 8.1 Values of σ_{ϕ} and $\sigma_{\phi,t}$ for Each Transport Equations

	σ_{ϕ}	$\sigma_{\phi,t}$
Momentum equations	1.00	1.00
Energy equation	0.72	0.90
k-equation (standard model)	—	1.00
ε-equation (standard model)	—	1.30
k-equation (extended model)	—	0.8927
ε-equation (extended model)	—	1.15
Species equations	1.00	0.90

$$J = \frac{\partial(\xi,\eta,\varsigma)}{\partial(x,y,z)} = \left[x_\xi (y_\eta z_\varsigma - y_\varsigma z_\eta) - x_\eta (y_\xi z_\varsigma - y_\varsigma z_\xi) + x_\varsigma (y_\xi z_\eta - y_\eta z_\xi) \right]^{-1} \quad (8.2)$$

$$U^c = \vec{V} \cdot \frac{\partial \xi}{\partial \vec{X}} = \sum_{j=1}^{3} V^j \frac{\partial \xi}{\partial X^j}; \quad V^c = \vec{V} \cdot \frac{\partial \eta}{\partial \vec{X}} = \sum_{j=1}^{3} V^j \frac{\partial \eta}{\partial X^j}$$

$$\vec{V} = [V^1, V^2, V^3] = [u, v, w]; \quad \vec{X} = [X^1, X^2, X^3] = [x, y, z]$$

where

$$\xi_x = J(y_\eta z_\xi - y_\xi z_\eta); \quad \eta_x = J(y_\varsigma z_\xi - y_\xi z_\varsigma); \quad \varsigma_x = J(y_\xi z_\eta - y_\eta z_\xi)$$
$$\xi_y = J(z_\eta x_\xi - z_\xi x_\eta); \quad \eta_y = J(z_\varsigma x_\xi - z_\xi x_\varsigma); \quad \varsigma_y = J(z_\xi x_\eta - z_\eta x_\xi)$$
$$\xi_z = J(x_\eta y_\xi - x_\xi y_\eta); \quad \eta_z = J(x_\varsigma y_\xi - x_\xi y_\varsigma); \quad \varsigma_z = J(x_\xi y_\eta - x_\eta y_\xi)$$

The source terms for the transport equations can be expressed as

$$\frac{S_\phi}{J} = \frac{1}{J}\begin{cases} 0 \\ -\dfrac{\partial p}{\partial X^i} + \dfrac{\partial \Gamma_\phi}{\partial X^j}\dfrac{\partial V^j}{\partial X^i} - \dfrac{2\lambda}{3}\dfrac{\partial \Gamma_\phi}{\partial X^i}\dfrac{\partial V^j}{\partial X^j} + \dfrac{\lambda}{3}\dfrac{\partial}{\partial X^i}\left(\Gamma_\phi \dfrac{\partial V^j}{\partial X^j}\right) + F_i \\ \dfrac{Dp}{Dt} + \Phi + Q_t \\ \rho(P_r - \varepsilon) \\ \rho\dfrac{\varepsilon}{k}\left[(C_1 + C_3 P_r /\varepsilon)P_r - C_2\varepsilon \right] \\ (S_\omega), \quad k = 1,2,3,\ldots, NS \end{cases} \quad (8.3)$$

where

Φ, Q_t, and $(S_\omega)_k$ represent energy dissipation, heat source, and species source terms, respectively

λ is the unity for compressible flows and zero for incompressible flows

NS is the number of chemical species involved

and

$$P_r = \frac{\Gamma_\phi}{\rho}\left[\begin{array}{l} 2(u_x^2 + v_y^2 + w_z^2) + (u_y + v_x)^2 + (v_z + w_y)^2 \\ + (w_x + u_z)^2 - \dfrac{2\lambda}{3}(u_x + v_y + w)_z^2 \end{array} \right] \quad (8.4)$$

Values of modeling constants are given in Table 8.2.

TABLE 8.2 Turbulence Modeling Constants

	C_μ	C_1	C_2	C_3
Standard model	0.09	1.43	1.92	0.00
Extended model	0.09	1.15	1.90	0.25

The first term on the left-hand side of Equation 8.1 is a temporal term, the next two terms represent convection terms and the last two terms are the orthogonal part of the diffusion terms. The first term on the right-hand side is a source term and the next two terms represent the nonorthogonal part of the diffusion terms. 2-D equations are derived here for simplicity in the present numerical formulation. Generalization to 3-D space can be done following the same procedures. For planar or axisymmetric 2-D flow problems, the Jacobian is modified to reflect the volumetric effects as

$$J = y^{(\text{IAX}-1)} z^{(2-\text{IAX})} \frac{\partial(\xi, \eta)}{\partial(x, y)}; \quad \text{IAX} = \begin{cases} 1, \text{ for 2-D planar flows} \\ 2, \text{ for axisymmetric flows} \end{cases}$$

Next, let us define the grid index systems. Referring to the grid system described in Section 8.1, the I- and J-coordinate are coincident with the ξ- and η-coordinate in the above equation, respectively. The temporal coordinate direction is denoted as N-coordinate with discrete grid index as $n-1$, n, and $n+1$ for previous, current, and next time steps, respectively. With these coordinates defined, it is straightforward to discretize the above transport equation term by term. Apply first-order backward difference scheme and second-order central difference scheme for the temporal and convection terms, respectively, these terms can be discretized as

$$\frac{\partial}{\partial t}\left(\frac{\rho\phi}{J}\right) + \frac{\partial}{\partial\xi}\left(\frac{\rho U^c\phi}{J}\right) + \frac{\partial}{\partial\eta}\left(\frac{\rho V^c\phi}{J}\right)$$

$$= \frac{\left(\dfrac{\rho\phi}{J}\right)^{n+1} - \left(\dfrac{\rho\phi}{J}\right)^{n}}{\Delta t} + \frac{\left(\dfrac{\rho U^c\phi}{J}\right)_{i+\frac{1}{2},j} - \left(\dfrac{\rho U^c\phi}{J}\right)_{i-\frac{1}{2},j}}{\Delta\xi}$$

$$+ \frac{\left(\dfrac{\rho V^c\phi}{J}\right)_{i,j+\frac{1}{2}} - \left(\dfrac{\rho V^c\phi}{J}\right)_{i,j-\frac{1}{2}}}{\Delta\eta} \tag{8.5}$$

$$
= \frac{\left[\left(\frac{\rho\phi}{J}\right)^{n+1} - \frac{\rho^{n}}{J}\phi^{n+1} + \frac{\rho^{n}}{J}\phi^{n+1} - \left(\frac{\rho\phi}{J}\right)^{n}\right]}{\Delta t}
$$

$$
+ \frac{\left(\frac{\rho U^{c}}{J}\right)_{i+\frac{1}{2},j}\left(\frac{\phi_{i+1,j} + \phi_{i,j}}{2}\right) - \left(\frac{\rho U^{c}}{J}\right)_{i-\frac{1}{2},j}\left(\frac{\phi_{i-1,j} + \phi_{i,j}}{2}\right)}{\Delta\xi}
$$

$$
+ \frac{\left(\frac{\rho V^{c}}{J}\right)_{i,j+\frac{1}{2}}\left(\frac{\phi_{i,j+1} + \phi_{i,j}}{2}\right) - \left(\frac{\rho V^{c}}{J}\right)_{i,j-\frac{1}{2}}\left(\frac{\phi_{i,j-1} + \phi_{i,j}}{2}\right)}{\Delta\eta} \tag{8.6}
$$

$$
= \left(\frac{\rho}{J}\right)^{n}\frac{(\phi^{n+1} - \phi^{n})}{\Delta t} + \frac{\left(\frac{\rho U^{c}}{2J}\right)_{i+\frac{1}{2},j}\phi_{i+1,j} - \left(\frac{\rho U^{c}}{2J}\right)_{i-\frac{1}{2},j}\phi_{i-1,j}}{\Delta\xi} + \frac{\left(\frac{\rho V^{c}}{2J}\right)_{i,j+\frac{1}{2}}\phi_{i,j+1} - \left(\frac{\rho V^{c}}{2J}\right)_{i,j-\frac{1}{2}}\phi_{i,j-1}}{\Delta\eta}
$$

$$
+ \left\{ \frac{\rho^{n+1} - \rho^{n}}{J\Delta t}\phi^{n+1} + \left[\frac{\left(\frac{\rho U^{c}}{J}\right)_{i+\frac{1}{2},j} - \left(\frac{\rho U^{c}}{J}\right)_{i-\frac{1}{2},j}}{\Delta\xi} + \frac{\left(\frac{\rho V^{c}}{J}\right)_{i,j+\frac{1}{2}} - \left(\frac{\rho V^{c}}{J}\right)_{i,j-\frac{1}{2}}}{\Delta\eta}\right]\left(\frac{\phi_{i,j}}{2}\right)\right\} \tag{8.7}
$$

The last term in the braces of Equation 8.7 is approaching zero and can be neglected in steady state applying the mass conservation condition (i.e., the continuity equation). Next, the orthogonal part of the diffusion terms is discretized using the central differencing scheme. That is,

$$
-\frac{\partial}{\partial\xi}\left(\frac{\xi_{x}^{2} + \xi_{y}^{2}}{J}\Gamma_{\phi}\frac{\partial\phi}{\partial\xi}\right) - \frac{\partial}{\partial\eta}\left(\frac{\eta_{x}^{2} + \eta_{y}^{2}}{J}\Gamma_{\phi}\frac{\partial\phi}{\partial\eta}\right)
$$

$$
= -\frac{1}{\Delta\xi}\left[\left(\frac{\xi_{x}^{2} + \xi_{y}^{2}}{J}\Gamma_{\phi}\right)_{i+\frac{1}{2},j}\frac{\phi_{i+1,j} - \phi_{i,j}}{\Delta\xi} - \left(\frac{\xi_{x}^{2} + \xi_{y}^{2}}{J}\Gamma_{\phi}\right)_{i-\frac{1}{2},j}\frac{\phi_{i,j} - \phi_{i-1,j}}{\Delta\xi}\right]
$$

$$
-\frac{1}{\Delta\eta}\left[\left(\frac{\eta_{x}^{2} + \eta_{y}^{2}}{J}\Gamma_{\phi}\right)_{i,j+\frac{1}{2}}\frac{\phi_{i,j+1} - \phi_{i,j}}{\Delta\eta} - \left(\frac{\eta_{x}^{2} + \eta_{y}^{2}}{J}\Gamma_{\phi}\right)_{i,j-\frac{1}{2}}\frac{\phi_{i,j} - \phi_{i,j-1}}{\Delta\eta}\right] \tag{8.8}
$$

Similarly, central difference scheme is also applied to the source terms and the nonorthogonal terms of the diffusion terms of the general conservation equation. Finally, the discretized conservation equation can be written as

$$
\left(\frac{\rho}{J}\right)^n \frac{(\phi^{n+1} - \phi^n)}{\Delta t} + \frac{\left(\frac{\rho U^c}{2J}\right)_{i+\frac{1}{2},j} \phi_{i+1,j} - \left(\frac{\rho U^c}{2J}\right)_{i-\frac{1}{2},j} \phi_{i-1,j}}{\Delta \xi}
$$

$$
+ \frac{\left(\frac{\rho V^c}{2J}\right)_{i,j+\frac{1}{2}} \phi_{i,j+1} - \left(\frac{\rho V^c}{2J}\right)_{i,j-\frac{1}{2}} \phi_{i,j-1}}{\Delta \eta}
$$

$$
- \frac{1}{\Delta \xi}\left[\left(\frac{\xi_x^2 + \xi_y^2}{J}\Gamma_\phi\right)_{i+\frac{1}{2},j} \frac{\phi_{i+1,j} - \phi_{i,j}}{\Delta \xi} - \left(\frac{\xi_x^2 + \xi_y^2}{J}\Gamma_\phi\right)_{i-\frac{1}{2},j} \frac{\phi_{i,j} - \phi_{i-1,j}}{\Delta \xi}\right]
$$

$$
- \frac{1}{\Delta \eta}\left[\left(\frac{\eta_x^2 + \eta_y^2}{J}\Gamma_\phi\right)_{i,j+\frac{1}{2}} \frac{\phi_{i,j+1} - \phi_{i,j}}{\Delta \eta} - \left(\frac{\eta_x^2 + \eta_y^2}{J}\Gamma_\phi\right)_{i,j-\frac{1}{2}} \frac{\phi_{i,j} - \phi_{i,j-1}}{\Delta \eta}\right]
$$

$$
= \left(\frac{S_\phi}{J}\right)_{i,j} + \frac{1}{\Delta \xi}\left[\left(\frac{\xi_x \eta_x + \xi_y \eta_y}{J}\Gamma_\phi\right)_{i+\frac{1}{2},j} \frac{\phi_{i+\frac{1}{2},j+1} - \phi_{i+\frac{1}{2},j-1}}{2\Delta \eta}\right.
$$
$$
\left. - \left(\frac{\xi_x \eta_x + \xi_y \eta_y}{J}\Gamma_\phi\right)_{i-\frac{1}{2},j} \frac{\phi_{i-\frac{1}{2},j+1} - \phi_{i-\frac{1}{2},j-1}}{2\Delta \eta}\right]
$$

$$
+ \frac{1}{\Delta \eta}\left[\left(\frac{\eta_x \xi_x + \eta_y \xi_y}{J}\Gamma_\phi\right)_{i,j+\frac{1}{2}} \frac{\phi_{i+1,j+\frac{1}{2}} - \phi_{i-1,j+\frac{1}{2}}}{2\Delta \xi}\right.
$$
$$
\left. - \left(\frac{\eta_x \xi_x + \eta_y \xi_y}{J}\Gamma_\phi\right)_{i,j-\frac{1}{2}} \frac{\phi_{i+1,j-\frac{1}{2}} - \phi_{i-2,j-\frac{1}{2}}}{2\Delta \xi}\right] \tag{8.9}
$$

8.3 UPWIND AND DISSIPATION SCHEMES

The left-hand side terms of Equation 8.9 is arranged in matrix form so that the variable is solved implicitly using iterative matrix solvers. The right-hand side terms are evaluated explicitly. This approach makes the overall numerical scheme semi-implicit in obtaining a steady-state solution. For numerical stability, it is required to have diagonal dominance of the final matrix equation. The objective is to have the coefficients associated with the central node, (i, j), to be equal or larger than that of its surrounding nodes, i.e., coefficients of the off-diagonal terms. For most flow problems, the convection terms are the dominant terms in the conservation equation and the first-order gradient terms contribute to the possibility of off-diagonal dominance of the discrete equation. To achieve this stability requirement, one extra term related to the convection term is therefore added to both sides of Equation 8.9. The net effect is to have diagonal dominance of the matrix coefficients on the left-hand side and keeping the equation essentially the same as the original form. This extra term takes the form of

$$
\begin{aligned}
&\frac{-\left|\dfrac{\rho U^{c}}{2J}\right|_{i+\frac{1}{2},j}(\phi_{i+1,j}-\phi_{i,j})+\left|\dfrac{\rho U^{c}}{2J}\right|_{i-\frac{1}{2},j}(\phi_{i,j}-\phi_{i-1,j})}{\Delta\xi}\\
&+\frac{-\left|\dfrac{\rho V^{c}}{2J}\right|_{i,j+\frac{1}{2}}(\phi_{i,j+1}-\phi_{i,j})+\left|\dfrac{\rho V^{c}}{2J}\right|_{i,j-\frac{1}{2}}(\phi_{i,j}-\phi_{i,j-1})}{\Delta\eta}
\end{aligned}
\tag{8.10}
$$

When the above extra term is added to both sides of Equation 8.9, the numerical scheme is still keeping its original second-order central difference form. For many complex flow problems, a relaxation factor between zero and unity can be applied to the extra term added on the right-hand side of the equation to allow more damping and stability for the numerical solutions. This method is also employed to obtain solutions near flow regions with large gradients or discontinuities, e.g., across shock waves. Also, a relaxation factor, 1-REC, is applied to the term added to the right-hand side for better stability in solving complex flow problems, where REC represents a dissipation parameter. A typical value of 0.25 for REC is used for turbulent or reacting flows. Larger REC can also be used for problems with higher stiffness in obtaining a stable solution. It can be observed that when zero value of the relaxation factor is applied, or REC = 1, the scheme is essentially a first-order upwind scheme for the convection terms. Since only the first-order upwind scheme is stable across shock waves, the following shock monitoring parameter in I-direction is also employed to gradually switch-off high-order terms when second-order pressure derivatives is large.

$$
\begin{aligned}
\alpha_{d} &= \max\left\{0, 1-25\max\left(\psi_{i+1}, \psi_{i}, \psi_{i-1}\right)\right\}\\
\psi_{i} &= \left|p_{i+1}-2p_{i}+p_{i-1}\right|/\left(p_{i+1}+2p_{i}+p_{i-1}\right)
\end{aligned}
\tag{8.11}
$$

J- and K-direction shock monitoring parameters are formulated in the similar way. To formulate other high-order upwind schemes for the convection terms, different form of the extra term is added to the right-hand side of Equation 8.9. For example, a high-order upwind scheme or a total variation diminishing (TVD) scheme, etc. can be resulted while keeping the final convection terms on the left-hand side of the equation retain the same diagonal dominant first-order upwind scheme. For a second-order upwind scheme, the extra term added to the right-hand side takes the following form:

$$
\frac{\left.\left|\frac{\rho U^c}{2J}\right|\right|_{i+\frac{1}{2},j}\left[a(\phi_{i,j}-\phi_{i-1,j})+(1-a)(\phi_{i+2,j}-\phi_{i+1,j})\right]-\left.\left|\frac{\rho U^c}{2J}\right|\right|_{i-\frac{1}{2},j}\left[b(\phi_{i-1,j}-\phi_{i-2,j})+(1-b)(\phi_{i+1,j}-\phi_{i,j})\right]}{\Delta\xi}
$$

$$
+\frac{\left.\left|\frac{\rho V^c}{2J}\right|\right|_{i,j+\frac{1}{2}}\left[c(\phi_{i,j}-\phi_{i,j-1})+(1-c)(\phi_{i,j+2}-\phi_{i,j+1})\right]-\left.\left|\frac{\rho V^c}{2J}\right|\right|_{i,j-\frac{1}{2}}\left[d(\phi_{i,j-1}-\phi_{i,j-2})+(1-d)(\phi_{i,j+1}-\phi_{i,j})\right]}{\Delta\eta}
$$

$$(8.12)$$

For a third-order upwind scheme, the extra term added to the right-hand side is expressed as

$$
-\frac{\left.\left|\frac{\rho U^c}{6J}\right|\right|_{i+\frac{1}{2},j}\left[a(2\phi_{i+1,j}-\phi_{i,j}-\phi_{i-1,j})+(1-a)(-2\phi_{i,j}+\phi_{i+1,j}+\phi_{i+2,j})\right]}{\Delta\xi}
$$

$$
+\frac{\left.\left|\frac{\rho U^c}{6J}\right|\right|_{i-\frac{1}{2},j}\left[b(2\phi_{i,j}-\phi_{i-1,j}-\phi_{i-2,j})+(1-b)(-2\phi_{i-1,j}+\phi_{i,j}+\phi_{i+1,j})\right]}{\Delta\xi}
$$

$$
+\frac{-\left.\left|\frac{\rho V^c}{6J}\right|\right|_{i,j+\frac{1}{2}}\left[c(2\phi_{i,j+1}-\phi_{i,j}-\phi_{i,j-1})+(1-c)(-2\phi_{i,j}+\phi_{i,j+1}+\phi_{i,j+2})\right]}{\Delta\eta}
$$

$$
+\frac{\left.\left|\frac{\rho V^c}{6J}\right|\right|_{i,j-\frac{1}{2}}\left[d(2\phi_{i,j}-\phi_{i,j-1}-\phi_{i,j-2})+(1-d)(-2\phi_{i,j-1}+\phi_{i,j}+\phi_{i,j+1})\right]}{\Delta\eta}
$$

$$(8.13)$$

where

$$
\left\{
\begin{array}{llll}
a = 1, & \text{if } U^c_{i+\frac{1}{2},j} \geq 0; & a = 0, & \text{if } U^c_{i+\frac{1}{2},j} < 0 \\
b = 1, & \text{if } U^c_{i-\frac{1}{2},j} \geq 0; & b = 0, & \text{if } U^c_{i-\frac{1}{2},j} < 0 \\
c = 1, & \text{if } V^c_{i,j+\frac{1}{2}} \geq 0; & c = 0, & \text{if } V^c_{i,j+\frac{1}{2}} < 0 \\
d = 1, & \text{if } V^c_{i,j-\frac{1}{2}} \geq 0; & d = 0, & \text{if } V^c_{i,j-\frac{1}{2}} < 0
\end{array}
\right\}
$$

For a TVD upwind scheme, the extra term added to the right-hand side is expressed as

$$
\frac{\left.\dfrac{\rho U^c}{J}\right|_{i+\frac{1}{2},j} A_{i+\frac{1}{2},j} - \left.\dfrac{\rho U^c}{J}\right|_{i-\frac{1}{2},j} A_{i-\frac{1}{2},j}}{\Delta \xi} + \frac{\left.\dfrac{\rho V^c}{J}\right|_{i,j+\frac{1}{2}} A_{i,j+\frac{1}{2}} - \left.\dfrac{\rho V^c}{J}\right|_{i,j-\frac{1}{2}} A_{i,j-\frac{1}{2}}}{\Delta \eta} \tag{8.14}
$$

$$
A_{i+\frac{1}{2},j} = \frac{1}{4}
\begin{cases}
\Psi^+_{i+\frac{1}{2}} + \Psi^-_{i-\frac{1}{2}} + \alpha_t(\Psi^+_{i+\frac{1}{2}} - \Psi^-_{i-\frac{1}{2}}), & \text{if } U^c_{i+\frac{1}{2},j} > 0 \\
\Psi^-_{i+\frac{1}{2}} + \Psi^+_{i+\frac{3}{2}} + \alpha_t(\Psi^-_{i+\frac{1}{2}} - \Psi^+_{i+\frac{3}{2}}), & \text{if } U^c_{i+\frac{1}{2},j} < 0
\end{cases}
$$

$$
A_{i-\frac{1}{2},j} = \frac{1}{4}
\begin{cases}
\Psi^+_{i-\frac{1}{2}} + \Psi^-_{i-\frac{3}{2}} + \alpha_t(\Psi^+_{i-\frac{1}{2}} - \Psi^-_{i-\frac{3}{2}}), & \text{if } U^c_{i-\frac{1}{2},j} > 0 \\
\Psi^-_{i-\frac{1}{2}} + \Psi^+_{i+\frac{1}{2}} + \alpha_t(\Psi^-_{i-\frac{1}{2}} - \Psi^+_{i+\frac{1}{2}}), & \text{if } U^c_{i-\frac{1}{2},j} < 0
\end{cases}
$$

$$
A_{i,j+\frac{1}{2}} = \frac{1}{4}
\begin{cases}
\Psi^+_{j+\frac{1}{2}} + \Psi^-_{j-\frac{1}{2}} + \alpha_t(\Psi^+_{j+\frac{1}{2}} - \Psi^-_{j-\frac{1}{2}}), & \text{if } V^c_{i,j+\frac{1}{2}} > 0 \\
\Psi^-_{j+\frac{1}{2}} + \Psi^+_{j+\frac{3}{2}} + \alpha_t(\Psi^-_{j+\frac{1}{2}} - \Psi^+_{j+\frac{3}{2}}), & \text{if } V^c_{i,j+\frac{1}{2}} < 0
\end{cases}
$$

$$
A_{i,j-\frac{1}{2}} = \frac{1}{4}
\begin{cases}
\Psi^+_{j-\frac{1}{2}} + \Psi^-_{j-\frac{3}{2}} + \alpha_t(\Psi^+_{j-\frac{1}{2}} - \Psi^-_{j-\frac{3}{2}}), & \text{if } V^c_{i,j-\frac{1}{2}} > 0 \\
\Psi^-_{j-\frac{1}{2}} + \Psi^+_{j+\frac{1}{2}} + \alpha_t(\Psi^-_{j-\frac{1}{2}} - \Psi^+_{j+\frac{1}{2}}), & \text{if } V^c_{i,j-\frac{1}{2}} < 0
\end{cases}
$$

$$
\Psi^\pm_{i+\frac{1}{2}} = \text{sign}(\Delta \phi_{i+\frac{1}{2}}) \max\{0, \min [|\Delta \phi_{i+\frac{1}{2}}|, \alpha_c \text{sign}(\Delta \phi_{i+\frac{1}{2}}) \Delta \phi_{i+\frac{1}{2}\mp 1}]\}
$$

$$
\Psi^\pm_{j+\frac{1}{2}} = \text{sign}(\Delta \phi_{j+\frac{1}{2}}) \max\{0, \min [|\Delta \phi_{j+\frac{1}{2}}|, \alpha_c \text{sign}(\Delta \phi_{j+1/2}) \Delta \phi_{j+\frac{1}{2}\mp 1}]\}
$$

and $\quad \alpha_c = \dfrac{3 - \alpha_t}{1 - \alpha_t};\quad$ where $\alpha_t = \begin{cases} -1, & \text{for second-order upwind TVD} \\ 1/3, & \text{for third-order upwind TVD} \end{cases}$

$$
\Delta \phi_{i+\frac{1}{2}} = \phi_{i+1,j} - \phi_{i,j}; \quad \Delta \phi_{j+\frac{1}{2}} = \phi_{i,j+1} - \phi_{i,j}
$$

An upwind switching parameter, IREC, is employed for the user to select what high-order upwind scheme to be used to discretized the convection terms. This is summarized in Table 8.3.

TABLE 8.3   CTP Code Upwind Scheme Options

	Second-Order Upwind	Third-Order Upwind	Second-Order Central	Second-Order Upwind TVD	Third-Order Upwind TVD
IREC	0	1	2	3	4

To provide background smoothing of the solutions while keeping second-order accuracy of the baseline numerical schemes, a fourth-order dissipation term is incorporated as a source term to the transport equations. This term takes the following form:

$$0.05\alpha_d \left(\frac{\left|\frac{\rho U^c}{J}\right|_{i+\frac{1}{2},j} D_{i+\frac{1}{2},j} - \left|\frac{\rho U^c}{J}\right|_{i-\frac{1}{2},j} D_{i-\frac{1}{2},j}}{\Delta\xi} + \frac{\left|\frac{\rho V^c}{J}\right|_{i,j+\frac{1}{2}} D_{i,j+\frac{1}{2}} - \left|\frac{\rho V^c}{J}\right|_{i,j-\frac{1}{2}} D_{i,j-\frac{1}{2}}}{\Delta\eta} \right) \quad (8.15)$$

where

$$D_{i+\frac{1}{2},j} = -(\phi_{\xi\xi\xi})_{i+\frac{1}{2}} = 2\Delta\phi_{i+\frac{1}{2}} - \Delta\phi_{i+\frac{3}{2}} - \Delta\phi_{i-\frac{1}{2}}$$

$$D_{i-\frac{1}{2},j} = -(\phi_{\xi\xi\xi})_{i-\frac{1}{2}} = 2\Delta\phi_{i-\frac{1}{2}} - \Delta\phi_{i+\frac{1}{2}} - \Delta\phi_{i-\frac{3}{2}}$$

$$D_{i,j+\frac{1}{2}} = -(\phi_{\eta\eta\eta})_{j+\frac{1}{2}} = 2\Delta\phi_{j+\frac{1}{2}} - \Delta\phi_{j+\frac{3}{2}} - \Delta\phi_{j-\frac{1}{2}}$$

$$D_{i,j-\frac{1}{2}} = -(\phi_{\eta\eta\eta})_{j-\frac{1}{2}} = 2\Delta\phi_{j-\frac{1}{2}} - \Delta\phi_{j+\frac{1}{2}} - \Delta\phi_{j-\frac{3}{2}}$$

and α_d is defined in Equation 8.11

8.4 SOLUTION STRATEGY

To obtained solutions of the flow variables, discretized conservation equations in the form described above are solved using iterative matrix solvers in sequence. The source terms on the right-hand side of the conservation equations are calculated explicitly based on known flow variables at the current time level. This constitutes a semi-implicit approach to obtain numerical solutions. The main advantage of semi-implicit method is in the programming simplicity, development efficiency, maintenance robustness and matrix solver efficiency, and stability. This approach only suffers in the time-step size limit that cannot be as large as a fully implicit method.

Since the pressure gradient terms in the momentum equations are first order, which is the same order as the convection terms, the velocity vectors and pressure field tend to decouple as the flowfield solution evolves. One way to circumvent the decoupling problem is to store the velocity vectors at staggered spatial locations with respect to where the pressure variable is stored around a control volume. Then pressure and velocity corrections are made based on a pressure Poisson equation. However, this approach makes the programming more complicated, especially for general curvilinear coordinates mesh systems.

To use collocated variable approach (i.e., velocity vectors and other flow variables including pressure are stored at the center of control volumes), for solution accuracy and programming consistency, velocity–pressure coupling formulation is incorporated at control volume interfaces. First of all, the discretized momentum equation in algebraic equation form can be written as

$$A_p \vec{V}_p + \sum_{n=1}^{N} A_n \vec{V}_n = -\frac{1}{J} \nabla p + \vec{S}_V \tag{8.16}$$

Assuming that velocity perturbations in space is a function of pressure perturbations only and can be estimated based on Equation 8.16. That is,

$$f(A_p, A_n) \vec{V}' = -\frac{1}{J} \nabla p' \tag{8.17}$$

or

$$\vec{V}' = -\frac{1}{J f(A_p, A_n)} \nabla p' = -D_u \nabla p' \tag{8.18}$$

Next, the continuity equation is employed to derive a pressure correction equation that is solved to update the pressure and velocity fields such that the mass conservation condition is enhanced at the end of each time-marching step. The continuity equation can be written as

$$\frac{\partial \rho}{\partial t} + \nabla \cdot (\rho \vec{V}) = 0 \tag{8.19}$$

The density and velocity fields are perturbed as $\rho = \rho^* + \rho'$ and $\vec{V} = \vec{V}^* + \vec{V}'$, respectively, where ρ^* and $\vec{V}^*$ are density and velocity fields of the current time level, respectively. Substitute these perturbed fields into the continuity equation, the following expression is obtained.

$$\frac{\partial(\rho* + \rho')}{\partial t} + \nabla \cdot (\rho* \vec{V}*) + \nabla \cdot (\vec{V}* \rho') + \nabla \cdot (\rho* \vec{V}') + \nabla \cdot (\rho' \vec{V}') = 0 \quad (8.20)$$

Then, by assuming small perturbations and omitting high-order perturbation terms, i.e., the last term on the left-hand side of Equation 8.20

$$\frac{\partial \rho'}{\partial t} + \nabla \cdot (\vec{V}* \rho') + \nabla \cdot (\rho* \vec{V}') = -\left[\frac{\partial \rho*}{\partial t} + \nabla \cdot (\rho* \vec{V}*) \right] \quad (8.21)$$

By substituting the velocity–pressure coupling expression and equation of state (EOS) in Equation 8.21, a pressure correction Poisson equation is obtained. That is,

$$\frac{1}{RT} \frac{\partial p'}{\partial t} + \nabla \cdot \left(\frac{\vec{V}*}{RT} p' \right) - \nabla \cdot (\rho* D_u \nabla p') = -\left[\frac{\partial \rho*}{\partial t} + \nabla \cdot (\rho* \vec{V}*) \right] \quad (8.22)$$

Here, to evaluate the last term on the right-hand side of Equation 8.22, the velocity vectors at control volume interfaces are calculated based on the nodal values and extrapolated to the interface locations. An extrapolation scheme is employed based on the discretized momentum equation shown above. The application of this scheme ends up with a fourth-order pressure dissipation term on the right-hand side of Equation 8.22 and gives smooth pressure field solutions. That is,

$$u^*_{i+\frac{1}{2}} = 0.5(u_{i+1} + u_i) - D_u \left\{ (p_x)_{i+\frac{1}{2}} - 0.5\left[(p_x)_{i+1} + (p_x)_i \right] \right\}$$

$$v^*_{i+\frac{1}{2}} = 0.5(v_{i+1} + v_i) - D_u \left\{ (p_y)_{i+\frac{1}{2}} - 0.5\left[(p_y)_{i+1} + (p_y)_i \right] \right\} \quad (8.23)$$

$$w^*_{i+\frac{1}{2}} = 0.5(w_{i+1} + w_i) - D_u \left\{ (p_z)_{i+\frac{1}{2}} - 0.5\left[(p_z)_{i+1} + (p_z)_i \right] \right\}$$

and

$$U^c_{i+\frac{1}{2}} = u^*_{i+\frac{1}{2}} \xi_x + v^*_{i+\frac{1}{2}} \xi_y + w^*_{i+\frac{1}{2}} \xi_z$$

$$V^c_{j+\frac{1}{2}} = u^*_{j+\frac{1}{2}} \eta_x + v^*_{j+\frac{1}{2}} \eta_y + w^*_{j+\frac{1}{2}} \eta_z \quad (8.24)$$

$$W^c_{k+\frac{1}{2}} = u^*_{k+\frac{1}{2}} \varsigma_x + v^*_{k+\frac{1}{2}} \varsigma_y + w^*_{k+\frac{1}{2}} \varsigma_z$$

The above pressure corrected contravariant velocities at control volume interfaces are used for the convection terms of all the conservation equations discussed above. It is clear that the above pressure correction equation is also in the form of a general conservation equation and can be discretized and solved the same way as other transport equations described previously. For good numerical stability in solving the above pressure correction equation, a first-order upwind scheme is applied to discretize the convection term, i.e., the second term on the right-hand side of the equation. It is worth to note that the temporal and convection terms of this equation become dominant for high Mach number flow regions and vanish when local flow Mach number is approaching zero. This property agrees with the gas dynamics of compressible flows and gives physically correct pressure correction solutions based on the mass conservation condition. After the perturbed pressure field is solved, the velocity fields are corrected using the velocity–pressure coupling expression. This correction cycle can be repeated several times for each time step to enhance mass conservation at the end of the solution procedure. Then, other transport equations are solved based on the mass conserved flow-field for good overall conservation solutions.

8.5 TIME-MARCHING SCHEME

For time accuracy, an efficient noniterative time-centered time-marching scheme with a multicorrector solution algorithm is employed. First of all, the governing equations are linearized, by applying the aforementioned finite difference discretization schemes to the flux and source terms. A system of linear algebraic equations is obtained as a result of the linearization. A relaxation solution procedure (i.e., the linearized algebraic equations are solved sequentially with an iterative full matrix solver) is employed for the solutions of the governing equations. For convenience, the conservation equation can be written as

$$\frac{1}{J}\frac{\partial \rho \phi}{\partial t} = \frac{\partial F_i}{\partial \xi_i} + S_\phi = R_\phi \qquad (8.25)$$

or, in finite difference form,

$$\frac{1}{J \Delta t}\left[(\rho \phi)^{n+1} - (\rho \phi)^n \right] = \theta R_\phi^{n+1} + (1-\theta) R_\phi^n \qquad (8.26)$$

where

superscripts n and $n+1$ denote the current and the next time levels, respectively

θ is a time-marching control parameter, which is specified in the input data file

$\theta = 1$ and $\theta = 1/2$ are for an implicit Euler time-marching and a time-centered time-marching schemes, respectively

The following linearization is then incorporated.

$$(\rho\phi)^{n+1} = (\rho\phi)^n + \rho^n \Delta\phi^n$$

$$R_\phi^{n+1} = R_\phi^n + \left(\frac{\partial R_\phi}{\partial\phi}\right)^n \Delta\phi^n \tag{8.27}$$

With these relations, a delta form of the time-marching equation can be written as

$$\left(\frac{\rho}{J\Delta t}\theta\frac{\partial R_\phi}{\partial\phi}\right)^n \Delta\phi^n = R_\phi^n \tag{8.28}$$

This system of equations is solved numerically, by using the matrix solver specified by the user through input data selections.

8.6 BOUNDARY CONDITIONS

The CTP code provides options (mostly through input data specifications) for treating various types of boundaries (e.g., inlet, outlet, symmetry, periodic, freestream, singularity lines, and solid-wall boundaries with or without blowing) and the location of each boundary. User-defined boundary conditions for overriding the specified ones can be provided in one of the include files (i.e., fmain02) with proper FORTRAN programming. The input data controlled boundary conditions are described below.

8.6.1 INLET FLOW BOUNDARIES

For incompressible flow inlet boundaries, only the pressure waves are extrapolated upstream. For subsonic inlet boundaries, two types of inlet boundary conditions can be specified. They are (1) fixed inlet total conditions (i.e., type-1 in the input data); and (2) fixed mass flow rate inlet condition (i.e., type-1 in the input data). The second subsonic inlet boundary condition (type-1) is usually used for mass injected inlet boundaries such as near the propellant burning-surface of a solid rocket motor combustion chamber. For supersonic inlet boundaries, all the flow variables are fixed at specified values (i.e., type 0 inlet).

8.6.2 EXIT FLOW BOUNDARIES

For outlet boundaries, all variables are extrapolated downstream in the first step. A special adaptive gradient detection extrapolation method is employed for the velocity vectors and pressure and temperature fields to provide smooth wave propagation out of the outlet boundaries. For subsonic or incompressible outlet boundaries, two options are provided: (1) outlet velocity vectors are corrected based on the global mass conservation conditions (i.e., through an exit boundary condition control parameter, PRAT = 0.0 in the input data); and (2) outlet pressure profile is updated such that the ratio of the pressure at an outlet pressure reference point (IPEX, JPEX in the input data) to the atmospheric pressure (14.7 psi) is kept at a specified value (i.e., PRAT = pressure ratio). IPEX represents the global grid number of the selected grid point in zones JPEX. The zonal and global grid node numbering system used throughout the CTP code is calculated using the following formulas, respectively.

Zonal: $III = I + (J - 1) \times JZS(NZ) + (K - 1) \times KZS(NZ)$, for zone NZ
Global: $IJK = IZS(NZ) + III$

where $IZS(NZ) = \sum_{i=1}^{NZ-1} \left(I_{max} \times J_{max} \times K_{max}\right)_i$

and IZS(NZ), JZS(NZ), and KZS(NZ) are the grid index incremental counts for zone number NZ, which are calculated based on the input grid sizes, $I_{max} = IZT(NZ)$, $J_{max} = JZT(NZ)$, and $K_{max} = KZT(NZ)$.

The first method is mainly used for incompressible or subsonic internal flow problems. The second method can be used for external and internal subsonic outlet boundaries. It is important that for incompressible flow cases, the global pressure reference point (IPC, JPC) must be an interior point (i.e., not a boundary point), where IPC represents the global grid number of the selected grid point in zones JPC. For supersonic outlet boundaries, only extrapolated conditions are employed (i.e., PRAT = –1.0). For scalar variables such as turbulence quantities and species concentrations, two kinds of outlet extrapolation methods are available. They are zero gradient extrapolation (i.e., extrapolation control parameter, IEXX = 1); and linear extrapolation (i.e., IEXX = 2). The zero gradient scalar extrapolation method is recommended for reacting flow applications to provide better solution stability near the outlet boundaries.

8.6.3 SYMMETRY BOUNDARIES

For symmetry boundary conditions on symmetry planes, or symmetry lines of 2D cases, or inviscid slip boundaries, zero normal gradients for all scalar quantities are specified. Special treatments for the velocity vectors are provided to reflect zero mass flux condition on the symmetry planes. The velocity vectors are first projected onto the symmetry planes. The surface normal components of the vectors are then assigned to be zero such that the resultant vectors are tangent to the symmetry planes.

8.6.4 ZONAL INTERFACE BOUNDARIES

For multizone interface boundary conditions, two methods are available. In the first method, grid lines must be continuous across patched zonal interface. The zonal patching index specification is given through the input data file. The second zonal method allows the use of noncontinuous patched and nonoverlaid mesh systems. In this case, the user needs to implement the subroutine INFACE (in f1.f module) to conduct the zonal interpolation index identification, grid movement, and zonal interpolation. The subroutine INFACE is currently empty in the CTP code. For turbomachinery applications using periodic boundaries, special input data specification must be used (i.e., ICYC = 3 and IGEO = 9). Multistage turbomachinery with interstage motion or for rocket stage separation applications, special user-defined zonal boundary conditions can be implemented in subroutine INFACE.

8.6.5 SINGULARITY BOUNDARIES

An averaging procedure along singularity lines is provided in the CTP code to circumvent possible numerical difficulties for resolving the flowfield near singularity lines. The flow solutions on the singularity lines are assumed to take the averaged values of the surrounding points. Additional conditions applied to the singularity lines can then be treated explicitly in the include file fmain02.

Additional boundary conditions such as freestream inlet flow angle extrapolation, jet outlet pressure condition updating and time-dependent inlet and/or wall boundary conditions can be implemented by adding program coding in one of the include files (i.e., fmain02) in the main program. For flowfield conditions modifications during restart of a run, another include file (i.e., fmain01) of the main program can be utilized to implement user-defined settings.

8.6.6 WALL BOUNDARIES

The CTP code provides a multiple solid-wall blocking feature within a mesh, which allows the user to specify wall blocks/elements anywhere inside the flow domain. The wall surface orientations (direction cosine) are also calculated and used for turbulence wall functions modeling and wall extrapolation purpose. For viscous flow computations, nonslip boundary conditions are employed for the momentum equations. A standard wall function approach with a modified universal velocity profile is employed for turbulent flow computations. Fixed wall temperature distributions or adiabatic wall boundary conditions are the two wall boundary conditions available for the energy equation (i.e., IWTM = −1 and 1, respectively). Since the current version provides the option of running conjugate heat transfer between solid-wall and fluid, fixed wall temperature condition (IWTM = −1) is set by the code when conjugate heat transfer option is activated (i.e., IWALL = 1 to activate and IWALL = 0 to deactivate). Pressure along the wall is evaluated by using extrapolation. Since the CTP code is designed for generalized coordinate systems, nonorthogonal boundary grid effects are also taken into account when normal gradients at the solid-wall surface are evaluated. Let us consider a wall surface on the i–k plane and at $j = 1$ with local direction cosines of the normal vector defined as: $\cos\alpha$, $\cos\beta$, and $\cos\gamma$. The local zero normal gradient condition for the flow variable q can be written as

$$\frac{\partial\phi}{\partial n} = 0 = \phi_x\cos\alpha + \phi_y\cos\beta + \phi_z\cos\gamma \qquad (8.29)$$

where

$$\cos\alpha = \frac{\eta_x}{\sqrt{A}}; \quad \cos\beta = \frac{\eta_y}{\sqrt{A}}; \quad \cos\gamma = \frac{\eta_z}{\sqrt{A}}; \quad A = \eta_x^2 + \eta_y^2 + \eta_z^2$$
$$\phi_x = \phi_\xi\xi_x + \phi_\eta\eta_x + \phi_\zeta\zeta_x; \quad \phi_y = \phi_\xi\xi_y + \phi_\eta\eta_y + \phi_\zeta\zeta_y; \quad \phi_z = \phi_\xi\xi_z + \phi_\eta\eta_z + \phi_\zeta\zeta_z \qquad (8.30)$$

With these relations, the following equation is obtained.

$$\phi_\eta = -\left[(\xi_x\eta_x + \xi_y\eta_y + \xi_z\eta_z)\phi_\xi + (\eta_x\zeta_x + \eta_y\zeta_y + \eta_z\zeta_z)\phi_\zeta\right]/A$$
$$= -\frac{1}{2}(3\phi_{i,1,k} - 4\phi_{i,2,k} + \phi_{i,3,k}) \qquad (8.31)$$

The wall boundary values are calculated using following expression derived based on Equation 8.31. That is,

$$\phi_{i,1,k} = \phi_{i,2,k} + \frac{1}{3}(\phi_{i,2,k} - \phi_{i,3,k} - 2\phi_\eta) \qquad (8.32)$$

The quantities, ϕ_ξ and ϕ_ς, are evaluated on the $j = 1$ plane. This method is a good approximation providing that the grid variation away from the wall surfaces is smooth. The above zero normal gradient treatment is performed explicitly at the end of each time step.

8.7 INITIAL CONDITIONS

The mesh systems and flowfield initial guesses for running the CTP code can be prepared in two ways. The first method is used in setting up some sample cases that are included in the compact disk (CD) with this book. Using this method, grid and initial flow generation FORTRAN codes are written to generate grid and initial flowfield data files (in file unit fort.10 format for reading sample cases). These data files are then read in from the example include file, fexmp01, when the calculation is started using the example start option (IDATA = 2). The second method involves the preparation of grid and flowfield restart files (fort.12 and fort.13, respectively). Then, the CTP code is started using the restart option (IDATA = 1 or 0). To use the second method, the following restart file data formats (Tables 8.4 and 8.5) must be used in preparing the data files of grid mesh and the initial flowfield.

TABLE 8.4   Data Format of the Restart Grid File

```
    WRITE(12,1) IZON
    DO IZ=1,IZON
       WRITE(12,1) IZT(IZ),JZT(IZ),KZT(IZ)
    ENDDO
    DO IZ=1,IZON
       I2=IZT(IZ)
       J2=JZT(IZ)
       K2=KZT(IZ)
       WRITE(12,2)(((X(I,J,K,IZ),I=1,I2),J=1,J2),K=1,K2)
       WRITE(12,2)(((Y(I,J,K,IZ),I=1,I2),J=1,J2),K=1,K2)
       WRITE(12,2)(((Z(I,J,K,IZ),I=1,I2),J=1,J2),K=1,K2)
    ENDDO
 1  FORMAT(15I5)
```

Notes: IZON: the total number of grid zones; IZT(IZ), JZT(IZ), and KZT(IZ): maximum grid numbers in the I-, J-, and K-direction for zone #IZ. (For 2D flows, KZT(IZ) = 1); For 2D axisymmetric flows, Z(I,J,K,IZ) = 1.0; while for 2D planar flows, Z(I,J,K,IZ) equals to 1.0 or the depth of the flow domain; If IDATA = 0 (unformatted file option), then the format in each write statement should be eliminated.

TABLE 8.5 Data Format of the Restart Flow File

```
WRITE(13,3) INSO(1),INSO(4),INSO(5),INSO(7),NGAS
DO IZ=1,IZON
  I2=IZT(IZ)
  J2=JZT(IZ)
  K2=KZT(IZ)
  WRITE(13,2)(((DEN(I,J,K,IZ),I=1,I2),J=1,J2),K=1,K2)
  WRITE(13,2)(((U(I,J,K,IZ),I=1,I2),J=1,J2),K=1,K2)
  WRITE(13,2)(((V(I,J,K,IZ),I=1,I2),J=1,J2),K=1,K2)
  WRITE(13,2)(((W(I,J,K,IZ),I=1,I2),J=1,J2),K=1,K2)
  WRITE(13,2)(((P(I,J,K,IZ),I=1,I2),J=1,J2),K=1,K2)
  IF(INSO(4) .EQ. 1)
&    WRITE(13,2)(((TM(I,J,K,IZ),I=1,I2),J=1,J2),K=1,K2)
  IF(INSO(5) .EQ. 1) THEN
    WRITE(13,2)(((DK(I,J,K,IZ),I=1,I2),J=1,J2),K=1,K2)
    WRITE(13,2)(((DE(I,J,K,IZ),I=1,I2),J=1,J2),K=1,K2)
  ENDIF
  IF(AMC.GT.0.)WRITE(13,2)(((AM(I,J,K,IZ),I=1,I2),J=1,J2),K=1,K2)
  IF(INSO(7).GE.1)WRITE(13,2)(((Q(I,J,K,IZ),I=1,I2),J=1,J2),K=1,K2)
  IF(NGAS.GT.0) THEN
    DO KK=1,NGAS
      WRITE(13,2)(((FM(I,J,K,IZ,KK),I=1,I2),J=1,J2),K=1,K2)
    ENDDO
  END IF
ENDDO
2  FORMAT(5(1P,E16.8))
3  FORMAT(8I5)
```

Notes: Definitions of NGAS and INSO are detailed in Chapter 3.

DEN, U, V, W, P, TM, DK, DE, AM, Q, and FM are the flow density, flow velocities in X-, Y- and Z-axis, static pressure, static temperature, turbulent kinetic energy and its dissipation rate, local flow Mach numbers, fluid qualities, and species mass fractions, respectively. The flow Mach number and quality can have the value of unity as the initial guess. For 2D axisymmetric flows, W (azimuthal velocity) can have nonzero value if the swirling component exists.

The CTP code uses grid systems that follow the right-hand rule for the I-, J- and K-line orientations such that the cell volumes are always positive.

The grid cell volumes are calculated and checked in subroutine TRANF. On detection of any zero or negative volumes, a warning message is printed and the program stops. To avoid this message for singularity lines where surfaces collapse into lines, small finite radii (1E-06 for instance) must be used for generating the singularity line surfaces (e.g., centerlines of pi segments). It is also recommended that the grid used should have smooth Jacobian variations so that unfavorable grid effects can be minimized.

8.7.1 REFERENCE CONDITIONS

The initial flow variables should be nondimensionalized by the reference conditions specified (depending on the parameter IUNIT which can be either 1 or 2) in the input data file (fort.11). If the reference quantities were chosen to be unity, then the system of the governing equations would be solved in dimensional form. The user must make sure that the nondimensionalization process is consistent throughout the initial flow data preparation process. Either SI or English units can be used based on the user's preference. The reference conditions are defined below:

SI unit: (for IUNIT = 1)
    Density: kg/m^3
    Velocity: m/s
    Temperature: K
    Length: m

English unit: (for IUNIT = 2)
    Density: slugs/ft^3
    Velocity: ft/s
    Temperature: R
    Length: ft

If the compressibility option, AMC, in the input data file is specified to be greater than zero (e.g., specified as unity), then the compressible flow option is activated and the flow properties of the initial data need to be normalized by the thermal properties of air at the reference conditions listed as follows, as well as those specified in the input data file.

SI unit: (for IUNIT = 1)
    Reference temperature (T_{ref}): TREF = 300 K
    Reference gas constant (R_{ref}): RMXBAR = 288.5939026 J/kg·K
    Reference specific heat (Cp_{ref}): CPBAR = 1012.790527 J/kg·K
    Reference viscosity (μ_{ref}): VISC = user specified (in N·s/m^2)
    Reference pressure (P_{ref}): PREF = user specified (in Pa)
    Reference density (ρ_{ref}): DENREF = calculated (in kg/m^3)
    Reference velocity (U_{ref}): UREF = user specified (in m/s)

English unit: (for IUNIT = 2)
    Reference temperature (T_{ref}): TREF1 = 540 R
    Reference gas constant (R_{ref}): RMXBAR = 53.62715 ft-lb$_f$/lb$_m$·R
    Reference specific heat (Cp_{ref}): CPBAR = 188.199 ft-lb$_f$/lb$_m$-R

Reference viscosity (μ_{ref}): VISC = user specified (in $lb_f \cdot s/ft^2$)
Reference pressure (P_{ref}): PREF = user specified (in psi)
Reference density (ρ_{ref}): DNREF1 = calculated (in slugs/ft³)
Reference velocity (U_{ref}): UREF = user specified (in ft/s)

The corresponding values of other reference variables are calculated in the code based on the following relations.

Reference density (ρ_{ref}) = $P_{ref}/(R_{ref} T_{ref})$
Reference Mach number (M_{ref}) = AMC = U_{ref}/a_{ref}
a_{ref} = reference speed of sound = $[\gamma_{ref} R_{ref} T_{ref}]^{0.5}$; and $R_{ref} = (R_u/M_w)_{ref}$

where γ_{ref} (GAMA) = 1.3985, and R_u and $(M_w)_{ref}$ are the universal gas constant and the molecular weight of air, respectively. However, if the reference Mach number (AMC) in the input data files is set to 0, then the incompressible flow calculation will be activated and all the reference properties will be set to unity. And, the reference viscosity is set equal to the inverse of the reference Reynolds number. In addition, the reference viscosity (μ_{ref}) is used to calculate local fluid viscosity based on a power Law:

$$\frac{\mu}{\mu_{ref}} = \left(\frac{T}{T_{ref}}\right)^{0.7} \tag{8.33}$$

8.7.2 NORMALIZATION OF FLOW VARIABLES

The flow variables and grid coordinates are nondimensionalized as

Density: Density/ρ_{ref}
Velocity: Velocity/U_{ref}
Pressure: Pressure/$(\rho_{ref} U_{ref}^2)$
Temperature: Temperature/T_{ref}
Length: Length/X_{ref}
Viscosity: Viscosity/$(\rho_{ref} U_{ref} X_{ref})$
Turbulence kinetic energy (k): $k/(U_{ref}^2)$
Turbulence dissipation rate (ε): $\varepsilon/(U_{ref}^3/X_{ref})$

Consistent units (either SI or English unit) must be used throughout the nondimensionalization process. For turbulent flow applications, the turbulence kinetic energy (k) and its dissipation rate (ε) can be initialized using the nominal nondimensional values when no measured data are available. That is,

$k = (4/3) \times$ (turbulence intensity)^2

$\varepsilon = 0.09 \times k^{1.5}/(0.03 \times$ characteristics length)

where the characteristics length can be a channel width or twice a boundary layer thickness, etc.

Besides the above reference quantities, there are two additional reference values, HHREF and HHREF1, which are important for calculating wall heat fluxes. The wall heat fluxes are calculated using a nondimensional heat-flux variable, HTWN(I), stored at near-wall points (or wall function points). That is,

Wall Heat Flux = −HTWN(IJLO(J))*HHREF—for SI unit, Watts/cm^2

or

Wall Heat Flux = −HTWN(IJLO(J))*HHREF1—for English unit, Btu/s-ft^2

where IJLO(J) is a wall function point indices bookkeeping array.

8.8 CTP CODE FEATURES

The CTP code is a fully transparent and user-friendly computational fluid dynamics code which is used to analyze a wide variety of fluid dynamics-related engineering problems (e.g., internal and/or external flows with complex geometries, cases with laminar or turbulent flow conditions, and flows with ideal, real or reacting gas effects for all speed range—incompressible to hypersonic flow regimes). For programming simplicity and computational efficiency, all the flow variables, except those in the subroutines for calculating chemical reaction source terms, are stored using COMMON blocks. There are 18 COMMON blocks include files: fdns01; fdns02; fdns03;...; fdns17; and fluid.inc. 1D arrays are used for all flow variables representing 2D or 3D flow problems using structured single or multiple-block grids. The conversion between the (i, j, k) indices and the global 1D indices is described in Section 8.6.2.

Before compiling the code, one must make sure that proper array dimensions are set in the first COMMON block include file, fdns01. An example of the fdns01 include file is shown below.

\<List of fdns01\>

```
PARAMETER (IIQMAX = 110000, IWP = 11000,  ISLMAX = 1)
PARAMETER (NSPM   = 11,  ISPMAX = IIQMAX)
PARAMETER (NPMAX  = 1,   IJKPMX = 1)
```

```
PARAMETER (NPOROX = 1,   IJKVMX = 1,   IJKWMX = 1)
PARAMETER (MSP = NSPM, MST = 100, MEL = 10)
PARAMETER (MZON = 200, MBIF = 200, MBIO = 100, MBWA = 200,
&          MBSN = 6,  MZPO = 20)
```

In this fdns01 example, the total grid size of the problem is IIQMAX = 110000, total number of wall function points is IWP = 11000; total number of grid points for the sliding boundary, ISLMAX, is set to be 1; maximum number of chemical elements is MEL = 10 (**do not change**); maximum number of chemical species is NSPM = 11; the multispecies thermodynamics data dimension, ISPMAX, must be equal to IIQMAX to activate the multispecies option; the maximum number of particle trajectories, NPMAX, is set to be 1; the particle property dimension, IJKPMX, must be set to be IIQMAX in order to activate the Lagrangian particle tracking option; the maximum number of porous volumes is NPOROX = 1; the porosity dimension, IJKVMX, must be set to IIQMAX to activate the porosity option; the maximum number of surface porosity is IJKWMX = 1; MST is the maximum number of chemical reaction steps; MZON is the maximum number of blocks (zones); MBIF is the maximum number of zonal interfaces; MBIO is the maximum number of flow boundaries; and MBWA, MBSN, and MZPO are the maximum numbers of wall segments, singularity lines, and porosity zones in the input data file, fort.11, respectively.

The main program of the present code, which is the main driver of other subroutines, defines input/output units and control parameters, provides problem restart modifications (through include file fmain01), defines the solution calling sequence, and provides time-marching control and timely input of run-time problem modifications and data output (through include file fmain02). Input and output units, IR1, IR2, IR3, IW1, IW2, and IW3, are assigned in the main program. Unit IR1 (input data file fort.11) is for setting up the flow domain and problem control parameters. Flow domains sizes, zonal interfaces locations, boundaries locations and types, wall block locations, job control parameters, upwind scheme selections, turbulence model selections, setting the reference conditions (viscosity = 1/Re, Mach number, reference density, reference velocity, reference temperature, reference length, etc.), and thermodynamics and reaction data, are included in unit IR1. Units IR2 and IR3 are assigned for restart grid and flowfield data files (fort.12 and fort.13), respectively. The flow solutions convergence history and evolution of the monitoring point flow variables are printed out to unit 6 (fort.6 or nohup. out). IW1 or unit 21 is not used in the current version. Units IW2 and IW3 are used for the grid data and flowfield solutions output. The current version also provides PLOT3D format grid and flow outputs through file units fort.91, fort.92, and fort.93 for postprocessing. These files units are assigned to their specific files names as listed in the main program. The file unit fort.91 contains the grid data, fort.92 contains five variables (i.e., density, u-velocity, v-velocity, w-velocity, and total pressure), and fort.93 also contains five flow variables (i.e., density, pressure, temperature, Mach number, and species 1 concentration). In addition, unit fort.94 contains both grid and flow outputs (including density, velocities, pressure, temperature, and Mach numbers), in the format of commercial graphics software "TECPLOT," for the purpose of postprocessing. For making output of different variables or using different data format (e.g., a binary data option is given in the source code), user-defined specific coding can be implemented by modifying the subroutine DATAIO (only the PLOT3D section which is located near the end of DATAIO) in f1.f. The program start or restart status is defined by setting IDATA = 2 or IDATA = 1 in the input data (fort.11), respectively. When IDATA = 2 is selected, grid and initial flowfield data must be created and made available for data read-in from one of the include files, fexmp01, for a fresh start.

Grid and initial flowfield data files can be generated using known preprocessor grid generation codes such as Gridgen, Umesh or EAGLE, etc., and user-developed initial flowfield generation codes. A grid/initial flow generator included in Appendix C provides an example of how these preprocessing tools can be constructed. The user may follow these examples for the generations of other new cases.

To present the flowfield solutions graphically using the output files of the CTP code, graphics packages using C++ or open-GL utilities such as PLOT3D (developed at NASA/Ames Research Center) or a commercial software—TECPLOT can be used to plot the grid, velocity vectors, and contour lines of selected field quantities. Other flow parameter outputs such as surface pressure, skin friction distributions, and heat transfer coefficient distributions (which are output from the include file fmain02) can also be plotted.

The basic code structure of the CTP code is further elaborated in the following section. This section provides a general outline of the code. However, it is thought to be detailed enough for the CTP users to perform daily engineering analysis applications.

8.8.1 CTP CODE STRUCTURE

The CTP code is designed for robustness and user-friendliness. The entire program is written in standard FORTRAN 77 language. The code is generally not machine dependent. It has been successfully tested on the personal computers with Windows-XP, Windows-2000, Linux clusters, IBM workstation, IBM clustered computers, HP computers, etc.

Besides the main program, there are 91 subroutines and 3 entries in the CTP code. They are grouped in seven files namely "CTP.f," "f1.f," "f2.f," "f3.f," "f4.f," "f5.f," and "f6.f." In addition, there is a database file, "dbase.dat," which contains thermodynamics data of various species for the real-fluid EOS. The subroutines' names and their major functions are summarized below.

< In CTP.f >

CTP	The main program which is the main driver for other sub-routines, defines input/output units and control parameters, provides problem restart modifications (through include file fmain01), defines the solution calling sequence, and provides time-marching control and timely input of run-time problem modifications and data output (through include file fmain02).
EXAMP	To allow the user to generate (or read in) grid and prepare initial flowfield data using an include file, fexmp01, for a fresh start.
XISENT	For calculating nozzle Mach number, pressure, density, and temperature variations (as a function of local versus throat area ratio) based on isentropic relations.
USUBIO	To calculate and store initial flowfield total pressure, total enthalpy, and mass flux conditions and provide inlet boundary conditions based on total conditions or mass conservation.
CHOEQ	To calculate equilibrium species concentrations for the wet-CO mechanism by solving a set of algebraic equations.

< In f1.f >

ZONCHK	To check the grid indices at the zonal interface, to calculate as well as exchange the grid spacing between the grid point at the zonal interface and its neighboring point, and also to compute the angle between cyclic boundaries.
CYCANG	To calculate the angle between each pair of cyclic boundaries.
BCCOND	To provide implicit and explicit boundary conditions for all flow variables that includes mass conservation conditions, pressure conditions, and pressure equation boundary condition setting.

DATINN	To read in CTP restart grid and flowfield data files (IR2—fort.12 and IR3—fort.13).
DATOUT	To output files CTP restart grid and flowfield data files (IW2—fort.22 and IW3—fort.23).
INFACE	To perform multiple-block, zonal-interface interpolation boundary conditions that requires overlaid zonal interfaces. Interface identification, interface grid movement, and interface interpolations are handled in this subroutine. Make sure that the statement {include 'inface.inc'} in INFACE is activated.
INFINT	This is a tool for 2D interpolations using bilinear interpolation scheme.
INIT	To provide initialization of problem control parameters, model constants and zero-out flowfield variables (not for initial flow conditions).
WALLFN	To calculate near-wall velocity profiles, heat flux to the wall, static enthalpy on the wall, and near-wall turbulence quantities by using wall function models for turbulent flow boundary conditions. The heat of pyrolysis model for hybrid fuel regression is also evaluated here.

< In f2.f >

SOLVEU	To provide solutions for the momentum and energy equations using high-order upwind, TVD, or central difference schemes plus adaptive dissipation terms.
SOLVEQ	To provide solutions for the turbulence model transport equations.
SOLVES	To provide solutions for the chemical species mass fraction transport equations.
SOLVET	To provide solutions for the thermal conduction equation for block of solid-wall points.
SOLVEP	To solve the pressure correction equation and perform pressure, temperature, velocity, and density field updating.

< In f3.f >

| AINDEX | To convert the global 1D indices, IJK, into the multiple-zone 3D indices (NZ, I, J, K). |
| AREAIO | To calculate density-weighted inlet and outlet areas. |

BCCHAR	To provide outlet properties extrapolations and perform outlet and symmetry plane velocity vector conditions.
BOUNC	This is a driver for wall function points by calling WALLFN.
DIRCOS	To register wall element identification and wall function control parameters and calculate wall surface orientations.
FLOWIO	To calculate the mass flow rates at the outlet boundaries.
LINER0	To perform 1D regular TDMA matrix inversion.
LINER1	To perform 1D periodic TDMA matrix inversion.
LINERA	To perform global point-by-point, L-U iterative, conjugate gradient, or GMRES matrix solution.
CGSOLV	A driver for choosing either conjugate gradient or GMRES method.
PNTBYPNT	Point-by-point solver in the zone-by-zone matrix solver.
PTCR1	Preconditioned zone-by-zone conjugate residual matrix solver with DKR and DD factorization.
PGMRES1	Preconditioned zone-by-zone GMRES matrix solver.
PTCR2	Preconditioned multizone conjugate residual matrix solver with DKR and DD factorization.
PGMRES2	Preconditioned multizone GMRES matrix solver.
BNDCOEF	To reset the link coefficients of the zonal boundary points matrix to zero.
DOTSUM	To calculate inner product of two vector arrays in the zone-by-zone matrix solver.
MATRIX	To perform multiplication of matrix in the zone-by-zone matrix solver.
PRINTN	To print out a matrix.
DECOM1	Dupont–Kendall–Rachford (DKR) factorization in the zone-by-zone matrix solver.
DECOM2	Double decomposition (DD) in the zone-by-zone matrix solver.
INVER1	Inverse LDU for DKR decomposition in the zone-by-zone matrix solver.
INVER2	Inverse LDU for DD decomposition in the zone-by-zone matrix solver.

PTBYPT	Point-by-point solver in the zone-by-zone matrix solver.
IJKXYZ	To calculate total grid number in each zone.
DOTSUM2	To calculate inner product of two vector arrays in the multizone matrix solver.
MATRIX3	To perform multiplication of matrix in the multizone matrix solver.
DECOM3	Dupont–Kendall–Rachford (DKR) factorization in the multizone matrix solver.
DECOM4	Double decomposition (DD) in the multizone matrix solver.
INVER3	Inverse LDU for DKR decomposition in the multizone matrix solver.
INVER4	Inverse LDU for DD decomposition in the multizone matrix solver.
PTBYPT2	Point-by-point solver in the multizone matrix solver.
UPBCPT	To update values for boundary conditions and zonal interface points.
PRINTM	To print out a matrix field.
EQUAL1	To assign a constant to a vector array.
EQUAL2	To assign the values of one vector array to another vector array.
DIVID1	To divide a vector array with a constant.
NORM2	To calculate the norm of a vector array.
RVA4	To assign four real constants.
LINKFA	To identify the indices for the grid points at the zonal interface boundaries.

< In f4.f >

SOURCE	To evaluate source terms for all transport equations. Multiple-phase interphase source terms are also calculated here.
SOURCX	Three entries are included: NEWVIS—To calculate turbulence eddy viscosity. BBLOCK—To assign wall element identifications. PROPTY—To calculate thermodynamics data and chemical reaction source terms (use include file propty.inc).
TRANF	To calculate the Jacobian of the coordinate transformation and grid spacing variations.

UNEWIO   To perform outlet velocity corrections based on mass conservation conditions.

UVCON   This subroutine is currently null.

WALVAL   For assigning boundary values for the wall surface points, inlet planes, outlet planes, symmetry planes, and singularity lines.

HEAT00   A general fluid thermodynamics subroutines driver which calls HEAT2A, HEAT2C, HEAT2D, FLINT, NBSLOX, and NBSH2O.

HEAT2A   To calculate the thermodynamics properties and to assemble finite-rate chemistry source terms.

SOOTOX   To calculate the reaction rate for the soot oxidation chemistry model.

HEAT2B   To find ideal gas temperature (in K) based on the given enthalpy and gas species concentrations using Newton's method.

GAUSS   To solve a matrix by using Gaussian elimination with the pivoting method.

CPHG   To calculate specific heat (C_p), enthalpy (h/R), and Gibb = s free energy at a given temperature based on CEC thermal data.

HEAT2C   To calculate pressure and enthalpy based on the density and temperature for either ideal gas or real fluid.

HEAT2D   To find ideal gas temperature (in K) based on the given enthalpy and gas species concentrations by using Newton's method, and calculate C_p and γ.

NBSLOX   To provide table look-up of LOX properties (H–P–T diagram) based on the National Bureau of Standard data.

NBSH2O   To provide table look-up of water properties (H–P–T diagram) for pressure less than 1.5 atm.

< In f5.f >

LPTSD   This is a main driver for setting up the particulate phase integration scheme that includes the particle initial conditions, Lagrangian integration of the particle trajectories and the assembly of the interphase source terms.

HPTDAT   To calculate temperatures of Al_2O_3 particles based on their enthalpy.

SEARCH   To locate particle inside the computational domain. A multiple zone algorithm is included.

PSOURC   To calculate the drag forces and heat transfer rates for the particle based on the interphase slip conditions.

POROST   This is a main driver for a porosity model.

DRAGHT   To calculate the drag forces and heat transfer based on the assigned porosity.

< In f6.f >

FLINT   The interface with real-fluids model.

ACENF   To calculate acentric factor using Lee–Kesler model.

DBASE   Routine containing database for fluids thermodynamics properties (see file "dbase.dat" for species represented).

IDMIX   Control routine for ideal mixture using HBMS EOS.
              Mode = 1: Initialization
                     2: Solve pressure, enthalpy based on density and temperature
                     3: Solve density, enthalpy based on pressure and temperature
                     4: Solve temperature, enthalpy based on density and pressure
                     5: Solve pressure, temperature based on density and enthalpy
                     6: Solve density, temperature based on pressure and enthalpy

HBMS   Routine to compute real-fluid properties using HBMS EOS.

SATLINE   To compute saturation line conditions.

THERMAL   To calculate ideal gas thermodynamic properties.

VFROMPT   Routine to determine volume from given pressure and temperature.

LAGRAN   Lagrange interpolation module.

< In io.f >

READ_RESTRAT_RFV   To read in CTP (RFV version format) restart grid and flow files from the working directory of each host machine.

WRITE_RESTART_RFV	To print out CTP (RFV version format) restart grid and flow files to the working directory of each host machine.
READ_GRID_RFV	To read in CTP (RFV version format) grid file.
READ_FLOW_RFV	To read in CTP (RFV version format) flow file.
WRITE_GRID_RFV	To print out CTP (RFV version format) grid file.
WRITE_FLOW_RFV	To print out CTP (RFV version format) flow file.
CHECK_IIQMAX	To check whether the total number of grid points exceeds the maximum memory allocation or not.
READ_PLOT3DG	To read in PLOT3D grid file.
WRITE_PLOT3DG	To print out PLOT3D grid file.
READ_PLOT3DQ	To read in PLOT3D flow file.
WRITE_PLOT3DQ	To print out PLOT3D flow file.

< In flib.f >

GETWORD	To convert an input character string into a character array without blank space.
GETNUM	To get the processor index from an input line.
FIND_WORD	To find the key word for an entry in the input data file.
INUMBER_CNUMBER	To convert a number character to an integer.
LENGTH	To obtain the length of a character string not including the blank space.
READCARD1	To read in a dummy line from the input data file.
IVA4	To assign four integer constants.

The basic structure of the CTP code is depicted in Chart 8.1, which is basically a functional flow chart for the main program.

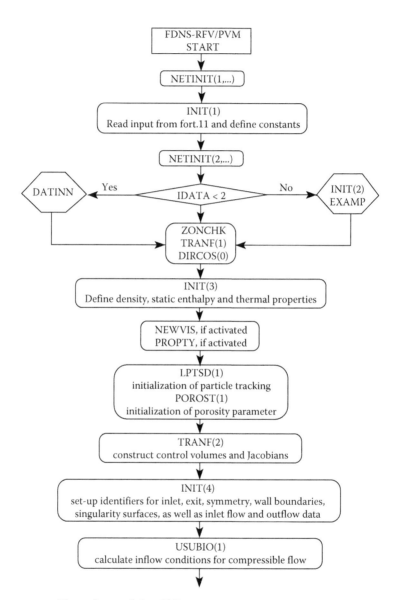

CHART 8.1 Flow chart of the CTP main program.

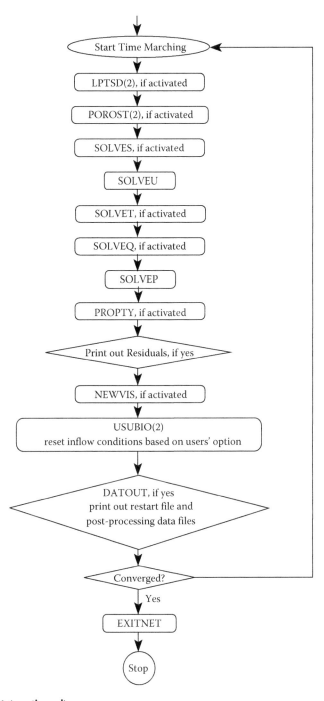

CHART 8.1 (continued)

8.9 USER'S GUIDE

8.9.1 INPUT DATA (FORT.11) DEFINITION

The input data file (fort.11) of the CTP code consists of 19 card groups and 2 entries. Definitions of these input data are given below. A sample input file is listed in Appendix A.

Card Group #1	Defines the case title and whether the problem is 2D or 3D
Format (1 line)	Title (Put title of the problem here—maximum 60 characters)
Format (1 line)	IDIM
Definition	IDIM = 2: For 2D planar or axisymmetric flow problems
	= 3: For 3D flow problems

Card Group #2	Specifies zonal information and number of flow and wall boundaries
Format (1 line)	IZON, IZFACE, IBND, ID, ISNGL
Definition	IZON: Number of zones or mesh blocks
	IZFACE: Number of patched interfaces
	IBND: Number of flow boundaries (e.g., inlet, outlet, or symmetry planes)
	ID: Number of wall elements (blocks)
	ISNGL: Number of singularity lines/surfaces

Card Group #3	Specifies zonal grid size and zonal rotational/translational speeds
Format (1 line*IZON)	IZT, JZT, KZT, CBG1, CBG2, CBG3, CBV1, CBV2 CBV3
Definition	IZT(II): *I*-max in zone II
	JZT(II): *J*-max in zone II
	KZT(II): *K*-max in zone II

Card Group #3 (continued)

CBG1(II): Coordinate rotation speed $(X_{ref}\Omega_x/U_{ref})$ of zone II about X-axis

CBG2(II): Coordinate rotation speed $(X_{ref}\Omega_y/U_{ref})$ of zone II about Y-axis

CBG3(II): Coordinate rotation speed $(X_{ref}\Omega_z/U_{ref})$ of zone II about Z-axis

CBV1(II): Coordinate translation speed of zone II in X-direction

CBV2(II): Coordinate translation speed of zone II in Y-direction

CBV3(II): Coordinate translation speed of zone II in Z-direction

---

Card Group #4	Identifies the zonal interface matching indices
Format	NNBC, IZB1, IZF1, IJZ11, IJZ12, JKZ11, JKZ12,
(2 lines*IZFACE)	IZB2, IZF2, IJZ21, IJZ22, JKZ21, JKZ22

Definition	NNBC:	$> = 0$, IZFACE counter
	IZB1:	Zonal index of interface plane # 1
	IZF1:	Interface plane identifier for plane #1
		$= 1$: $I = I$-max or east boundary
		$= 2$: $I = 1$ or west boundary
		$= 3$: $J = J$-max or north boundary
		$= 4$: $J = 1$ or south boundary
		$= 5$: $K = K$-max or top boundary= 6: $K = 1$ or bottom boundary
	IZB2:	Zonal index of interface plane #2
	IZF2:	Interface plane identifier for plane #2
	IJZ11, IJZ12:	The starting and ending points of the first running index on the interface plane #1

(continued)

Card Group #4 (continued)

JKZ11, JKZ12:   The starting and ending points of the second running index on the interface plane #1

IJZ21, IJZ22:   The starting and ending points of the first running index on the interface plane #2

JKZ21, JKZ22:   The starting and ending points of the second running index on the interface plane #2

Example: If IZF1 or IZF2 is either 1 or 2 then IJZ11, IJZ12, IJZ21, and IJZ22 are the indices in J-direction, and JKZ11, JKZ12, JKZ21, and JKZ22 are the indices in K-direction.

If IZF1 or IZF2 is either three or four then IJZ11, IJZ12, IJZ21, and IJZ22 are the indices in I-direction, and JKZ11, JKZ12, JKZ21, and JKZ22 are the indices in K-direction.

If IZF1 or IZF2 is either five or six then IJZ11, IJZ12, IJZ21, and IJZ22 are the indices in I-direction, and JKZ11, JKZ12, JKZ21, and JKZ22 are the indices in J-direction.

Note: The interface patching surface indices for planes #1 and #2 (i.e., IJZ11 → IJZ12 and IJZ21 → IJZ22, JKZ11 → JKZ12 and JKZ21 → JKZ22 must have consistent running order). Also, IJZ12 > IJZ11 and JKZ12 > JKZ11 (but not necessary for IJZ21, IJZ22, JKZ21, and JKZ22)

Card Group #5	Specifies flow boundaries (inlet, outlet, symmetry)	
Format	IBCZON, IDBC, ITYBC, IJBB, IJBS, IJBT, IKBS, IKBT	
(1 line*IBND)		
Definition	IBCZON	Zonal index for the flow boundary
	IDBC	Boundary facing index:
		= 1: $I = I$-max or east
		= 2: $I = 1$ or west
		= 3: $J = J$-max or north
		= 4: $J = 1$ or south
		= 5: $K = K$-max or top
		= 6: $K = 1$ or bottom
	ITYBC	Identifies boundary type:
		= −2: Inlet with mass flow rate and velocities fixed
		= −1: Inlet with mass flow rates conserved (e.g., solid fuel blowing surfaces)
		= 0: Inlet with all variables fixed (e.g., supersonic)
		= 1: Inlet with constant total pressure (compressible flow only)
		= 2: Outlet boundary
		= 3: Symmetry plane (can also be regarded as slip/inviscid wall boundary conditions, but should not be combined with wall block in Card Group #6)
	IJBB	$I, J,$ or K location (depends on IDBC) of the boundary
	IJBS, IJBT	Boundary starting and ending indices (for I or J)
	JKBS, JKBT	Boundary starting and ending indices (for J or K)

Card Group #6   Specifies wall block indices

Format (1 line*ID)   IWBZON, L1, L2, M1, M2, N1, N2, IWTM, HQDOX, IWALL, DENNX, VISWX

Definition

IWBZON:	Zonal index for the wall block
L1, L2:	Starting and ending indices in the I-direction
M1, M2:	Starting and ending indices in the J-direction
N1, N2:	Starting and ending indices in the K-direction
IWTM:	Solid-wall thermal boundary condition options

         $= -1$:   For fixed temperature wall boundary

         $= 1$:   For heat-flux ($=$ HQDOX) wall boundary

HQDOX:	Nondimensional wall heat flux when IWTM $= 1$, the value is positive if it is from wall to fluid

Normalization for $\dot{Q}$:

$$\text{SI unit} = \dot{Q}/(\rho_{ref}U_{ref}Cp_{ref}T_{ref})$$
$$\text{English units} = \dot{Q}/(32.174\rho_{re}{}_{f1}U_{ref1}Cp_{ref1}T_{ref1})$$

IWALL:	Solid-wall heat conduction option

         $= 0$:   To deactivate

         $= 1$:   To activate

DENNX:	Nondimensional solid-wall density (wall density/ρ_{ref})
VISWX:	Nondimensional solid-wall thermal conductivity

$$= \kappa/(X_{ref}\rho_{ref}U_{ref}Cp_{ref})$$

Note: The specified values of DENNX and VISWX will be meaningful only when IWALL $= 1$ is selected, and the program will set IWTM $= -1$, since this is a correct combination

Card Group #7   Specifies the singularity lines
Format (1 line*ISNGL)   ISNZON, ISNBC, ISNAX, ISNBS, ISNBT
Definition

	ISNZON	Zonal index for the singularity lines
	ISNBC	Singularity line boundary facing index
		= 1: $I = I$-max or east
		= 2: $I = 1$ or west
		= 3: $J = J$-max or north
		= 4: $J = 1$ or south
		= 5: $K = K$-max or top
		= 6: $K = 1$ or bottom
	ISNAX	Orientation of the singularity line axis
		On I–J plane (ISNBC= 5 or 6)
		ISNAX = 1 for I-axis
		ISNAX = 2 for J-axis
		On J–K plane (ISNBC = 1 or 2)
		ISNAX = 1 for J-axis
		ISNAX = 2 for K-axis
		On K–I plane (ISNBC = 3 or 4)
		ISNAX = 1 for I-axis
		ISNAX = 2 for K-axis
	ISNBS, ISNBT	Starting and ending indices along ISNAX

Card Group #8   I/O parameters and problem control parameters
Format (1 line)   IDATA, IGEO, ITT, ITPNT, ICOUP, NLIMT, IAX, ICYC
Definition   IDATA   Restart options
                                 = 1:   For regular restart runs. Restart grid and flow files, fort.12 and

(continued)

Card Group #8 (continued)

		fort.13, must be made available. The format of the restart files can be specified by IOFINN in Card Group #15
	= 2:	For example start run. Users must implement the subroutine EXAMP in the fexmp01 include file to properly read in the pregenerated grid and flow data.
IGEO		Geometry parameter (for user applications)
	= 1:	Is specifically for problems without inlets and outlets (e.g., cavity flows)
	= 19:	Is reserved for linear cascades applications
ITT		Number of time-steps limit
ITPNT		The frequency on printing out solutions (through files fort.22, fort.23, fort.91, fort.92, fort.93, and fort.94)
ICOUP		number of pressure correctors (typically 1 for steady-state applications and 3–6 for transient or rough initial start applications)
NLIMT	= 1:	For regular run
	= 0:	For printing out the initial or restart files without going through solution procedures
IAX	= 1:	For 2D planar or 3D flows
	= 2:	For 2D axisymmetric flow problems
ICYC		Cyclic or periodic boundary conditions identifier. Currently, only ICYC = 3 (in K-direction) is active for turbomachinery applications, where all flow variables at $K = 1$ are the same as those at $K = K$-max in the corresponding zones.

Card Group #9	Time-step size, upwind schemes and time-marching scheme selections.		
Format (1 line)	DTT, IREC, REC, THETA, BETAP, IEXX, PRAT		
Definition	DTT		Nondimensional time-step size, $\Delta t(U_{ref}/X_{ref})$
	IREC		Selects upwind scheme options
		= 0:	For second-order upwind scheme
		= 1:	For third-order upwind scheme
		= 2:	For second-order central scheme
		= 3:	For second-order TVD scheme
		= 4:	For third-order TVD scheme
	REC		Upwind damping parameter (0.1 recommended)
		= 0.0:	For second-order accuracy
		= 1.0:	For first-order upwind scheme
	THETA		Time-marching scheme parameter
		= 1.0:	For steady-state applications
		= .99:	For implicit-Euler transient applications
		= 0.5:	For Crank–Nicholson second-order accurate transient applications
	BETAP	≤1.0	Pressure updating under-relaxation parameter, typically 1.0; small values can be used to reduce the amount on pressure corrections for rough start initial runs (in this case, choose ISWP ≥ 80 as explained in Card Group #12)
		>1.0	Factor for the diagonal term of matrix coefficients in the pressure correction equation to maintain stability of matrix solver (typically 1.01 for incompressible flows, single species, or premixed multispecies flows)

(continued)

Card Group #9 (continued)

        IEXX   Outlet extrapolation parameter for scalar quantities

              = 1:   For zero-gradient extrapolation

              = 2:   For linear extrapolation

        PRAT   Specifies outlet boundary condition (BC) options

              = −1:   For supersonic outlet BC.

              = 0.0:   For outlet mass conservation BC.

              > 0:   For outlet fix pressure BC. The outlet pressure reference point (IPEX, JPEX) is used here. Pressure at this point is maintained at a value of PRAT*PPCN, where PPCN is the atmospheric pressure (1 atm).

Card Group #10	Specifies inlet, outlet pressure points and data monitoring point
Format (1 line)	IPC, JPC, IPEX, JPEX, IMN, JMN
Definition	IPC, JPC: Flowfield reference point at the grid index of IPC in zone JPC (not the global grid index)
	IPEX, JPEX: Outlet pressure reference point (same way of indexing as IPC, JPC)
	IMN, JMN: Solution monitoring point

Card Group #11	Gives reference viscosity, Mach number and options of turbulence models	
Format (1 line)	VISC, IG, ITURB, AMC, GAMA, CBE, CBH, EREXT	
Definition	VISC	Dimensional fluid viscosity of air at the sea level (the unit has to be consistent with that specified by IUNIT)
	IG	= 1: For laminar flow option
		= 2: For turbulent flow option
	ITURB	For turbulence model selection
		= 1: For standard high-Re k–ε model
		= 2: For extended high-Re k–ε model
		= 3: for Lam-Bremhorst low-Re k–ε model[19]
		= 4: for H-G low-Re k–ε model
	AMC	= 0.0: For incompressible flow calculation
		> 0.0: For compressible flow calculation
	GAMA	Reference-specific heat ratio (not used in the code)
	CBE	= 0: No buoyancy effect
		> 0.5: Include buoyancy effect for compressible flow
		< 0: Nondimensional buoyancy force parameter for incompressible flow, $|CBE| = Gr/Re^2$, where Gr is the Grashof number and Re is the Reynolds number
	CBH	Select compressibility corrections for the k–ε turbulence model
		= 0.0: No compressibility correction
		= −1.0: For k-corrected model
		= −2.0: For ε-corrected model
		< −3.0: For T-corrected model where $C_3(T/T_a)\lambda$, $\lambda = *CBH*-3$
	EREXT	Convergence criterion (typically 5.0E-05 for steady-state solutions)

Card Group #12   Specifies number of zonal iterations in the matrix
                 solver when (INFACE > 0) is used for overlaid
                 grid zonal interface interpolations and indicates
                 orthogonal or nonorthogonal grid options

Format (1 line) ISWU, ISWP, ISWK, ISKEW

Definition   ISWU   For the momentum and energy
                         equations

                         ≤80   Number of iterations for the over-
                               laid zonal boundaries by using
                               point-by-point matrix solver

                         >80   Conjugate gradient matrix solver
                               (solving multizones as a whole) is
                               used to solve the matrices until its
                               residuals drop (ISWU-80) orders

                         >85   GMRES matrix solver (solving
                               multizones as a whole) is used to
                               solve the matrices until its resid-
                               uals drop (ISWU-85) orders

                         >90   Conjugate gradient matrix solver
                               (solving zone-by-zone) is used to
                               solve the matrices until its resid-
                               uals drop (ISWU-90) orders

                         >95   GMRES matrix solver (solving
                               zone-by-zone) is used to solve
                               the matrices until its residuals
                               drop (ISWU-95) orders

                 ISWP   For the pressure correction equations

                         ≤80   Number of iterations for the over-
                               laid zonal boundaries by using
                               point-by-point matrix solver

                         >80   Same as above for ISWU

                 ISWK   For the scalar equations (e.g., k, ε, and
                         species equations)

                         ≤80   Number of iterations for the over-
                               laid zonal boundaries by using
                               point-by-point matrix solver

                         >80   Same as above for ISWU

Card Group #12 (continued)

        ISKEW   Nonorthogonal grid viscous flux option
- = 0: For orthogonal grid
- = 1: For nonorthogonal grid

Card Group #13	Specifies which equations are to be solved
Format (1 line)	INSO(IEQ): U, V, W, TM, DK, DE, FL, 8, EQ, VS, FM, SP

Definition (0 to deactivate; 1 to activate)			
	U, V, W	= 1:	Solving the momentum equations
	TM	= 1:	Solving the energy equation
	DK, DE	= 1:	Solving the turbulence model
	FL	= 0:	For ideal gas flow model
		= $n > 0$:	For real-fluid flow model with the quality of the n-th species being saved (In this case, a "FLUID" entry at the end of the input file is needed to identify the species which properties will be calculated through real-fluid model.)
	8	Not used	
	EQ	= 0:	No equilibrium chemistry
		= 1:	H_2/O_2 equilibrium chemistry
		= 2:	Wet-CO equilibrium chemistry
	VS	= 1:	For updating the turbulence eddy viscosity

(continued)

Card Group #13 (continued)

FM	= 0:	Deactivate the species mass-fraction equations (e.g., single species or air flows)
	= 1:	Activate the species mass-fraction equations
SP		For calculating the gas thermal properties, and selecting various treatment for species production term
	= 1:	Explicit chemistry model (penalty function)
	= 11 or 12:	Implicit chemistry model (first or second-order) with pseudotime-step size
	= 21 or 22:	Implicit chemistry model (first or second-order) with real time-step size
	= 31 or 32:	First or second-order implicit chemistry model with time integration (constant T, P)
	= 33:	Fourth-order PARASOL chemistry model with time integration (constant T, P)
	= 41 or 42:	First- or second-order implicit chemistry model with time integration (constant H, P)
	= 43:	fourth-order PARASOL chemistry model with time integration (constant H, P)

Card Group #14	Specifies number of gas species and reactions, and gives the reference conditions	
Format (1 line)	NGAS, NREACT, IUNIT, DENREF, UREF, TREF, XREF	
Definition	NGAS	Number of chemical species which thermal properties in CEC tables will be read in
		= 0: For single species, ideal gas flow
		> 0: For multiple chemical species flow
		= −1: For LOX flow calculation where its thermodynamics properties are calculated from NBS table look-up
		= −3: For water flow calculation where its thermodynamics properties are calculated from table look-up
	NREACT	Number on reaction steps to be used
		= 0: For nonreacting flow
		> 0: For finite-rate reacting flow
	IUNIT	= 1: For SI-unit reference conditions
		= 2: For English-unit reference conditions
	DENREF	Reference density (in kg/m^3 or $slug/ft^3$), not used in the code
	UREF	Reference velocity (in m/s or ft/s)
	TREF	Reference temperature (in K or R), not used in the code
	XREF	Reference length (in m or ft)
	PREF	Reference pressure (in psi or N/m^2)

Card Group #15	If NGAS > 0, include the CEC thermodynamics data here	
Format	SPECIE, WTMOLE, <==== (1 line)	
(4 lines*NGAS)	HF(7,2) <==== (3 lines)	
Definition	SPECIE	Name of the chemical species (20 characters)
	WTMOLE	Molecular weight of the chemical species
	HF(7,2)	Polynomial coefficients of CEC thermodynamics data of the species

Card Group #16	If NREACT > 0, specifies the finite-rate reaction steps	
Format	REACTION: Species names, N = 1, NGAS (only 1 line as a title)	
(2 or 3 lines*	IREACT, A, B, E/RT, ITHIRD, IGLOB,	
NREACT)	(STOCEF(N, IREACT), N = 1, NGAS),	
	(STOCEG(N, IREACT), N = 1, NGAS)* *needed if IGLOB = 2	
Definition	IREACT	Reaction step counter
	A	Reaction rate leading constant ($A = 0$ is designated for the soot-oxidation chemistry)
	B	Reaction rate temperature exponent
	E/R	Reaction rate activation energy constant (if $A = 0$, then E/R is the assumed diameter of the soot particle in meter, typically 4×10^{-6} m)
	ITHIRD	Third-body reaction indicator
	= 0:	Deactivated
	= n:	For using the N-th species as third body
	= 999:	For global (every species) third body
	IGLOB	Global reaction model indicator

Card Group #16 (continued)

= 0:		Elementary reactions with the rate of backward reaction is calculated from equilibrium constant and forward reaction rate
= 1:		One-way reaction (either forward or backward reaction controlled by the sign of STCOEF); need only one input line of STCOEF
= 2:		One-way reaction with power dependency; need input line for STCOEF and STCOEG
	STCOEF	Stoichiometric coefficients of elementary reactions (negative signs apply to reactants and positive signs are for the products)
	STCOEG	Power dependency coefficients

Card Group #17	If IJKPMX= IIQMAX in the parameter specification, then read in the following particle input control		
Format (1 line)	IDPTCL, IPREAD		
Definition	IDPTCL	Number on particle initial condition input lines	
		= 0:	To deactivate particulate phase calculation
		= 1:	To activate particulate phase calculation
	IPREAD	= 0:	Read in particle inlet conditions in next card group

(continued)

Card Group #17 (continued)

| | = 1: | For reading in particle data (fort.14) from upstream domain (this allows transferring the outlet particle data from the upstream domain solutions to the inlet boundary for succeeding domain computations— especially useful for multiphase rocket plume simulations) |

Card Group #18   Particle initial conditions (for steady state runs only)

Format   IPTZON, IDBCPT, LPTCL1, LPTCL2, MPTCL1, MPTCL2, NPTCL1, NPTCL2

(2 lines*IDPTCL)   ITPTCL, DDPTCL, DNPTCL, WDMASS, UUPTCL, HTPTCL

Definition   IPTZON   Zonal index for the particle initial position

IDBCPT   I-, J- or K-plane identifier

	= 1:	For I-plane (plane normal to I lines)
	= 2:	For J-plane (plane normal to J lines)
	= 3:	For K-plane (plane normal to K lines)

LPTCL1, LPTCL2   I-interval for the particle initial position

MPTCL1, MPTCL2   J-interval for the particle initial position

NPTCL1, NPTCL2   K-interval for the particle initial position

ITPTCL   Number of particle groups (trajectories) starting from each grid cell

Card Group #18 (continued)

DDPTCL		Particle diameter in μm
DNPTCL		Particle density in lbm/ft^3
WDMASS		Particle mass flow rates for the current particle group and area involve the current input line
UUPTCL		Particle/gas velocity ratio at the initial positions
HTPTCL		Particle initial enthalpy in ft^2/s^2

8.9.2 USER-DEFINED RUN-TIME MODIFICATIONS

Three run-time modification include files are used for the current version of the CTP code. They are: fmain01, fmain02 in the main program, and fexmp01 in the subroutine EXAMP. All these include files are entered in the CTP.f file. Only this file needs to be recompiled after any of these three include files are changed. However, if any of the COMMON block include files, fdns01 through fdns17, and fluid.inc, is changed, the entire code must be recompiled. Samples of the usage of these include files for given example problems are listed in Appendix C.

8.9.3 MAIN PROGRAM INCLUDE FILES (FMAIN01 AND FMAIN02)

The include file fmain01 is only used after the restart files are read. Any modifications to the restarted data such as the inlet pressure level setting, inlet velocity profile modification, reinitializing part of the flowfield, and/or wall temperature resetting, etc. FORTRAN statement numbers between 7000 and 7900 can be used in the coding of this include file.

The second include file fmain02 is entered after every time step. Any runtime modification such as boundary condition adjustment, grid modification (note that DIRCOS and TRANF must be called after grid modification), and run-time printing of any data of interest (file unit numbers between 30 and 89 are recommended for printing out user's data) can be added in this include file. FORTRAN statement numbers between 8000 and 8900 can be used in the coding of fmain02.

8.9.4 EXAMPLE SUBROUTINE INCLUDES (FEXMP01)

In the example start include file, fexmp01 (which is included in the subroutine EXAMP), one can include simple grid generation and initial flowfield specification FORTRAN coding to start a problem. Subroutines IVA4 and RVA4 are used to simplify the coding and subroutine XISENT can be employed to generate initial nozzle flowfield based on 1-D isentropic relations. Another way of using the fexmp01 file, such as the one used for some of the example problems is to write a simple but general grid and flow data read-in code in the fexmp01 file. Then, a separate grid generation and flowfield initialization program is written for generating the grid and flow data file, which is then used by the fexmp01 data input code.

8.9.5 RESTART/OUTPUT FILES (IN MAIN, DATINN, AND DATOUT)

The grid and flowfield restart input files (fort.12 and fort.13) and output files (fort.22 and fort.23) are handled in the subroutines DATINN and DATOUT. The subroutine DATP3D prints out one PLOT3D gridfile (fort.91) and two PLOT3D q-files (through fort.92 and fort.93). The PLOT3D grid data are not rescaled for general cases, and except for 3-D pump problems where grid data are normalized by the pump tip diameter. The first PLOT3D q-file, fort.92, includes the following five variables:

q1   Density in lb_m/ft^3 or unity
q2   U-velocity in ft/s or normalized by the pump tip speed
q3   V-velocity in ft/s or normalized by the pump tip speed
q4   W-velocity in ft/s or normalized by the pump tip speed
q5   Nondimensional total pressure.

The second PLOT3D q-file, fort.93, includes the following five variables:

q1   Density in lb_m/ft^3 or unity
q2   Pressure in psia
q3   Temperature in EK
q4   Mach number
q5   First species mass-fraction, or quality of the specified species

Units of the PLOT3D outputs can be modified, by editing the DATAIO subroutine. Unformatted data are used in DATAIO subroutine. Other formats can be incorporated, by modifying DATAIO. The TECPLOT data file contains both grid coordinates and flow variables, which are listed as: X, Y, Z (3D only), U, V, W (3D only) for pressure, density, temperature, Mach number, quality (if real-fluid model is activated), and species mass fractions, respectively. The units of all variables are determined through the input file (fort.11). Note that in the main program, CTP.f, the names of the data files described above are assigned according to the following definitions:

fort.2 → dbase. dat
fort.9 → fluid.inp
fort.11 → input
fort.12 → restart.x
fort.13 → restart.q
fort.22 → output.x
fort.23 → output.q
fort.91 → plot.x
fort.92 → plot.q1
fort.93 → plot.q2
fort.94 → tecplot.dat

8.10 NOMENCLATURE

See Section 6.5, exceptions are the replacement of the symbols: g with ϕ and ϕ with Γ. Other symbols are used, but they are not blamed as they only have meaning in the equations in which they are used.

CHAPTER 9

Multiphase Phenomena

9.1 SCOPE

Multiphase flows are important in numerous industrial and environmental situations. The chemical process industries, aerospace propulsion systems, and sediment transport by rivers and ocean currents generally involve dispersed solid particulate and bubbly flows in liquid and gaseous carrier streams. An equally huge variation of scale is associated with these phenomena. From the measurement of single particle fall velocity measurements in graduated cylinders, to the transport of tons of sediment into the Gulf of Mexico by the Mississippi River each year, these multiphase flows demand our attention. Since the current computational power of today's computers allows investigators to provide high-quality approximate solutions to the transient, three-dimensional conservation equations, much progress has been made in predicting and understanding many of these complex transport processes. Perhaps, the most astounding feature of these analyses is that the conservation equation solutions apply to laboratory experiments on a scale of a few millimeters to environmental flows, which extend for hundreds of miles.

Particulate pollutants carried by the atmosphere and sediment carried by rivers were among the first multiphase flows to receive serious study. Such flows were rather dilute suspensions and were modeled as such. The analysis of dilute particulate transport was given a major boost by NASA and the Air Force's interests in the flow of aluminum oxide particles in solid rocket motors and their plumes. Since the aerospace industry pioneered the use of computers for solving transport phenomena problems because laboratory experiments could not duplicate space conditions, more reliance was forced on numerical simulation. The process industries utilized laboratory experiments, which simplified the flow phenomena as much as possible to create phase interfaces for which the multitude of chemical systems could be studied. Interphase transport was measured primarily in falling film devices for which the fluid velocity fields were assumed to be known. Practical processes were analyzed by empirical extrapolation of such test data. Later, fluidized bed technology was utilized for reactor design, which forced the process industries into exploring the dense suspension flows involved in such devices. Environmental scientists had the full-scale transport processes at hand and consequently spent their efforts on measuring and collecting experimental data. Once again such data were interpreted by empirical correlations and dimensional analysis. Numerical simulations of the large-scale flows involved had to await the development of more computer power and of innovative averaging techniques.

These brief introductory remarks do not do justice to the major efforts that have been undertaken to study multiphase flows. Since the intent of this work is to introduce the application of computational technology to the analysis of transport processes, several examples will be given to suggest how multiphase transport may be simulated. Be warned that the mechanics of particulate and bubble transport has received far more attention than the equally important thermodynamic and chemical aspects of interfacial phenomena.

The examples included in this overview are as follows:

- Dilute suspensions of solid particulates and bubbles in liquids. The effect of dispersed phase properties and several comprehensive analyses will be discussed.
- The capabilities of the CTP code for analyzing multiphase transport will be explained. These capabilities include the use of real-fluid chemical and physical properties to treat otherwise extremely complex flows.
- The importance and description of interphase transport. Phase boundaries will be described as boundary conditions separating the phases and also as an integral part of the solution of the conservation equations.

- Dense suspensions typical of fluidized bed operation will be described. The conservation equations which apply to such flows require extensive modification from those previously mentioned in this work. These modified equations will be presented and discussed herein. Specialized codes capable of solving the transport equations for these systems will be illustrated.

9.2 DILUTE SUSPENSIONS

Sediment (Yalin, 1972; Waldrop and Farmer, 1976) and atmospheric particulate pollutants (Zannetti, 1990; Seinfeld and Pandis, 1998) transport has been simulated as flow of dilute suspensions for some time. When metallized propellants were introduced into rocket motors, a lower combustion efficiency was produced than had been previously experienced. Kliegel (1963) attributed this effect to be solid and/or liquid metal oxide particles in the exhaust causing lower volumetric flowrates than expected. By postulating drag and heat transfer coefficients and particle size distributions, the flowfield was calculated and the loss effect was explained. A similar, but less extensive, modeling study was accomplished by Ishii et al. (1989). Over the years, NASA (Smith, 1984) and the Air Force (Simmons, 2000) have refined the analyses of these dilute particulate flows extensively to better predict the performance and plumes of rocket propulsion systems. Various subsonic and supersonic gas dynamic analyses were used to simulate the carrier fluid. The physical effects considered and modeled include particle collisions and agglomeration, solidification, crystallization, fragmentation, nucleation, condensation, vaporization, sublimation, and melting.

Dilute suspensions use an Eulerian analysis for the carrier (continuum) fluid with a Lagrangian description of the suspended particulates. Iteration is performed to couple the two flows. The Eulerian step solves the continuum equations for a fixed grid. The Lagrangian step accounts for the energy and momentum exchange with the dispersed phase. A host of effects other than particle size and drag and heat transfer coefficients have been evaluated but few of these are yet to be included in comprehensive flowfield codes. A recent review of effects such as virtual mass to simulate acceleration forces, interaction of condensed phases with the turbulent field, particle shape, mixture sound speed, etc. is given by Brennen (2005).

The analyses just mentioned for dilute suspensions have tacitly assumed that there is no exchange of mass from the dispersed to the continuum phase. This is usually an excellent assumption for the examples mentioned. However, if condensation or evaporation or chemical reaction occurs, interfacial conditions must be specified more completely.

9.3 INTERPHASE MASS TRANSFER

When interphase mass transfer is not important, modeling the exchange of heat and momentum and the accompanying swelling or shrinkage of the particulates is adequate to describe the process. When mass is exchanged, both physical and chemical processes may occur. Such systems defy generalization. Several specific, important phenomena will be described to illustrate the level of detail required to simulate a multiphase process.

9.3.1 INTERFACIAL EQUILIBRIUM

Consider a multicomponent gas–liquid. If the phases are separated by a simple geometric interface, the following conditions may occur. If the liquid is dispersed as droplets, such conditions may apply locally around each drop.

Let y_i be the mass (or mole) fraction in the gas phase and x_i be the mass (or mole) fraction in the liquid phase. For example, place a gaseous mixture of ammonia and air along with an amount of water in a closed container. Shake the container and let it sit. The pressure and temperature will become constant. The ammonia will be distributed between the gas (air) and liquid (water) phases. The amount of air dissolved in the water, and the amount of water which humidifies the air will be and are assumed negligible. The distribution of the ammonia will constitute an equilibrium split in the two phases. The thermodynamic statement of this equilibrium condition is that the chemical potential (or, equivalently, Gibbs free energy or activity) of the ammonia in the air and in the water will be equal.

Now inject additional ammonia, and repeat the experiment. A new set of equilibrium values will be obtained. The results of these experiments can be plotted as an equilibrium curve, i.e., the y_A vs. x_A values (where A represents ammonia). The volume of the container and heat exchange to or from it is to be adjusted as required to maintain the original temperature and pressure. In general,

1. Equilibrium conditions for other systems can be established in a similar manner.
2. There is no net diffusion (of the solute) for equilibrium conditions, assuming that initially the mixture was shaken sufficiently to make the concentration uniform in the each phase.
3. For a static system not in equilibrium, equilibrium will exist at the interface and laminar and turbulent diffusion will occur until the entire system reaches equilibrium.
4. For a wetted wall column in either cocurrent or countercurrent laminar or turbulent flow in which the gas stream and the liquid stream are in fully developed flow, i.e., there is negligible velocity component normal to the wall; conditions are the same as for the static system at each elevation in the column.

Statements 3 and 4 are not absolutely true, as some interfacial resistance to mass transfer may exist. Such resistance may be estimated with kinetic theory by calculating a maximum transfer rate accounting for molecular strikes on the gas side and an accommodation coefficient for them sticking to the surface. Not considering this effect might introduce errors of the order of a few percent, under normal processing conditions. More serious deviation from this equilibrium condition may be caused by their being significant surface tension effects, having turbulent flow in the system which produces surface ripples, and/or chemical reactions occurring near the interface.

9.3.2 TWO-FILM THEORY

For the wetted wall column just described, there is a resistance in either phase for producing the interfacial compositions from the bulk composition in the phases. Whitman introduced this idea in 1923 (Sherwood et al., 1975) by suggesting

$$N_A = k_y (y_{AG} - y_{Ai}) = k_x (x_{Ai} - x_{AL}) \tag{9.1}$$

where N_A is the mass flux of A for the condition of no average velocity. The fully developed flow assumption in this system implies no average velocity normal to the column wall at each local elevation. The flux indicated by N_A is a combination of convection and diffusion. Frequently, it is used when the transport effect being considered does not change the velocity field (a very severe restriction). The k's are film coefficients which convert the concentration difference driving forces to fluxes. They are obtained from experiments. The subscripts G, i, and L indicate gas, interface, and liquid values, respectively. Hence,

$$\frac{y_{AG} - y_{Ai}}{x_{AL} - x_{Ai}} = -\frac{k_y}{k_x} \tag{9.2}$$

Whitman's work predated application of boundary layer theory to mass transfer, but the concept is still used in the chemical industry. The use of bulk values implies that boundary layer behavior is not being considered. The analyses using such methodology are referred to as film theories. Since compositions at the interface are difficult to measure, they are usually replaced by assuming equilibrium conditions at the interface. The superscript asterisk indicates equilibrium conditions corresponding to bulk values in the gas or liquid phase.

$$N_A = K_y(y_{AG} - y_A^*) = K_x(x_A^* - x_{AL}) \tag{9.3}$$

Combining the two resistances is accomplished by defining

$$m' \equiv \frac{y_{Ai} - y_A^*}{x_{Ai} - x_{AL}} \quad \text{and} \quad m'' \equiv \frac{y_{AG} - y_{Ai}}{x_A^* - x_{Ai}} \tag{9.4}$$

The overall resistances become

$$\frac{1}{K_y} = \frac{1}{k_y} + \frac{m'}{k_x} \tag{9.5}$$

or based on the liquid side,

$$\frac{1}{K_x} = \frac{1}{m''k_y} + \frac{1}{k_x} \tag{9.6}$$

These relationships are used extensively to analyze mass transfer process, as described in the texts Treybal (1968) and Sherwood et al. (1975).

The film concept for predicting mass transfer rates is not directly useful in computational analyses of these processes, but the physics and thermodynamics involved indicate how boundary conditions and internal jump conditions may be introduced into computational analysis.

9.3.3 SIMULTANEOUS HEAT AND MASS TRANSFER

To relate humidity to wet- and dry-bulb temperature measurements, the following analysis is made. The wick around the wet bulb will exchange heat and mass with the ambient air until a steady-state equilibrium is reached, whereas the dry bulb simply measures the air temperature. Let α designate the diffusing species water vapor and utilize the definitions of average velocities and fluxes for a two-component air–water system:

$$c_\alpha \vec{V} + \vec{J}_\alpha = \vec{N}_\alpha \tag{9.7}$$

Assume the diffusion and flow are one-dimensional, the total molar concentration is constant for this ideal gas mixture of known temperature and pressure ($c = P/RT$).

$$c_\alpha V = N_\alpha - J_\alpha = c_\alpha \sum_\beta y_\beta W_\beta = y_\alpha \sum_\beta c_\beta W_\beta = y_\alpha (N_\alpha + N_\beta) \tag{9.8}$$

Assume N_β is zero for air not diffusing into the water contained in the wick. This means that the water vapor is diffusing from the wet surface through a stagnant film. Upon replacing the diffusion flux with a coefficient and concentration gradient,

$$J_\alpha = (1 - y_\alpha) N_\alpha = -cD_{\alpha\beta} \frac{\partial y_\alpha}{\partial z} \tag{9.9}$$

Integrating over a thickness δ

$$N_\alpha = \frac{D_{\alpha\beta} P}{RT\delta} \ln\left(\frac{1 - y_{\alpha 2}}{1 - y_{\alpha 1}}\right) \tag{9.10}$$

Also,

$$\left(\frac{1 - y_\alpha}{1 - y_{\alpha 1}}\right) = \left(\frac{1 - y_{\alpha 2}}{1 - y_{\alpha 1}}\right)^{y/\delta} \tag{9.11}$$

The concentrations correspond to the partial pressure of water at the wick surface (location 1) and zero at the film edge δ where the air stream contains only ambient moisture.

$$N_\alpha = \frac{D_{\alpha\beta} P}{RT\delta} \ln\left(\frac{1 - (p_{\alpha 2}/P)}{1 - (p_{\alpha 1}/P)}\right) \tag{9.12}$$

An analogous equation can be written for energy transfer.

$$q_T = \frac{N_\alpha C_\alpha}{1 - \exp\left\{-(N_\alpha C_\alpha / \hbar)\right\}} (T_2 - T_1) - \lambda_\alpha N_\alpha = q_{sen} + q_{vap} \tag{9.13}$$

For the boundary conditions that $T = T_1$ at the interface and $T = T_2$ at the edge of the film. Since no heat is added to the wet wick, q_T is zero. The film coefficient for heat transfer is $\hbar$; C_α and λ_α are the heat capacity of water vapor and heat of vaporization of water, respectively.

The problem of determining the simultaneous transfer of mass and energy has been solved. Unfortunately, the film thickness is not known and the use of the thermal conductivity κ, film coefficient $\hbar$, and diffusion coefficient $D_{\alpha\beta}$ imply that the results apply only to laminar transport. The analysis shown is crude because the species conservation equation is not directly solved, but if it were for the same conditions the answer would be the same (see Bird et al., 2002). So much for trying to solve the conservation equations. Simply resorting to empiricism, let the mass and heat flux of α diffusing into stagnant β be represented by a film coefficient times a driving force:

$$N_\alpha \approx +k_G \left(p_{\alpha 1} - p_{\alpha 2}\right) \tag{9.14}$$

$$q_{sen} = \frac{N_\alpha C_\alpha}{1 - \exp\left\{-(N_\alpha C_\alpha / \hbar)\right\}} (T_2 - T_1) \approx h_G (T_G - T_1) \tag{9.15}$$

$$q_T \approx h_G \left(T_G - T_1\right) + \lambda_\alpha k_G \left(p_{\alpha 2} - p_{\alpha 1}\right) = 0 \tag{9.16}$$

$$\left(T_G - T_1\right) = \lambda_\alpha k_G \left(p_{\alpha 1} - p_{\alpha 2}\right) / h_G \tag{9.17}$$

The film coefficients are related to diffusion coefficients for mass and heat transfer, respectively:

$$k_y = cD_{\alpha\beta} / \delta \quad \hbar = \kappa / \delta_T \tag{9.18}$$

The film thicknesses must be known to use these equations. The k_G and h_G are film coefficients for a low rate of mass and heat transfer and are evaluated from experimental data. They are defined such that the film thickness does not have to be defined. Other nomenclatures for these coefficients are for gases

$$N_\alpha = k_c \left(c_{\alpha 1} - c_{\alpha 2} \right) = k_G \left(p_{\alpha 1} - p_{\alpha 2} \right) = k_y \left(y_{\alpha 1} - y_{\alpha 2} \right) \qquad (9.19)$$

and for liquids:

$$N_\alpha = k_c \left(c_{\alpha 1} - c_{\alpha 2} \right) = k_L \left(c_{\alpha 1} - c_{\alpha 2} \right) = k_x \left(x_{\alpha 1} - x_{\alpha 2} \right) \qquad (9.20)$$

Using these approximations, and since both heat and mass transfer coefficients are included in this heat flux equation, one coefficient can be eliminated by the analogy between heat and mass transfer. If the configuration of the wick is cylindrical,

$$\frac{h_G}{k_G} = C_s \left(\frac{Sc}{Pr} \right)^{0.56} = C_s \left(Le \right)^{0.56} \qquad (9.21)$$

C_s is the humid heat required to raise one pound of water and whatever water vapor it contains to 1°F. This correlation gives $C_s = 0.24$ for air–water system compared to measured values of 0.227. For the air–water system, the constant C_s is essentially the same as that for the adiabatic saturation temperature. This is purely fortuitous because the ratio Sc/Pr (i.e., the Lewis number) is approximately one. In this example, the Lewis number is evaluated for the major species, which is dry air. These approximate mass and heat transfer correlations are very satisfactory to explain the operation of the wet- and dry-bulb temperature determination of humidity.

Correlations for the k_G and h_G coefficients abound in the literature for various binary fluid mixtures and geometrically simple configurations (Treybal, 1968; Skelland, 1974; Sherwood et al., 1975; White, 1988; Bird et al., 2002; Geankoplis, 2003). For the wet- and dry-bulb system, the thermometer bulb may be cylindrical with cross-flow or spherical. Frequently, the correlations are given with Colburn's j-factors.

$$j_D = \left(k_c / V \right) Sc^{2/3} = fn\{Re\}; \quad j_H = \left(h / \rho C_p V \right) Pr^{2/3} = gn\{Re\} \qquad (9.22)$$

It was stated that these approximations are valid if the mass flux of the diffusing species is small. Less obvious is the fact that the bulk velocity has been neglected. Consider what the definition of N_α being constant means. For one-dimensional flow, the stream moves inviscidly through an imaginary, constant diameter tube. If the flux of a species is constant and the species concentration decreases along the tube, how is the mass balance satisfied? This can be only approximately true when the concentration of the diffusing species is very small. But the diffusion velocity is also very small, how can one be sure that the diffusive velocity exceeds the convective velocity sufficiently to make a meaningful analysis?

Nevertheless, extensive use has been made of film theory and mass flux methodology. The literature also refers to mass transfer coefficients for "high mass flux" conditions. These coefficients are the same as for low mass flux conditions after they have had the driving force modified by accounting for the $y_\alpha N_\alpha$ term. This term represents the bulk velocity for diffusion through a stagnant film (or other diffusion flow analyses). The use of film theory eliminates the need for defining a film thickness and linearizes the driving force term. Do both of these factors need to be modified to treat high mass flux conditions? Consider first the logarithmic form of the driving force.

Defining log–mean values by

$$LM\{w_2, w_1\} = \frac{w_2 - w_1}{\ln(w_2/w_1)} = p\beta M \tag{9.23}$$

For example,

$$N_\alpha = \frac{D_{\alpha\beta}P}{RT\delta}\left(\ln\frac{P - p_{\alpha2}}{P - p_{\alpha1}}\right) = \frac{D_{\alpha\beta}P}{RT\delta}\left(\frac{p_{\alpha1} - p_{\alpha2}}{LM\{p_{\beta2}, p_{\beta1}\}}\right) \tag{9.24}$$

where $P = p_{\alpha1} + p_{\beta1} = p_{\alpha2} + p_{\beta2}$. The manipulations to produce this equation are as follows:

$$\left(\frac{p_{\beta2} - p_{\beta1}}{\ln\{p_{\beta2}/p_{\beta1}\}}\right) = \frac{P - p_{\alpha2} - P + p_{\alpha1}}{\ln\left[\frac{P - p_{\alpha2}}{P - p_{\alpha1}}\right]} = \frac{+p_{\alpha1} - p_{\alpha2}}{\ln\left[\frac{P - p_{\alpha2}}{P - p_{\alpha1}}\right]} \quad \text{or}$$

$$\ln\left[\frac{P - p_{\alpha2}}{P - p_{\alpha1}}\right] = \frac{p_{\alpha1} - p_{\alpha2}}{\left(\frac{p_{\beta2} - p_{\beta1}}{\ln\{p_{\beta2}/p_{\beta1}\}}\right)} = \frac{p_{\alpha1} - p_{\alpha2}}{LM\{p_{\beta2}/p_{\beta1}\}} = \frac{p_{\alpha1} - p_{\alpha2}}{p\beta M} \tag{9.25}$$

It does not appear that much has been accomplished by the introduction of the log–mean term, but it preserves the appearance of the linear term as the driving force.

$$N_\alpha = +k_G^\bullet \left(p_{\alpha 1} - p_{\alpha 2} \right) = \left[\frac{+k_G}{p \beta M} \right] \left(p_{\alpha 1} - p_{\alpha 2} \right) \tag{9.26}$$

The dot exponent on the film coefficient indicates that it is for the high mass flux conditions. The literature suggests that the film coefficient has been changed to give high mass flux conditions. As is evident by this example, it is the driving force that has been redefined, not the film coefficient. The film coefficients are obtained from the low mass flux conditions and then modified with the log–mean term. Such corrections might look different depending on the concentration units used to describe the driving force, but the method of making the correction remains the same. This also removes the confusion resulting from finding that the high flux film coefficient is smaller than the low flux coefficient (unless the fluid properties vary significantly).

9.3.4 TURBULENT FILM COEFFICIENTS FOR MASS TRANSFER

Now it is argued that the ratio k_G/h_G can be evaluated for either laminar or turbulent flow by using j-factors. The log–mean correction factors must be used for this interpretation to make any sense. Otherwise, one is assuming the flow to be turbulent but with a zero bulk velocity. With this proviso, turbulent mass transfer is addressed as follows.

In Chapter 1, simple examples were given to relate momentum, heat, and mass transfer. The examples were simple with respect to geometry; fully developed pipe flow was described. It was stated that such flow was like other near wall flows such that the variations were local and normal to the wall. In this respect they became similar to the mass flux analyses presented in the previous section. Chapter 1 examples for heat and mass transfer were also simple in that they were assumed to be of such small values that the velocity profiles were not changed by their presence. Such a prediction is only a slight improvement over the mass flux analyses which neglect the bulk flow normal to a wall altogether. The next level of improvement would be to account for the influence of momentum, heat, and mass transport on the near wall velocity profiles. To make such analyses, the impact on the entire flowfield should be determined. Such accounting must be done with a fully coupled CTP solution. However, the empiricism developed from test data will be reviewed so that a more rigorous set of boundary conditions can be determined for use in numerical solutions.

For near wall flows, a fluid which is the same composition as the free stream can be blown into or sucked from the boundary layer. Practically, such flows can be used to control friction or heating on a surface. Studies of these flows were reviewed by White (2006). The effect of blowing and suction on turbulent convective heat transfer boundary layers has been frequently measured and is well correlated by the following equation (Schetz, 1993). The correlation is for a single species, i.e., the injected species is the same as that in the free stream.

$$\frac{St}{St_0} = \frac{C_f}{C_{f0}} = \left[\frac{\ln(1+B_{f,h})}{B_{f,h}}\right]^{5/4} (1+B_{f,h})^{1/4} \tag{9.27}$$

where $B_f = -(\rho_w \upsilon_w / \rho_c U_c)/(C_f/C_{f0})$ $\quad B_h = (\rho_w \upsilon_w / \rho_c U_c)/St$

The Stanton number is $St = \hbar/\rho U_c C_p$. The subscript c represents centerline or free stream value. The subscript 0 represents the value without injection or suction at the wall. The two flows involved are to be compared at the same Reynolds number based on momentum thickness, Re_θ. The Stanton number for heat transfer is related to that for mass transfer (denoted by the subscript m) by the Chilton-Colburn analogy:

$$St = \frac{\hbar}{\rho U_c C_p} = \frac{C_f}{2} Pr^{-2/3} \tag{9.28}$$

$$St_m = \frac{k_y}{U_c} = \frac{C_f}{2} Sc^{-2/3} \tag{9.29}$$

The discussion has been relative to near wall flows. To calculate numerical values for these correlations, one must utilize data for the specific geometries of interest from standard texts, for example, those cited at the beginning of the discussion on wet- and dry-bulb temperature measurements for the determination of humidity. These correlations give a good approximate indication of the interaction of mass, heat, and momentum transfer at an interface or wall. More accurate analyses would involve detailed predictions within the flowfield using these approximations to construct boundary conditions.

To gain more insight into turbulent mass transfer, the analogy between the transport processes needs to be reconsidered. Consider the transport processes to be described by

$$N_{\alpha y} = c_\alpha V_y - J_{\alpha y} = c_\alpha (N_\alpha + N_\beta) + (D_{\alpha m} + E_\alpha) \frac{\partial c_\alpha}{\partial y} \tag{9.30}$$

$$q_y = V_y \, \rho C_p T - \frac{(\kappa + E_h)}{\rho C_p} \frac{\partial (\rho C_p T)}{\partial y} \tag{9.31}$$

$$\tau_{yx} = (\nu + E_\nu) \frac{\partial \rho V_x}{\partial y} \tag{9.32}$$

These equations are written for constant density, steady-state, one-dimensional flow in the x-direction, and one-dimensional heat and mass transfer in the y-direction. The conserved quantities are assumed time averaged. The turbulent transport properties are described only in terms of the eddy diffusion, conductivity, and viscosity. The species continuity equation is written for α diffusing into a mixture m. These equations are the starting point for most mass transfer analyses (e.g., Sherwood et al., 1975). Usually, the statements are made that the analogy between the processes is not exact because the shear stress is a tensor. This restriction is not precise. Usually, the velocity parallel to a surface or interface is much larger than the velocity normal to the surface. Hence, the momentum transport is definitely a two-dimensional process (at least). Worse still, the bulk velocity term (V_z) is frequently omitted in the mass transport equation. As previously mentioned, this is the reason that the mass transport film coefficients are limited to small transfer rates when the driving forces are approximated as linear differences of concentration gradients. These things being said, we will make the conventional sin and proceed assuming that all of these restrictions apply. Most specifically, the bulk velocity in the species equation will be assumed zero.

To investigate the description of mass transfer with these restricted transport equations, the fully developed pipe flow example from Chapter 1 will be revisited. Using inner law dimensionless variables, utilizing the linear variation of shear-stress across the pipe, and the mass average velocities,

$$u^+ = \frac{U}{U_{av}} \sqrt{\frac{2}{f}} \quad y^+ = \frac{(R-r)U_{av}}{\nu} \sqrt{\frac{f}{2}} \tag{9.33}$$

$$\tau = \frac{r}{R} \tau_w \quad \frac{E_\nu}{\nu} = \frac{r}{R} \frac{dy^+}{du^+} - 1 \tag{9.34}$$

Assume that the laminar transport is important only very near the wall, that in this near wall region N_α is constant and that $r = R$. It follows that

$$u^+ = \int_0^{y_1^+} \frac{dy^+}{(E_v/v)+1} + \int_{y_1^+}^{y^+} \frac{dy^+}{(E_v/v)} \qquad (9.35)$$

Assume that the flux equation omits the bulk velocity term, that the concentration of α and the velocity equal free stream values at y_1^+, and that $E_D = E_v$ at y_1^+ it follows that

$$\frac{(c_w - c_{av})U_{av}}{N_\alpha} = \frac{2}{f} + \sqrt{\frac{2}{f}} \int_0^{y_1^+} \left(\frac{v}{E_D + D} - \frac{v}{E_D + 1} \right) dy^+ = \frac{1}{St_m} = \frac{U_{av}}{k_c} \qquad (9.36)$$

Now if E_D is specified as a function of y^+, the function *gn* can be determined:

$$\frac{1}{St_m} = \frac{2}{f} + \sqrt{\frac{f}{2}} gn\{Sc\} \qquad (9.37)$$

This equation indicates that if two resistances are present, the first term is a resistance outside of the wall layer and second is effective inside the wall layer where both turbulent and laminar effects are present. If the *Sc* is unity, the second resistance is zero and the Reynolds analogy results, which is,

$$St_m = \frac{k_c}{U_{av}} = \frac{f}{2} \qquad St = \frac{\hbar}{\rho U_{av} C_p} = \frac{f}{2} \qquad (9.38)$$

Several attempts have been made to use a different "universal" velocity function very near the wall, resulting in various degrees of success. Also, different models for representing mass diffusion very near the wall have been used. These do not use the analogy in this region. Several such functions use E_D proportional to y^{+3} in this region. Skelland (1974) reviewed several of these modified velocity profile approaches and found that some of the better ones predicted the mass transfer rates rather well, but produced kinks around the point where the wall layer was fit to the inner law region. Skelland reviewed Gowariker and Garner's velocity profile model which eliminated the kink and gave comparable predictions to the other models.

$$y^+ = \left(\frac{F_1\{Re\}F_2\{Re\}-1}{F_1\{Re\}-1}\right)u^+ + 3\times10^{-7}\left(u^+\right)^{F_1\{Re\}}$$

$$F_1\{Re\} = 7 - \exp\{-2.3\times10^{-8}\left(Re\right)^2\} \qquad (9.39)$$

$$F_2\{Re\} = 1 + \exp\{-3.83\left(Re\right)^{0.05}\}$$

Such models have been criticized as too complex, but the ready availability of PCs has rendered such criticism mote.

A purely empirical attempt to predict better eddy mass transport near the wall is

$$E_D/\nu = 1.77\left(f/2\right)^{3/2}\left(y^+\right)^3 \qquad (9.40)$$

which yields the Chilton-Colburn analogy. Regardless of all of these attempts and numerous publications, the film coefficients for small rates of mass and heat transfer are still usually predicted using this 1934 analogy.

These results indicate that the velocity, Sc number, and Pr number profiles discussed in Chapter 1 apply to conditions where the mass and heat transfer rates are small and may be used to evaluate k_c and h_G film coefficients. Corrections for turbulent high mass and heat flux conditions are yet to be described.

To describe high mass transfer from an interface into a turbulent flow, Equation 9.26 can be used if the film coefficient is evaluated from a turbulent flow Reynolds number correlation or if the transfer from the interface is by diffusion only. The log–mean correction term in this equation accounts for the bulk flow in the direction of diffusion.

A further analysis developed by Mickley et al. (1954) and reviewed by Sherwood et al. (1975) was described as an air stream passing through the porous wall of a pipe into an air stream. The interior of the porous surface was to be maintained wet with water. The water was assumed to vaporize and be carried into the main air stream by diffusion. Such conditions can almost be duplicated experimentally without physically blowing water off of the surface. The analysis should be very appropriate for describing the injected flow to be air with some other gas or vapor component. The primary air stream would then have this additional species flow and diffuse into it. The injected mass flux (N_T) would still be a total value representing bulk velocity and set as a boundary condition. The mass balance describing this flow is

$$N_\alpha = N_T y_\alpha + J_\alpha \tag{9.41}$$

and the species continuity equation is

$$\frac{d^2 y_\alpha}{dy^2} - \frac{N_T RT}{D_{\alpha\beta} P} \frac{dy_\alpha}{dy} = 0 \tag{9.42}$$

boundary conditions: $y_\alpha = y_{\alpha 1}$ at $y = 0$ and $y_\alpha = y_{\alpha 2}$ at $y = \delta$

The solution is

$$J_\alpha = -\frac{D_{\alpha\beta} P}{RT}\left(\frac{dy_\alpha}{dy}\right)_0 = \frac{N_T(y_{\alpha 1} - y_{\alpha 2})}{\left[\exp\{N_T RT \delta/D_{\alpha\beta} P\}\right] - 1} \tag{9.43}$$

The driving force is linear in mole fraction, therefore no log–mean correction is needed. The result is a determination of the diffusion flux, but it is still in terms of an undefined film thickness, δ. The term involving the fictitious film thickness can be evaluated in terms of the low mass transfer rate film coefficient (k_y) to completely specify J_α, which is then used to determine N_T.

$$J_\alpha = k_y^\bullet \left(y_{\alpha 1} - y_{\alpha 2}\right) = \frac{N_T(y_{\alpha 1} - y_{\alpha 2})}{\left[\exp\{N_T/k_y\}\right] - 1} \tag{9.44}$$

The simplification in defining the high mass transfer film coefficient arises because the total mass flux (or in other words bulk velocity) is set as a boundary condition.

This brief discussion of film coefficients to describe mass transfer provides the background to help specify boundary conditions for a CTP solution. The emphasis herein has been on film coefficients and one-dimensional transport. Film coefficients for other geometries and two-way diffusion (usually equal molal counter diffusion) can readily be found in mass transfer texts. The discussion herein also serves as a primer to prepare one for reading such texts. This is useful because so many detailed summaries of test data and perturbations of the chemical systems involved the obscure basic transport mechanisms involved.

For a computational analysis of simultaneous heat and mass transfer, the one-dimensional analyses are only used to provide information for designing boundary conditions at the interfaces. Temperature and species distributions in the field are calculated with the computational code used. An even simpler situation exists when surface reactions occur. The energy effect matching the mass transfer of reaction products into the flowfield is specified as a boundary condition. This simple calculation is often clouded by the fact that the surface reaction rate and participating species must be determined by experiment. When such data are not available, only poor estimates must be used. Sometimes availability of good validation test data for other flowfield properties mitigates this problem.

9.4 MULTIPHASE EFFECTS INCLUDED IN THE CTP CODE

The CTP code was intended primarily for a single phase multicomponent flow simulator. However, it was found to be capable of simulating several multiphase phenomena with little modification and/or with inventive use of its single-phase capability.

9.4.1 DILUTE PARTICULATE CLOUD TRACKING

The CTP code solves the particle equations of motion and the particle energy equation on the Lagrangian framework. Each particle group trajectory is tracked by integrating the particle equations of motion. The interphase drag forces and heat transfer fluxes are stored and included in the gas phase governing equations. Only steady-state flows can be simulated. The particle momentum and energy equations are written as (Al_2O_3 particle properties are used in the present model)

$$\frac{DV_i}{Dt} = \frac{(U_i - V_i)}{t_d} \quad \text{and} \quad \frac{Dh_p}{Dt} = C_p \frac{(T_{aw} - T_p)}{t_H} - \frac{6\sigma \varepsilon f T^4}{(\rho d)_p}$$

$$t_d = \frac{4\rho_p d_p}{3 C_d \rho_g |U_i - V_i|} \quad t_H = \frac{\rho_p d_p P_r}{12 N u \mu}$$

(9.45)

9.4.2 CONJUGATE HEAT TRANSFER

Analysis of convective heat transfer to pipe flows frequently uses constant wall temperature of constant wall heat flux as boundary conditions. What if neither is appropriate for the problem at hand? Often a more realistic solution can be obtained if the wall boundary condition is moved outside of the pipe where insulation could make the outside condition adiabatic or one of natural convection to the environment. This type flow can be analyzed using the CTP code by considering grid points within the wall of the pipe as nonflow points. The code would still solve the energy equation at such points; hence, the flow along the inside pipe wall would become like an interface across which heat could flow. Thus, the inside wall temperature be predicted as part of the simulation.

9.4.3 REACTING WALL BOUNDARY CONDITIONS

Ablating surfaces utilize the endothermic reactions of pyrolizing polymers as a heat sink to protect structures. Hybrid rocket motors flow oxidizer over polymeric fuel surfaces to provide the working fluid for propulsion. In the hybrid motor, the fuel might react at the surface or might pyrolize and react in the gas phase. Which case actually occurs is difficult to ascertain, but either might be modeled by the assumption that one of the processes occurs and set the reaction rates to match overall process measurements. The CTP code simulates such processes by treating the products from the receding surface as flow through a fictitious inlet to the flowfield. The calculation is rigorous, but its accuracy depends on correctly modeling the surface reactions. Analysis of hybrid three-dimensional grain geometries has been successfully simulated using this procedure. The surface reactions were empirically determined by calibration with subscale, geometrically simple test motor data (Cheng et al., 1998).

9.4.4 REAL-FLUID PROPERTY FOR REACTING SPRAY SIMULATIONS

Spray combustion in liquid rocket engines has proven to be very difficult to measure or to simulate. Many injector elements are utilized to ensure efficient mixing of the propellant streams. The resulting high temperatures and pressures of the combustion process present too hostile an environment to interrogate with probes. Studying a single injector element has proven to be likewise difficult to measure. Laser Doppler methods are valuable for dilute flows, but near the element exit, the liquid stream is too dense for the laser beam to penetrate. Even if measurements downstream were available, no information is provided on how the initial stream atomizes and spreads. Such information is essential if the injector element characteristics are to be coupled to the efficiency of the spray formation. Much effort has been devoted to two-phase flow analysis of such spray combustion processes. Although the mechanics of the two-phase, non-reacting portion of the flowfield have been predicted, these simulations cannot be matched to the resulting initial spread and droplet formation processes.

The CTP code was utilized to address this problem by a different route. The real-fluid properties in the CTP code allow vapors and liquids to be treated simultaneously by simply accounting for the mixture density of the flow. The mixing and reactions accompanying the cocurrent flow of liquid oxygen surrounding by a gaseous (or liquid) hydrogen stream could be simulated with the CTP code. The combustion would be simulated as an equilibrium or finite-rate reaction process. The mixing would be described with a two-equation turbulence model. Such a simulation was conducted (Cheng and Farmer, 2006). It was anticipated that a modification to the turbulence model could be made to fit available test data. Surprisingly enough, no modification was required for the model to fit available data rather well. Such a model is not only fortuitous for subcritical combustion but also should be rigorously correct for supercritical combustion simulation. Basically, the fluid mechanics of the spray formation process were overwhelmed by the highly exothermic reactions of the propellant combustion. The conclusion to be drawn from the success of this modeling methodology is that all of the process must be considered, one cannot always simply solve a succession of simple processes to get to the final result desired. Some of the very complex intermediate processes might prove to be unnecessary, if the more important steps are addressed first (not last) in developing the simulation.

9.5 POPULATION BALANCE MODELS

The population balance is a methodology adapted from statistical mechanics which is most useful for describing particulate systems. Hulbert and Katz (1964) and Randolph (1964) independently reported the development of a Liouville type equation from statistical mechanics to describe the number density of particles in phase space. The particles in statistical mechanics are hard spheres or molecules. The particulate processes likewise modeled become aerosols, comminution products, crystallizers, granulators, flocculation systems, combustion, polymerization, biomedical cell states, and even shrimp distributions in bays. The distribution of the particulates and the distribution of their properties are the results of applying this methodology to a system of interest. Texts which describe this modeling methodology are Randolph and Larson (1971), Himmelblau and Bischoff (1968), and Ramkrishna (2000).

The population balance takes the fluid continuum conditions of temperature, velocity, and composition and the dependence of particle nucleation and growth on its local environment as known. Thus, the continuum flowfield and algebraic rate equations must be solved concurrently with the population balance. The population balance is formulated as a phase space with external geometric coordinates (x_i) to locate the particle and internal coordinates to describe its interaction with the environment (r_j). The x_i's may be taken as the Cartesian coordinates (x, y, z). The r_j's are a characteristic of the particles; for the present say a characteristic length, like an effective radius. The total phase coordinates are $\xi = \{x, r\}$, which describe the state of the fluid with first-order ordinary differential equations (ODEs) as $d\xi_i/dt = u_i\{\xi, t\}$. The particles flow with their local velocity. The change of any particle property (taken here to be particle size) with time is a "velocity" like quantity. Here the change is taken to be a growth rate (G), which depends on the local concentration (c) and temperature (θ), these ODEs may be represented by

$$\frac{dx_i}{dt} = V_i\{x, t\} \quad \frac{dr_j}{dt} = G_j\{c(x, t), \theta(x, t), r\} \tag{9.46}$$

The particle velocity (V) is related to the local fluid velocity with a drag law. The determination of the x_i's locates the position of the particles. It is assumed that there are a sufficient number of particles to form a continuum. For convenience, a vector particle phase–space velocity is defined by

$$\vec{W} = u\vec{\delta}_x + v\vec{\delta}_y + w\vec{\delta}_z + G_1\vec{\delta}_1 + G_2\vec{\delta}_2 + \cdots + G_m\vec{\delta}_m \quad \text{or}$$
$$\vec{W} = \vec{V} + \vec{G} \tag{9.47}$$

The deltas are unit vectors all of which are orthogonal to one another. Herein, such equations are termed mathematical vectors. They are best considered as relationships between arrays of terms. See Appendix B for further details.

Introduce a distribution function for the number density of particles, $f\{x, r, t\}$. A particle population balance for a control volume moving in phase–space, i.e., a Lagrangian viewpoint is

$$\frac{d}{dt}\int_{R_1} f \, dR = \int_{R_1} (B - D)\, dR = \int_{R_1}\left[\frac{\partial f}{\partial t} + \nabla \cdot f\left(\frac{d\vec{\xi}}{dt}\right)\right] dR \qquad (9.48)$$

$B - D$ is the birthrate minus the death rate of particles; a net generation type term.

$$\int_{R_1}\left[\frac{\partial f}{\partial t} + \nabla \cdot f\vec{V} + \nabla \cdot f\vec{G} + D - B\right] dR = 0 \qquad (9.49)$$

Since the integration region is arbitrary it follows that

$$\left[\frac{\partial f}{\partial t} + \nabla \cdot f\vec{V} + \nabla \cdot f\vec{G} + D - B\right] = 0 \qquad (9.50)$$

Alternatively, a function (h) defining the net number of particles introduced into the system per unit time and per unit phase–space can be defined to represent $B - D$. A differential particle balance can be written directly:

$$\frac{\partial f}{\partial t} + \sum_{i=1}^{3}\frac{\partial fV_i\{x,t\}}{\partial x_i} + \sum_{j}\frac{\partial\left(fG_j\left[c\{x,t\},\theta\{x,t\},r\right]\right)}{\partial r_j} = h\{x,r,t\} \qquad (9.51)$$

This is the partial differential equation which represents the population balance. It will be termed the *micro-distributed population balance*. Notice that this equation contains no unit vectors. The independent variables in the population balance are one time variable, up to three geometry variables, and j internal variables. This equation along with equations to define G and h and the conservation equations for representing the continuum phase must be solved to represent the particulate flow. Appropriate initial and boundary conditions are also needed. Even though only one internal variable (usually a length scale) is frequently used to describe the particulates, the number of independent variables is different from that for the continuum phase conservation equations. The variation in dimensionality makes this system of equations much more difficult to solve. Although this is the most general form of the population balance, various approximations to simplify obtaining a solution are frequently made.

The general population balance resembles the basic equation of statistical mechanics, but there are significant differences. In statistical mechanics applications:

- No distinction is made between the spatial and internal coordinates, hence only ξ is used to indicate the coordinates.

$$\frac{\partial f}{\partial t} + \sum_i \frac{\partial \left(u_i\{\xi,t\}f\right)}{\partial \xi_i} = h\{\xi,t\} \tag{9.52}$$

- The particles are considered to be molecules which are neither created nor destroyed, so h is zero. The divergence of the generalized velocity field is assumed to be zero; this is by analogy to the incompressible flow equations. These conditions result in the statistical mechanics equations becoming

$$\frac{\partial f}{\partial t} + \sum_i u_i\{\xi,t\}\frac{\partial f}{\partial \xi_i} = 0 \tag{9.53}$$

- This equations implies that a cluster of particles that move in phase space always occupy the same volume.

Returning to determining a solution to the micro-population balance, for a very simple system, let $j = 1$ and G and h be defined as simple functions. Examples of such simplification will be presented subsequently.

If one is not interested in the spatial distribution of particulates within a process vessel, i.e., reactor, crystallizer, etc., an average of external coordinates over the vessel volume, $\forall$, with inlet and outlet flow rates, Q_k, the macroscopic population balance is

$$\frac{\partial f}{\partial t} + \nabla \cdot f\vec{G}_i + f\frac{d\{\ln\forall\}}{dt} = B - D - \sum_k \frac{Q_k f_k}{\forall} \tag{9.54}$$

For a steady-state process, this form of the balance reduces to an ODE. No information on the spatial variation within the process vessel is obtained. The birth and death functions may be very general.

One might be satisfied with an approximate description of the particulate phase. This can be accomplished by describing the particulate distribution in terms of its moments. An infinite number of moments must be defined to describe a function exactly. However, if only the leading moments are used, a practical solution can be obtained. It must be determined how many "leading moments" are needed to give the accuracy desired. To illustrate the procedure, consider a phase space defined by only one internal coordinate, the particulate size (L), and one particle growth rate ($G = G_0\{1 + aL\}$). The jth moment of the distribution is

$$m_j\{\vec{x}, t\} + \int_0^\infty fL^j dL \tag{9.55}$$

The population balance becomes

$$\int_0^\infty L^j \left[\frac{\partial f}{\partial t} + \nabla \cdot f\vec{V} + \frac{\partial\{fG\}}{\partial L} - B + D \right] dL = 0 \tag{9.56}$$

The first two terms are evaluated by reversing the order of integration.

$$\int_0^\infty L^j \left[\frac{\partial f}{\partial t} + \nabla \cdot f\vec{V} \right] dL = \frac{\partial m_j}{\partial t} + \nabla \cdot m_j\vec{V} \tag{9.57}$$

assuming $\vec{V} \neq \vec{V}\{L\}$. The B and D functions are described by k-moments, as

$$\int_0^\infty \left[B\{\vec{x}, L, f, t\} - D\{\vec{x}, L, f, t\} \right] L^j dL = \bar{B}\{\vec{x}, t, m_k\} - \bar{D}\{\vec{x}, t, m_k\} \tag{9.58}$$

where $j = 0, 1, 2, \ldots \geq k$

The third term in the balance equation is integrated by parts to give

$$\int_0^\infty L^j \frac{\partial\{fG\}}{\partial L} dL = -0^j \cdot B^0 - jG_0(m_{j-1} - am_j) \tag{9.59}$$

B^0 is the number flux entering the internal coordinate region at $L = 0$. Finally, the micro-moment population balance becomes

$$\partial m_j/\partial t + \nabla \cdot m_j\vec{V} = 0^j \cdot B^0 + jG_0(m_{j-1} - am_j) + \bar{B} - \bar{D} \tag{9.60}$$

where $j = 0, 1, 2, \ldots \geq k$

The micro-moment population balance has the same dimensionality as the transport equations for the continuum phase, hence it is easier to solve. The external coordinates might also be averaged using moments, but this is not frequently done.

Using the same assumptions on G, B, D, and V, the macro-moment population balance may be obtained from the micro-moment balance:

$$dm_j/dt + m_j \frac{d\ln\forall}{dt} = 0^j \cdot B^0 + jG_0 (m_{j-1} - am_j) + \bar{B} - \bar{D} - \sum_k \frac{Q_k m_{j,k}}{\forall} \qquad (9.61)$$

where $j = 0, 1, 2, \ldots \geq k$

This form reduces the balance equation to be an ODE in time. Such a model is useful for stability and tracer studies.

Population balance models do not have to be formulated in terms of number density. The particle volume formulation is very useful when mass balances are to be determined. For dN particles in a particle volume range of $\upsilon + d\upsilon$, $dN = f\{\upsilon\}d\upsilon$. $[B\{\upsilon\} - D\{\upsilon\}]d\upsilon$ is the net creation of particles of size υ, $(G_\upsilon f)$ is the convection in υ direction, $G_\upsilon = d\upsilon/dt$ is the rate of change of particle volume. $\Sigma Q_k f_k d\upsilon$ is the inlet and outlet flow of specific volume $d\upsilon$. The macro-population balance becomes

$$\frac{\partial(f\forall)}{\partial t} + \frac{\partial\left[f\vec{G}_\upsilon \forall \right]}{\partial \upsilon} = [B - D]\forall - \sum_k Q_k f_k \qquad (9.62)$$

Verkoeijen, Pouw, Meesters, and Scarlett (2002) recommended the volume formulation because (1) when two particles identified by υ_1 and υ_2 agglomerate, their specific volumes are additive. Had the particulates been identified by a linear dimension, such dimensions would not be additive, and (2) instead of a number distribution function, a mass distribution function associated with the particle specific volumes will convert the population balance to a mass conservation equation for disperse flows.

It is argued that the birth and death terms are not rate processes for breakage and agglomeration because these are instantaneous events. However, if a model is used to represent the process over some average period of time, as it usually is, then using rate expressions for the generation and depletion terms is reasonable and proper.

In general, the population balance is difficult to solve because more independent variables are used to describe the system. Not only are temporal and spatial variables used, but internal coordinates are also used. The simplifications mentioned are made to make the solution more tractable. However, the method is far more powerful than this implies.

Recent examples indicate the extent to which such generalizations have been developed. All of these methods use the volume of the particle as its measure.

- Kumar et al. (2008) investigated two solution methods for solving a system of ODEs to represent aggregation and breakage of a distribution of particles. This is described as a population balance equation (PBE) model, but it is limited to a well-mixed process with no spatial velocity component. Only calculated values are compared and discussed.
- Darelius et al. (2006) used the volume-based, matrix representation of a discretized set of particle sizes suggested by Verkoeijen et al. (2002) to analyze the coalescence and compaction occurring in a wet granulation process. The particle properties were the volume of solid and liquid (both in pores and on the surface). Three particle sizes were analyzed simultaneously. Rates for the coalescence and compaction were specified. Eighty particle size cuts were analyzed. The particles were saturated such that no air was in the pores and a layer of liquid surrounded the particles. The resulting system of ODEs was stated to be stiff, but no difficulty was reported in obtaining a solution with Matlab®. The simulation was for well-stirred conditions, such that fluid velocity was not considered. This was a well-conceived and reported study. It is typical of what is reported to be PBE model simulations. The work reported by Verkoeijen et al. was a useful analysis, but no data comparisons accompanied it.
- Silva, Damian, and Lage (2007) implemented an analysis which coupled a CFD solver and a PBE model to represent two-phase flow of a particulate suspension. Two flow solvers—The ANSYS CFX code by ANSYS, Inc. and the Open FOAM C + + code (Weller et al. 1998)—were used to represent the continuum phase. The direct quadrature method of moments (DQMOM) code (Fox, 2003; Marchisio and Fox, 2005) is a quadrature closure approximation for the integrals of the distribution function for internal variables in terms of Dirac delta functions. The CFD and PBE codes were coupled to describe the laminar flow of water in oil suspension over a back step (presumably, a two-dimensional analysis). Break up and coalescence of the water drops were simulated with an analytical model. Extensive numerical simulations were

compared and reasonable results were obtained. Details of the method of coupling the two codes were not discussed in the publication. No tests were available for comparison. This is one of the few attempts to utilize a coupled CFD and PBE analysis to solve the general PBE equation for particulate flow.

The article just referred to and the text of Fox is a good first step in producing this important simulation technology. Such technology should be developed and validated for general purpose use in the near future.

9.6 DENSE PARTICULATE FLOWS

Fluidized bed chemical reactors, sediment transport, boiling heat transfer, crystallizers, pulverized-coal furnaces, and pyroclastic flows involve dense particulate flows. Fluidized bed reactors have received the most investigation by computational analysis. Yet the shear scope of interests and the wealth of the detailed available information collected to understand and predict these phenomena provide impetus for comprehensive synergistic study of the mechanics of dense particulate flows. Lagrangian analyses of dilute particulate flows are adequate if only a modest number of particulate groups must be tracked and coupled to the carrier flow. For dense particulate flows, the particles can interact with each other and form a fluid-like continuum of their own. For these conditions a different modeling strategy must be utilized. This methodology is defined and its application is presented in this section.

9.6.1 LOCAL SPATIAL AVERAGING TO DESCRIBE MULTIPHASE FLOWS

A practical approach to describing the mechanics of dense particulate flow should include no more than a few partial differential equations. For complex thermodynamics and chemical reactions, more equations may be required. Two approaches have been suggested, one by Anderson and Jackson (1967) and Jackson (1997, 1998) which modified the single phase conservation equations to treat particulate clouds. The second used an analogy to the Boltzmann equation by approximating the particulates as molecules (Murray, 1966). These methods are quite general, and unfortunately contain many defined terms which are still lacking in quantitative evaluation. Subsequent review and discussion of these models are available in Drew (1983), Faghri and Zhang (2006), Delhaye (1970), and Ishii (1975). Professor J.S. Curtis from the University of Florida and her colleagues have made extensive contributions to the application of these models by computational analyses.

The approach of Anderson and Jackson is termed a two-phase model. The dense particle cloud is analyzed as a continuum by a local averaging process similar to the Reynolds averaging of the turbulent flow equations. Their original derivation is rather lengthy and obscures the simplicity of the approach. Whitaker (1969, 1973) presents a clearer derivation by considering three length scales: $d \ll \ell \ll L$. The particle size is typically d, the volume over which the variables in the conservation laws are to be averaged is ℓ, and the characteristic size of the process is L. If q represents a variable to be conserved for the α or β phase, where these symbols are used as subscripts to denote which phase is being considered, three averages can be defined. Broken brackets are used to denote average particle quantities.

$$\langle q \rangle = \frac{1}{\forall} \int_{\forall} q \, d\forall \quad \langle q_\alpha \rangle = \frac{1}{\forall} \int_{\forall_\alpha} q_\alpha \, d\forall \quad \langle q_\alpha \rangle^\alpha = \frac{1}{\forall_\alpha} \int_{\forall_\alpha} q_\alpha \, d\forall \quad (9.63)$$

where $\forall_\alpha$ can be function of time. The first average is for both phases over the volume; the second is the α phase averaged over the volume; and the third is the α intrinsic phase averaged over the volume occupied by the α phase. Gray (1975) reexamined Whitaker's work to make a clearer distinction between the convection and diffusion terms. Drew also reviewed this methodology and introduced further terms:

$$\langle q \rangle \{\vec{x}, t\} = \frac{1}{\ell^3} \int_{x_1 - 0.5\ell}^{x_1 + 0.5\ell} \int_{x_2 - 0.5\ell}^{x_2 + 0.5\ell} \int_{x_3 - 0.5\ell}^{x_3 + 0.5\ell} q\{\vec{x}', t\} dx_1' dx_2' \, dx_3' \quad (9.64)$$

and a weighted space average:

$$\langle q \rangle \{\vec{x}, t\} = \iiint_{\ell^3} g\{\vec{x} - \vec{x}'\} q\{\vec{x}', t\} \, d\vec{x}' \quad (9.65)$$

Whitaker also provided the following definitions:

Averages of the velocity property product

$$\langle \vec{U}_\alpha q_\alpha \rangle = \langle \vec{U}_\alpha \rangle \langle q_\alpha \rangle + \langle \tilde{\vec{U}}_\alpha \tilde{q}_\alpha \rangle \quad \text{where} \quad \langle \tilde{\vec{U}}_\alpha \rangle = \langle \tilde{q}_\alpha \rangle = 0 \quad (9.66)$$

relationship of the phase average to the intrinsic phase average:

$$\langle q_\alpha \rangle = \varepsilon_\alpha \langle q_\alpha \rangle^\alpha \tag{9.67}$$

where ε is volume fraction; the average of the derivative is related to the derivative of the average by

$$\langle \nabla q_\alpha \rangle = \nabla \langle q_\alpha \rangle + \frac{1}{\forall} \int_{A_{\alpha\beta}} q_\alpha \vec{n} \, dA \tag{9.68}$$

where $A_{\alpha\beta}$ is the interfacial area between phases α and β; it may be a function of time. Whitaker included some additional definitions for the purpose of applying the results to chemically reacting flow systems.

Anderson and Jackson (1967) and Jackson (1997) utilized these averaged parameters to construct continuity and momentum equations for gas–solid multiphase flows. Ishii (1975) did the same for liquid–droplet gas flows. The conservation laws resulting from these analyses were not closed because constitutive relations for most of the particle–fluid interaction effects were not quantitatively defined. There is still controversy over their values. An example to show what such effects are and some relations used for their evaluation is presented in the next section.

9.6.2 MODELS FOR DENSE PARTICULATE FLOWS

A comparative study of dense particle models has been reported by van Wachem et al. (2001).

The study mainly addresses solid–gas flows, but includes some pertinent droplet-gas observations. The models compared are the two-phase analyses of Jackson and Ishii. Since these, as most models are, have evolved through the years, the only form reported in this comparison paper will be discussed.

The averaging function used was

$$4\pi \int_0^\infty g\{r\} r^2 \, dr = 1 \tag{9.69}$$

with an effective averaging volume defined by

$$\int_0^\ell g\{r\} r^2 \, dr = \int_\ell^\infty g\{r\} r^2 \, dr \tag{9.70}$$

The gas-phase volume fraction and particle number density are

$$\varepsilon\{\vec{x}\}_g = \int g\{|\vec{x} - \vec{y}|d\forall_y \quad \text{and} \quad n\{\vec{x}\} = \sum_p g\{|\vec{x} - \vec{y}|\}$$

(9.71)

The local mean gas-phase flow properties are

$$\varepsilon\{\vec{x}\}_g \langle q \rangle_g \{\vec{x}\} = \int_{\forall_g} q\{\vec{y}\}g\{|\vec{x} - \vec{y}|d\forall_y$$

(9.72)

The mean value of the solid-phase properties is not averaged like the gas, they are averaged by

$$n\{\vec{x}\}\langle q \rangle_s \{\vec{x}\} = \sum_p q_s\{|\vec{x} - \vec{y}|\}$$

(9.73)

Applying these averaging rules to the point-value continuity equations for each phase:

$$\frac{\partial \varepsilon_g}{\partial t} + \nabla \cdot \left(\varepsilon_g \vec{v}_g \right) = 0 \quad \text{and} \quad \frac{\partial \varepsilon_s}{\partial t} + \nabla \cdot \left(\varepsilon_s \vec{v}_s \right) = 0$$

(9.74)

The gas-phase momentum equation of Jackson is

$$\rho_g \varepsilon_g \left[\frac{\partial \vec{v}_g}{\partial t} + \vec{v}_g \cdot \nabla \vec{v}_g \right] = \varepsilon_g \nabla \cdot \vec{\tau}_g - \varepsilon_g \nabla P - \beta(\vec{v}_g - \vec{v}_s) + \varepsilon_g \rho_g \vec{g}$$

(9.75)

and the solid-phase momentum equation of Jackson is

$$\rho_s \varepsilon_s \left[\frac{\partial \vec{v}_s}{\partial t} + \vec{v}_s \cdot \nabla \vec{v}_s \right] = \varepsilon_s \nabla \cdot \vec{\tau}_g - \varepsilon_s \nabla P + \nabla \cdot \vec{\tau}_s - \nabla P_s - \beta(\vec{v}_g - \vec{v}_s) + \varepsilon_s \rho_s \vec{g}$$

(9.76)

The gas-phase momentum equation of Ishii is

$$\rho_g \varepsilon_g \left[\frac{\partial \vec{v}_g}{\partial t} + \vec{v}_g \cdot \nabla \vec{v}_g \right] = \nabla \cdot \varepsilon_g \vec{\tau}_g - \varepsilon_g \nabla P - \beta(\vec{v}_g - \vec{v}_s) + \varepsilon_g \rho_g \vec{g}$$

(9.77)

and the solid-phase momentum equation of Ishii is

$$\rho_s \varepsilon_s \left[\frac{\partial \vec{v}_s}{\partial t} + \vec{v}_s \cdot \nabla \vec{v}_s \right] = \nabla \cdot \varepsilon_s \vec{\tau}_s - \varepsilon_s \nabla P_s - \beta(\vec{v}_g - \vec{v}_s) + \varepsilon_s \rho_s \vec{g}$$

(9.78)

In the form presented, the two approaches yield very similar equations. In Jackson's solid-phase momentum equation, pressure and shear stress terms for both phases are included. In Ishii's dispersed-phase momentum equation, the volume fraction is taken inside the gradient operator. There are some other differences in the two approaches not evident in the formulated equations. Ishii used a different spatial averaging technique. Namely, a phase indicator function was used to which phase was present at a specific point. Jump conditions were used to cross the phase boundaries. These differences made the Ishii analysis more appropriate for the simulation of liquid droplets as the condensed phase.

The continuity and momentum equations for the dense two-phase system are not closed. Constitutive equations are needed to express the solid-phase stress term as a function of the velocity field. Analogy to kinetic theory was used to evaluate the particle collision stress term. A granular "temperature" was defined to evaluate this effect. A transport equation for energy is required to evaluate the granular temperature (Θ), but this equation was approximated as an algebraic balance between the generation and dissipation terms. The result is

$$\left(-\nabla P_s \vec{\vec{I}} + \vec{\vec{\tau}}_s\right) : \nabla \vec{v}_s - \gamma_s = 0 \tag{9.79}$$

where γ_s is the dissipation of granular energy. The solid-phase pressure is given by

$$P_s = \rho_s \varepsilon_s \Theta \left[1 + 2(1+e)g_o \varepsilon_s\right] \tag{9.80}$$

where
  e is the coefficient of restitution
  g_o is the radial distribution function
  d_s is the particle size

The bulk viscosity is

$$\lambda_s = (4/3)\,\varepsilon_s^2 \rho_s d_s g_o (1+e)\sqrt{\Theta/\pi} \tag{9.81}$$

Several shear/viscosity correlations have been reported; the one suggested is

$$\mu_s = \frac{5\sqrt{\pi\Theta}}{96}\rho_s d_s \left[\left(\frac{2R}{(R+\lambda_{mfp})(1+e)} + \frac{8\varepsilon_s}{5} \right) \left(\frac{1+8/5\eta(3\eta-2)\varepsilon_s g_o}{2-\eta} \right) + \frac{768}{25\pi}\eta\varepsilon_s^2 g_o \right] \tag{9.82}$$

Most viscosity expressions fit available data, but this one produced the gas-phase viscosity when no particles were present.

The remaining parameters which must be evaluated to complete the closure of the momentum equations are as follows. The conductivity of the granular energy

$$\kappa = \frac{15d_s\rho_s\varepsilon_s\sqrt{\Theta/\pi}}{4(41-33\eta)}\left[1+\frac{12}{5}\eta^2(4\eta-3)\varepsilon_s g_o + \frac{16}{15\pi}(41-33\eta)\eta\varepsilon_s g_o\right] \tag{9.83}$$

The dissipation due to particle–particle collisions is approximated as

$$\gamma_s = 12(1-e^2)\frac{\varepsilon_s^2\rho_s g_o}{d_s\sqrt{\pi}}\Theta^{1.5} \tag{9.84}$$

The dissipation due to the fluctuating force exerted by the gas through the fluctuating velocity of the particles is

$$J_s = \beta(\langle \vec{v}_s \cdot \vec{v}_s \rangle - \langle \vec{v}_g \cdot \vec{v}_s \rangle) \tag{9.85}$$

which is empirically evaluated as

$$J_s = \left[\frac{3\mu_s\varepsilon_s\Theta}{d_s^2}R_{diss} - \frac{\beta^2 d_s\left(\left|\vec{v}_g - \vec{v}_s\right|\right)^2}{4\varepsilon_s\rho_s\sqrt{\pi\Theta}}S^* \right] \tag{9.86}$$

where
   β is the interphase drag coefficient
   R_{diss} is a fudged drag coefficient to fit test data
   S^* is an energy source given by

$$s^* = \left(1/2\sqrt{\pi}\right)R_s\beta^2 \tag{9.87}$$

where R_s represents a mean force acting on the particles; it too is obtained by fitting test data. The radial distribution function is

$$g_o = \left[1 - \left(\varepsilon_s / \varepsilon_{s,max}\right)^{1/3}\right]^{-1} \tag{9.88}$$

For very dense particulate flows, the kinetic theory analogy is not sufficient. The particles apparently stay in contact with one another long enough to cause additional friction. Attempts to model this phenomenon have generally used the conventional Newtonian stress rate-of-strain equations for a separate solid phase to develop an additional correlation. The lower dense phase and gas pressure and viscosity is increased by adding additional terms. One model of these additional pressure and viscosity terms are

$$P_f = A\left(\varepsilon_s - \varepsilon_{s,min}\right)^n \approx 10^{25} \left(\varepsilon_s - 0.59\right)^{10} \tag{9.89}$$

$$\mu_f = \cfrac{P_f \sin\Phi}{\varepsilon_s \sqrt{\cfrac{1}{6}\left[\left(\cfrac{\partial u_s}{\partial x} - \cfrac{\partial \upsilon_s}{\partial y}\right)^2 + \left(\cfrac{\partial u_s}{\partial x}\right)^2 + \left(\cfrac{\partial \upsilon_s}{\partial y}\right)^2\right] + \cfrac{1}{4}\left(\cfrac{\partial u_s}{\partial y} + \cfrac{\partial \upsilon_s}{\partial x}\right)^2}} \tag{9.90}$$

where Φ is typically 25°. The very magnitude of the powers in the pressure term indicates that the prediction would yield results which would vary orders of magnitude. Other models for this phenomenon involve even more outlandish variation. The effect is apparently real, but one must conclude that a reliable model is not yet available.

The article by van Wachem et al. (2001) applies this dense phase model, along with variations resulting from other suggested constitutive submodels to three fluidized bed configurations: (1) for bubbling fluidized beds, (2) slugging fluidized beds, and (3) bubble injection fluidized beds. Gravity and drag terms in the momentum equations were found to be the most dominating. The Jackson vs. Ishii formulations and the other submodel variations were found to have only a modest effect on the results. The frictional stress had a major effect on the results for very dense beds. This is far from surprising, since the predictions of this effect varied by many orders of magnitude, depending on the submodel chosen to represent it.

The review of this article proves that the very comprehensive dense phase computational model accounts for many of the factors controlling the bed operation. It shows what parameters are adequately represented and which need further improvement. It shows the frictional stress phenomena for very dense beds is not sufficiently understood or modeled. Above all, it shows that evaluation of the many simultaneously occurring factors which control the operation of dense fluidized beds cannot be reasonably evaluated and understood without using computation methodology.

9.7 NOMENCLATURE

9.7.1 NOMENCLATURE FOR SECTIONS 9.1 THROUGH 9.4

9.7.1.1 English Symbols

B_f	blowing rate for mass transfer
B_h	blowing rate for energy transfer
c	driving force in mass transfer coefficients is concentration
c_i	concentration of i
C_f	skin friction
C_i	heat capacity of species i
C_p	constant pressure heat capacity
C_s	humid heat
D_{ij}	binary diffusion coefficient
E_k	eddy transport coefficient; $k=$D, h, v diffusion, heat, and momentum
f	Fanning friction factor
G	driving force in mass transfer coefficients is partial pressure
h_i	film coefficient for heat transfer; i indicates phase
H, w	dummy variable to indicate mathematical operation
H_0	parameter without mass transfer
j_i	mass diffusion flux
j_D	Colburn's j-factor for mass transfer
j_H	Colburn's j-factor for heat transfer
J_i	molar diffusion flux
K_x	film coefficient based on bulk and equilibrium liquid composition
k_x	film coefficient on liquid side
K_y	film coefficient based on bulk and equilibrium gas composition
k_y	film coefficient on gas side
L	same as c for liquids
Le	Lewis number
m	type of heat flux

m'	slope of operating point and bulk liquid composition
m''	slope of operating point and bulk gas composition
N_i	one-dimensional mass (or molar) flux of i
N_T	total mass flux
P	pressure
p_{in}	partial pressure of species i at location n
Pr	Prandtl number
q_m	heat flux; m indicates type of exchange
R	universal gas constant; pipe radius
Re_θ	Reynolds number based on momentum thickness
Sc	Schmidt number
St	Stanton number
T	temperature
u^+	inner dimensionless variable for velocity
u_I	velocity; I indicates centerline, free-stream, and average value
V_z	z velocity component
w	wall value
w_i	arbitrary property
W_i	magnitude of velocity of species i
x_i	mass (or mole) fraction in liquid phase
y^+	inner dimensionless variable for distance from the wall
y_i	mass (or mole) fraction in gas phase

9.7.1.2 Greek Symbols

δ	film thickness
κ	thermal conductivity
λ_i	heat of vaporization of species i
ν	kinematic viscosity
τ	shear stress

9.7.1.3 Subscripts

A	diffusing species
G, i, L	indicates gas, interface, and liquid conditions
α	species is water
β	species is air

9.7.1.4 Superscripts

*	equilibrium conditions

9.7.1.5 Mathematical Symbols

$fn\{\}$, $gn\{\}$	arbitrary functions
$F_1\{\}$, $F_2\{\}$	empirical functions of the Reynolds number; Equation 9.39
$\hbar$	heat transfer coefficient
$\dot{k}_G$	film coefficient for high rate of mass transfer
$LM\{H\}$	log mean value of H; also indicated by HM

9.7.2 NOMENCLATURE FOR SECTION 9.5

9.7.2.1 English Symbols

B, D	birth and death functions, respectively, of particles
B^0	number of particles entering vessel
c	concentration
G	growth rate
G, a, L	growth rate parameters
H	dummy variable to illustrate mathematical operation
L	particle size
m_j	the jth moment
N	number of particles per volume of particle
Q_r	mass flowrate into vessel
R	Lagrangian mass of fluid
r_i	internal coordinates; such as particle size r
t	time
u_i	solution vector
V	particle velocity vector
W	particle phase-space velocity
x_i	external coordinates

9.7.2.2 Greek Symbols

θ	temperature
ξ_i	phase coordinates
υ	particle specific volume

9.7.2.3 Mathematical Symbols

$f\{\vec{x}, r, t\}$	number density of particles
$h\{\}$	function to represent birth and death of particles
$\bar{H}$	average value of H

$\forall$ volume of process vessel

$\vec{W}$ velocity in phase space

9.7.2.4 Acronyms

ANSYS	name of a company
CFX	name of a computer code
DQMOM	direct quadrature method of moments
FOAM	name of a computer code
PBE	population balance equation

9.7.3 NOMENCLATURE FOR SECTION 9.6

9.7.3.1 English Symbols

A, η	parameters used to represent P_f
$A_{\alpha,\beta}$	interface between phases α and β
d	characteristic particle size
e	coefficient of restitution
g_o	radial distribution function
J_s	dissipation fluctuating force correlation
L	characteristic size of process vessel
ℓ	grid space over which particles are to be averaged
n	particle number density
P, P_s	pressure of gas and solid continuum, respectively
P_f	pressure effect cause by particles rubbing together
q	property to be averaged
r	particle size
R_{diss}	fictitious drag coefficient to fit experimental data
R_s	fictitious force operating on particles obtained by fitting test data
S^*	energy source operating on particles
t	time
u_s, υ_s	magnitude of velocity components of particle cloud
x, y	coordinates within particle cloud

9.7.3.2 Greek Symbols

β	interphase drag coefficient
γ	dissipation of granular energy
ε	volume fraction

η	parameter to define e
Θ	grandular temperature parameter
θ	temperature of solid
κ	conductivity of grandular energy
λ_s	viscosity of solid phase continuum
μ_f	effective viscosity attributed to P_f
Φ	angle that particles rub together—not measurable

9.7.3.3 Mathematical Symbols

$\vec{g}$	gravity
$\vec{H}$	average over particle class
$\vec{n}$	normal to a surface
$g\{\vec{r}\}$	weighting function
$\vec{x}$	position vector
$\vec{x}'$	dummy position variable
$\{x-y\}$	increment of distance to be used in averaging process
$\vec{\vec{\tau}}$	shear stress tensor
$\forall$	volume

9.7.3.4 Subscripts

g, s	gas and solid phases, respectively
α, β	denote two phases

REFERENCES

Anderson, T. B. and R. Jackson. 1967. A fluid mechanical description of fluidized beds. *I&EC Fundamentals.* 6:527–539.

Bird, R. B., W. E. Stewart, and E. N. Stewart. 2002. *Transport Phenomena*, 2nd edn. New York: McGraw-Hill.

Brennen, C. E. 2005. *Fundamentals of Multiphase Flow.* Cambridge: Cambridge University. Press.

Cheng, G. C., R. C. Farmer, S. H. Jones, and J. P. Alves. 1998. A Practical CFD Model for Simulating Hybrid Motors. Paper presented at JANNAF 35th Combustion, Airbreathing Propulsion and Propulsion Systems Conference. Tucson.

Cheng, G. C. and R. C. Farmer. 2006. Real fluid modeling of multiphase flows in liquid rocket engine combustors. *J. of Propulsion and Power.* 22:1373–1381.

Darelius, A., H. Brage, A. Rasmuson, I. Niklasson, and S. Folestad. 2006. A volume-based multi-dimensional population balance approach for modelling high shear granulation. *Chem. Eng. Sci.* 61:2482–2493.

Delhaye, J.-M. 1970. Fundamental Equations of Two-Phase Flow. Part I. General Equations of Conservation. Paper presented before the Franco-Soviet Seminar on Heat Transfer, Grenoble. Translated by the NTIS as PB193989.

Drew, D. A. 1983. Mathematical modeling of two-phase flow. *Ann. Rev. Fluid Mech.* 15:261–291.

Faghri, A. and Y. Zhang. 2006. *Transport Phenomena in Multiphase Systems.* Boston: Elsevier.

Farmer, R. C. and W. R. Waldrop. 1977. Model for sediment transport and delta formation. *Proceedings of the Conference: Coastal Sediments 77.* New York: American Society of Civil Engineers.

Fox, R. O. 2003. *Computational Models for Turbulent Reacting Flows.* Cambridge: Cambridge Press.

Geankoplis, C. J. 2003. *Transport Processes and Separation Principles,* 4th edn. Upper Saddle River, NJ: Prentice Hall.

Gray, W. G. 1975. A derivation of the equations for multi-phase transport. *Chem. Eng. Sci.* 30:229–233.

Himmelblau, D. M. and K. B. Bischoff. 1968. *Process Analysis and Simulation: Deterministic Systems.* New York: John Wiley & Sons.

Hulbert, H. M. and S. Katz. 1964. Some problems in particle technology. *Chem. Eng. Sci.* 19:555–574.

Ishii, M. 1975. *Thermo-Fluid Model and Hydrodynamic Constitutive Relations.* Diretion des Etudes et Recherches d'Electricite' de France. Eyrolles, Paris.

Ishii, R., Y. Umeda, and M. Yuhi. 1989. Numerical analysis of gas-particle two-phase flows. *J. Fluid Mech.* 203:475–515.

Jackson, R. 1997. Locally averaged equations of motion for a mixture of identical spherical particles and a Newtonian fluid. *Chem. Eng. Sci.* 52:2457–2469.

Jackson, R. 1998. Erratum: Locally averaged equations of motion for a mixture of identical spherical particles and a Newtonian fluid. *Chem. Eng. Sci.* 53:1955.

Kleigel, J. R. 1963. Gas particle nozzle flows. *Ninth Symposium (International) of Combustion.* New York: Academic Press.

Kumar, J., G. Warnecke, M. Peglow, and S. Heinrich. 2009. Comparison of numerical methods for solving population balance equations incorporating aggregation and breakage. *Powder Technology.* 189(1):1810-1830.

Marchisio, D. and R. O. Fox. 2005. Solution of the population balance equation using direct quadrature method of moments. *J. Aerosol Sci.* 36:43–73.

Mickley, H. S., R. C. Ross, A. L. Squyers, and W. E. Stewart. 1954. Heat, Mass, and Momentum Transfer for Flow over a Flat Plate with Blowing or Suction. NACA TN 3208. National Advisory Committee for Aeronautics.

Murray, J. D. 1966. On Equations of Motion for a Particle-Fluid Two Phase Flow. Report No. AP 00455-01. University of Michigan, Ann Arbor, Michigan.

Ramkrishna, D. 2000. *Population Balances. Theory and Applications to Particulate Systems.* New York: Academic Press.

Randolph, A. D. and M. A. Larson. 1971. *Theory of Particulate Processes: Analysis and Techniques of Continuous Crystallization,* 2nd edn. San Diego, CA: Academic Press.

Randolph, A. D. 1964. A population balance for countable entities. *Can. J. Chem. Eng.* 42:280–281.

Schetz, J. A. 1993. *Boundary Layer Analysis.* Englewood Cliffs, NJ: Prentice Hall.

Seinfeld, J. S. and S. N. Pandis. 1998. *Atmospheric Chemistry and Physics.* New York: John Wiley and Sons.

Sherwood, T. K., R. L. Pigford, and C. R. Wilke. 1975. *Mass Transfer.* New York: McGraw-Hill.

Silva, L. F. L. R., R. B. Damian, and P. L. C. Lage. 2007. Implementation and analysis of numerical solution of the population balance equation in CFD packages. *Comput. Chem. Eng.* 32(12):2933–2945.

Simmons, F. S. 2000. *Rocket Exhaust Plume Methodology.* Reston: AIAA.

Skelland, A. H. P. 1974. *Diffusional Mass Transfer.* New York: John Wiley and Sons.

Smith, S. D. 1984. High Altitude Chemically Reacting Gas Particle Mixtures Vol. I-A Theoretical Analysis and Development of the Numerical Solution. LMSC-HREC TR D867 400-1. Lockheed Missiles and Space Co. Huntsville.

Treybal, R. E. 1968. *Mass-Transfer Operations.* New York: McGraw-Hill.

Verkoeijen, D., G. A. Pouw, G. M. H. Meesters, and B. Scarlett. 2002. *Chem. Eng. Sci.* 57:2287–2303.

Wachem, B. G. M. van, J. C. Schouten, C. M. van Bleek, R. Krishna, and J. L. Sinclair. 2001. Comparative analysis of CFD models of dense gas-solid systems. *AIChE J.* 47:1035–1051.

Waldrop, W. R. and R. C. Farmer. 1976. Three-dimensional computation of buoyant plumes. *J. Geophys. Res.* 79:1269–1276.

Weller, H., G. Jasak, and C. Fureby. 1998. A tensorial approach to continuum mechanics using object-oriented techniques. *Comput. Phys.* 12:620–631.

Whitaker, S. 1969. Advances in theory of fluid motion in porous media. *Ind. Eng. Chem.* 61:14–28.

Whitaker, S. 1973. The transport equations for multi-phase systems. *Chem. Eng. Sci.* 28:139–147.

White, F. M. 1988. *Heat and Mass Transfer.* Reading, MA: Addison Wesley.

White, F. M. 2006. *Viscous Fluid Flow,* 3rd edn. New York: McGraw-Hill.

Yalin, M. S. 1972. *Mechanics of Sediment Transport.* Oxford: Pergamon Press.

Zannetti, P. 1990. *Air Pollution Modeling: Theories: Computational Methods and Available Software.* New York: Van Nostrand Reinhold.

CHAPTER 10

Closure

There are three sources of information available to engineers to aid in solving a complex transport phenomena problem. Full-scale experiments, subscale or laboratory experiments, and analysis. Computational transport phenomena is now an accepted methodology for providing dependable analysis. Production quality computer codes are not something that one undertakes to write when a specific problem arises. There are several companies which provide computational services of a high quality, at a price. There are many codes which have been generated and may be available as shareware to investigate a particular process. This chapter is designed to provide one with the background to assess when a computational analysis would be beneficial to a project and what type of computational tool would be appropriate for such an analysis. Finally, mastery of the material presented herein would allow one to make an informed decision as to whether the individual could perform the required computational analysis, or whether a code or vendor could provide the analysis, or whether the problem is so complex that the current state of the art of CTP is insufficient to even attempt using such methodology.

The CTP code described herein has been used successfully by these authors and the supporters of the code development, primarily NASA, the Air Force, and various aerospace companies, for the past 20 years. The validation process involved not only basic flows like boundary layers, backward facing steps, and round jets, but also analyses of real devices. The performance of the Space Shuttle Main Engine (SSME) was predicted with the CTP code described in Chapter 8 (Wang and Chen, 1993) to be well within the accuracy that it could be measured on a test stand. This analysis involved calculating the interior flowfield of the SSME and the plume emitted by the engine. Similar analyses and performance comparisons were also made for hybrid rocket engines under development by NASA and Martin-Marietta Corp. and AMROC (Cheng et al., 1994).

The major advantage that computational analyses of transport phenomena offers the engineer is that processes which are highly three-dimensional may be simulated with good accuracy. High performance turbopumps require effective and robust analytical tools to optimize their operation. The complex interaction of the inducer, impeller, and diffuser flows were analyzed with early versions of the CTP code (Cheng et al., 1993). Another three-dimensional flow of critical interest to launch vehicle design is the base-flow and associated heating caused by the separated flow region and pumping effect caused by the supersonic flow from the rocket engines. It is not uncommon for the base heat shield to weigh as much as the payload of the vehicle. Lacking computational computer power and the ability to conduct full-scale ground testing simulations of flight conditions, base flows were estimated from subscale model test for the Saturn vehicles (Brewer and Craven, 1969). Twenty-seven years later computational analysis and computers had improved sufficiently that a simulation could be made of this subscale experiment (Wang, 1996). Of course, the advantage of the simulation is that numerically the process could be scaled-up from subscale to prototype.

Computational analyses of transport phenomena involving chemical reactions and the associated heating characteristics can be realistically analyzed. The predecessors of the CTP code have been used to analyze a dump combustor of burning hydrogen and oxygen (Wang et al., 1989). Predictions compared very well to the experimental data obtained by Smith and Giel (1980). Many more recent experiments have been reported and analyzed which were designed to provide design information on single injector elements of rocket engines (Farmer et al., 2000). Since the experiments were to study conditions typical of rocket operation the pressures involved were supercritical and near supercritical. The concept of such experiments was to relate the geometry of the element with the flowfield and combustion efficiency of the device. The simulations (made with the CTP code methodology) compared well with the test data. Unfortunately, detailed flowfield data collected were not definitive, so comprehensive comparisons were seldom obtained. The propellants studied were hydrogen and RP-1 and oxygen. Both gaseous and liquid propellants were investigated and simulated. Liquid propellant elements are difficult to characterize because initial drop size and droplet spreading have not been successfully measured. A major advantage of the CTP code simulations was that the real fluid characterization and modeling of the spray as a multicomponent, single phase consisting of both gaseous and liquid components proved to be a good characterization of the spray flame (Cheng and Farmer, 2006). Describing spray flames at near supercritical conditions computationally has proven to be a most difficult task, and the simplicity and accuracy offered by the CTP approach is a most useful engineering analysis.

Accurate simulations of the combustion devices just mentioned also imply that the associated heat transfer phenomena could also be accurately predicted. This indeed proved to be the case. A classic heat transfer experiment for rocket combustion chamber type conditions was conducted by Bartz (1965). Subsequent experiments of this type have also been conducted (Elam, 1991) and have been simulated with the CTP code (Wang and Luong, 1994). A further test of the computational analysis was conducted for two film cooling tests in which a film of hydrogen was injected under a supersonic stream of air (Chen et al., 1992). These simulations showed the finite-rate combustion reactions had little effect at the test conditions, i.e., the combustion was fast and in a near equilibrium condition. However, the turbulence model had to be adjusted to match the test data. Analyses of these film cooling experiments also provided validation of the conjugate heat transfer analysis included in the CTP code. This is the basis for the temperature corrections in the TKE model which were discussed in Chapter 4. The CTP code was also used to analyze the heat transfer region in the LOX post region of the SSME. This region resembles the tube side of a shell-and-tube in which there are hundreds of posts (tubes) with cross-flow entering from two sides. The analysis was successfully completed by assuming the computational field was a porous media (Cheng et al., 1995). The distribution of posts was used to evaluate the local porosity. This analysis was then used to estimate the distribution of inlet flow rates into the combustion chamber through the injector.

These practical reacting, multiphase and highly three-dimensional engineering analyses serve as an impressive validation for the CTP code. Hundreds of combustion reactions and turbulence models were more complex that the TKE model were not needed to match test data from a variety of sources. Turbulence and combustion researchers, prompted mainly by the goal of producing publications, constantly suggest that their methodology is necessary for analyzing transport phenomena. Such technology is decades away, if ever, from making the type analyses just summarized. Also, significant problems can be analyzed with employing huge PC clusters and parallel processing. These comments are not meant to defame elaborate computational analysis efforts, but to emphasize that less elaborate methods can be very educational and useful. There is still nothing wrong with making intelligent approximations to expedite a study.

One must always be aware of the capabilities and limitations of computational analysis. Some time ago a major power producer was seeking renewal of discharge permits from EPA. The power company maintained an instrumented boat, a hydraulics laboratory, and computer modeling capabilities to monitor the hot water discharges from their power plants. The estimated worst-case scenarios of maximum power production at minimum river stages were estimated with computational solutions of transport equations. The government official wanted measurements made to verify the CTP predictions for the worst cases which would occur over the next 10 years. The company was asked to provide this verification within the next 30 days. The company engineer responded "I will be glad to take the instrument boat out and make such measurements. You just tell me which day will have the worst conditions for the next 10 years so I will know when to make the measurements." Computational analysis must be relied upon to play its proper role as an engineering tool.

Thirty-two years ago, a computational study of the Mississippi River flowing into the Gulf of Mexico was made. Computers were slower and computational grids were coarser, but a two-component fresh water/salt-water analysis was made (Waldrop and Farmer, 1976). The analysts took their solution to their colleagues who were researchers in coastal studies. The colleagues remarked upon looking at the computed results: "We made a field study to measure the diffusion across the free-shear layer separating the blue gulf water from the brown river water. Dye was periodically injected to be monitored as it crossed the interface. We put the dye in, but could not find where it went. Thirty days later after repeating the test many times, we finally discovered that the dye came up 3 miles downstream and in the center of the river plume—exactly where your computer simulation showed it to be!" Yes, the gulf flow was in a westerly direction and the river flow in a southerly direction. The buoyant river plume flowed over the dense gulf water, but it also was turned by the gulf water which pushed on the side of the plume. The turn caused the curvature which set up a vortex that carried the dye down into the standing vortex and finally back up to surface where it was found. The computer simulation was crude by today's standards, but modest physics even in a coarse grid with a simple turbulence model produced very realistic results. A more elaborate model with a fine grid would not have produced a solution with the older computers. But such refinements were not necessary to produce a valuable engineering simulation. Modern computational tools with today's fast PCs can yield amazing results.

REFERENCES

Bartz, D. R. 1965. Turbulent boundary-layer heat transfer from rapidly accelerating flow of rocket combustion gases and of heated air. In *Advances in Heat Transfer*, Vol. 2. T. P. Hartnett and T. F. Irvine Jr. (Eds.), New York: Academic Press.

Brewer, E. B. and C. E. Craven. 1969. Experimental investigation of base flow field at high altitude for a four-engine clustered nozzle configuration. NASA TND-5164. National Aeronautics and Space Administration: Huntsville, AL.

Chen, Y. S., G. C. Cheng, and R. C. Farmer. 1992. Reacting and non-reacting flow simulation for film cooling in 2-D supersonic flows. AIAA 92-3602. AIAA/ASME/SAE/ASEE 28th Joint Propulsion Conference and Exhibit, Nashville, TN.

Cheng, F. C. and R. C. Farmer. 2006. Real fluid modeling of multiphase flows in liquid rocket engine combustors. *J. Propul. Power* 22:1373–1381.

Cheng, G. C., Y.-S. Chen, and T.-S. Wang. 1995. Flow Distribution within the SSME Main Injector Assembly Using Porosity Formulation. AIAA 95-0350. 33rd Aerospace Sciences Meeting and Exhibit, Reno, NV.

Cheng, G. C., R. C. Farmer, S. H. Jones, and J. S. McFarlane. 1994. Numerical Simulation of the Internal Ballistics of a Hybrid Rocket Motor. AIAA 94-0554. 32nd Aerospace Sciences Meeting and Exhibit, Reno, NV.

Cheng, G. C., Y.-S. Chen, R. Garcia, and R. W. Williams. 1993. Numerical Study of 3-D Inducer and Impeller for Pump Model Development. AIAA 93-3003. AIAA 24th Fluid Dynamics Conference, Orlando, FL.

Elam, S. K. 1991. Subscale LOX/Hydrogen Testing with a Modular Chamber and a Swirl Coaxial Injector. *AIAA*. Paper 91-1874.

Farmer, R., G. Cheng, H. Trinh, and K. Tucker. 2000. A Design Tool for Liquid Rocket Engine Injectors. AIAA 2000-3499. 36th AIAA/ASME/SAE/ASEE Joint Propulsion Conference and Exhibit, Huntsville, AL.

Smith, G. D. and T. V. Giel. 1980. An Experimental Investigation of Reactive, Turbulent, Recirculating jet Mixing. AEDC-Tr-79-79. Arnold Engineering Development Center, Tullahoma, TN.

Waldrop, W. R. and R. C. Farmer. 1976. Three-dimensional computation of buoyant plumes. *J. Geophys. Res.* 79:1269–1276.

Wang, T.-S. 1996. Grid-resolved analysis of base flowfield for four-engine clustered nozzle configuration. *J. Spacecraft Rockets* 33:22–29.

Wang, T.-S. and Y.-S. Chen. 1993. Unified Navier-Stokes flowfield and performance analysis of liquid rocket engines. *J. Propul. Power* 9:678–685.

Wang, T.-S. and V. Luong. 1994. Hot-gas-side and coolant-side heat transfer in liquid rocket engine combustors. *J. Thermophys. Heat Trans.* 8:524–530.

Wang, T., Y. Chen, and R. Farmer. 1989. Numerical study of reactive ramjet dump combustor flowfields with a pressure based CFD method. *AIAA* 89–2798. AIAA/ASME/SAE/ASEE 25th Joint Propulsion Conference, Monterey, CA.

Grid Stencils and Example Problems

Tutorial example cases for preparing and running the CTP code are presented in this section. These example cases cover a wide range of applications of internal, external, ideal-gas, real-fluid, single-species, and multispecies flow conditions. The users can construct their specific applications based on these examples by modifying the grid generation and flow initialization codes and the input files provided. These cases are:

1. Boundary layer flow over a flat plate (working directory: z01-BL)
2. Developing and fully developed pipe flow (working directory: z02-DPF)
3. Flow over a backstep (working directory: z03-BStep)
4. A cylinder in cross-flow (working directory: z04-Cylinder)
5. Flow over an airfoil (working directory: z05-Airfoil)
6. Shell and tube heat exchanger (working directory: z06-HeatEx)
7. Converging–diverging nozzle flow (working directory: z07-Nozzle)
8. Orifice flow and an ejector pump (working directory: z08-Orifice)
9. Flow through a pipe elbow (working directory: z09-PipeElbow)
10. Flow through a pipe tee (working directory: z10-PipeT)
11. Free-surface flow in an open duct (working directory: z11-FreeSurface)
12. Flow in a stirred-tank (working directory: z12-StirredTank)

To test these cases, the user is to open a DOS Command Prompt in a PC Windows operating environment and change directory to the specific working directory. The grid generation and flow initialization operations are performed in a FORTRAN code, *grid.f*. This code generates *restrt.x* and *restrt.q* for the mesh and flow variables. Input parameters that control the running options of the CTP code are specified in the input data file, *input*. The users can refer to the input parameter definition section for details. After running the CTP code, by executing *xctp.exe* on the Command Prompt, either when the convergence criterion is satisfied or the time steps, ITT, are completed, *output.x* and *output.q* (these output files are generated for every ITPNT steps) are generated in the working directory. To continue running the case, copy *output.q* to *restrt.q*, modify input if necessary, and execute *xctp.exe* again. Other output files, *tecplot.dat*, *plot.q1*, and *plot.q2* are available for plotting graphical representations of the flow solutions. For running real-fluid cases, *dbase.dat* and *fluid.inp* (input files for the real-fluid module) have to be present in the working directory. In CTP code, nondimensional time step size, $DTT = dt * U\text{-ref}/X\text{-ref}$ (where dt is physical time step size in second), is specified in the input data file. Also, in the input data file, $VISC = \text{viscosity}/(\text{Density-ref} * U\text{-ref} * X\text{-ref})$. Where Density-ref, U-ref, and X-ref are reference density, velocity, and length, respectively.

A.1 BOUNDARY LAYER FLOW OVER A FLAT PLATE

This example problem simulates a turbulent boundary layer flow development over a flat plate. A two-dimensional (2-D) rectangular domain of $5.1\,\text{m} \times 0.1\,\text{m}$ is used for numerical computation. The flat plate is located at $y = 0$ and extends from $x = 0\,\text{m}$ to $x = 5\,\text{m}$. A uniform flow of 1.0 m/s speed enters the domain from the left at $x = -0.1\,\text{m}$ (with inlet fixed mass flow rate condition) and exits at $x = 5\,\text{m}$ (with outlet mass conservation condition). The top boundary at $y = 0.1\,\text{m}$ is assigned as a symmetry boundary. Part of the bottom boundary from $x = -0.1\,\text{m}$ to $x = 0\,\text{m}$ is also assigned as a symmetry boundary. A mesh of $121 \times 81 \times 1$ is generated in *grid.f* to represent the computational domain. Initial flow properties are also specified in *grid.f*. Input parameters of this case are provided in an input file, *input*, which is read by the present CFD code at startup. The following specifications are pertinent to this example problem:

- Mesh (1 block): $121 \times 81 \times 1$ (wall: $I = 21, 121; J = 1$)
- Working directory: z01-BL (contains grid generation code, *grid.f*, and input file, *input*)

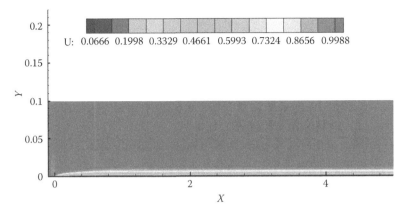

FIGURE A.1 (See color insert following page 294.) Boundary layer flow over a flat plate (working directory: z01-BL).

- VISC = 1.5E-05, U-inlet = 1.0 m/s, U-ref = 100 m/s, X-ref = 0.1 m, Turbulent flow (Dk-in = 0.005 * (U-inlet/U-ref)^2, DE-in = 0.09 * DK-in$^{1.5}$/(0.03 * 2.0))
- Reynolds number, Re = (1/VISC) (1/X-ref) (U-inlet/U-ref) = 6666.7 (1/m)

To run this case, compile *grid.f* to create *xgrid.exe* using a command prompt on a PC Windows system. Next, execute *xgrid.exe* to generate *restrt.x* for the grid file and *restrt.q* for the flow variables file. Then, examine the input file, *input*, to check that the input parameters are set properly. Finally, execute *xcpt.exe* to solve this case. Figure A.1 shows the predicted u-velocity contours of a converged solution of this case. The boundary layer development is clearly shown.

A.2 DEVELOPING AND FULLY DEVELOPED PIPE FLOW

This test case is designed to simulate a developing turbulent flow in a circular pipe. A 2-D axisymmetric rectangular domain of 5 m × 0.1 m is used for numerical computation. The pipe wall is located at $y = 0.1$ m and extends from $x = 0$ m to $x = 5$ m. A uniform flow of 1.0 m/s speed enters the domain from the left at $x = 0$ m (with inlet fixed mass flow rate condition) and exits at $x = 5$ m (with outlet mass conservation condition). The bottom boundary at $y = 0$ m is assigned as a symmetry boundary. A mesh of 121 × 81 × 1 is generated in *grid.f* to represent the computational domain. Initial flow properties are also specified in *grid.f*. Input parameters of this case are provided in an input file, input, which is read by the present CFD code at startup. The following specifications are pertinent to this example problem:

- Mesh (1 block): 121 × 81 × 1 (wall: I = 1, 121; J = 81)
- Working directory: z02-DPF (contains grid generation code, *grid.f*, and input file, *input*)
- VISC = 2.0E-05, U-inlet = 1.0 m/s, U-ref = 100 m/s, X-ref = 0.1 m, Turbulent flow (Dk-in = 0.005 * (U-inlet/U-ref)^2, DE-in = 0.09 * DK-in$^{1.5}$/(0.03 * 2.0))
- Reynolds number, Re = (1/VISC) (0.2/X-ref) (U-inlet/U-ref) = 4000

To run this case, compile *grid.f* to create *xgrid.exe* using a command prompt on a PC Windows system. Next, execute *xgrid.exe* to generate *restrt.x* for the grid file and restrt.q for the flow variables file. Then, examine the input file, *input*, to check that the input parameters are set properly. Finally, execute *xcpt.exe* to solve this case. Figure A.2 shows the predicted

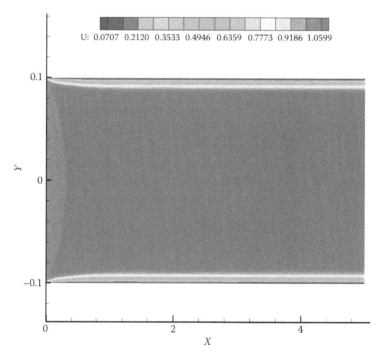

FIGURE A.2 (See color insert following page 294.) Developing and fully developed pipe flow (working directory: z02-DPF).

u-velocity contours of a converged solution of this case. The pipe entrance flow development that reached a fully developed pipe flow downstream is clearly shown.

A.3 FLOW OVER A BACKSTEP

This example case is used to simulate a turbulent recirculating flow behind a backward-facing step. The experimental investigation of this case was conducted by Kim et al. (1978). A 2-D backward-facing step domain of a 0.4 m × 0.2 m inlet duct (with top and bottom walls) plus a 2 m × 0.3 m downstream duct (with top and bottom walls) is used for numerical computation. The backward-facing step is located at $x = 0$ m and extends from $y = -0.1$ m to $y = 0.2$ m. A uniform flow of 10 m/s speed enters the domain from the left at $x = -0.4$ m (with inlet fixed mass flow rate condition) and exits at $x = 2.4$ m (with outlet mass conservation condition). A two-block mesh of 121 × 41 × 1 and 101 × 21 × 1 is generated in *grid.f* to represent the computational domain. Initial flow properties are also specified in *grid.f*. Input parameters of this case are provided in an input file, *input*, which is read by the present CFD code at startup. The following specifications are pertinent to this example problem:

- Mesh (2 blocks): 121 × 41 × 1 and 101 × 21 × 1
- Working directory: z03-BStep (contains grid generation code, *grid.f*, and input file, *input*)
- VISC = 3.0E-08, U-inlet = 10.0 m/s, U-ref = 300 m/s, X-ref = 0.1 m, Turbulent flow (Dk-in = 0.005 * (U-inlet/U-ref)^2, DE-in = 0.09 * DK-in$^{1.5}$/(0.03 * 2.0))
- Reynolds number, Re = (1/VISC) (0.2/X-ref) (U-inlet/U-ref) = 2.2222E + 06

To run this case, compile *grid.f* to create *xgrid.exe* using a command prompt on a PC Windows system. Next, execute *xgrid.exe* to generate *restrt.x* for the grid file and *restrt.q* for the flow variables file. Then, examine the input file, *input*, to check that the input parameters are set properly. Finally, execute *xcpt.exe* to solve this case. Figure A.3 shows the predicted u-velocity contours of a converged solution of this case. The recirculating flow downstream of the backward-facing step shows that the length of the recirculation zone is about seven times the step height, which is very close to the experimental measurement.

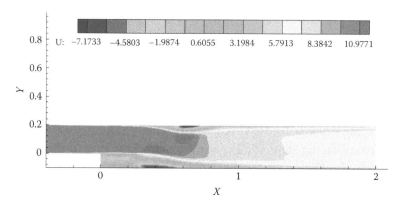

FIGURE A.3 (See color insert following page 294.) Flow over a backstep.

A.4 A CYLINDER IN CROSS-FLOW

This example case is used to simulate a laminar flow past a circular cylinder. Experimental investigations of this type of flow are summarized in Schlichting (1979) and also studied by Coutanceau and Bouard (1977). A 2-D circular domain of 0.2 m radius is used for numerical computation. The circular cylinder radius is 0.005 m, which is the inner boundary of the computational domain. A uniform flow of 10 m/s speed is specified at the outer boundary (with mass conservation condition). A mesh of $181 \times 121 \times 1$ is generated in *grid.f* to represent the computational domain. Initial flow properties are also specified in *grid.f*. Input parameters of this case are provided in an input file, *input*, which is read by the present CFD code at startup. The following specifications are pertinent to this example problem:

- Mesh (1 block): $181 \times 121 \times 1$ (wall: $I = 1, 181; J = 1$)
- Working directory: z04-Cylinder (contains grid generation code, *grid.f*, and input file, *input*)
- VISC = 8.3333E-04, U-inlet = 10.0 m/s, U-ref = 300 m/s, X-ref = 0.01 m, Laminar flow
- Reynolds number, $Re = (1/\text{VISC})(0.01/\text{X-ref})(\text{U-inlet}/\text{U-ref}) = 40$

To run this case, compile *grid.f* to create *xgrid.exe* using a command prompt on a PC Windows system. Next, execute *xgrid.exe* to generate *restrt.x* for the grid file and *restrt.q* for the flow variables file. Then, examine the input file, *input*, to check that the input parameters are set properly. Finally, execute *xcpt.exe* to solve this case. Figure A.4 shows the predicted

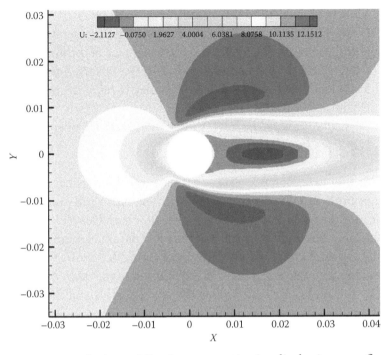

FIGURE A.4 (See color insert following page 294.)   A cylinder in cross-flow.

u-velocity contours of a converged solution of this case. The recirculating flow downstream of the circular cylinder shows that the length of the
recirculation zone is about 2.2 times the cylinder diameter, which is very
close to the experimental measurement.

A.5 FLOW OVER AN AIRFOIL

This example case is used to simulate a turbulent flow past an NACA-
0012 airfoil. A 2-D C-grid domain is used for numerical computation.
The airfoil coordinates are read to form part of the inner boundary of
the computational domain. A uniform flow of 68 m/s speed (about Mach
number 0.2) is specified at the outer boundary (with mass conservation
condition). A mesh of 401 × 81 × 1, with airfoil wall located at part of the
bottom boundary, is generated in *grid.f* to represent the computational
domain. Initial flow properties are also specified in *grid.f*. Input parameters of this case are provided in an input file, *input*, which is read by the
present CFD code at startup. The following specifications are pertinent to
this example problem:

- Mesh (1 block): $401 \times 81 \times 1$ (wall: $I = 81, 321; J = 1$)
- Working directory: z05-Airfoil (contains grid generation code, *grid.f*, and input file, *input*), the angle of attack of the airfoil is $2°$
- VISC = 1.0E-06, U-inlet = 68.0 m/s, U-ref = 300 m/s, X-ref = 1.0 m, Turbulent flow (Dk-in = 0.005 * (U-inlet/U-ref)^2, DE-in = 0.09 * DK-in$^{1.5}$/(0.03 * 1.0))
- Reynolds number, Re = (1/VISC) (1.0/X-ref) (U-inlet/U-ref) = 2.2667E + 05

To run this case, compile *grid.f* to create *xgrid.exe* using a command prompt on a PC Windows system. Next, execute *xgrid.exe* to generate *restrt.x* for the grid file and *restrt.q* for the flow variables file. Then, examine the input file, *input*, to check that the input parameters are set properly. Finally, execute *xcpt.exe* to solve this case. Figure A.5 shows the predicted Mach number contours of a converged solution of this case. At $2°$ angle of attack, the flow is accelerated along the upper surface and the stagnation point is moved toward the lower surface of the leading edge. An oscillating wake flow is also clearly shown from the solution.

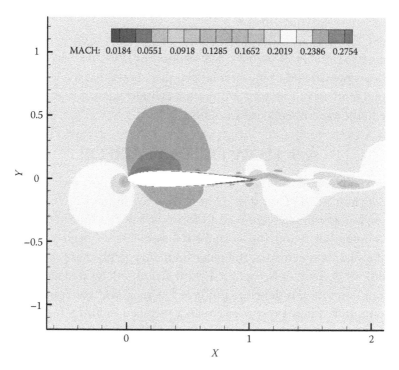

FIGURE A.5 (See color insert following page 294.) Flow over an airfoil.

A.6 CROSS-SECTION OF A SHELL AND TUBE HEAT EXCHANGER

This example case is used to simulate a laminar flow in a heat exchanger. A 2-D O-grid domain is used for numerical computation. The inner and outer shells of the computational domain also include the wall thickness. The initial flow is quiescent and is developed into recirculating flow pattern due to thermal buoyancy effects. A mesh of 121 × 81 × 1, with inner and outer wall segments, is generated in *grid.f* to represent the computational domain. Initial flow properties are also specified in *grid.f*. Input parameters of this case are provided in an input file, *input*, which is read by the present CFD code at startup. The following specifications are pertinent to this example problem:

- Mesh (1 block): 121 × 81 × 1 (walls: $I = 1, 121; J = 1, 11$ and $I = 1, 121; J = 71, 81$)
- Working directory: z06-HeatEx (contains grid generation code, *grid.f*, and input file, *input*), turn on buoyancy effect (CBE = 1.0) and conjugate heat transfer model
- VISC = 1.0E-04, U-initial = 0 m/s, U-ref = 300 m/s, X-ref = 1.0 m, Laminar flow
- Reynolds number, $Re = (1/VISC) (2.0/X\text{-ref}) = 20,000$

To run this case, compile *grid.f* to create *xgrid.exe* using a command prompt on a PC Windows system. Next, execute *xgrid.exe* to generate *restrt.x* for the grid file and *restrt.q* for the flow variables file. Then, examine the input file, *input*, to check that the input parameters are set properly. Finally, execute *xcpt.exe* to solve this case. Figure A.6 shows the predicted temperature contours of a converged solution of this case. Due to temperature difference between the inner and outer shells and the buoyancy effects, recirculating flow pattern is developed to assist the heat exchange between the hot inner shell and the cold outer shell.

A.7 CONVERGING–DIVERGING NOZZLE FLOW

This example case is used to compute a turbulent flow inside a chamber and conical nozzle. A 2-D domain is used for numerical computation. A two-block grid domain is used to describe the geometry. A uniform flow of 100 m/s speed is specified at the chamber inlet. The two-block

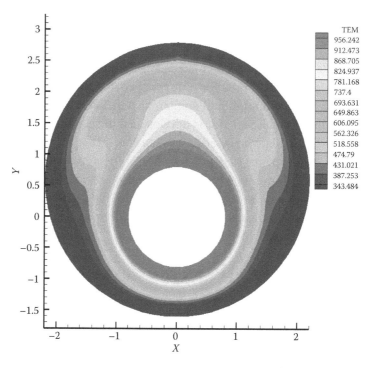

FIGURE A.6 (See color insert following page 294.) Shell and tube heat exchanger.

mesh, of $61 \times 61 \times 1$ and $121 \times 61 \times 1$, with outer wall boundaries, is generated in *grid.f* to represent the computational domain. Initial flow properties are also specified in *grid.f*. Input parameters of this case are provided in an input file, *input*, which is read by the present CFD code at startup. The following specifications are pertinent to this example problem:

- Mesh (2 blocks): $61 \times 61 \times 1$ and $121 \times 61 \times 1$
- Working directory: z07-Nozzle (contains grid generation code, *grid.f* with data file *nozzle.dat*, and input file, *input*)
- VISC = 1.65E-06, U-inlet = 100.0 m/s, U-ref = 300 m/s, X-ref = 0.3048 m, Turbulent flow (Dk-in = 0.005 * (U-inlet/U-ref)^2, DE-in = 0.09 * DK-in$^{1.5}$/(0.03 * 0.5))
- Reynolds number, Re = (1/VISC) (0.4/X-ref) (U-inlet/U-ref) = 8.0808E + 04

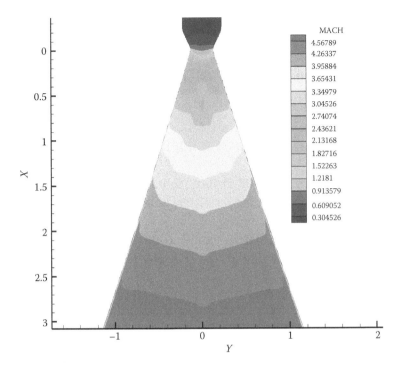

FIGURE A.7 (See color insert following page 294.) Converging–diverging nozzle flow.

To run this case, compile *grid.f* to create *xgrid.exe* using a command prompt on a PC Windows system. Next, execute *xgrid.exe* to generate *restrt.x* for the grid file and *restrt.q* for the flow variables file. Then, examine the input file, *input*, to check that the input parameters are set properly. Finally, execute *xcpt.exe* to solve this case. Figure A.7 shows the predicted Mach number contours of a converged solution of this case. Sonic flow conditions are predicted near the nozzle throat. At the nozzle exit, Mach number around 5 is predicted for the present ideal-gas model.

A.8 ORIFICE FLOW AND AN EJECTOR PUMP

This example case simulates a turbulent flow inside an orifice and ejector pump. A 2D axisymmetric domain is used for numerical computation. A ten-block grid domain is used to describe the geometry, which is created using a separate grid generator. A uniform flow of 5 m/s speed is specified at the orifice inlet. The ten-block mesh, with wall boundaries form the orifice and ejector pump geometry, is generated in *grid.f* to represent the computational domain. Initial flow properties are also specified in *grid.f*. Input parameters of this case are provided in an input file, *input*, which is read by the present CFD code at startup. The following specifications are pertinent to this example problem:

- Mesh (10 blocks): $21 \times 21 \times 1$; $31 \times 41 \times 1$; $16 \times 21 \times 1$; $11 \times 41 \times 1$; $41 \times 21 \times 1$; $41 \times 9 \times 1$; $41 \times 41 \times 1$; $101 \times 21 \times 1$; $101 \times 9 \times 1$; and $101 \times 41 \times 1$
- Working directory: z08-Orifice (contains grid generation code, *grid.f* with grid data file *fort12.fmt* generated by a separate grid generator, and input file, *input*)
- VISC = 1.65E-06, U-inlet = 5.0 m/s, U-ref = 300 m/s, X-ref = 0.1 m, Turbulent flow (Dk-in = 0.005 * (U-inlet/U-ref)^2, DE-in = 0.09 * DK-in$^{1.5}$/(0.03 * 0.1))
- Reynolds number, Re = (1/VISC) (0.01/X-ref) (U-inlet/U-ref) = 1010

To run this case, compile *grid.f* to create xgrid.exe using a command prompt on a PC Windows system. Next, execute *xgrid.exe* to generate restrt.x for the grid file and restrt.q for the flow variables file. Then, examine the input file, *input*, to check that the input parameters are set properly. Finally, execute *xcpt.exe* to solve this case. Figure A.8 shows the predicted

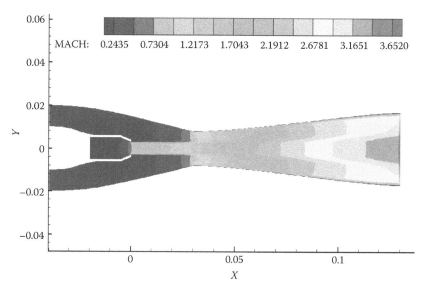

FIGURE A.8 (See color insert following page 294.) Orifice flow and an ejector pump.

Mach number contours of a converged solution of this case. The orifice flow is choked with Mach unity predicted at the orifice. The pumping effects due to the orifice flow are clearly shown from the solution.

A.9 FLOW THROUGH A PIPE ELBOW

This example case simulates a turbulent flow developing through a 90° circular pipe elbow. A 3-D pipe elbow domain is used for numerical computation. A three-block grid domain is used to describe the geometry, which represents an inlet pipe section, a 90° turning section and an exit pipe section. A uniform flow of 100 m/s speed is specified at the pipe inlet with mass conservation condition. Outlet mass conservation condition is also applied. The three-block mesh, with wall boundaries at the lateral boundaries of the geometry, is generated in *grid.f* to represent the computational domain. Initial flow properties are also specified in *grid.f*. Input parameters of this case are provided in an input file, *input*, which is read by the present CFD code at startup. The following specifications are pertinent to this example problem:

- Mesh (3 block2): $41 \times 21 \times 21$; $41 \times 21 \times 21$; and $41 \times 21 \times 21$
- Working directory: z09-PipeElbow (contains grid generation code, *grid.f*, and input file, *input*)
- VISC = 1.65E-06, U-inlet = 100.0 m/s, U-ref = 300 m/s, X-ref = 0.1 m, Turbulent flow (Dk-in = 0.005 * (U-inlet/U-ref)^2, DE-in = 0.09 * DK-in$^{1.5}$/(0.03 * 1.0))
- Reynolds number, Re = (1/VISC) (0.1/X-ref) (U-inlet/U-ref) = 2.0202E + 05

To run this case, compile *grid.f* to create *xgrid.exe* using a command prompt on a PC Windows system. Next, execute *xgrid.exe* to generate *restrt.x* for the grid file and *restrt.q* for the flow variables file. Then, examine the input file, *input*, to check that the input parameters are set properly. Finally, execute *xcpt.exe* to solve this case. Figure A.9 shows the predicted pressure contours of a converged solution of this case. The flow enters from the lower-left pipe inlet boundary and going through the 90° turn section to develop the high and low pressure region along the outer and inner pipe surfaces of the elbow, respectively. The flow then exits the pipe vertically from the outlet boundary at the upper-right of the domain shown.

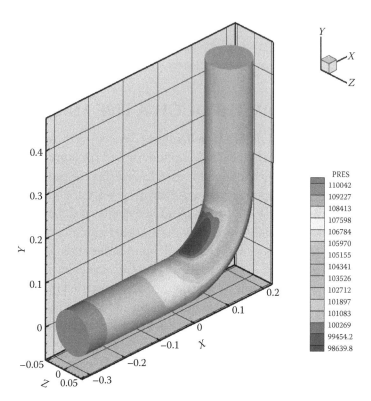

PRES
110042
109227
108413
107598
106784
105970
105155
104341
103526
102712
101897
101083
100269
99454.2
98639.8

FIGURE A.9 (See color insert following page 294.) Flow through a pipe elbow.

A.10 FLOW THROUGH A PIPE TEE

This example case simulates a turbulent flow developing through a square pipe tee geometry. A 3-D pipe tee domain is used for numerical computation. A four-block grid domain is used to describe the geometry, which represents an outlet section, a pipe tee midsection, a second outlet section, and a pipe inlet section. A uniform flow of 100 m/s speed is specified at the pipe inlet with mass conservation condition. Outlet mass conservation conditions are also applied for both outlets. The four-block mesh, with wall boundaries at the lateral boundaries of the geometry, is generated in *grid.f* to represent the computational domain. Initial flow properties are also specified in *grid.f*. Input parameters of this case are provided in an input file, *input*, which is read by the present CFD code at startup. The following specifications are pertinent to this example problem:

- Mesh (4 blocks): 41 × 21 × 21; 21 × 21 × 21; 41 × 21 × 21; and 21 × 41 × 21
- Working directory: z10-PipeT (contains grid generation code, *grid.f*, and input file, *input*)
- VISC = 1.65E-06, U-inlet = 100.0 m/s, U-ref = 300 m/s, X-ref = 0.1 m, Turbulent flow (Dk-in = 0.005 * (U-inlet/U-ref)^2, DE-in = 0.09 * DK-in$^{1.5}$/(0.03 * 1.0))
- Reynolds number, *Re* = (1/VISC) (0.1/X-ref) (U-inlet/U-ref) = 2.0202E + 05

To run this case, compile *grid.f* to create *xgrid.exe* using a command prompt on a PC Windows system. Next, execute *xgrid.exe* to generate *restrt.x* for the grid file and *restrt.q* for the flow variables file. Then, examine the input file, *input*, to check that the input parameters are set properly. Finally, execute *xcpt.exe* to solve this case. Figure A.10 shows the predicted pressure contours of a converged solution of this case. The flow enters from the lower

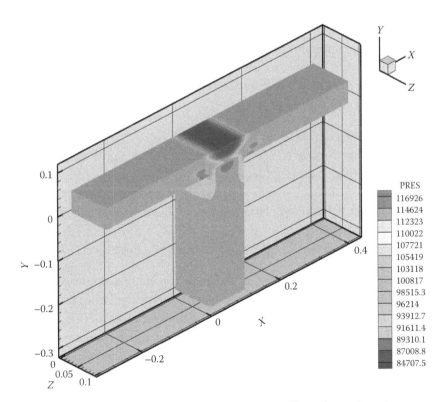

FIGURE A.10 (See color insert following page 294.) Flow through a pipe tee.

pipe inlet boundary and going through the pipe tee junction to develop the high and low pressure region through the midsection of the pipe tee. The flow then exits the pipe tee horizontally from the outlet boundaries at the upper-left and upper-right of the domain shown in the figure.

A.11 FREE-SURFACE FLOW IN AN OPEN DUCT

This example case is designed to simulate a laminar free-surface flow developing through an open duct. A 2-D open duct domain is used for numerical computation. A single-block grid domain is used to describe the geometry, which represents an outlet section, a midsection with mild slope, and a duct outlet section. A uniform flow of 1 m/s speed is speci-fied at the duct inlet with mass conservation condition. Gas flow enters from the upper half of the inlet boundary and liquid water enters from the bottom half. Outlet mass conservation conditions are also applied at the outlet boundary. The real-fluid model (RF = 2) with two species (nitro-gen, N_2, and water, H_2O, NGAS = 2) is employed to simulate the liquid water flow conditions. The single-block mesh, with wall boundaries at the bottom boundary of the geometry, is generated in *grid.f* to represent the computational domain. Initial flow properties are also specified in *grid.f*. Input parameters of this case are provided in an input file, *input*, which is read by the present CFD code at startup. The following specifications are pertinent to this example problem:

- Mesh (1 block): $141 \times 81 \times 1$
- Working directory: z11-FreeSurface (contains grid generation code, *grid.f*, and input file, *input*)
- VISC = 5.0E-06, U-inlet = 1.0 m/s, U-ref = 300 m/s, X-ref = 0.1 m, Laminar flow, RF = 2, NGAS = 2 and CEC thermodynamics data specified
- Reynolds number, $Re = (1/\text{VISC}) (0.1/\text{X-ref}) (\text{U-inlet}/\text{U-ref}) = 666.67$

To run this case, compile *grid.f* to create *xgrid.exe* using a command prompt on a PC Windows system. Next, execute *xgrid.exe* to generate *restrt.x* for the grid file and *restrt.q* for the flow variables file. Then, exam-ine the input file, *input*, to check that the input parameters are set properly. Finally, execute *xcpt.exe* to solve this case. Figure A.11 shows the predicted density contours of a converged solution of this case. The flow enters from the left boundary of the open duct and accelerates through the middle sec-tion with minor slope, then exits from the outlet boundaries at the right

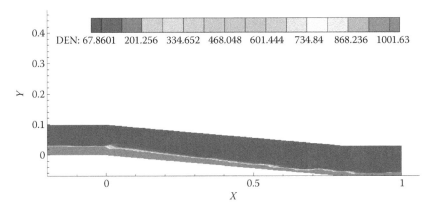

FIGURE A.11 (See color insert following page 294.) Free-surface flow in an open duct.

side of the domain shown in the figure. It is clearly seen that the water layer is thinning due to acceleration through the middle section.

A.12 FLOW IN A STIRRED-TANK

This example case simulates a turbulent flow developing in a stirred-tank geometry. A 3-D stirred-tank domain is used for numerical computation. A four-block grid domain is used to describe the geometry, which represents a circular cylinder stirred-tank section, two rectangular inlet pipes, and a square outlet pipe. A uniform flow of 10 m/s speed is specified at both rectangular pipe inlets (oxygen gas enters from the left boundary and hydrogen enters from the right boundary) with mass conservation condition. Outlet mass conservation condition is applied at the outlet boundary. The four-block mesh, with wall boundaries at the lateral boundaries of the geometry, is generated in *grid.f* to represent the computational domain. Initial flow properties are also specified in *grid.f*. Input parameters of this case are provided in an input file, *input*, which is read by the present CFD code at startup. The following specifications are pertinent to this example problem:

- Mesh (4 blocks): $41 \times 41 \times 41$; $21 \times 11 \times 11$; $21 \times 11 \times 11$; and $15 \times 21 \times 15$
- Working directory: z12-StirredTank (contains grid generation code, *grid.f*, and input file, *input*)
- VISC = 1.0E-06, U-inlet = 10.0 m/s, U-ref = 300 m/s, X-ref = 1 m/s, Turbulent flow (Dk-in = 0.005 * (U-inlet/U-ref)^2, DE-in = 0.09 *

DK-in$^{1.5}$/(0.03 $*$ 1.0)), reacting flow model (SP = 32), NGAS = 6 and a single-step reaction

- Reynolds number, Re = (1/VISC) (1.0/X-ref) (U-inlet/U-ref) = 3.3333E + 04

To run this case, compile *grid.f* to create *xgrid.exe* using a command prompt on a PC Windows system. Next, execute *xgrid.exe* to generate *restrt.x* for the grid file and *restrt.q* for the flow variables file. Then, examine the input file, *input*, to check that the input parameters are set properly. Finally, execute *xcpt.exe* to solve this case. Figure A.12 shows the predicted oxygen (O$_2$) mass-fraction contours of a converged solution of this case. The flow enters from the lower-left and lower-right pipe inlet boundaries and going through the circular cylinder stirred-tank to provide mixing of the oxygen and hydrogen gases. Due to low temperature conditions, no apparent reaction occurred in the flow domain. The flow then exits

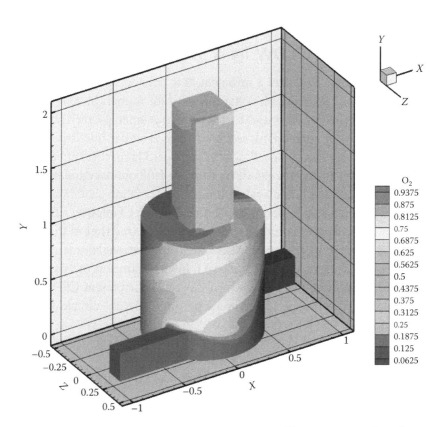

FIGURE A.12 (See color insert following page 294.) Flow in a stirred-tank.

the upper pipe vertically from the outlet boundaries at the upper part of the domain shown in the figure. Mixing effects are shown through the flow domain.

A.13 ADDITIONAL COMMENTS

A few additional comments on the files in the working directory are given here. Generally, the files found in the working directory (folder) are shown in Table A.1. The text in this appendix describes most of these files. The output file in Tecplot format is the final result of the simulation. If one does not have access to the Tecplot code, the following example would enable one to read and/or plot the results with another code.

Example of a Tecplot data file:
TITLE = PROPERTIES
VARIABLES = "X" "Y" "Z" "U" "V" "W" "PRES" "DEN" "TEM"
"MACH" "..."
ZONE I = 31, J = 21, K = 11, F = BLOCK
-0.250000E-01 -0.248516E-01 -0.246905E-01 -0.245159E-01 -0.243269E-01

TABLE A.1 Summary of File Types for the CTP Code

File	Type	Purpose
example name	Lists files in folder	Identify working directory
dbase.dat	DAT file	See text and z12 stierred tank for data format
fluid.inp	INP file	See text
fort.59	59 file	User specified output data
grid.f	Fortran source	See txt
grid.out	inter. file	Provide grid to solver
grid00.f	Fortran source	A specific grid generator
input	File	Input for case to be run
make.bat	MS-DOS Batch	Written to work on a variety of computers.
make01	File	These make files should require no user modification
PO.lay	tecplot document	Prepares data for ploting
restrt.q	Q file	See text
restrt.x	X file	See text
tecplot.dat	DAT file	Output from simulation in tecplot format
tecplot.phy	PHY file	Names file of the example
U-01.tif	MS-office imaging	Graphics figure of results

-0.241225E-01 -0.239018E-01 -0.236638E-01 -0.234076E-01 -0.231323E-01

...

...

...

ZONE I= 61, J= 41, K = 11, F = BLOCK

-0.250000E-01 -0.248516E-01 -0.246905E-01 -0.245159E-01

-0.243269E-01

-0.241225E-01 -0.239018E-01 -0.236638E-01 -0.234076E-01 -0.231323E-01

...

...

...

1. The first line is a header. If a key word "TITLE" present then the following words will be ignored and used as a record.
2. The second line is also a header. If a key word "VARIABLES," then all the variables contained in this data file need to be listed sequentially after the key word. For 2-D flow, the first two variables will be the X- and Y-coordinates of the grid points; whereas for 3-D flow, the first three variables are the X-, Y-, and Z-coordinates of the grid points as shown in the example file. The values of all flow variables such as velocity components, pressure, temperature, etc. will be written to the data file in the same sequence as that listed in the second line.
3. The third line is a zone record, which consists a keyword "ZONE" followed by a set of numerical data called the zone data, which lists the number of grid points in the I-, J-, and K-directions in that zone. There are two data formats, "POINTS" or "BLOCK" can be specified after the record of grid point numbers. In the "POINT" format, the values for all variables are given for the first point, then the second point, and so on. In "BLOCK" format, all of the values for the first variable are given in a block, then all the values for the second variable, then all the values for the third variable, and so forth.
4. The fourth line and onwards are the values of all variables in either the "POINT" or "BLOCK" format.
5. If multizone mesh system is used, then Items 3 and 4 will be repeated for each zones.

For additional details, documentation of a Tecplot data file can be found at the webpage (http://www.tecplot.com).

Further details on the files in the working directory should be obtained by study of the examples described in this appendix. The examples and their solution and the CTP code can be obtained from the publishers webpage.

A.14 NOMENCLATURE

See Section 6.5 and Chapter 8.

REFERENCES

Coutanceau, M. and R. Bouard. 1977. Experimental determination of the main features of the viscous flow in the wake of a circular cylinder in uniform translation. Part 1, Steady flow. *J. Fluid Mech.* 79:231–256.

Kim, J., S. Kline, and J. Johnson. 1978. Investigation of separation and reattachment of a turbulent shear layer. Flow over a backward facing step. Rpt. MD, 37. Mechanical Engineering, Stanford University, CA.

Schlichting, H. 1979. *Boundary Layer Theory*, 7th ed. New York: McGraw-Hill.

Rudiments of Vector and Tensor Analysis

B.1 OVERVIEW

Equations which represent physical processes must be consistent with respect to the mathematical operations involved and the units employed. The partial differential equations which describe complex transport phenomena must be solved numerically on a computational grid. Both vectors and second-order tensors must be used to represent the physical phenomena of interest. Tensor analysis conveniently describes the coordinates and operations which are used to formulate the conservation equations. Unfortunately, there are several dialects of tensor terminology in use. Scalars, vectors, and tensors may be described from a purely geometric viewpoint, or as objects arising from linear transformations of matrix algebra. The geometric objects have evolved from those described with Cartesian coordinates. The approach based on linear transformations may have no reference whatsoever to geometric base vectors or directions. Details of this twofold development of tensor analysis are described by Sokolnikoff (1964) and Lanczos (1961). This appendix introduces tensor analysis in sufficient detail to describe industrial transport processes, without addressing the full ramifications of tensor methodology as used in other branches of physics and mathematics.

The presentation will proceed in three steps:

- Tensor objects in orthogonal Cartesian coordinates will be discussed as it is the simplest methodology.
- Tensor objects in curvilinear, nonorthogonal coordinates will be discussed.
- Linear transformations without geometric interpretations will be discussed.

The order of presentation roughly covers the historical development and generalization of the concepts involved. Be advised, there is no standard nomenclature in use for describing the various dialects of tensor methodology. Also, many authors do not carefully distinguish to which type of tensor analysis their work applies.

B.2 CARTESIAN TENSORS

Consider an orthogonal Cartesian coordinate (OCC) system consisting of base vectors, coordinates and an origin, such that the position vector is defined by

$$\vec{R} = X_1\vec{\chi}_1 + X_2\vec{\chi}_2 + X_3\vec{\chi}_3 \tag{B.1}$$

The location of R is given by the three scalar components (X_i) measured in length along the three directions $(\vec{\chi}_i)$. The base vectors are of unit magnitude. Unless stated otherwise, the coordinate system shall be right handed. This means that the three, positive-directions denoted by 1, 2, and 3 shall point in the direction of the thumb, index finger, and middle finger of the right hand. A plane or surface vector in OCC is identified as being directed normal to the surface and positive in the outward direction of the volume it encloses.

For two points, $\vec{R}\{1\}$ and $\vec{R}\{2\}$ in the $\vec{\chi}_3$ plane, $\vec{R}\{2\}-\vec{R}\{1\}$ would be a vector. If a linear transformation were performed to create a second OCC system, the two position points would be invariant. That is, a new set of X_i's (say X_j's) to go with the new set of specified $\vec{\chi}_i$'s (say $\vec{\chi}_j$'s) would produce the same two $\vec{R}$'s. The new directions would be chosen by the coordinate transformation; the distances along the new directions would be determined such that the original points were not moved. The straight line between these two points would constitute a vector (say $\vec{A}$); in this case from point 1 to point 2. The length of the vector, its magnitude, would be

$$\left|\vec{A}\right| = A = +\sqrt{(X_2\{2\} - X_2\{1\})^2 + (X_1\{2\} - X_1\{1\})^2}$$
$$= +\sqrt{(X_2^{\cdot}\{2\} - X_2^{\cdot}\{1\})^2 + (X_1^{\cdot}\{2\} - X_1^{\cdot}\{1\})^2} \tag{B.2}$$

Both the magnitude and direction of the vector quantity are invariant with coordinate transformations. Note, this example is for a two-dimensional situation. If the coordinate lengths were scaled to match another variable, say velocity, the vector would have the scaled units. The vector is defined by its three scalar components (or elements), once a coordinate system is chosen.

$$\vec{A} = A_1\vec{\chi}_1 + A_2\vec{\chi}_2 + A_3\vec{\chi}_3 \tag{B.3}$$

The position point could also be scaled to represent some nondirectional quantity (a scalar) such as temperature. The scalar is defined to be a single value at a point located by the chosen coordinate system.

A second-order tensor in OCC is defined by nine scalar components associated with two directions simultaneously:

$$\vec{\vec{T}} \equiv T_{11}\vec{\chi}_1\vec{\chi}_1 + T_{12}\vec{\chi}_1\vec{\chi}_2 + T_{13}\vec{\chi}_1\vec{\chi}_3$$
$$+ T_{21}\vec{\chi}_2\vec{\chi}_1 + T_{22}\vec{\chi}_2\vec{\chi}_2 + T_{23}\vec{\chi}_2\vec{\chi}_3$$
$$+ T_{31}\vec{\chi}_3\vec{\chi}_1 + T_{32}\vec{\chi}_3\vec{\chi}_2 + T_{33}\vec{\chi}_3\vec{\chi}_3 \tag{B.4}$$

This is the point where complexity enters into the description of tensors. Just as with a vector, a coordinate system is first chosen and then nine scalar components are specified to define the second-order tensor. If the coordinate system is redefined, a new set of nine components must be evaluated to redefine the second-order tensor. The rules and identities developed for defining vectors and second-order tensors in OCC must be completely changed to define these quantities in curvilinear coordinate systems. Furthermore, tensor is a term which has been "generalized" to the point where one is not sure of its meaning in a particular instance. Scalars, vectors, and tensors have been used to mean zero-, first-, and second-order tensors. Second-order tensors have also been termed a dyad or outer products of two vectors. Tensor has been used to indicate a particular type of linear transformation, i.e., objects which transform by a certain set of equations (rules) are termed certain types of tensors. One must know how the author uses the term in its local context. In this appendix, "tensor" shall be used exclusively to mean a second-order tensor. Any additional meaning will be stated as a modification to the term as it is used.

B.2.1 SCALAR, VECTOR, AND TENSOR ALGEBRA IN OCC

Physical quantities needed to describe transport phenomena can be placed in the following three categories:

- Scalars (zero-order tensors)—temperature, pressure, volume
- Vectors (first-order tensors)—velocity, momentum, force
- Tensors (second-order tensors)—stress tensor, rate-of-strain tensor

Physical laws must be independent of any particular coordinate system used to describe them mathematically if they are to be valid. The physical requirements place mathematical restrictions on the elements which define vectors and tensors. The relationship between these elements and the physical and flow properties of the fluid are termed the constitutive equations of continuum mechanics. These include the principle of material indifference which states that the response of a material must be the same to all observers, no matter their coordinate system of reference (Brodkey, 1967). This analysis leads to the rheological equations of state. Rheology is the science of deformation and flow. Turbulence models are required to describe the constitutive equations when necessary.

Vectors and tensors will be represented with the component and base vector nomenclature introduced in Sections B.2 and B.3. This nomenclature was devised by Gibbs during 1881–1884 (Gibbs and Wilson, 1960) and is most often used to represent engineering equations. Vectors are sometimes said to be represented by a 3×1 column matrix and tensors by a 3×3 matrix. This is not generally true. Usually, such a statement means that the vector and tensor components may be represented by matrix nomenclature. The usual rules for matrix manipulation apply to such components. Occasionally, base vectors are included as components in a matrix representation of a vector or tensor manipulation. Such practice is legitimate and offers no complications. Section B.6 will describe the use of matrices to describe vectors and tensors as linear transformations.

Algebraic operations involving scalars, vectors, and tensors are summarized in Table B.1. The sum of two vectors is a vector whose components are the sum of the corresponding components of the two summed vectors. The dot (·) product of two vectors is a scalar, the product of their magnitudes times the cosine of the included angle between them. Thus, it is the projection of either of these vectors on the other. The cross (×) product of two vectors is a vector of magnitude equal to the product of the magnitudes of the two vectors times the sine of their included angle. This magnitude is equal to the area of the parallelogram formed by the two product vectors. The direction of the vector product is normal to the surface defined by the two vectors and is normal to that surface in a positive direction indicated by the positive advancement of a screw rotating from the first vector toward the second vector. For three vectors, consider first: $\vec{A} \cdot \vec{B} \times \vec{C}$. This is a scalar of magnitude equal to the volume of the parallelepiped of which $\vec{A}$, $\vec{B}$, $\vec{C}$ coterminous edges. Next consider: $\vec{A} \times \vec{B} \cdot \vec{C} = \vec{A} \cdot \vec{B} \times \vec{C}$ which is equal to the determinant formed by the three coefficients

TABLE B.1 Cartesian Vector and Tensor Algebra

Properties of the unit (base) vectors

$$\vec{\chi}_1 \cdot \vec{\chi}_1 = \vec{\chi}_2 \cdot \vec{\chi}_2 = \vec{\chi}_3 \cdot \vec{\chi}_3 = 1$$
$$\vec{\chi}_1 \cdot \vec{\chi}_2 = \vec{\chi}_2 \cdot \vec{\chi}_3 = \vec{\chi}_3 \cdot \vec{\chi}_1 = 0 \tag{A}$$

$$\vec{\chi}_1 \times \vec{\chi}_1 = \vec{\chi}_2 \times \vec{\chi}_2 = \vec{\chi}_3 \times \vec{\chi}_3 = 0$$
$$\vec{\chi}_1 \times \vec{\chi}_2 = \vec{\chi}_3 = -\vec{\chi}_2 \times \vec{\chi}_1$$
$$\vec{\chi}_2 \times \vec{\chi}_3 = \vec{\chi}_1 = -\vec{\chi}_3 \times \vec{\chi}_2$$
$$\vec{\chi}_3 \times \vec{\chi}_1 = \vec{\chi}_2 = -\vec{\chi}_1 \times \vec{\chi}_3 \tag{B}$$

Vector $\vec{A}$

$$\vec{A} = A_1 \vec{\chi}_1 + A_2 \vec{\chi}_2 + A_3 \vec{\chi}_3 \tag{C}$$

$$|A| = A = \sqrt{A_1^2 + A_2^2 + A_3^2} \tag{D}$$

Dot product of two vectors

$$\vec{A} \cdot \vec{B} = |\vec{A}| \cos\{\vec{A}, \vec{B}\} \, |\vec{B}| = A \cos\{\vec{A}, \vec{B}\} \, B = \sum_{i=1}^{3} A_i B_i \tag{E}$$

Cross product of two vectors

$$\vec{A} \times \vec{B} = \vec{V} = \begin{vmatrix} \vec{\chi}_1 & \vec{\chi}_2 & \vec{\chi}_3 \\ A_1 & A_2 & A_3 \\ B_1 & B_2 & B_3 \end{vmatrix} = (A_2 B_3 - A_3 B_2)\vec{\chi}_1 + (A_3 B_1 - A_1 B_3)\vec{\chi}_2 + (A_1 B_2 - A_2 B_1)\vec{\chi}_3 \tag{F}$$

where $|\vec{V}| = |\vec{A}| \sin\{\vec{A}, \vec{B}\} \, |\vec{B}|$

Vector resolute is the projection of one vector on another

$$(|\vec{B}| \cos\theta)(\vec{A}/A) = \vec{C} \tag{G}$$

Scalar resolute is the projection of one vector on the unit vector of another

$$\vec{B} \cdot (\vec{A}/A) = |\vec{B}| \cos\theta = D \tag{H}$$

(continued)

TABLE B.1 (continued) Cartesian Vector and Tensor Algebra

Tensor $\vec{\vec{T}}$

$$
\begin{aligned}
\vec{\vec{T}} \equiv\ & T_{11}\vec{\chi}_1\vec{\chi}_1 + T_{12}\vec{\chi}_1\vec{\chi}_2 + T_{13}\vec{\chi}_1\vec{\chi}_3 \\
& + T_{21}\vec{\chi}_2\vec{\chi}_1 + T_{22}\vec{\chi}_2\vec{\chi}_2 + T_{23}\vec{\chi}_2\vec{\chi}_3 \\
& + T_{31}\vec{\chi}_3\vec{\chi}_1 + T_{32}\vec{\chi}_3\vec{\chi}_2 + T_{33}\vec{\chi}_3\vec{\chi}_3
\end{aligned}
\tag{I}
$$

Properties of the unit base vectors using cyclic permutations

$$
\begin{aligned}
&(\vec{\chi}_1\vec{\chi}_1)\cdot\vec{\chi}_1 = \vec{\chi}_1(\vec{\chi}_1\cdot\vec{\chi}_1) = \vec{\chi}_1 \qquad \vec{\chi}_1\cdot(\vec{\chi}_1\vec{\chi}_1) = (\vec{\chi}_1\cdot\vec{\chi}_1)\vec{\chi}_1 = \vec{\chi}_1 \\
&(\vec{\chi}_1\vec{\chi}_2)\cdot\vec{\chi}_1 = \vec{\chi}_1(\vec{\chi}_2\cdot\vec{\chi}_1) = 0 \qquad \vec{\chi}_1\cdot(\vec{\chi}_1\vec{\chi}_2) = (\vec{\chi}_1\cdot\vec{\chi}_1)\vec{\chi}_j = \vec{\chi}_j \\
&(\vec{\chi}_2\vec{\chi}_1)\cdot\vec{\chi}_1 = \vec{\chi}_2(\vec{\chi}_1\cdot\vec{\chi}_1) = \vec{\chi}_2 \qquad \vec{\chi}_1\cdot(\vec{\chi}_2\vec{\chi}_1) = (\vec{\chi}_1\cdot\vec{\chi}_2)\vec{\chi}_1 = 0 \\
&\text{repeat for } \vec{\chi}_1,\vec{\chi}_2 \Rightarrow \vec{\chi}_2,\vec{\chi}_3 \Rightarrow \vec{\chi}_3,\vec{\chi}_1
\end{aligned}
\tag{J}
$$

$$
\begin{aligned}
&(\vec{\chi}_1\vec{\chi}_1)\times\vec{\chi}_1 = \vec{\chi}_1(\vec{\chi}_1\times\vec{\chi}_1) = 0 \qquad \vec{\chi}_1\times(\vec{\chi}_1\vec{\chi}_1) = (\vec{\chi}_1\times\vec{\chi}_1)\vec{\chi}_1 = 0 \\
&(\vec{\chi}_1\vec{\chi}_2)\times\vec{\chi}_1 = \vec{\chi}_1(\vec{\chi}_2\times\vec{\chi}_1) = -\vec{\chi}_1\vec{\chi}_3 \qquad \vec{\chi}_1\times(\vec{\chi}_1\vec{\chi}_2) = (\vec{\chi}_1\times\vec{\chi}_1)\vec{\chi}_2 = 0 \\
&(\vec{\chi}_2\vec{\chi}_1)\times\vec{\chi}_1 = \vec{\chi}_2(\vec{\chi}_1\times\vec{\chi}_1) = 0 \qquad \vec{\chi}_1\times(\vec{\chi}_2\vec{\chi}_1) = (\vec{\chi}_1\times\vec{\chi}_2)\vec{\chi}_1 = \vec{\chi}_3\vec{\chi}_1 \\
&\text{repeat for } \vec{\chi}_1,\vec{\chi}_2,\vec{\chi}_3 \Rightarrow \vec{\chi}_2,\vec{\chi}_3,\vec{\chi}_1 \Rightarrow \vec{\chi}_3,\vec{\chi}_1,\vec{\chi}_2
\end{aligned}
\tag{K}
$$

The repetitive substitutions are cyclic permutation.

Dyad, a product of two vectors without a × or · contraction

$$
\begin{aligned}
\vec{A}\vec{B} \equiv\ & A_1B_1\vec{\chi}_1\vec{\chi}_1 + A_1B_2\vec{\chi}_1\vec{\chi}_2 + A_1B_3\vec{\chi}_1\vec{\chi}_3 \\
& + A_2B_1\vec{\chi}_2\vec{\chi}_1 + A_2B_2\vec{\chi}_2\vec{\chi}_2 + A_2B_3\vec{\chi}_2\vec{\chi}_2 \\
& + A_3B_1\vec{\chi}_3\vec{\chi}_1 + A_3B_2\vec{\chi}_3\vec{\chi}_2 + A_3B_3\vec{\chi}_3\vec{\chi}_2
\end{aligned}
\tag{L}
$$

The product of two vectors is a dyad, but a dyad is not necessarily the product of two vectors, the dyad is a (second-order) tensor.

Properties of unit dyads

$$
\begin{aligned}
&(\vec{\chi}_i\vec{\chi}_j : \vec{\chi}_k\vec{\chi}_\ell) = (\vec{\chi}_j\cdot\vec{\chi}_k)(\vec{\chi}_i\cdot\vec{\chi}_\ell) = \delta_{jk}\delta_{i\ell} \\
&(\vec{\chi}_i\vec{\chi}_j\cdot\vec{\chi}_k) = \vec{\chi}_i(\vec{\chi}_j\cdot\vec{\chi}_k) = \vec{\chi}_i\delta_{jk} \\
&(\vec{\chi}_i\cdot\vec{\chi}_j\vec{\chi}_k) = (\vec{\chi}_i\cdot\vec{\chi}_j)\vec{\chi}_k = \delta_{ij}\vec{\chi}_k \\
&(\vec{\chi}_i\vec{\chi}_j\cdot\vec{\chi}_k\vec{\chi}_\ell) = \vec{\chi}_i(\vec{\chi}_j\cdot\vec{\chi}_k)\vec{\chi}_\ell = \delta_{jk}\vec{\chi}_i\vec{\chi}_\ell
\end{aligned}
\tag{M}
$$

$$
\vec{\chi}_i\vec{\chi}_j\times\vec{\chi}_k = \vec{\chi}_i(\vec{\chi}_j\times\vec{\chi}_k) = \sum_{\ell=1}^{3}\varepsilon_{jkl}\vec{\chi}_i\vec{\chi}_\ell
$$

$$
\vec{\chi}_i\times\vec{\chi}_j\vec{\chi}_k = (\vec{\chi}_i\times\vec{\chi}_j)\vec{\chi}_k = \sum_{\ell=1}^{3}\varepsilon_{ijk}\vec{\chi}_\ell\vec{\chi}_k
$$

Double dot product of two tensors

$$
\vec{\vec{T}}:\vec{\vec{S}} = \sum_{i=1}^{3}\sum_{j=1}^{3}T_{ij}S_{ji} = \vec{\vec{S}}:\vec{\vec{T}} = E, \quad \text{a scalar}
\tag{N}
$$

TABLE B.1 (continued) Cartesian Vector and Tensor Algebra

		Multiplication Table		
Operation	**Order**	**Contraction**	**Exponent**	**Terms in Product**
S	0	0	0	1
V	1	0	0	3
T	2	0	0	9
$T * T$	4	0	0	81
$T * V$	3	0	3	27
$T \times T$	4	−1	3	27
$T \cdot T$	4	−2	2	9
$V * V = T$	2	0	2	9
$V \div V = T$	2	0	2	9
$V \times V$	2	−1	1	3
$V \cdot V$	2	−2	0	1
$T : T$	4	−4	0	1
$T \cdot V$	3	−2	1	3

Notes: Order is the total number of unit vectors or subscripts in the terms being multiplied.
S is a scalar, V is a vector, and T is a tensor.
The asterisk used as a multiplication symbol is the same as a blank space.

from each of the three vectors. These operations are shown in Table B.1, as well as other vector functions derived from those just mentioned.

A scalar times a tensor is a new tensor with each component multiplied by the scalar.

$$s\vec{\vec{T}} = \vec{\vec{W}} \quad \text{where} \quad sT_{ij} = W_{ij} \tag{B.5}$$

The sum of two tensors is a new tensor whose components are the sum of the components of the two individual tensors:

$$\vec{\vec{T}} = \vec{\vec{U}} + \vec{\vec{V}} \quad \text{where} \quad T_{ij} = U_{ij} + V_{ij} \tag{B.6}$$

For a tensor dotted with a vector and for a vector dotted with a tensor

$$\vec{\vec{T}} \cdot \vec{A} = \vec{B} \quad \text{where} \quad \sum_{j=1}^{3} T_{ij} A_j = B_i \quad \text{for} \quad i = 1,2,3 \tag{B.7}$$

$$\vec{A} \cdot \vec{\vec{T}} = \vec{C} \quad \text{where} \quad \sum_{j=1}^{3} A_j T_{ji} = C_i \quad \text{for} \quad i = 1,2,3 \tag{B.8}$$

$$\text{note} \quad \vec{A} \cdot \vec{\vec{T}} \neq \vec{\vec{T}} \cdot \vec{A}$$

The $\vec{\vec{T}} \cdot \vec{A} = \vec{B}$ expression could be read that the tensor T operates on the vector A to produce a new vector B. This is the connection that will be expanded further in Section B.6 to discuss linear transformations. In dealing with tensors, the dot and cross products of the unit vectors are also needed. These and other tensor operations are also summarized in Table B.1. Symbols-used only to define generic operations included in this section are not included in the nomenclature.

A tensor is symmetric if $T_{ij} = T_{ji}$; it is asymmetric if $T_{ij} = -T_{ji}$ for $i \neq j$. A tensor may be expressed as the sum of a symmetric part and an asymmetric part by

$$\vec{\vec{T}} = \vec{\vec{T}}_s + \vec{\vec{T}}_a \quad \text{where} \quad T_{ij} = \tfrac{1}{2}\left(T_{ij} + T_{ji}\right) + \tfrac{1}{2}\left(T_{ij} - T_{ji}\right) \tag{B.9}$$

B.2.2 SCALAR, VECTOR, AND TENSOR DIFFERENTIAL OPERATORS IN OCC

The differential operator for scalars, vectors, and tensors is denoted by the symbol del, ∇. These operations are shown in Table B.2 for OCC. The *substantial derivative* operator contains the local fluid velocity, $\vec{U}$, and is defined as

$$\frac{D\{\}}{Dt} = \frac{\partial\{\}}{\partial t} + \left(\vec{U} \cdot \nabla\{\}\right) \tag{B.10}$$

The substantial derivative operator applied to a scalar $a\{x, y, z\}$ in OCC coordinates gives

$$\frac{Da}{Dt} = \frac{\partial a}{\partial t} + \left(\vec{U} \cdot \nabla a\right) = \frac{\partial a}{\partial t} + \sum_1^3 U_i \frac{\partial a}{\partial X_i} \tag{B.11}$$

The substantial derivative operator applied to a vector $\vec{A}\{t, X_1, X_2, X_3\}$ in OCC coordinates gives

$$\frac{D\vec{A}}{Dt} = \frac{\partial \vec{A}}{\partial t} + \left(\vec{U} \cdot \nabla \vec{A}\right) = \frac{\partial \vec{A}}{\partial t} + \sum_{i=1}^3 U_i \frac{\partial \vec{A}}{\partial X_i} \tag{B.12}$$

TABLE B.2 Differential Operators in OCC

Del operator

$$\nabla\{\ \} = \frac{\partial\{\ \}}{\partial \vec{R}} = \sum_{i=1}^{3} \vec{\chi}_i \frac{\partial\{\ \}}{\partial X_i} \tag{A}$$

Gradient of a scalar (S)

$$\nabla\{S\} = \frac{\partial\{S\}}{\partial \vec{R}} = \sum_{i=1}^{3} \vec{\chi}_i \frac{\partial\{S\}}{\partial X_i} \tag{B}$$

Divergence of a vector ($\vec{A}$)

$$\nabla \cdot \vec{A} = \sum_{i=1}^{3} \vec{\chi}_i \cdot \frac{\partial \vec{A}}{\partial X_i}$$

$$\nabla \cdot \vec{A} = \frac{\partial A_1}{\partial X_1} + \frac{\partial A_2}{\partial X_2} + \frac{\partial A_3}{\partial X_3} \tag{C}$$

Curl of a vector $\vec{A}$

$$\nabla \times \vec{A} = \sum_{j=1}^{3}\sum_{k=1}^{3} (\vec{\chi}_j \times \vec{\chi}_k)\left(\frac{\partial A_k}{\partial X_j}\right) \tag{D}$$

$$\nabla \times \vec{A} = \vec{\chi}_1\left(\frac{\partial A_3}{\partial X_2} - \frac{\partial A_2}{\partial X_3}\right) + \vec{\chi}_2\left(\frac{\partial A_1}{\partial X_3} - \frac{\partial A_3}{\partial X_1}\right) + \vec{\chi}_3\left(\frac{\partial A_2}{\partial X_1} - \frac{\partial A_1}{\partial X_2}\right)$$

Other operations

$$\nabla\vec{A} = \sum_{i=1}^{3}\sum_{j=1}^{3} \vec{\chi}_i\vec{\chi}_j \frac{\partial A_j}{\partial X_i} \quad \text{and} \quad \vec{A}\nabla\{\ \} = \sum_{i=1}^{3}\sum_{j=1}^{3} \vec{\chi}_i\vec{\chi}_j A_i \frac{\partial\{\ \}}{\partial X_j} \tag{E}$$

Tensor operator

$$\nabla\nabla \equiv \vec{\chi}_1\vec{\chi}_1 \frac{\partial^2}{\partial X_1^2} + \vec{\chi}_1\vec{\chi}_2 \frac{\partial^2}{\partial X_1 \partial X_2} + \vec{\chi}_1\vec{\chi}_3 \frac{\partial^2}{\partial X_1 \partial X_3}$$

$$+ \vec{\chi}_2\vec{\chi}_1 \frac{\partial^2}{\partial X_2 \partial X_1} + \vec{\chi}_2\vec{\chi}_2 \frac{\partial^2}{\partial X_2^2} + \vec{\chi}_2\vec{\chi}_3 \frac{\partial^2}{\partial X_2 \partial X_3}$$

$$+ \vec{\chi}_3\vec{\chi}_1 \frac{\partial^2}{\partial X_3 \partial X_1} + \vec{\chi}_3\vec{\chi}_2 \frac{\partial^2}{\partial X_3 \partial X_2} + \vec{\chi}_3\vec{\chi}_3 \frac{\partial^2}{\partial X_3^2} \tag{F}$$

Let $\vec{\vec{T}} = \sum_{i=1}^{3}\sum_{j=1}^{3} \vec{\chi}_i\vec{\chi}_j T_{ij}$

$$\vec{\vec{T}} \cdot \nabla\{\ \} = \vec{\chi}_1 \sum_{i=1}^{3} T_{1i} \frac{\partial\{\ \}}{\partial X_i} + \vec{\chi}_2 \sum_{i=1}^{3} T_{2i} \frac{\partial\{\ \}}{\partial X_i} + \vec{\chi}_3 \sum_{i=1}^{3} T_{3i} \frac{\partial\{\ \}}{\partial X_i} \tag{G}$$

$$\nabla \cdot \vec{\vec{T}} = \vec{\chi}_1 \sum_{i=1}^{3} \frac{\partial}{\partial X_i} T_{i1} + \vec{\chi}_2 \sum_{i=1}^{3} \frac{\partial}{\partial X_i} T_{i2} + \vec{\chi}_3 \sum_{i=1}^{3} \frac{\partial}{\partial X_i} T_{i3} \tag{H}$$

B.2.3 INTEGRAL EXPRESSIONS IN OCC

Gauss' divergence theorem is proven in Sokolnikoff and Redheffer (1966, p. 397) for vectors. This theorem is stated to be true for tensors in Morse and Feshbach (1953, p. 66, Part I) and in Korn and Korn (1968, p. 560). These theorems are stated as

$$\iiint \left(\nabla \cdot \vec{A} \right) d\forall = \iint \vec{A} \cdot d\vec{S} \tag{B.13}$$

$$\iiint \left(\nabla \cdot \overset{\rightrightarrows}{A} \right) d\forall = \iint \overset{\rightrightarrows}{A} \cdot d\vec{S} \tag{B.14}$$

or

$$\iiint \left(\frac{\partial A_1}{\partial X_1} + \frac{\partial A_2}{\partial X_2} + \frac{\partial A_3}{\partial X_3} \right) dX_1 dX_2 dX_3$$
$$= \iint \left(A_1 dX_2 dX_3 + A_2 dX_3 dX_1 + A_3 dX_1 dX_2 \right) \tag{B.15}$$

$$\iiint \left(\vec{\chi}_1 \sum_{i=1}^{3} \frac{\partial}{\partial X_i} A_{i1} + \vec{\chi}_2 \sum_{i=1}^{3} \frac{\partial}{\partial X_i} A_{i2} + \vec{\chi}_3 \sum_{i=1}^{3} \frac{\partial}{\partial X_i} A_{i3} \right) d\forall$$
$$= \iint \overset{\rightrightarrows}{Q} \cdot \left[\left(\vec{\chi}_1 dX_2 dX_3 \right) + \left(\vec{\chi}_2 dX_3 dX_1 \right) + \left(\vec{\chi}_3 dX_1 dX_2 \right) \right] \tag{B.16}$$

where

$$\overset{\rightrightarrows}{Q} = A_{11} \vec{\chi}_1 \vec{\chi}_1 + A_{12} \vec{\chi}_1 \vec{\chi}_2 + A_{13} \vec{\chi}_1 \vec{\chi}_3 + A_{21} \vec{\chi}_2 \vec{\chi}_1 + A_{22} \vec{\chi}_2 \vec{\chi}_2$$
$$+ A_{23} \vec{\chi}_2 \vec{\chi}_3 + A_{31} \vec{\chi}_3 \vec{\chi}_1 + A_{32} \vec{\chi}_3 \vec{\chi}_2 + A_{33} \vec{\chi}_3 \vec{\chi}_3$$

Although this expression is lengthy, the final result indicates vectors are under the integral signs. Since the OCC unit vectors are constant in direction and magnitude, they can be factored through the integral signs. This means that the integrals are vectors, such that three scalar equations can be extracted and evaluated.

Several other integral theorems are used in transport phenomena, for instance, to change the order of differentiation and integration, the partial differentiation of an integral is given by the Leibnitz formula (Brodkey, 1967):

$$\frac{d}{dt} \int_{a(t)}^{b(t)} f(x,t) dx = \int_{a(t)}^{b(t)} \frac{\partial f(x,t)}{\partial t} dx + f(b,t) \frac{db(t)}{dt} - f(a,t) \frac{da(t)}{dt} \tag{B.17}$$

The Leibnitz formula for differentiating a triple integral with time with a volume $\forall$ surrounded by a closed surface S is given by the following equation:

$$\frac{d}{dt}\int_{\forall} a(X_1, X_2, X_3, t)d\forall = \int_{\forall} \frac{\partial a}{\partial t}\,d\forall + \int_S a(\vec{U}_s \cdot \vec{n})\,dS \qquad (B.18)$$

where

$a\{X_1, X_2, X_3, t\}$ is a scalar function
$\vec{n}$ is the unit normal vector to the surface
$\vec{U}_s$ is the velocity of a surface element

The above equation has the following form when the velocity, $\vec{U}_s$, is the local fluid velocity, $\vec{U}$. The use of the continuity equation is required to obtain this equation.

$$\frac{d}{dt}\int_{\forall} \rho a(X_1, X_2, X_3, t)d\forall = \int_{\forall} \rho\frac{Da}{Dt}\,d\forall \qquad (B.19)$$

where
the substantial derivative of the scalar, a, appears on the right-hand side
ρ is the fluid density

This form of the theorem is called the Reynolds transport theorem (Bird and Stewart, 2002).

The line integral of a vector A bounded by a closed curve is related to the surface integral of the curl of the vector A by Stokes theorem given below:

$$\oint_C \vec{A} \cdot d\vec{L} = \int_S (\nabla \times \vec{A}) \cdot d\vec{S} \qquad (B.20)$$

$$\oint_C \vec{\vec{A}} \cdot d\vec{L} = \int_S (\nabla \times \vec{\vec{A}}) \cdot d\vec{S} \qquad (B.21)$$

Again the tensor form of this integral equation depends on the unit vector being factored through the integral signs to obtain a solution. These relationships are used to determine the strength of a fluid vortex (Brodkey, 1967).

B.3 SCALARS, VECTORS, AND TENSORS IN NONORTHOGONAL CURVILINEAR COORDINATES

B.3.1 OVERVIEW

Cartesian tensors describe only spaces in which the spatial coordinates are straight and never change direction. The various generalizations of tensors are useful and should not be a problem. However, when authors are not careful, the reader might not be informed when certain stated relationships are only valid in a limited type domain. There are other limitations which might be imposed, for example, vector spaces or affine spaces. In vector space, coordinates are defined so that the difference between two close position points is an increment of length, i.e., a metric is defined. In affine space, the difference between two close position points may have no physical meaning. For example, a P–$\forall$–T plot represents an affine space (Brillouin and Brennan, 1964); the difference between two points does not represent a length. Coordinates in tensor analysis are generalized to the point that some coordinates might have linear dimensions and some angular dimensions, for example, cylindrical coordinates. Base vectors are used to indicate direction. If the base vectors in a coordinate system are of unit magnitude, we shall term the coordinates physical. We shall use the term mathematical to represent coordinates, vectors, and tensors in general. Finally, some literature has generalized the term coordinates to mean locations in n-dimensional spaces. We do not need n dimensions to describe transport phenomena. Three dimensions are adequate, and four is the maximum needed when adaptive grids are utilized. Vinokur (1974) formally used time to cast the conservation equations as a four-dimensional system, in order to state the equations in a strong conservation form. This appendix is limited to the description of three-dimensional nonadaptive coordinates and tensor relations pertinent thereto.

The grid will be considered fixed relative to an inertial reference. Fictitious Coriolis and centrifugal forces will be utilized to approximate moving coordinates. Adaptive grids are used to reposition grid points during the course of a computational simulation. However, grid locations do not need to be coupled in real time or iteration level to provide useful solutions. More complicated noninertial flow analyses are not considered herein; for example, see complex gravitational analysis (Misner et al., 1973).

Many books on tensor analysis do not clearly address the issues and generalizations just mentioned. The texts by Margenau and Murphy (1956), Brillouin and Brennan (1964), Hawkins (1963), Sokolnikoff (1964), and Borisenko and Tarapov (1968) are recommended for their clarity and completeness.

B.3.2 TYPES OF COORDINATE SYSTEMS

We have stated that the value of scalars, vectors, and tensors in analyzing physical phenomena is that they do not change in magnitude or direction regardless of the coordinate system being used. Obviously, a scalar does not change because the coordinates change. The transformation of a scalar from one coordinate system to another is called an invariant transformation, i.e., the scalar does not change its value. Vectors indicated with an arrow over them or tensors with two arrows over them do not change with coordinate transformations. However, we have to play lawyer at this point and weasel out of this simple claim. The components used to define vectors and tensors certainly do change when the coordinate system is changed. In fact, for some coordinate systems, some of the coordinates are in angles and some are in lengths. To obtain a clear understanding of geometric tensors and vectors, six coordinate systems will be defined and given their individual nomenclature. Then more complex geometric objects will be described. These six coordinate systems are defined in Table B.3.

Orthogonal Cartesian and curvilinear coordinates and physical nonorthogonal curvilinear coordinates are used extensively in solving the transport equations. The other three types of coordinate systems must be understood to read the basic tensor literature. Nonorthogonal Cartesian coordinates is another type of coordinate system, but they will not be considered. Any use of the term Cartesian coordinates herein shall refer only to orthogonal Cartesian coordinates.

TABLE B.3 Six Basic Coordinate Systems

Orthogonal (rectangular) Cartesian coordinates (OCC)

$$d\vec{R} = \sum_{i=1}^{3} \vec{\chi}_i * dX_i \tag{A}$$

Orthogonal curvilinear coordinates (NCC)

$$d\vec{R} = \sum_{i=1}^{3} \vec{\upsilon}_i * H_i * dY_i \tag{B}$$

Mathematical (tangential) general curvilinear coordinates (MTC)
(Covariant base vectors and contravariant coordinate lines)

$$d\vec{R} = \sum_{i=1}^{3} \frac{\partial \vec{R}}{\partial Z^i} * dZ^i = \sum_{i=1}^{3} \vec{\zeta}_i * dZ^i \tag{C}$$

Physical (tangential) general curvilinear coordinates (PTC)

$$d\vec{R} = \sum_{i=1}^{3} \frac{\partial \vec{R}}{\partial Z^{(i)}} * dZ^{(i)} = \sum_{i=1}^{3} \vec{\zeta}_{(i)} * dZ^{(i)} $$

$$\text{where } \vec{\zeta}_{(i)} = \frac{1}{\sqrt{g_{ii}}} \vec{\zeta}_i \quad \text{and} \quad \Delta Z^{(i)} = \sqrt{g_{ii}} \Delta Z^i \tag{D}$$

(continued)

TABLE B.3 (continued) Six Basic Coordinate Systems

Mathematical (normal) general curvilinear coordinates (MNC)
(Contravariant base vectors and covariant coordinate lines)

$$d\vec{R} = \sum_{i=1}^{3} \frac{\partial \vec{R}}{\partial Z_i} * dz_i = \sum_{i=1}^{3} \vec{\zeta}^i * dZ_i \tag{E}$$

Physical (normal) general curvilinear coordinates (PNC)

$$d\vec{R} = \sum_{i=1}^{3} \frac{\partial \vec{R}}{\partial Z_{(i)}} * dZ_{(i)} = \sum_{i=1}^{3} \vec{\zeta}^{(i)} * dZ_{(i)} \tag{F}$$

$$\text{where } \vec{\zeta}^{(i)} = \frac{1}{\sqrt{g^{ii}}} \vec{\zeta}^i \quad \text{and} \quad \Delta Z_{(i)} = \sqrt{g^{ii}} \Delta Z_i$$

$$\text{where } |\vec{\chi}_i| = |\vec{v}_i| = |\vec{\zeta}_{(i)}| = |\vec{\zeta}^{(i)}| = 1 \quad \text{and} \quad |\vec{\zeta}_i| = \sqrt{g_{ii}} \quad \text{and} \quad |\vec{\zeta}^i| = \sqrt{g^{ii}}$$

Six kinds of coordinate systems are defined by the local value of an increment of the position vector. The asterisk indicates simple multiplication which is neither a dot nor a cross operation.

B.3.2.1 Distances Associated with $\overrightarrow{dR}$

For OCC system, $\vec{R} = X_1\vec{\chi}_1 + X_2\vec{\chi}_2 + X_3\vec{\chi}_3$. For the other coordinate systems, the integral value of $\vec{R}$ is not as easy to evaluate, since the base vectors $\left(\dfrac{\partial \vec{R}}{\partial \ell}\right)$, where ℓ represents any of the coordinate lines, are not always oriented in the same direction, nor are they necessarily constant. Base vectors may be defined to have a constant magnitude of one, in which case they are termed unit vectors. However, the incremental arc length is the same in each coordinate system. The strategy is that we know the behavior of the OCC system, therefore the other coordinates systems will be defined in terms of the known OCC system.

Arc length squared is

$$(ds)^2 = d\vec{R} \cdot d\vec{R} = dX_i dX_i = \sum_{i=1}^{3} (dX_i)^2 \tag{B.22}$$

Let

$$\vec{\zeta}_i = \frac{\partial \vec{R}}{\partial Z^i} = \sum_{j=1}^{3} \vec{\chi}_j \frac{\partial X_j}{\partial Z^i} = \vec{\chi}_j \frac{\partial X_j}{\partial Z^i} \tag{B.23}$$

This relationship may be described as a transformation of the base vector in the OCC system to one in the MTC system. This form of the transformation law is termed covariant and the $\vec{\zeta}_i$ are the covariant vector components whereas the Z^i are the contravariant coordinates. The contravariant law is an alternative transformation; it reads

$$\vec{\zeta}^i = \frac{\partial \vec{R}}{\partial Z_i} = \sum_{j=1}^{3} \vec{\chi}_j \frac{\partial Z_i}{\partial X_j} = \vec{\chi}_j \frac{\partial Z_i}{\partial X_j} \tag{B.24}$$

The $\vec{\zeta}^i$ are the contravariant vector components and the Z_i are the covariant coordinates. The superscripts and subscripts used in defining the MTC and MNC systems are not needed in the OCC system because the normals and tangents are in the same directions when the coordinates are orthogonal. Returning to the calculation of the arc length,

$$d\vec{R} = \sum_{i=1}^{3} \frac{\partial \vec{R}}{\partial Z^i} dZ^i = \sum_{i=1}^{3} \vec{\zeta}_i dZ^i = \vec{\zeta}_i dZ^i \tag{B.25}$$

$$d\vec{R} = \sum_{i=1}^{3} \frac{\partial \vec{R}}{\partial Z_i} dZ_i = \sum_{i=1}^{3} \vec{\zeta}^i dZ_i = \vec{\zeta}^i dZ_i \tag{B.26}$$

Arc length is

$$(ds)^2 = d\vec{R} \cdot d\vec{R} = \left(\sum_{i=1}^{3} \sum_{j=1}^{3} \vec{\zeta}_i \cdot \vec{\zeta}_j dZ^i dZ^j \right) = g_{ij} dZ^i dZ^j \tag{B.27}$$

$$(ds)^2 = d\vec{R} \cdot d\vec{R} = \left(\sum_{i=1}^{3} \sum_{j=1}^{3} \vec{\zeta}^i \cdot \vec{\zeta}^j dZ_i dZ_j \right) = g^{ij} dZ_i dZ_j \tag{B.28}$$

Note, the second dummy index has been changed from i to j which is legitimate.

Frequently the summation sign is omitted and the summation is indicated by the repeated subscript, this is termed the Einstein summation convention. Also, note that the relationship between the g's with superscripts and subscripts has not yet been established.

B.3.2.2 Metric Tensor

For nonorthogonal coordinate systems, the coordinate surfaces are defined as surfaces for which one of the coordinates is constant. The intersection of two coordinate surfaces defines a coordinate line. Two coordinate systems result from this observation. The coordinates may be tangential to the coordinate lines ($\vec{\zeta}_i, Z^i$) or normal to the coordinate surfaces ($\vec{\zeta}^i, Z_i$). The superscript or subscript position in nonorthogonal coordinates is significant and must be properly placed, as this positioning distinguishes between these two types of coordinate systems. The use of the index position to distinguish between coordinate systems is awkward to use, but it is the convention employed in the literature. In an orthogonal system, the tangential and normal directions are the same so no indicial distinction is needed.

For the tangential coordinate system, let $\vec{\zeta}_i \cdot \vec{\zeta}_j = g_{ij}$ and form the matrix

$$\hat{g} = \begin{bmatrix} g_{ij} \end{bmatrix} = \begin{bmatrix} g_{11} & g_{12} & g_{13} \\ g_{21} & g_{22} & g_{23} \\ g_{31} & g_{32} & g_{33} \end{bmatrix} \tag{B.29}$$

$$g_{ij} = \frac{\partial X_1}{\partial Z^i}\frac{\partial X_1}{\partial Z^j} + \frac{\partial X_2}{\partial Z^i}\frac{\partial X_2}{\partial Z^j} + \frac{\partial X_3}{\partial Z^i}\frac{\partial X_3}{\partial Z^j} \quad \text{for} \quad i, j = 1, 2, 3 \tag{B.30}$$

Note $g_{ij} = g_{ji}$. The "for $i, j = 1, 2, k$ (shorthand for 9 terms) or $i, j, k = 1, 2, 3$ (shorthand for 27 terms)" is referred to as indicating "cyclic permutation of the indices."

For the tangential coordinate system, the base vectors are termed covariant and the coordinate lines are termed contravariant. The elements g_{ij} are of a covariant second-order tensor called the fundamental or metric tensor. Notice that it has two directions (i and j) associated with each of its components. Due to the definitions of the $\vec{\zeta}_i$-vectors, g_{ij} may be evaluated by a transformation of an increment of arc length from a rectangular Cartesian coordinate system to a general (tangential) coordinate system. The metric tensor is termed a tensor because of the way it is transformed, but it does not include a definition of base vectors. A more appropriate definition would be to term it a transformation matrix for differential lengths.

Transformation equations are written as

$$dX_i = \sum_{m=1}^{3} M_{im} dZ^m, \quad \text{for} \quad i = 1, 2, 3 \tag{B.31}$$

$$d\tilde{X} = \hat{M} d\tilde{Z}$$

$$
\begin{pmatrix} dX_1 \\ dX_2 \\ dX_3 \end{pmatrix} =
\begin{bmatrix}
\dfrac{\partial X_1}{\partial Z^1} & \dfrac{\partial X_1}{\partial Z^2} & \dfrac{\partial X_1}{\partial Z^3} \\[8pt]
\dfrac{\partial X_2}{\partial Z^1} & \dfrac{\partial X_2}{\partial Z^2} & \dfrac{\partial X_2}{\partial Z^3} \\[8pt]
\dfrac{\partial X_3}{\partial Z^1} & \dfrac{\partial X_3}{\partial Z^2} & \dfrac{\partial X_3}{\partial Z^3}
\end{bmatrix}
\begin{pmatrix} dZ_1 \\ dZ_2 \\ dZ_3 \end{pmatrix}
\tag{B.32}
$$

Mathematicians term the single-subscripted terms "vectors" and the double-subscripted terms "matrices." The tilde will be used to denote vector-valued functions, and the circumflex, matrices. These terms are also called arrays. The term array is most meaningful since this is the manner that it will be treated in a computer code.

Denote J as the determinate of $\hat{M}: J = |\hat{M}|$.

Since curvilinear coordinate systems are only locally linear, differential rather than integral position vectors are required and defined:

$$d\vec{R} = \left(\frac{\partial \vec{R}}{\partial Z^1}\right) dZ^1 + \left(\frac{\partial \vec{R}}{\partial Z^2}\right) dZ^2 + \left(\frac{\partial \vec{R}}{\partial Z^3}\right) dZ^3 \tag{B.33}$$

but $\vec{R}$ is defined in the $\vec{\chi}$, X coordinate system. Hence,

$$\frac{\partial \vec{R}}{\partial Z^i} = \vec{\chi}_1 \frac{\partial X_1}{\partial Z^i} + \vec{\chi}_2 \frac{\partial X_2}{\partial Z^i} + \vec{\chi}_3 \frac{\partial X_3}{\partial Z^i} \quad \text{for} \quad i = 1,2,3 \tag{B.34}$$

Since $d\vec{R} \cdot d\vec{R}$ must be the same in any coordinate system and it is known to be

$$d\vec{R} \cdot d\vec{R} = \left(dX_1\right)^2 + \left(dX_2\right)^2 + \left(dX_3\right)^2 = dR * dR = \left(dR\right)^2 \tag{B.35}$$

$$
\left(dR\right)^2 = \left[\left(\frac{\partial X_1}{\partial Z^1}\right) dZ^1 + \left(\frac{\partial X_1}{\partial Z^2}\right) dZ^2 + \left(\frac{\partial X_1}{\partial Z^3}\right) dZ^3\right]^2
$$

$$
+ \left[\left(\frac{\partial X_2}{\partial z^1}\right) dZ^1 + \left(\frac{\partial X_2}{\partial Z^2}\right) dZ^2 + \left(\frac{\partial X_2}{\partial Z^3}\right) dZ^3\right]^2
$$

$$
+ \left[\left(\frac{\partial X_3}{\partial Z^1}\right) dZ^1 + \left(\frac{\partial X_3}{\partial Z^2}\right) dZ^2 + \left(\frac{\partial X_3}{\partial Z^3}\right) dZ^3\right]^2 \tag{B.36}
$$

$$(dR)^2 = \left[\left(\frac{\partial X_1}{\partial Z^1}\right)dZ^1\right]^2 + \left[\left(\frac{\partial X_1}{\partial Z^2}\right)dZ^2\right]^2 + \left[\left(\frac{\partial X_1}{\partial Z^3}\right)dZ^3\right]^2 + 2\left(\frac{\partial X_1}{\partial Z^1}\right)\left(\frac{\partial X_1}{\partial Z^2}\right)dZ^1 dZ^2$$

$$+2\left(\frac{\partial X_1}{\partial Z^1}\right)\left(\frac{\partial X_1}{\partial Z^3}\right)dZ^1 dZ^3 + 2\left(\frac{\partial X_1}{\partial Z^2}\right)\left(\frac{\partial X_1}{\partial Z^3}\right)dZ^2 dZ^3 + \left[\left(\frac{\partial X_2}{\partial Z^1}\right)dZ^1\right]^2$$

$$+\left[\left(\frac{\partial X_2}{\partial Z^2}\right)dZ^2\right]^2 + \left[\left(\frac{\partial X_2}{\partial Z^3}\right)dZ^3\right]^2 + 2\left(\frac{\partial X_2}{\partial Z^1}\right)\left(\frac{\partial X_2}{\partial Z^2}\right)dZ^1 dZ^2$$

$$+2\left(\frac{\partial X_2}{\partial Z^1}\right)\left(\frac{\partial X_2}{\partial Z^3}\right)dZ^1 dZ^3 + 2\left(\frac{\partial X_2}{\partial Z^2}\right)\left(\frac{\partial X_2}{\partial Z^3}\right)dZ^2 dZ^3$$

$$+\left[\left(\frac{\partial X_3}{\partial Z^1}\right)dZ^1\right]^2 + \left[\left(\frac{\partial X_3}{\partial Z^2}\right)dZ^2\right]^2 + \left[\left(\frac{\partial X_3}{\partial Z^3}\right)dZ^3\right]^2 + 2\left(\frac{\partial X_3}{\partial Z^1}\right)\left(\frac{\partial X_3}{\partial Z^2}\right)dZ^1 dZ^2$$

$$+2\left(\frac{\partial X_3}{\partial Z^1}\right)\left(\frac{\partial X_3}{\partial Z^3}\right)dZ^1 dZ^3 + 2\left(\frac{\partial X_3}{\partial Z^2}\right)\left(\frac{\partial X_3}{\partial Z^3}\right)dZ^2 dZ^3 \tag{B.37}$$

$$(dR)^2 = \sum_{m=1}^{3}\sum_{n=1}^{3}\sum_{i=1}^{3} M_{im} dZ^m M_{in} dZ^n = \sum_{m=1}^{3}\sum_{n=1}^{3} g_{mn} dZ^m dZ^n \tag{B.38}$$

since the M's do not depend on the dZ^i's. Also,

$$(dR)^2 = \sum_{m=1}^{3}\sum_{n=1}^{3} g_{mn} dZ^m dZ^n = \sum_{m=1}^{3}\sum_{n=1}^{3}\left(\sum_{i=1}^{3}\frac{\partial X_i}{\partial Z^m}\frac{\partial X_i}{\partial Z^n}\right)dZ^m dZ^n \tag{B.39}$$

Elements of several $\hat{g}$ matrices for different coordinate transformations are given in Table B.4.

An element of surface, dS_i, where Z^i being constant is given by

$$dS_i = \sqrt{g_{jj}g_{kk}-(g_{jk})^2}\,dZ^j dZ^k \quad \text{for} \quad i,j,k = 1,2,3 \tag{B.40}$$

An increment of volume, $d\forall$

$$d\forall = \sqrt{g}\,dZ^1 dZ^2 dZ^3 \quad \text{where} \quad g = \det\left|g_{ik}\right| \tag{B.41}$$

The Jacobian (J) of the incremental coordinate transformation is equal to the square root of g.

The relationship between the two types of base vectors is given by

$$\vec{\zeta}^i = \frac{(\vec{\zeta}_j \times \vec{\zeta}_k)}{\forall} = \frac{(\vec{\zeta}_j \times \vec{\zeta}_k)}{\left(\det\left|g_{jk}\right|\right)\left(|\vec{\zeta}_1||\vec{\zeta}_2||\vec{\zeta}_3|\right)} \tag{B.42}$$

TABLE B.4 Properties of the Metric

Metric for the transformation of curvilinear mathematical tangential coordinates to rectangular Cartesian coordinates

$$\sum (M_{im})(M_{ni}) \equiv \hat{g} = \begin{bmatrix} \left(\dfrac{\partial X_1}{\partial Z^1}\dfrac{\partial X_1}{\partial Z^1} + \dfrac{\partial X_2}{\partial Z^1}\dfrac{\partial X_2}{\partial Z^1} + \dfrac{\partial X_3}{\partial Z^1}\dfrac{\partial X_3}{\partial Z^1} \right) & \left(\dfrac{\partial X_1}{\partial Z^1}\dfrac{\partial X_1}{\partial Z^2} + \dfrac{\partial X_2}{\partial Z^1}\dfrac{\partial X_2}{\partial Z^2} + \dfrac{\partial X_3}{\partial Z^1}\dfrac{\partial X_3}{\partial Z^2} \right) & \left(\dfrac{\partial X_1}{\partial Z^1}\dfrac{\partial X_1}{\partial Z^3} + \dfrac{\partial X_2}{\partial Z^1}\dfrac{\partial X_2}{\partial Z^3} + \dfrac{\partial X_3}{\partial Z^1}\dfrac{\partial X_3}{\partial Z^3} \right) \\ \left(\dfrac{\partial X_1}{\partial Z^2}\dfrac{\partial X_1}{\partial Z^1} + \dfrac{\partial X_2}{\partial Z^2}\dfrac{\partial X_2}{\partial Z^1} + \dfrac{\partial X_3}{\partial Z^2}\dfrac{\partial X_3}{\partial z^1} \right) & \left(\dfrac{\partial X_1}{\partial Z^2}\dfrac{\partial X_1}{\partial Z^2} + \dfrac{\partial X_2}{\partial Z^2}\dfrac{\partial X_2}{\partial Z^2} + \dfrac{\partial X_3}{\partial Z^2}\dfrac{\partial X_3}{\partial Z^2} \right) & \left(\dfrac{\partial X_1}{\partial Z^2}\dfrac{\partial X_1}{\partial Z^3} + \dfrac{\partial X_2}{\partial Z^2}\dfrac{\partial X_2}{\partial Z^3} + \dfrac{\partial X_3}{\partial Z^2}\dfrac{\partial X_3}{\partial Z^3} \right) \\ \left(\dfrac{\partial X_1}{\partial Z^3}\dfrac{\partial X_1}{\partial Z^1} + \dfrac{\partial X_2}{\partial Z^3}\dfrac{\partial X_2}{\partial Z^1} + \dfrac{\partial X_3}{\partial Z^3}\dfrac{\partial X_3}{\partial Z^1} \right) & \left(\dfrac{\partial X_1}{\partial Z^3}\dfrac{\partial X_1}{\partial Z^2} + \dfrac{\partial X_2}{\partial Z^3}\dfrac{\partial X_2}{\partial Z^2} + \dfrac{\partial X_3}{\partial Z^3}\dfrac{\partial X_3}{\partial Z^2} \right) & \left(\dfrac{\partial X_1}{\partial Z^3}\dfrac{\partial X_1}{\partial Z^3} + \dfrac{\partial X_2}{\partial Z^3}\dfrac{\partial X_2}{\partial Z^3} + \dfrac{\partial X_3}{\partial Z^3}\dfrac{\partial X_3}{\partial Z^3} \right) \end{bmatrix} \quad \text{(A)}$$

$$= \begin{bmatrix} g_{11} & g_{12} & g_{13} \\ g_{21} & g_{22} & g_{23} \\ g_{31} & g_{32} & g_{33} \end{bmatrix} \quad Note: \quad g_{nm} = g_{mn}, \text{ i.e., the matrix is symmetric} \quad \text{(B)}$$

Incremental distance

$$\left| d\vec{R} \right| = [g_{11}dZ^1 dZ^1 + g_{12}dZ^1 dZ^2 + g_{13}dZ^1 dZ^3 + g_{21}dZ^2 dZ^1 + g_{22}dZ^2 dZ^2 + g_{23}dZ^2 dZ^3$$
$$+ g_{31}dZ^3 dZ^1 + g_{32}dZ^3 dZ^2 + g_{33}dZ^3 dZ^3]^{0.5} \quad \text{(C)}$$

$$d\vec{R} = \sum_{i=1}^{3} \vec{\zeta}_{(i)} * \left[\left(\dfrac{\partial X_1}{\partial Z^i} \right)^2 + \left(\dfrac{\partial X_2}{\partial Z^i} \right)^2 + \left(\dfrac{\partial X_3}{\partial Z^i} \right)^2 \right]^{0.5} \Delta Z^i \quad \text{(D)}$$

(continued)

TABLE B.4 (continued) Properties of the Metric

Determinant of $\hat{g} \equiv g = J^2$, the Jacobian squared of the transformation

$$\det \hat{g} = g_{11}g_{22}g_{33} + g_{12}g_{23}g_{31} + g_{13}g_{21}g_{32} - g_{13}g_{22}g_{31} - g_{11}g_{23}g_{32} - g_{12}g_{21}g_{33} \equiv \left[\frac{\partial(X_1,X_2,X_3)}{\partial(Z^1,Z^2,Z^3)}\right]^2 = J^2 \tag{E}$$

Volume increments

$$= \mathrm{d}P\left(\frac{\partial(X_1,X_2,X_3)}{\partial(Z^1,Z^2,Z^3)}\right)\mathrm{d}Z^1\mathrm{d}Z^2\mathrm{d}Z^3 = \sqrt{g}\,\mathrm{d}Z^1\mathrm{d}Z^2\mathrm{d}Z^3 = H_1H_2H_3\,\mathrm{d}Y_1\mathrm{d}Y_2\mathrm{d}Y_3 = \mathrm{d}X_1\mathrm{d}X_2\mathrm{d}X_3 \tag{F}$$

The metric for the transformation of curvilinear physical tangential coordinates to rectangular Cartesian coordinates-

$$\sum(N_{im})(N_{ni}) \equiv \hat{g}_0 = \left[g_{(i)(j)}\right] \equiv \hat{n} = \left[\sum_{m=1}^{3}\sum_{n=1}^{3}\left(\sum_{i=1}^{3}\frac{\partial X_i}{\partial Z^{(m)}}\frac{\partial X_i}{\partial Z^{(n)}}\right)\right] \tag{G}$$

$$= \begin{bmatrix} 1 & \cos\{\vec{\zeta}_{(1)},\vec{\zeta}_{(2)}\} & \cos\{\vec{\zeta}_{(1)},\vec{\zeta}_{(3)}\} \\ \cos\{\vec{\zeta}_{(2)},\vec{\zeta}_{(1)}\} & 1 & \cos\{\vec{\zeta}_{(2)},\vec{\zeta}_{(3)}\} \\ \cos\{\vec{\zeta}_{(3)},\vec{\zeta}_{(1)}\} & \cos\{\vec{\zeta}_{(3)},\vec{\zeta}_{(2)}\} & 1 \end{bmatrix} \tag{H}$$

If $g_{(nm)} = g_{(mn)}$, the matrix is symmetric.

Incremental distance

$$|d\vec{R}| = [dZ^{(1)}dZ^{(1)} + dZ^{(2)}dZ^{(2)} + dZ^{(3)}dZ^{(3)} + 2\cos\{\vec{\zeta}_{(1)}, \vec{\zeta}_{(2)}\}dZ^{(1)}dZ^{(2)}$$
$$+ 2\cos\{\vec{\zeta}_{(2)}, \vec{\zeta}_{(3)}\}\, dZ^{(2)}dZ^{(3)} + 2\cos\{\vec{\zeta}_{(3)}, \vec{\zeta}_{(1)}\}\, dZ^{(3)}dZ^{(1)}]^{0.5}$$

(I)

Determinant of $\hat{g}_0 \equiv g_0$

$$\det \hat{g}_0 = 1 + 2\cos\{1,2\}\cos\{2,3\}\cos\{3,1\} - [\cos\{1,3\}]^2 - [\cos\{2,3\}]^2 - [\cos\{1,2\}]^2 = \left[\frac{\partial(X_1, X_2, X_3)}{\partial(Z^{(1)}, Z^{(2)}, Z^{(3)})}\right]^2 = J_0^2 = g_0$$

(K)

where $\cos\{I, J\} = \cos\{\vec{\zeta}_{(I)}, \vec{\zeta}_{(J)}\}$ etc.

Volume increments

$$\Delta P = \left(\frac{\partial(X_1, X_2, X_3)}{\partial(Z'^{(1)}, Z'^{(2)}, Z'^{(3)})}\right) dP^{(1)}dP^{(2)}dP^{(3)} = \sqrt{g_0}\, dP^{(1)}dP^{(2)}dP^{(3)}$$

(L)

Relationships between physical covariant and covariant base vectors

$$Z^{(i)} = Z^{(i)}\{Z^i\}; \quad \Delta Z^{(i)} = \sqrt{g_{ii}} \Delta Z^i = \left[\left(\frac{\partial X_1}{\partial Z^i}\right)^2 + \left(\frac{\partial X_2}{\partial Z^i}\right)^2 + \left(\frac{\partial X_3}{\partial Z^i}\right)^2\right]^{0.5} \Delta Z^i$$

(B.43)

$$\vec{\zeta}_{(i)} \cdot \vec{\zeta}_{(j)} = g_{(ij)} = \frac{1}{\sqrt{g_{ii}} \sqrt{g_{jj}}} g_{ij} = \cos\{\vec{\zeta}_{(i)}, \vec{\zeta}_{(j)}\}$$

$$\vec{\zeta}_{(i)} \cdot \vec{\zeta}_{(i)} = 1.0$$

(B.44)

The physical coordinates are those for which the base vectors have been defined as unit vectors without changing their direction. M and P in Table B.3 refer to the coordinates being mathematical with base vectors not necessarily unit in length and physical with base vectors of unity, respectively.

For orthogonal curvilinear coordinates (NCC), the metric tensor as represented by its matrix components is considerably simplified. The metric coefficients for cylindrical coordinates $\{r, \theta, z\}$ are defined as

$$g_{ij} = \begin{pmatrix} 1 & 0 & 0 \\ 0 & r^2 & 0 \\ 0 & 0 & 1 \end{pmatrix} = \begin{pmatrix} (H_1)^2 & 0 & 0 \\ 0 & (H_2)^2 & 0 \\ 0 & 0 & (H_3)^2 \end{pmatrix}$$

(B.45)

$$(ds)^2 = (dr)^2 + r^2(d\theta)^2 + (dz)^2$$

(B.46)

For the nomenclature in Tables B.3 and B.5, $(y_1, y_2, y_3) \Rightarrow (r, \theta, z)$ and $(H_1, H_2, H_3) \Rightarrow (1, r, 1)$ and $g_{ii} = (H_i)^2$.

B.3.2.3 Conjugate Metric Tensor

The metric tensor was obtained by requiring that a transformation from orthogonal Cartesian coordinates to tangential curvilinear coordinates maintained an increment of length (ds) as an invariant. This was represented as

$$(ds)^2 = \sum_i (dX_i)^2 = \sum_\ell \sum_n g_{\ell m} dZ^\ell dZ^m$$

(B.47)

The last pair of summed terms is called "a homogeneous quadratic form." The conjugate metric tensor $\hat{n}$ will be defined such that the homogeneous quadratic form is maintained when a normal curvilinear coordinate system is established. No longer will it be assumed that one of the coordinates systems

TABLE B.5 Tensor Operations in Orthogonal Curvilinear Coordinates

Vector: $\vec{A} = \vec{\upsilon}_1 A_{\upsilon1} + \vec{\upsilon}_2 A_{\upsilon2} + \vec{\upsilon}_3 A_{\upsilon3}$ (A)

Scalar or dot product: $\vec{A} \cdot \vec{B} = A_{\upsilon1} B_{\upsilon1} + A_{\upsilon2} B_{\upsilon2} + A_{\upsilon3} B_{\upsilon3}$ (B)

Vector or cross product: $\vec{A} \times \vec{B} = (A_{\upsilon2} B_{\upsilon3} - A_{\upsilon3} B_{\upsilon2})\vec{\upsilon}_1 + (A_{\upsilon3} B_{\upsilon1} - A_{\upsilon1} B_{\upsilon3})\vec{\upsilon}_2 + (A_{\upsilon1} B_{\upsilon2} - A_{\upsilon2} B_{\upsilon1})\vec{\upsilon}_3$ (C)

Gradient of a scalar: $\nabla U = \dfrac{\vec{\upsilon}_1}{H_1} \dfrac{\partial U}{\partial Y_1} + \dfrac{\vec{\upsilon}_2}{H_2} \dfrac{\partial U}{\partial Y_2} + \dfrac{\vec{\upsilon}_3}{H_3} \dfrac{\partial U}{\partial Y_3}$ (D)

Gradient of a vector: $\nabla \vec{A} = \overset{\leftrightarrow}{W}$ the elements of $\overset{\leftrightarrow}{W}$ are

$$W_{ii} = \frac{1}{H_i} \frac{\partial A_{\upsilon i}}{\partial Y_i} + \frac{1}{H_j} \frac{\partial H_i}{\partial Y_j} \frac{A_{\upsilon j}}{H_i} + \frac{1}{H_k} \frac{\partial H_i}{\partial Y_k} \frac{A_{\upsilon k}}{H_i} \quad \text{for} \quad i,j,k = 1,2,3; \quad 2,3,1; \quad 3,1,2$$ (E)

$$W_{ij} = \frac{1}{H_i} \frac{\partial A_{\upsilon i}}{\partial Y_j} - \frac{1}{H_j} \frac{\partial H_i}{\partial Y_i} \frac{A_{\upsilon j}}{H_j} \quad \text{for} \quad i,j = 1,2; \quad 1,3; \quad 2,1; \quad 2,3; \quad 3,1; \quad \text{and} \quad 3,2$$ (F)

Divergence of $\vec{A}$: $\nabla \cdot \vec{A} = \dfrac{1}{H_1 H_2 H_3} \left[\dfrac{\partial(H_2 H_3 A_{\upsilon1})}{\partial Y_1} + \dfrac{\partial(H_1 H_3 A_{\upsilon2})}{\partial Y_2} + \dfrac{\partial(H_1 H_2 A_{\upsilon3})}{\partial Y_3} \right]$ (G)

Curl of $\vec{A}$: $\nabla \times \vec{A} = \dfrac{1}{H_1 H_2 H_3} \begin{vmatrix} H_1 \vec{\upsilon}_1 & H_2 \vec{\upsilon}_2 & H_3 \vec{\upsilon}_3 \\ \partial/\partial Y_1 & \partial/\partial Y_2 & \partial/\partial Y_3 \\ H_1 A_{\upsilon1} & H_2 A_{\upsilon2} & H_3 A_{\upsilon3} \end{vmatrix}$ (H)

(continued)

TABLE B.5 (continued) Tensor Operations in Orthogonal Curvilinear Coordinates

Laplacian of U, a scalar:
$$\nabla^2 U = \nabla \cdot (\nabla U) = \frac{1}{H_1 H_2 H_3}\left[\frac{\partial}{\partial Y_1}\left(\frac{H_2 H_3}{H_1}\frac{\partial U}{\partial Y_1}\right) + \frac{\partial}{\partial Y_2}\left(\frac{H_3 H_1}{H_2}\frac{\partial U}{\partial Y_2}\right) + \frac{\partial}{\partial Y_3}\left(\frac{H_1 H_2}{H_3}\frac{\partial U}{\partial Y_3}\right)\right] \quad (I)$$

Laplacian of $\vec{A}$: $\nabla^2 \vec{A} = \nabla(\nabla\cdot\vec{A}) - \nabla\times(\nabla\times\vec{A})$

$$=\left\{\frac{1}{H_1}\frac{\partial}{\partial Y_1}(\nabla\cdot\vec{A}) + \frac{1}{H_2 H_3}\left[\frac{\partial}{\partial Y_3}\left\{\frac{H_2}{H_3 H_1}\left(\frac{\partial H_1 A_{v1}}{\partial Y_3} - \frac{\partial H_3 A_{v3}}{\partial Y_1}\right)\right\} - \frac{\partial}{\partial Y_2}\left\{\frac{H_3}{H_1 H_2}\left(\frac{\partial H_2 A_{v2}}{\partial Y_1} - \frac{\partial H_1 A_{v1}}{\partial Y_2}\right)\right\}\right]\right\}\vec{v}_1$$

$$+\left\{\frac{1}{H_2}\frac{\partial}{\partial Y_2}(\nabla\cdot\vec{A}) + \frac{1}{H_3 H_1}\left[\frac{\partial}{\partial Y_1}\left\{\frac{H_3}{H_1 H_2}\left(\frac{\partial H_2 A_{v2}}{\partial Y_1} - \frac{\partial H_1 A_{v1}}{\partial Y_2}\right)\right\} - \frac{\partial}{\partial Y_3}\left\{\frac{H_1}{H_2 H_3}\left(\frac{\partial H_3 A_{v3}}{\partial Y_2} - \frac{\partial H_2 A_{v2}}{\partial Y_3}\right)\right\}\right]\right\}\vec{v}_2 \quad (J)$$

$$+\left\{\frac{1}{H_3}\frac{\partial}{\partial Y_3}(\nabla\cdot\vec{A}) + \frac{1}{H_1 H_2}\left[\frac{\partial}{\partial Y_2}\left\{\frac{H_1}{H_2 H_3}\left(\frac{\partial H_3 A_{v3}}{\partial Y_2} - \frac{\partial H_2 A_{v2}}{\partial Y_3}\right)\right\} - \frac{\partial}{\partial Y_1}\left\{\frac{H_2}{H_3 H_1}\left(\frac{\partial H_1 A_{v1}}{\partial Y_3} - \frac{\partial H_3 A_{v3}}{\partial Y_1}\right)\right\}\right]\right\}\vec{v}_3 \quad (K)$$

Other operations: $(\vec{A}\cdot\nabla)U = \dfrac{A_{v1}}{H_1}\dfrac{\partial U}{\partial Y_1} + \dfrac{A_{v2}}{H_2}\dfrac{\partial U}{\partial Y_2} + \dfrac{A_{v3}}{H_3}\dfrac{\partial U}{\partial Y_3}$

$$(\vec{A}\cdot\nabla)\vec{B} = \frac{1}{H_1}\left\{\left(A_{v1}\frac{\partial B_{v1}}{\partial Y_1} + A_{v2}\frac{\partial B_{v1}}{\partial Y_2} + A_{v3}\frac{\partial B_{v1}}{\partial Y_3}\right) - \frac{A_{v2}}{H_2}\left(\frac{\partial H_2 B_{v2}}{\partial Y_1} - \frac{\partial H_1 B_{v1}}{\partial Y_2}\right) + \frac{A_{v3}}{H_3}\left(\frac{\partial H_1 B_{v1}}{\partial Y_3} - \frac{\partial H_3 B_{v3}}{\partial Y_1}\right)\right\}\vec{v}_1$$

$$+\frac{1}{H_2}\left\{\left(A_{v1}\frac{\partial B_{v2}}{\partial Y_1} + A_{v2}\frac{\partial B_{v2}}{\partial Y_2} + A_{v3}\frac{\partial B_{v2}}{\partial Y_3}\right) - \frac{A_{v3}}{H_3}\left(\frac{\partial H_3 B_{v3}}{\partial Y_2} - \frac{\partial H_2 B_{v2}}{\partial Y_3}\right) + \frac{A_{v1}}{H_1}\left(\frac{\partial H_2 B_{v2}}{\partial Y_1} - \frac{\partial H_1 B_{v1}}{\partial Y_2}\right)\right\}\vec{v}_2$$

$$+\frac{1}{H_3}\left\{\left(A_{v1}\frac{\partial B_{v3}}{\partial Y_1} + A_{v2}\frac{\partial B_{v3}}{\partial Y_2} + A_{v3}\frac{\partial B_{v3}}{\partial Y_3}\right) - \frac{A_{v1}}{H_1}\left(\frac{\partial H_1 B_{v1}}{\partial Y_3} - \frac{\partial H_3 B_{v3}}{\partial Y_1}\right) + \frac{A_{v2}}{H_2}\left(\frac{\partial H_3 B_{v3}}{\partial Y_2} - \frac{\partial H_2 B_{v2}}{\partial Y_3}\right)\right\}\vec{v}_3 \quad (L)$$

For $\vec{\vec{W}}$ being a symmetric tensor:

$$
\begin{aligned}
\vec{\vec{W}} : (\nabla \vec{B}) = {}& W_{v11}\left(\frac{1}{H_1}\frac{\partial B_{v1}}{\partial Y_1} + \frac{B_{v2}}{H_1 H_2}\frac{\partial H_1}{\partial Y_2} + \frac{B_{v3}}{H_1 H_3}\frac{\partial H_1}{\partial Y_3}\right) \\
&+ W_{v22}\left(\frac{1}{H_2}\frac{\partial B_{v2}}{\partial Y_2} + \frac{B_{v1}}{H_1 H_2}\frac{\partial H_2}{\partial Y_1} + \frac{B_{v3}}{H_2 H_3}\frac{\partial H_2}{\partial Y_3}\right) + W_{v33}\left(\frac{1}{H_3}\frac{\partial B_{v3}}{\partial Y_{32}} + \frac{B_{v1}}{H_1 H_2}\frac{\partial H_3}{\partial Y_1} + \frac{B_{v2}}{H_2 H_3}\frac{\partial H_3}{\partial Y_2}\right) \\
&+ W_{v12}\left(\frac{1}{H_2}\frac{\partial B_{v1}}{\partial Y_2} + \frac{1}{H_1}\frac{\partial B_{v2}}{\partial Y_1} - \frac{B_{v2}}{H_1 H_2}\frac{\partial H_2}{\partial Y_1} - \frac{B_{v1}}{H_1 H_2}\frac{\partial H_1}{\partial Y_2}\right) \\
&+ W_{v23}\left(\frac{1}{H_3}\frac{\partial B_{v2}}{\partial Y_3} + \frac{1}{H_2}\frac{\partial B_{v3}}{\partial Y_2} - \frac{B_{v3}}{H_2 H_3}\frac{\partial H_3}{\partial Y_2} - \frac{B_{v2}}{H_2 H_3}\frac{\partial H_2}{\partial Y_3}\right) \\
&+ W_{v31}\left(\frac{1}{H_3}\frac{\partial B_{v1}}{\partial Y_3} + \frac{1}{H_1}\frac{\partial B_{v3}}{\partial Y_1} - \frac{B_{v3}}{H_1 H_3}\frac{\partial H_3}{\partial Y_1} - \frac{B_{v1}}{H_1 H_3}\frac{\partial H_1}{\partial Y_3}\right)
\end{aligned} \tag{M}
$$

The components of $\vec{\vec{W}}$ are W_{ij} where i,j are: 1,2; 2,3; 3,1; 2,1; 3,2; 1,3; 1,1; 2,2; and 3,3.

being transformed will be orthogonal and Cartesian. Rather, one coordinate system $(\vec{\zeta}_i, Z^i)$ will be transformed to $(\vec{\zeta}^j, Z_j)$ by using the same homogeneous quadratic form. This line of thought is what leads to identifying covariant and contravariant quantities by being determined by how they transform.

The normal curvilinear coordinate system is related to the tangential curvilinear system by defining the conjugate metric tensor as the inverse of the metric tensor. New base vectors and coordinates are then named. The conjugate metric tensor becomes

$$
\hat{n} = \begin{bmatrix} g^{11} & g^{12} & g^{13} \\ g^{21} & g^{22} & g^{23} \\ g^{31} & g^{32} & g^{33} \end{bmatrix} = \hat{g}^{-1}
\tag{B.48}
$$

Taking the inverse of $\hat{g}$ is accomplished by

1. Exchange the rows and columns of $\hat{g}$ to form its transpose $\hat{g}^{\mathrm{T}}$
2. Compute the cofactors G^{ij} of the transposed matrix
3. The new matrix elements become

$$
g^{ij} = \frac{G^{ij}}{\det \hat{g}} \quad \text{or} \quad g_{ik}g^{kj} = \delta_i^j \quad \text{or} \quad g^{jk}g_{ki} = \delta_i^j
\tag{B.49}
$$

where the repeated index k indicates a summation and the Kronecker delta is

$$
\delta_i^j = 1, \quad \text{if } i = j \qquad \delta_i^j = 0, \quad \text{if } i \neq j
\tag{B.50}
$$

Note, $g^{ij} = 1/g_{ij}$ only if the curvilinear system is orthogonal. In which case the normal and tangential curvilinear systems are identical! This is the nomenclature of Hawkins (1963) and Borisenko and Tarapov (1968). Brillouin and Brennan (1964) uses the ij on G as a subscript to represent the cofactors of the transposed matrix.

To this point, the magnitude of the base vectors and the coordinate lines have been given a symbol. To proceed further, the coordinate lines must be defined. If curvilinear lines and surfaces are described analytically, empirically, or with tables of data points in an orthogonal Cartesian coordinate system, the independent variables in the conservation laws can be expressed in curvilinear coordinates and the scalar-dependent variables can be evaluated. This is the methodology described in Chapter 6.

B.3.3 EVALUATION OF BASE VECTORS

Given the components (U_i) of a vector in orthogonal Cartesian coordinates, they may be transformed into contravariant components (W^i) in a general curvilinear coordinate system by

$$
\begin{pmatrix} W^1 \\ W^2 \\ W^3 \end{pmatrix} = \begin{pmatrix} \dfrac{\partial Z^1}{\partial X_1} & \dfrac{\partial Z^1}{\partial X_2} & \dfrac{\partial Z^1}{\partial X_3} \\ \dfrac{\partial Z^2}{\partial X_1} & \dfrac{\partial Z^2}{\partial X_2} & \dfrac{\partial Z^2}{\partial X_3} \\ \dfrac{\partial Z^3}{\partial X_1} & \dfrac{\partial Z^3}{\partial X_2} & \dfrac{\partial Z^3}{\partial X_3} \end{pmatrix} \begin{pmatrix} U_1 \\ U_2 \\ U_3 \end{pmatrix}
\tag{B.51}
$$

Vectors which transform in this manner are termed contravariant vectors. Such vector components are parallel to the covariant base vectors.

Given the components (U_i) of a vector in orthogonal Cartesian coordinates, these may be transformed into covariant components (W_j) in a general curvilinear coordinate system by

$$
\begin{pmatrix} W_1 \\ W_2 \\ W_3 \end{pmatrix} = \begin{pmatrix} \dfrac{\partial X_1}{\partial Z^1} & \dfrac{\partial X_2}{\partial Z^1} & \dfrac{\partial X_3}{\partial Z^1} \\ \dfrac{\partial X_1}{\partial Z^2} & \dfrac{\partial X_2}{\partial Z^2} & \dfrac{\partial X_3}{\partial Z^2} \\ \dfrac{\partial X_1}{\partial Z^3} & \dfrac{\partial X_2}{\partial Z^3} & \dfrac{\partial X_3}{\partial Z^3} \end{pmatrix} \begin{pmatrix} U_1 \\ U_2 \\ U_3 \end{pmatrix}
\tag{B.52}
$$

The normal curvilinear coordinates are related to the tangential coordinates by requiring that the base vectors of the two systems be reciprocal to one another, or by transformations similar to Equations B.51 and B.52 (Hawkins, 1963). Mathematical base vectors are reciprocal if the following relations are true.

$$
\vec{\zeta}_i \cdot \vec{\zeta}^j = \left|\vec{\zeta}_i\right|\left|\vec{\zeta}^j\right| \cos\{\vec{\zeta}_j, \vec{\zeta}_i\} = 1 \quad \text{if } i = j \quad \text{otherwise} = 0
\tag{B.53}
$$

Reciprocal base vectors can be used to express multiplication as "raising and lowering" the indices. For physical problems using two coordinate systems when one is sufficient is awkward and unnecessary. The alternative is

$$
\vec{\zeta}^i = \frac{\vec{\zeta}_j \times \vec{\zeta}_k}{\forall}
\tag{B.54}
$$

where $\forall$ is the volume of the parallelepiped spanned by $\vec{\zeta}_1$, $\vec{\zeta}_2$, and $\vec{\zeta}_3$. This vector is orthogonal to the surface defined by the j and k base vectors.

B.3.3.1 Covariant and Contravariant Vectors and Tensors

For the general vector

$$\vec{A} = A^i \vec{\zeta}_i = A_i \vec{\zeta}^i = A^{(i)} \vec{\zeta}_{(i)} = A_{(i)} \vec{\zeta}^{(i)} \tag{B.55}$$

the index in the superscript position indicates that the quantity is contravariant; in the subscript position the quantity is covariant. A repeated index implies a summation over the three coordinates. Since our interest is in applications to transport analyses, only contravariant components and covariant base or unit vectors will be needed. The other operations are defined for completeness. Also, the covariant and contravariant label will not generally be repeated.

Indices will be used to indicate components of coordinates and dependent variables. This produces a problem since both coordinates and coordinate types would each require an indicial symbol. The value of vectors and tensors is that they are independent of coordinates, but herein different symbols will be used to indicate the coordinate system the variable is defined in, so that the double index is not needed. The double index will then be used only for defining second-order tensors. Higher order tensors will only be defined and used as they appear. Now to define the velocity vector we could use $\vec{A}$ and let it represent velocity in any coordinate system, but since we frequently need to talk about velocity components in a particular coordinate system, we shall use the following symbols for clarity:

$$
\begin{aligned}
\vec{A} &\equiv \text{velocity in any coordinate system} \\
\vec{U} &\equiv \text{velocity in OCC} \\
\vec{V} &\equiv \text{velocity in NCC} \\
\vec{W} &\equiv \text{velocity in MTC, PTC, MNC, PNC}
\end{aligned}
\tag{B.56}
$$

Components of these velocities are denoted by U_i, V_i, W^i, $W^{(i)}$, W_i, and $W_{(i)}$.

A tensor in general may have several directions associated with it. For example,

$$\vec{\vec{T}} = \sum_{i=1}^{3} \sum_{j=1}^{3} T^{ij} \vec{\zeta}_i \vec{\zeta}_j \tag{B.57}$$

for a second-order tensor in a three-dimensional field. Thus a vector is a first-order tensor. A zero-order tensor is a scalar. A second-order tensor is also referred to as a dyadic.

B.3.4 TENSOR OPERATIONS

The del operator is used to define differential operators in OCC systems, for example,

$$\nabla \cdot \vec{T} = \vec{\chi}_1 \sum_{i=1}^{3} \frac{\partial}{\partial X_i} T_{i1} + \vec{\chi}_2 \sum_{i=1}^{3} \frac{\partial}{\partial X_i} T_{i2} + \vec{\chi}_3 \sum_{i=1}^{3} \frac{\partial}{\partial X_i} T_{i3} \qquad \text{(B.58)}$$

When curvilinear coordinates are used, allowance must be made for the variation of the base vectors and metric coefficients in the field. The del operations can still be performed, but they do not produce tensors (Wylie, 1966). This situation is illustrated by the following. Let:

$$\vec{A} = A^j \vec{\zeta}_j = A_i \vec{\zeta}^i$$
$$\vec{\zeta}_j = \vec{\zeta}_j \{Z^1, Z^2, Z^3\} \quad \text{and} \quad \vec{\zeta}^i = \vec{\zeta}^i \{Z_1, Z_2, Z_3\} \qquad \text{(B.59)}$$
$$\vec{A} \cdot \vec{\zeta}^j = A^j \quad \text{and} \quad \vec{A} \cdot \vec{\zeta}_i = A_i$$

Repeated indices indicate summations.

The partial derivative of this vector with respect to Z^k is

$$\frac{\partial \vec{A}}{\partial Z^k} = \frac{\partial \left(A^j \vec{\zeta}_j \right)}{\partial Z^k} = \left(\frac{\partial A^j}{\partial Z^k} \right) \vec{\zeta}_j + A^j \left(\frac{\partial \vec{\zeta}_j}{\partial Z^k} \right)$$

$$\frac{\partial \vec{A}}{\partial Z^k} = \left[\frac{\partial A^m}{\partial Z^k} + \begin{Bmatrix} m \\ ij \end{Bmatrix} A^j \right] \vec{\zeta}_m = A^m_{,k} \vec{\zeta}_m = D_k \vec{\zeta}_m \qquad \text{(B.60)}$$

The Christoffel symbols are introduced as a shorthand representation for part of the partial differential operation. Two other nomenclature systems are also shown in Equation B.60. Only the symbols are different, the meaning is the same. The Christoffel symbol of the first kind is

$$[ik, j] \equiv \Gamma_{j,ik} = \frac{1}{2} \left(\frac{\partial g_{ij}}{\partial Z^k} + \frac{\partial g_{kj}}{\partial Z^i} - \frac{\partial g_{ik}}{\partial Z^j} \right) \qquad \text{(B.61)}$$

and of the second kind (denoted by the indices in braces or gamma) is

$$\left\{ {\ell \atop i\;k} \right\} \equiv \Gamma_{ik}^{\ell} \equiv \frac{1}{2} g^{\ell m} \left(\frac{\partial g_{im}}{\partial Z^k} + \frac{\partial g_{km}}{\partial Z^i} - \frac{\partial g_{ik}}{\partial Z^m} \right) = g^{\ell m} \left[ik,m \right] \tag{B.62}$$

Both the first and second Christoffel symbols are symmetric with respect to i and k.

Additional derivatives are obtained from these definitions (Hawkins, 1963) and (Margenau and Murphy, 1956).

$$\frac{\partial \vec{\zeta}_i}{\partial Z^j} = \left\{ {m \atop i\;j} \right\} \vec{\zeta}_m \tag{B.63}$$

$$\frac{\partial \vec{\zeta}^i}{\partial Z^j} = - \left\{ {i \atop m\;j} \right\} \vec{\zeta}^m \tag{B.64}$$

$$\frac{\partial g_{ij}}{\partial Z^k} = \left[ik,j \right] + \left[jk,i \right] \tag{B.65}$$

$$\frac{\partial g^{ij}}{\partial Z^k} = -g^{\alpha i} \left\{ {j \atop \alpha\;k} \right\} - g^{\alpha j} \left\{ {i \atop \alpha\;k} \right\} \tag{B.66}$$

The gradient of φ

$$\nabla \varphi = g^{ij} \vec{\zeta}_j \frac{\partial \varphi}{\partial Z^i} \tag{B.67}$$

The divergence of $\vec{A}$

$$\nabla \cdot \vec{A} = \frac{\partial A^i}{\partial Z^i} + A^j \left\{ {i \atop i\;j} \right\} = \frac{1}{\sqrt{g}} \frac{\partial \left(A^i \sqrt{g} \right)}{\partial Z^i} \tag{B.68}$$

The Laplacian of φ

$$\nabla^2 \varphi = \frac{1}{\sqrt{g}} \frac{\partial}{\partial Z^i} \left(g^{ij} \sqrt{g} \frac{\partial \varphi}{\partial Z^j} \right) \tag{B.69}$$

The curl of $\vec{A}$

$$\nabla \times \vec{A} = \frac{-1}{\sqrt{g}} \left[\left(\frac{\partial A_3}{\partial Z^2} - \frac{\partial A_2}{\partial Z^3} \right) \vec{\zeta}_1 + \left(\frac{\partial A_1}{\partial Z^3} - \frac{\partial A_3}{\partial Z^1} \right) \vec{\zeta}_2 + \left(\frac{\partial A_2}{\partial Z^1} - \frac{\partial A_1}{\partial Z^2} \right) \vec{\zeta}_3 \right] \tag{B.70}$$

B.3.4.1 Tensor Operations in Physical Curvilinear Coordinates

The following relationships and derivatives are derived by Lee (1997). The Γ terms appearing in these relationships are Christoffel symbols. As shown in Section B.3.4, either of several characters may be used to indicate Christoffel symbols. They arise when derivatives of base or general vectors and tensors are to be evaluated in nonorthogonal curvilinear coordinates:

$$W^{(i)} = U_m \left(\frac{\partial Z^i}{\partial X_m} \right) \sqrt{g_{ii}} \quad \text{sum on } m$$

$$U_m = W^{(i)} \frac{\partial X_m}{\partial Z^{(i)}} \quad \text{sum on } i \tag{B.71}$$

$$W^{(i)} = W^i \sqrt{g_{ii}} \quad \text{no sum on } i$$

$$\frac{\partial \vec{\zeta}_{(i)}}{\partial Z^{(j)}} = \left[\frac{\sqrt{g_{kk}}}{\sqrt{g_{ii}} \sqrt{g_{jj}}} \Gamma_{ij}^k - \delta_i^k \frac{g_{km}}{g_{ii} \sqrt{g_{jj}}} \Gamma_{ij}^m \right] \vec{\zeta}_{(k)} \quad \text{sum on } k \text{ and } m$$

$$= \Gamma_{(ij)}^{(k)} \vec{\zeta}_{(k)} \quad \text{sum on } k \tag{B.72}$$

$$\frac{\partial \vec{\zeta}_i}{\partial Z^j} = \Gamma_{ij}^k \vec{\zeta}_k \quad \text{and} \quad \Gamma_{(ij)}^{(k)} = \frac{\sqrt{g_{kk}}}{\sqrt{g_{ii}} \sqrt{g_{jj}}} \Gamma_{ij}^k - \delta_i^k \frac{g_{km}}{g_{ii} \sqrt{g_{jj}}} \Gamma_{ij}^m$$

Also, $\Gamma_{jk}^i = \Gamma_{jk}^i$ and $\Gamma_{(jk)}^{(i)} \neq \Gamma_{(kj)}^{(i)}$

The gradient of a scalar, Φ

$$\nabla \Phi = g^{(ik)} \frac{\partial \Phi}{\partial Z^{(k)}} \vec{\zeta}_{(i)} \quad \text{sum on } i \text{ and } k \tag{B.73}$$

$g^{(ik)}$ are elements of the conjugate metric tensor, $\hat{n}$. It is the conjugate metric tensor; it is the inverse of the metric tensor (both being for either the physical or mathematical coordinates).

The Laplacian of φ

$$\nabla^2 \Phi = \nabla \cdot (\nabla \Phi) = g^{(ij)} \frac{\partial^2 \Phi}{\partial Z^{(j)} \partial Z^{(i)}} - g^{(\ell j)} \Gamma_{(\ell j)}^{(i)} \frac{\partial \Phi}{\partial Z^{(i)}} \quad \text{sum on } i, j, \text{ and } \ell \tag{B.74}$$

The gradient of a vector $\vec{W}$

$$\nabla \vec{W} = g^{(ij)} \left(W_{(,i)}^{(k)} \right) \vec{\zeta}_{(k)} \vec{\zeta}_{(j)} \quad \text{sum on } i, j, \text{ and } k \tag{B.75}$$

where $W_{(,i)}^{(k)} \equiv \left(\frac{\partial W^{(k)}}{\partial Z^{(i)}} + W^{(m)} \Gamma_{(im)}^{(k)} \right) \quad \text{sum on } m$

The divergence of the vector $\vec{W}$

$$\nabla \cdot \vec{W} = W_{(,i)}^{(i)} = \frac{\sqrt{g_{ii}}}{J} \left[\frac{\partial}{\partial Z^{(i)}} \left(\frac{J}{\sqrt{g_{ii}}} W^{(i)} \right) \right] \quad \text{sum on } i \qquad (B.76)$$

The Laplacian of the vector $\vec{W}$

$$\nabla^2 \vec{W} = \nabla \cdot (\nabla \vec{W})$$

$$= g^{(jk)} \left[W_{(,j)}^{(m)} \Gamma_{(mk)}^{(i)} - W_{(,m)}^{(i)} \Gamma_{(jk)}^{(m)} + \frac{\partial W_{(,j)}^{(i)}}{\partial Z^{(k)}} \right] \vec{\zeta}_{(i)} \quad \text{sum on } i, j, k, \text{ and } m \quad (B.77)$$

The curl of the vector $\vec{W}$

$$\nabla \times \vec{W} = \frac{e^{ijk}}{J} \left(\frac{\partial W_k}{\partial z^j} - W_m \Gamma_{jk}^m \right) \vec{\zeta}_i \qquad (B.78)$$

Repeated indices are summed, and the permutation symbol is defined as

$$e^{ijk} = +1 \quad \text{if } ijk \text{ are } 123, 231, \text{ or } 312$$
$$e^{ijk} = -1 \quad \text{if } ijk \text{ are } 213, 132, \text{ or } 321 \qquad (B.79)$$
$$e^{ijk} = 0 \quad \text{if any two indicies are the same}$$

Indices may be superscripts, subscripts, or a combination thereof. The symbol e^{ijk} is used for OCC.

B.3.4.2 Vector and Tensor Operations in Orthogonal Curvilinear Coordinates (NCC)

For orthogonal curvilinear coordinates, the g_{ii}'s are evaluated as $g_{ii} = (H_i)^2$ and the g_{ij}'s as 0's. These relationships are shown in Table B.5. The base vectors ($\vec{v}_i$'s) are of unit magnitude.

B.4 VECTOR FORMS OF THE CONSERVATION LAWS

The Reynolds-averaged Navier–Stokes equations with body forces in OCC is:

$$\frac{\partial \vec{U}}{\partial t} + (\nabla \vec{U}) \cdot \vec{U} = -\nabla P_\text{T} + v_\text{E} \nabla^2 \vec{U} + \left[\nabla \vec{U} + (\nabla \vec{U})^\text{T} \right] \cdot \nabla v_\text{E} + \vec{F}$$

$$P_\text{T} = p + \frac{z}{Fr^2} + \frac{2}{3}k \quad \text{and} \quad v_\text{E} = v + \varepsilon \qquad (B.80)$$

where

v is the kinematic viscosity

ε is the eddy viscosity

v_E is the effective kinematic viscosity

Other forms of this equation do not show k explicitly, but there is no difference in these forms.

B.4.1 STATIONARY GENERAL (PHYSICAL) (TANGENTIAL) CURVILINEAR COORDINATES

Lee and Soni (1997) converted the scalar form of the incompressible conservation equations to

$$\frac{\partial}{\partial Z^i}\left(\frac{J}{\sqrt{g_{ii}}}W^{(i)}\right) = 0 \tag{B.81}$$

$$\frac{1}{v_E}\frac{\partial W^{(i)}}{\partial t} + \frac{1}{v_E}W^{(j)}\left(\frac{\partial W^{(i)}}{\partial Z^{(j)}} + W^{(k)}\Gamma^i_{kj}\right)$$

$$= -\frac{1}{v_E}\left[g^{(j)}\frac{\partial P}{\partial Z^{(j)}} - \frac{\partial v_E}{\partial Z^{(j)}}\left[g^{(jk)}\frac{\partial W^{(i)}}{\partial Z^k} + g^{(jk)}\Gamma^{(i)}_{(\ell k)}W^{(\ell)} + g^{(ik)}\frac{\partial W^{(i)}}{\partial Z^k} + g^{(ik)}\Gamma^{(j)}_{(\ell k)}W^{(\ell)}\right]\right]$$

$$+ g^{(jk)}\left[\frac{\partial^2 W^{(i)}}{\partial Z^{(k)}\partial Z^{(j)}} + \frac{\partial\left(W^{(\ell)}\Gamma^{(i)}_{(\ell j)}\right)}{\partial Z^{(k)}} + \Gamma^{(i)}_{(\ell k)}W^{(\ell)}_{,(j)} - \Gamma^\ell_{(jk)}W^i_{,(\ell)}\right] \tag{B.82}$$

The momentum equation may be written as

$$\frac{\partial W^{(i)}}{\partial t} + W^{(j)}\left(\frac{\partial W^{(i)}}{\partial Z^{(j)}} + W^{(k)}\Gamma^i_{kj}\right)$$

$$= -g^{(j)}\frac{\partial P_T}{\partial Z^{(j)}} - \frac{\partial v_E}{\partial Z^{(j)}}\left[g^{(jk)}\frac{\partial W^{(i)}}{\partial Z^k} + g^{(jk)}\Gamma^{(i)}_{(\ell k)}W^{(\ell)} + g^{(ik)}\frac{\partial W^{(i)}}{\partial Z^k} + g^{(ik)}\Gamma^{(j)}_{(\ell k)}W^{(\ell)}\right]$$

$$+ v_E g^{(jk)}\left[\frac{\partial^2 W^{(i)}}{\partial Z^{(k)}\partial Z^{(j)}} + \frac{\partial\left(W^{(\ell)}\Gamma^{(i)}_{(\ell j)}\right)}{\partial z^{(k)}} + \Gamma^{(i)}_{(\ell k)}W^{(\ell)}_{,(j)} - \Gamma^\ell_{(jk)}W^{(i)}_{,(\ell)}\right] \tag{B.83}$$

The comma in the subscript indicates partial differentiation with respect to the coordinate defined by the subscript.

$$W^{(\ell)}_{(,j)} = \frac{\partial W^{(\ell)}}{\partial Z^{(j)}} + \begin{Bmatrix} \ell \\ \alpha\, j \end{Bmatrix}W^{(\alpha)} \quad \text{and} \quad W^{(i)}_{(,\ell)} = \frac{\partial W^{(i)}}{\partial Z^{(\ell)}} + \begin{Bmatrix} i \\ \alpha\, \ell \end{Bmatrix}W^{(\alpha)} \tag{B.84}$$

Note that both the independent and dependent variables are transformed in these equations. As discussed by Lee and Soni (1997), the transformed velocity in curvilinear coordinates leads to a more accurate numerical solution when the curvilinear coordinates are highly skewed. This is not a problem if the skewness can be held to a moderate level.

B.4.2 UTILITY OF THE VECTOR FORM OF THE CONSERVATION LAWS

Having identified a meaningful average velocity for a fluid in motion, such a velocity can be described by a magnitude and direction. Mathematically, this velocity is represented as a vector. Considering only stationary (non-adaptive), orthogonal Cartesian coordinate system, unit vectors are defined to indicate direction and such vectors always point in the same direction. The direction of a vector is then the extent of the quantity (i.e., a velocity) which has fixed components in each of the three unit vector directions. Linear momentum conservation laws may be written for each of the three coordinate directions. Each of these equations may then be multiplied by the corresponding unit vector, and upon summing generate a vector equation for the momentum balance. Such an equation is neat and compact, but it can only be solved by evaluating each of the component balances. In curvilinear coordinates the base vectors are not constant; hence, vector forms of momentum equations are not used.

Upon writing each of the three momentum equations for a stationary, orthogonal Cartesian coordinate system, then multiplying each by the appropriate unit vector and adding

$$\frac{\partial}{\partial t}(\rho \vec{U}) + \nabla \cdot \rho \vec{U}\vec{U} = -\nabla P + \nabla \cdot \vec{\vec{\tau}} + \rho \vec{F}$$

$$\text{where } \nabla \cdot \vec{\vec{\tau}} = \nabla \cdot \left[\mu(\nabla \vec{U} + (\nabla \vec{U})^{\mathrm{T}}) + \left(\frac{2}{3}\mu - \kappa\right)\nabla \cdot \vec{U}\right]\vec{\vec{\delta}} \qquad (B.85)$$

Further complications also arise because some quantities require more elaborate definition than vectors. For example, shear-stress and rate-of-strain must be defined in terms of a magnitude and two directions. Tensor formalism is necessary for such definitions. The second complicating factor is that curvilinear coordinate systems are frequently needed to describe the geometry of the region of interest and its boundaries. The net results are that (1) vector formulation of the conservation equations serves only to be a neat, shorthand method of writing the equations, (2) various dialects of "tensor analysis" must be recognized and adhered to, and (3) the conservation laws are solved as partial differential equations (unless geometric idealizations are utilized to simplify the analysis). The implication is that integrals involving vectors which change direction within the flowfield do not have to be formally evaluated.

B.5 CONSERVATION EQUATIONS IN NONORTHOGONAL COORDINATE SYSTEMS

B.5.1 CONTINUITY EQUATION

Consider the parallelepiped spanned by $\vec{\zeta}_1, \vec{\zeta}_2, \vec{\zeta}_3$ (see Borisenko and Tarapov, 1968, p.24; Margenau and Murphy, 1968, p. 146). The volume of this parallelepiped is

$$\vec{\zeta}_1 \cdot (\vec{\zeta}_2 \times \vec{\zeta}_3) = \vec{\zeta}_2 \cdot (\vec{\zeta}_3 \times \vec{\zeta}_1) = \vec{\zeta}_3 \cdot (\vec{\zeta}_1 \times \vec{\zeta}_2) = \forall \qquad (B.86)$$

This is the scalar triple product rule. The cross products are the surface areas of the parallelograms which comprise the faces of the parallelepiped. The normals of these areas are directed positively in the direction that a right-handed screw would advance if the first vector in the product is twisted toward the second vector. The contravariant base vectors are related to these terms by

$$\vec{\zeta}^1 = \frac{\vec{\zeta}_2 \times \vec{\zeta}_3}{\forall}, \quad \vec{\zeta}^2 = \frac{\vec{\zeta}_3 \times \vec{\zeta}_1}{\forall}, \quad \vec{\zeta}^3 = \frac{\vec{\zeta}_1 \times \vec{\zeta}_2}{\forall} \qquad (B.87)$$

Compare the continuity equation in orthogonal Cartesian coordinates and in general tangential curvilinear coordinates.

$$\frac{\partial \rho}{\partial t} + \frac{\partial \rho U_1}{\partial X_1} + \frac{\partial \rho U_2}{\partial X_2} + \frac{\partial \rho U_3}{\partial X_3} = 0 \qquad (B.88)$$

$$\frac{\partial \rho}{\partial t} + \frac{1}{J} \left(\frac{\partial \rho W^1}{\partial Z^1} + \frac{\partial \rho W^2}{\partial Z^2} + \frac{\partial \rho W^3}{\partial Z^3} \right) = 0 \qquad (B.89)$$

Notice that the velocity components, coordinates, and area vectors are colinear in the rectangular Cartesian coordinate system. The contravariant velocity components and contravariant coordinates are also colinear, but the surface through which the velocity moves the fluid is directed in the covariant coordinate direction. This looks odd, so consider it further.

The net volumetric flow of fluid through the $\vec{\zeta}_2, \vec{\zeta}_3$ surface is $dW^1 dZ^2 dZ^3$, which is into the unit volume $d\forall = \sqrt{g}\,dZ^1 dZ^2 dZ^3$ or $d\forall = J dZ^1 dZ^2 dZ^3$. Introducing these terms into the mass balance yields

$$\frac{dW^1 dZ^2 dZ^3}{J dZ^1 dZ^2 dZ^3} = \frac{1}{J}\frac{\partial W^1}{\partial Z^1} \tag{B.90}$$

Including the flow through the other surfaces yields the final mass balance. This demonstrates that the flow through the parallelepiped is correctly represented by the stated continuity equation in tangential curvilinear coordinates.

B.5.2 MOMENTUM EQUATION

Recognizing that an integral form of the momentum equation cannot be readily integrated, a differential element suitable for use as a finite-difference or finite-volume representation of a computational element can be meaningfully analyzed. Yang and his colleagues have presented such a development in a series of papers (Yang and Lloyd, 1990; Yang et al., 1988, 1994). Their development is summarized as follows.

An elemental control volume for physical tangential curvilinear coordinates is shown in Figure B.1. This is the control volume analyzed by Yang and his colleagues. The incremental distance ds in this figure is the distance which is constructed to be the same between Cartesian coordinates and nonorthogonal curvilinear coordinates. The angle between

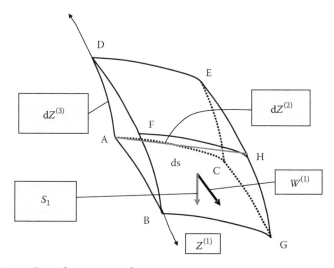

FIGURE B.1 Curvilinear coordinate system.

S_1 and $W^{(1)}$ is the result of the coordinates being nonorthogonal. Point A is the local origin. The line A–B is the increment $dZ^{(1)}$.

Continuity and Navier–Stokes equations

$$\frac{\partial \rho}{\partial t} + \nabla \cdot (\rho \vec{U}) = 0 \qquad (B.91)$$

$$\frac{\partial \rho}{\partial t} + \nabla \cdot (\rho \vec{U} \vec{U}) = -\nabla P + \nabla \cdot \left[\mu(\nabla \vec{U}) + \mu(\nabla \vec{U})^{\mathsf{T}} \right] + \rho \vec{F} \qquad (B.92)$$

To utilize tangential curvilinear coordinates, both the independent and dependent velocity variables are transformed.

The covariant base vectors in tangential curvilinear coordinates are

$$\vec{\zeta}_i = \sum_j \vec{\chi}_j \frac{\partial X_j}{\partial Z^i} \quad \text{and} \quad \vec{\zeta}_{(i)} = \frac{\vec{\zeta}_i}{\sqrt{g_{ii}}} \qquad (B.93)$$

The differential operators for scalar operations are

$$\nabla \varphi = \vec{\zeta}^{\,j} \frac{\partial \varphi}{\partial Z^j} = J \frac{\partial}{\partial Z^j} \left(\frac{\vec{\zeta}^{\,j}}{J} \varphi \right) \qquad (B.94)$$

$$\nabla^2 \varphi = J \frac{\partial}{\partial Z^j} \left(\frac{g^{jk}}{J} \frac{\partial \varphi}{\partial Z^k} \right) \qquad (B.95)$$

$$\nabla \cdot \vec{W} = J \frac{\partial}{\partial Z^j} \left(\frac{\vec{\zeta}^{\,j} \cdot \vec{W}}{J} \right) \qquad (B.96)$$

$$\frac{1}{J} = \frac{\partial \{X_1, X_2, X_3\}}{\partial \{Z^1, Z^2, Z^3\}} \qquad (B.97)$$

Options for representing the velocity vector are

$$\vec{W} = W^i \vec{\zeta}_i = W_i \vec{\zeta}^{\,i} = W^{(i)} \vec{\zeta}_{(i)} = W_{(i)} \vec{\zeta}^{\,(i)} \qquad (B.98)$$

These definitions imply the additional options:

$$W^{(i)} = W^i / \sqrt{g_{ii}}; \quad W_{(i)} = W_i / \sqrt{g_{ii}}; \quad W^i = \vec{W} \cdot \vec{\zeta}^{\,i}; \quad W_i = \vec{W} \cdot \vec{\zeta}_i \qquad (B.99)$$

Denote the velocity resolute by ω to obtain

$$\vec{\omega}_{(i)} = \left(\vec{W} \cdot \vec{\zeta}_{(i)}\right) \vec{\zeta}_{(i)} \tag{B.100}$$

The continuity and momentum equations in curvilinear coordinates become

$$\frac{\partial}{\partial t}\left(\frac{\rho}{J}\right) + \frac{\partial}{\partial Z^i}\left(\frac{\rho\vec{\zeta}^i \cdot \vec{W}}{J}\right) = 0 \tag{B.101}$$

$$\frac{\partial}{\partial t}\left(\frac{\rho\vec{W}}{J}\right) + \frac{\partial}{\partial Z^i}\left(\frac{\rho\vec{\zeta}^i \cdot \vec{W}\vec{W}}{J}\right) = -\frac{\vec{\zeta}^i}{J}\frac{\partial p}{\partial Z^i} + \frac{\partial}{\partial Z^i}\left(\frac{\mu g^{ik}}{J}\frac{\partial \vec{W}}{\partial Z^k}\right)$$
$$+ \frac{\partial}{\partial Z^i}\left(\frac{\mu\vec{\zeta}^i}{J} \cdot \frac{\partial \vec{W}}{\partial Z^k}\vec{\zeta}^k\right) + \frac{\vec{S}}{J} \tag{B.102}$$

The continuity equation can be integrated over the differential volume ($\forall$) surrounding the node point p.

$$\left[(\rho_p - \rho_p^o)\forall/\Delta t\right] + \sum_{f=1}^{3}\Delta M_f = 0 \tag{B.103}$$

ΔM_f is the net flowrate through the projected area of each of the cell faces. The physical velocity component is used to calculate this flowrate.

To avoid dealing with a vector in the PDEs to be solved, Yang and colleagues (1994) noticed that a base vector $\vec{b}$ could be dotted with the vector form of the momentum equation to eliminate vectors appearing in the momentum equation. Since the control volume was to be of differential dimensions, this vector could be assumed locally constant. The resulting equation for the unsteady, convection, and pressure terms in the inviscid momentum equation become

$$\frac{\partial}{\partial t}\left(\frac{\rho\vec{W} \cdot \vec{b}}{J}\right) + \frac{\partial}{\partial Z^i}\left(\frac{\rho W^i \vec{W} \cdot \vec{b}}{J}\right) - \frac{\partial\vec{b}}{\partial Z^i} \cdot \frac{\rho W^i}{J}\vec{W} = -\vec{b} \cdot \frac{\vec{\zeta}^i}{J}\frac{\partial P}{\partial Z^i} \tag{B.104}$$

Various choices of $\vec{b}$ resulted in various discretized forms of the momentum equation, many of which had been investigated by others. For instance, if $\vec{b}$ is the unit vector in the OCC system, the velocity components would be those from the OCC system. These are the same as those used in the conservation equations in Chapter 6 and in the CTP code. Other choices would result in various velocity components based on nonorthogonal curvilinear coordinates. Yang and colleagues (1994) presented several simulations using different $\vec{b}$ vectors. General conclusions could not be drawn, although some of the test cases showed improved computational performance.

B.6 LINEAR TRANSFORMATIONS

Matrices and column vectors are used to perform linear transformations as follows:

The matrix is indicated with a diacritic carat and the column vector with a tilde diacritic

$$\tilde{y} = \hat{M}\tilde{x} \quad \text{or} \quad \begin{bmatrix} y_1 \\ y_2 \\ y_3 \end{bmatrix} = \begin{bmatrix} M_{11} & M_{12} & M_{13} \\ M_{21} & M_{22} & M_{23} \\ M_{31} & M_{32} & M_{33} \end{bmatrix} \begin{bmatrix} x_1 \\ x_2 \\ x_3 \end{bmatrix} \quad \text{(B.105)}$$

The use of the term *column vector* is very unfortunate, as it is frequently shortened to just vector. Even worse both vectors and *column vectors* are frequently indicated with bold type so there is no apparent difference in their appearance. Then the *column vector* is confused with the physical vectors which were described in the previous sections. The matrix, $\hat{M}$, is said to operate on the (column) vector x to produce the solution vector y. Better terminology is that M is a multidimensional-array, and x and y are one-dimensional arrays. Unfortunately again, such terminology is not used, except in computer codes where such terms are so named and computer operations are so treated.

The inverse transformation is

$$\tilde{x} = \hat{M}^{-1}\tilde{y} = \frac{\hat{M}\tilde{y}}{|\hat{M}|} \quad \text{where} \quad |\hat{M}| \equiv \begin{vmatrix} M_{11} & M_{12} & M_{13} \\ M_{21} & M_{22} & M_{23} \\ M_{31} & M_{32} & M_{33} \end{vmatrix} = \det \hat{M}$$

$$|\hat{M}| = M_{11}M_{22}M_{33} + M_{12}M_{23}M_{31} + M_{13}M_{32}M_{21}$$
$$- M_{13}M_{22}M_{31} - M_{12}M_{21}M_{33} - M_{11}M_{32}M_{23}$$

$$\text{(B.106)}$$

The vertical bars indicate the determinant of the enclosed matrix. The transpose of a matrix is

$$\hat{M}^{\mathrm{T}} = \begin{bmatrix} M_{11} & M_{21} & M_{31} \\ M_{12} & M_{22} & M_{32} \\ M_{13} & M_{23} & M_{33} \end{bmatrix} \quad \text{(B.107)}$$

The Jacobian of the transformation from OCC to TCC was used in Chapter 6. In the nomenclature of the coordinates in this appendix,

$$
J = \frac{\partial(Z^1, Z^2, Z^3)}{\partial(X_1, X_2, X_3)} = \begin{vmatrix} \dfrac{\partial Z^1}{\partial X_1} & \dfrac{\partial Z^1}{\partial X_2} & \dfrac{\partial Z^1}{\partial X_3} \\[2mm] \dfrac{\partial Z^2}{\partial X_1} & \dfrac{\partial Z^2}{\partial X_2} & \dfrac{\partial Z^2}{\partial X_3} \\[2mm] \dfrac{\partial Z^3}{\partial X_1} & \dfrac{\partial Z^3}{\partial X_2} & \dfrac{\partial Z^3}{\partial X_3} \end{vmatrix} \tag{B.108}
$$

In Section B.3, this transformation was used twice since the increment of distance was ds^2 to obtain the metric coefficients. Since the metric tensor had elements of incremental distance, the Jacobian for the metric tensor is the product of two transformation matrices for the individual coordinates, the metric for the metric tensor is J^2. Or, in terms of the metric tensor proper it is $\det \hat{g} = g = J^2$.

More matrices properties can be found in Margenau and Murphy (1956) and Sokolnikoff et al. (1966), Lanczos (1961), and Gantmacher (1960).

The use of linear transformation laws allows the evaluation of vector and tensor components, but it does not include base or unit vectors. This observation should clarify the discussion in texts such as Brodkey's which uses column vectors and vectors interchangeably. This is not correct.

Chapter 6 uses these linear transformation laws to describe the conservation laws in PTC. This methodology also termed the use of linear vector spaces or the use of function spaces to represent transformations of variables. Notice again that the base or unit vectors are not used in such a development. This simplifies the coding necessary to obtain a solution and is what is used in the CTP code.

B.7 SUMMARY

This appendix presents vector and tensor analysis suitable for describing two- and three-dimensional geometries. The grids may be updated at intervals during the course of the calculations, but time (or iteration level) are not coupled to the grid geometry. More complex grid, vector, and tensor operations are discussed herein, but they are not needed to utilize the methodology presented in Chapters 6 and 8. Such discussion is included here so that other literature on computational transport analyses may be appreciated and so that a starting point for studying the vast literature on tensor analysis is established. Again, one is warned that the tensor literature is not consistent with nomenclature, definitions, and/or scope. The cited references are strongly recommended as a starting point to avoid hopeless confusion.

B.8 NOMENCLATURE

a	typical scalars
A, B, C	typical vectors
b	magnitude of arbitrary base vector
f	a function
Fr	Froude number
g, J^2	determinate of the metric tensor, i.e., a matrix
g^{ij}	component of the conjugate metric tensor, i.e., a matrix
g_{ij}	component of the metric tensor, i.e., a matrix
H_i	metric coefficients in NCC
J	Jacobian of the coordinate transformation matrix
J_{MT}	Jacobian of the metric tensor components
k	turbulent kinetic energy
p	a spatial location
P	pressure
R	magnitude of position vector
s	arc length
S	magnitude of surface vector
t	time
T	temperature
T_{ij}	second-order tensor components
U_s	surface velocity component
x_i, y_i	arbitrary column matrices
$X, \vec{\chi}, \vec{U}$	orthogonal Cartesian coordinate (OCC), unit vectors, velocities
$X', \vec{\chi}'$	a second OCC system
$Y, \vec{\upsilon}, \vec{V}$	orthogonal curvilinear coordinates (NCC), unit vectors, velocities
$Z^i, \vec{\zeta}_i, W^i$	mathematical contravariant curvilinear coordinates (MTC), covariant base vectors, contravariant l velocities
$Z_i, \vec{\zeta}^i, W_i$	mathematical covariant curvilinear coordinates (MNC), contravariant base vectors, covariant velocities
$Z^{(i)}, \vec{\zeta}_{(i)}, W^{(i)}$	physical contravariant curvilinear coordinates (PTC), covariant base vectors, covariant physical velocities
$Z_{(i)}, \vec{\zeta}^{(i)}, W_{(i)}$	physical covariant curvilinear coordinates (PNC), contravariant base vectors, covariant physical velocities

B.8.1 GREEK SYMBOLS

κ	second coefficient of viscosity
μ	viscosity
ν	kinematic viscosity
ρ	density
Φ, φ	scalar variables
ω	velocity resolute

B.8.2 SUBSCRIPTS

a	antisymmetric tensor
E	eddy transport term
i	covariant variable
s	symmetric tensor
T	a total value
$,i$	denotes partial differentiation WRT coordinate i

B.8.3 SUPERSCRIPTS

o	reference value
$\vec{\vec{T}}$	a second-order tensor
$\vec{V}$	a vector
$\forall$	volume
$\check{\forall}$	volume of a specific system
-1	inverse of a matrix

B.8.4 MATHEMATICAL SYMBOLS

$\tilde{C}$	column vector
e^{ijk}	permutation symbol
$\hat{g}$	covariant metric "tensor," i.e., a matrix
$\hat{g}^{T}$	transposed matrix
G^{ij}	cofactor of transposed metric "tensor"
$\hat{M}$	coordinate transformation matrix
$\hat{n}$	conjugate contravariant metric "tensor," i.e. a matrix
δ_i^j	Kronecker delta; δ_{ij} is used for OCC
Γ, { }	Christoffel symbol of the second kind

$[\ ]$ — Christoffel symbol of the first kind

$$\nabla\{\ \} = \frac{\partial\{\ \}}{\partial\vec{R}}$$ — del operator

$$\frac{D\{\ \}}{Dt} = \frac{\partial\{\ \}}{\partial t} + \vec{U}\cdot\nabla\{\ \}$$ — substantial derivative

REFERENCES

Bird, R. B., W. E. Stewart, and E. N. Lightfoot. 2002. *Transport Phenomena,* 2nd ed. New York: Wiley & Sons.

Borisenko, A. I. and I. E. Tarapov. 1968. *Vector and Tensor Analysis with Applications.* Translated by R. A. Silverman. Englewood Cliffs, NJ: Prentice-Hall.

Brillouin, L. and R. O. Brennan. 1964. *Tensors in Mechanics and Elasticity.* New York: Academic Press.

Brodkey, R. S. 1967. *The Phenomena of Fluid Motions.* Reading, MA: Addison-Wesley Publishing.

Gantmacher, F. R. 1960. *The Theory of Matrices.* New York: Chelsea Publishing.

Gibbs, J. W. and E. B. Wilson. 1960. *Vector Analysis.* New York: Dover Publications.

Hawkins, G. W. 1963. *Multilinear Analysis for Students in Engineering and Science.* New York: John Wiley & Sons.

Korn, G. A. and T. M. Korn. 1968. *Mathematical Handbook for Scientists and Engineers,* 2nd ed. New York: McGraw-Hill.

Lanczos, C. 1961. *Linear Differential Operators.* London: D. Van Nostrand.

Lee, S. 1997. Three-dimensional incompressible viscous solutions based on the physical curvilinear coordinate system. PhD disseartions, MS State University.

Lee, S. and B. K. Soni. 1997. Governing equations of fluid mechanics in physical curvilinear coordinate system. *Electronic Journal of Differential Equations, conference 01, 1997,* J. Graef, R. Shivaji, B. Soni, and J. Zhu, (Eds.) pp. 149–157. http://www.emis.ams.org/journals//conf/proc/01/toc.html

Margenau, H. and G. M. Murphy. 1956. *The Mathematics of Physics and Chemistry,* 2nd ed. Princeton: D. Van Nostrand.

Misner, C. W., K. S. Thorne, and J. A. Wheeler. 1973. *Gravitation.* San Francisco, CA: W. H. Freeman and Company.

Morse, P. M. and H. Feshbach. 1953. *Methods of Theoretical Physics, Part I,* 2nd ed. New York: McGraw-Hill.

Sokolnikoff, I. S. 1964. *Tensor Analysis: Theory and Applications to Geometry and Mechanics of Continua,* 2nd ed. New York: John Wiley & Sons.

Sokolnikoff, I. S. and R. M. Redheffer. 1966. *Mathematics of Physics and Modern Engineering,* 2nd ed. New York: Mc-Graw-Hill.

Vinokur, M. 1974. A New Formulation of the Conservation Equations of Fluid Mechanics. NASA TM X-62,415, National Aeronautics and Space Administration. Moffett Field.

Wylie, C. R., Jr. 1966. *Advanced Engineering Mathematics,* 3rd ed. New York: McGraw-Hill.

Yang, H. Q., K. T. Yang, and J. R. Lloyd. 1988. Buoyant flow calculations with non-orthogonal curvilinear co-ordinates for vertical and horizontal parallelepiped enclosures. *International Journal for Numerical Methods in Fluids*. 25: 331–345.

Yang, H. Q. and J. R. Lloyd. 1990. A control volume finite difference method for buoyant flow in three-dimensional curvilinear non-orthogonal co-ordinates. *International Journal for Numerical Methods in Fluids*. 10: 199–211.

Yang, H. Q., S. D. Habchi, and A. J. Przekwas. 1994. General strong conservation formulation of Navier-Stokes equations in nonorthogonal curvilinear coordinates. *AIAA Journal* 32: 936–841.

Fortran Primer

C.1 OVERVIEW

The Fortran (Formula translation) system consists of a mathematical language and a compiler. The mathematical language, called source code, allows one to use a set of statements to perform various functions much like algebra. The compiler converts these statements into machine language called the compiled code which runs on a computer. Fortran is about 50 years old, but it has been and still is very useful. This has been the choice of the engineers and scientists because of its high precision and other features which expedite solving numerical equations. There are literally hundreds of millions of dollars of Fortran codes which have been developed and most of them are still useful. These codes will not be replaced simply to convert them to another language. No one will provide the financing to perform such a task.

Fortran has evolved through the years. Fortran 77 was the major code for 25 years. Fortran 90, 95, 2002, etc., have been developed, but these codes offer marginal improvements. Fortran 77 can be learned from a good text, like *Fortran 77 for Engineers and Scientists* by L. Nyhoff and S. Leestma, 4th ed. from Prentice-Hall. Tutorials are also available on the Internet, for example, http://www.stanford.edu/class/me200c/tutorial_77/. The Internet is not a dependable source, because what is here today might be gone tomorrow. If you find something that you like—SAVE IT! Free Fortran 77 compilers are also available on the Internet for both Linux and Windows operating systems. Linux is the preferred computational operation system, but Windows is the most commonly available.

The CTP code described herein is provided as both source code and compiled code. The purpose of this text is to explain how the CTP code can be used to work transport problems; not to teach one how to write a Fortran code. One does not need the source code to perform a calculation. However, frequently one must have access to the source code to understand and fix a problem that arises in a given analysis. The more complex the analysis the more often this is the case. It is anticipated that once one masters the material herein, the user will want to know more about Fortran. The following is a brief outline of the Fortran language. Be warned! It does not replace texts and well-written tutorials.

The computer is a dumb beast. You must tell it exactly what to do in terms of its own rules, one step at a time. The language used for identifying the kinds of numbers and words used are called syntax, no exceptions. Assignment statements are sentences which instruct the machine to perform one specific operation. One must use a list of acceptable assignment statements to write a Fortran code—such a list amounts to a dictionary of terms which the Fortran code understands. The assignment statement is written on one line in plain text. The computer program consists of a structured algorithm which is the plan for combining these "one-liners" into a strategy for producing the desired analysis. The strategy may consist of (1) a sequence of calculations, (2) a selection among several routes depending on intermediate results, (3) a repetition of steps, and (4) a combination of these routes to arrive at the desired goal. The compilation step converts these instructions to machine language.

These restrictions appear severe and they are. But the result is extremely useful. That was the bad news, the good news is the following. Computers, even PCs, perform operations extremely rapidly and can handle and generate very large volumes of data. File structures of specified format are used to input required data and write out calculated data. While the calculation is in progress, arrays of data and variables are used to enable the immense numbers of calculations to be made in an organized fashion. Not only may data files become large, but the coding itself may be long and involved. Analysts have found it convenient to write and test blocks of code which is then collected for use with a "makefile" command. All of these operations are not described in basic Fortran literature, but many of them are used in the accompanying CTP source code. Learning and applying Fortran 77 coding is an evolutionary process. Hopefully, having the CTP source code and the examples supplied for its use will provide a "jump-start" for using the powerful computational tool.

Modern computer usage allows clusters of computers like PCs to be used simultaneously to analyze very large and complex problems. This practice is called parallel computing and it is beyond the scope of the present discussion.

The following is a brief outline of the Fortran language. Be warned! It does not replace texts and well-written tutorials. It is meant to give one an initial feel for Fortran coding and a convenient reference once one has ventured into the programming world.

C.2 LOOK OF FORTRAN 77

Fortran 77 is briefly described as follows. Dr. John Mathews provided the authors this material in a personal communication. Dr. Mathews is a retired mathematics professor at the California State University at Fullerton. We thank Professor Mathews for the use of this concise and informative summary.

FORTRAN Preliminaries

Specified Columns

Each line of FORTRAN code must be written using the following column conventions.

Column	Contents
1	"C" or "*" for a comment line
1–5	Statement number or label (non-signed integer).
6	Blank or character for continuation line.
7–72	FORTRAN statement
73–80	Sequence number (usually omitted).

Form of a FORTRAN Program

```
PROGRAM <program-name>
        {<specification-statements>}
        {<executable-statements>}
END
```

Form of a FORTRAN Subroutine

```
SUBROUTINE <subroutine-name>[<actual-argument-list>]
        {<specification-statements>}
        {<executable-statements>}
```

```
RETURN
END
```

Form of a Subroutine call.

```
CALL <subroutine-name>[<actual-argument-list>]
```

Form of a FORTRAN function

```
[<type>] FUNCTION <function-name> [(<dummy-assignment list>)]
        {<specification-statements>}
        {<executable-statements>}
RETURN
END
```

Specification Statements (or Type Declaration Statements)

```
INTEGER I,J,Row
REAL A0,B0,Max,X,Y
DOUBLE PRECISION
LOGICAL A,B,C
COMPLEX Z,W0
CHARACTER*10 Name
IMPLICIT F,DF
PARAMETER (Pi=3.1415926535)
DATA E/2.178281828/
COMMON A0,B0
COMMON /BlockA/ A,B,X,Y
```

Variables

Variable names can be from one to six characters long.
Variables beginning with I,J,K,L,M,N are presumed to be integers unless they are declared otherwise.

Assignment Statements

```
X = A(1) + 3
Y = A*X**2 + B*X + C
Flag = .TRUE.
Name = 'Fran'
```

Warning. Mixed Mode arithmetic might result in some frustrating surprises. For example, when two integers are divided the result is an unexpected integer value. To avoid errors use explicit decimal points for real numbers and the intrinsic functions REAL, INT, DBLE, and CMPLX to get the required type conversion.

Arrays

```
DIMENSION A(1:50), M(1:10,1:10)
```

Arithmetic Operations

+ Addition
− Subtraction
* Multiplication
/ Division
** Exponentiation

Relational Operators

.EQ. Equal to
.NE. Not equal to
.LT. Less than
.GT. Greater than
.LE. Less than or equal to
.GE. Greater than or equal to

Logical Operators

.NOT. Complement
.AND. True if both operands are true
.OR. True if either (or both) operands are true

Logical Constants

.TRUE.
.FALSE.

Remark Statement

```
C       This is a comment.
```

Place-holder Statement (or label)

```
10      CONTINUE
```

Unconditional Transfer

```
GOTO 10
```

Computed Control Statement

Transfers control to a specified statement, depending on the value of an integer expression, e.g., `<IJUMP>`.
```
GOTO(100,200,300,400), IJUMP
```
Warning. Do not use the `GOTO` statement to transfer into a `DO`, `IF`, `ELSE`, or `ELSEIF` block from outside the block.

IF (Arithmetic) Control Statement

Transfers control to a statement depending on whether the value of the (`<arithmetic-expression>`) is positive, negative or zero.
```
IF (<arithmetic-expression>) 100, 200, 300
```

IF (Logical) Control Statement

Executes a single statement only if the (`<logical-expression>`) is true.
```
IF (<logical-expression>) GOTO 100
IF (<logical-expression>) WRITE (*,*) 'Yes'
IF (<logical-expression>) X = A+B
```

IF (Block) Control Statement

Performs the series of {`<executable-statement>`} following it or transfers control to an `ELSEIF`, `ELSE`, or `ENDIF` statement, depending on the value of the (`<logical-expression>`).

```
IF (<logical-expression>) THEN
        {<executable-statements>}
ENDIF
```

```
IF (<logical-expression>) THEN
        {<executable-statements>}
ELSE
        {<executable-statements>}
ENDIF
```

```
IF (<logical-expression-#1>) THEN
        {<executable-statements>}
ELSEIF (<logical-expression-#2>) THEN
        {<executable-statements>}
ELSEIF (<logical-expression-#3>) THEN
        {<executable-statements>}
ELSE
        {<executable-statements>}
ENDIF
```

```
IF (ABS(P3-P1).LT.ABS(P3-P0)) THEN
  U=P1; P1=P0; P0=U
  V=Y1; Y1=Y0; Y0=V
ENDIF

IF (Df.EQ.0) THEN
  Dp=P1-P0
  P1=P0
ELSE
  Dp=Y0/Df
  P1=P0-Dp
ENDIF

IF (YC.EQ.0) THEN
  A=C
  B=C
ELSEIF ((YB*YC).GT.0) THEN
  B=C
  YB=YC
  KR=KR+1
ELSE
  A=C
  YA=YC
  KL=KL+1
ENDIF
```

DO (Block) Control Statement

```
DO K = M1, M2
       {<executable-statements>}
ENDDO

DO K = M1, M2, Mstep
       {<executable-statements>}
ENDDO

DO (M TIMES)
       {<executable-statements>}
ENDDO
SUM = 0
DO K=0,100,2
  SUM = SUM + K
ENDDO
```

```
 SUM = 0
 DO K=100,-1,1
   SUM = SUM + 1.0/REAL(K)
 ENDDO

 DO J=1,5
   DO K=1,5
       A(J,K) = 1.0/FLOAT(1+J+K)
   ENDDO
 ENDDO

 SUM = 0
 DO K=1,10000
   SUM = SUM + 1.0/REAL(K)
   IF (SUM.GT.5) EXIT
 ENDDO
```

WHILE (Block) Control Statement

```
 WHILE (<logical-expression>)
       {<executable-statements>}
 REPEAT
 SUM = 0
 WHILE (K.LT.10000)
   SUM = SUM + 1.0/REAL(K)
   IF (SUM.GT.5) EXIT
   K = K+1
 REPEAT
```

Input and Output

```
 READ *,<input-variable-name-list>
 READ <format>,<input-variable-name-list>
 READ <N>,<input-variable-name-list>
       where <N> is a FORMAT statement number, e.g.
```

```
111  FORMAT(5X,I10,4F15.5)
```

```
     PRINT *
     PRINT *,<output-expression-list>
     PRINT <format>,<output-expression-list>
     PRINT <N>,<output-expression-list>
         where <N> is a FORMAT statement number, e.g.
```

```
999  FORMAT(5X,'X = ',F15.5)

     WRITE (*,*)
     WRITE (*,*) <output-expression-list>
     WRITE (*,N) <output-expression-list>
         where <N> is a FORMAT statement number, e.g.

999  FORMAT(5X,'X = ',F15.5)
```

Pause Statement

```
     PAUSE
```

Stop Statement

```
     STOP
```

Mathematical Functions

`COS(X)`	cosine (radians)
`SIN(X)`	sine (radians)
`TAN(X)`	tangent (radians)
`EXP(X)`	exponential exp(x)
`ACOS(X)`	inverse cosine (radians)
`ASIN(X)`	inverse sine (radians)
`ATAN(X)`	inverse tangent (radians)
`ALOG(X)`	natural logarithm base e
`LOG10(X)`	common logarithm base 10
`SQRT(X)`	square root
`ABS(X)`	absolute value
`INT(X)`	conversion to integer
`FLOAT(I)`	conversion to real number type
`REAL(I)`	conversion to real number type
`DBLE(X)`	conversion to double precision type
`CMPLX(X)`	conversion to double precision type

C.3 OUTFITTING A PC FOR USING FORTRAN

There are free Fortran 77 compilers available for the three major operating systems (OS) used by a personal computer (PC): MS-Windows, Linux, and Mac OS-X. If you are using the Linux OS, the Fortran compiler (g77 or f77) generally comes with the OS. Even if you cannot find it on your system (most likely it did not get installed), you can install it from the CD-ROM of the OS that you purchased or comes with your system. For Mac OS-X, you can download the Fortran compiler from the Apple Web site (http://www.apple.com) under the "Development Tools" in the "Downloads" section. For the PC with the MS-Windows OS, there are various ways of getting a free Fortran compiler. You can go to http://kkourakis.tripod.com/ or http://www.geocities.com/Athens/Olympus/5564/ to download the Fortran g77 compiler, or http://ftp.g95.org/ for the Fortran g95 compiler. Another way of obtaining the Fortran compiler is to install a software package call "cygwin" (http://sources.redhat.com/cygwin/), which includes X-Windows (X11) emulator, text editors (such as "vi" or "emacs"), Fortran compiler (g77 or f77), networking software ("ssh", "sftp", …), etc. After downloading the software package from any of the above Web sites, you can follow the instruction to install the software. Additional sources of free Fortran compilers are listed at: http://www.thefreecountry.com/compilers/fortran. shtml

Index